ADVANCES IN RELIABILITY

ADVANCES IN RELIABILITY

Edited by

ASIT P. BASU

Department of Statistics
University of Missouri
Columbia, Missouri
USA

1993
NORTH-HOLLAND
AMSTERDAM • LONDON • NEW YORK • TOKYO

ELSEVIER SCIENCE PUBLISHERS B.V.
Sara Burgerhartstraat 25
P.O. Box 211, 1000 AE Amsterdam, The Netherlands

Library of Congress Cataloging-in-Publication Data

International Research Conference on Reliability (1991 : Columbia, Mo.)
Advances in reliability : a selection of papers from the International Research Conference on Reliability, Missouri, Columbia, 19-22 June, 1991 / edited by Asit P. Basu.
p. cm.
Includes bibliographical references and index.
ISBN 0-444-89645-7
1. Reliability (Engineering)--Congresses. I. Basu, Asit P. II. Title.
TA169.I584 1993
620'.00452--dc20 93-3047
CIP

ISBN: 0-444-89645-7

This book is printed on acid-free paper.

Printed in The Netherlands.

PREFACE

The fourth international research conference on Reliability was held at the University of Missouri-Columbia, Missouri, June 19-22, 1991. The purpose of the conference was to review recent developments in Reliability Theory and to stimulate interaction among leading researchers in the world and users from university, government and industry. After the conference each participant was asked to revise, update and submit a more complete paper so that a monograph consisting of papers on recent advances in Reliability Theory could be published.

This volume consists of updated versions of a selected number of papers presented at the conference and some additional invited papers. All papers were refereed. A number of distinguished workers presented important results on a broad spectrum of topics in Reliability. Because topics often overlap within a paper, instead of classifying according to topics, these papers are presented in alphabetical order by the first author. The only exception has been the after dinner nontechnical talk by Nozer Singpurwalla, which is the last paper in this volume.

I would like to thank a number of persons for their help in organizing the conference. Lee Bain, (University of Missouri-Rolla), Richard Barlow (University of California-Berkeley), Siddartha Dalal (Bell Communications Research), Robert Easterling (Sandia National Labs), Frank Guess (University of Tennessee), John Healey (Bellcore), Jeff Hooper (AT&T Bell Laboratories), Raghu Kacker (National Institute of Standards and Technology), John Kitchin (Digital Equipment Corporation), Robert Lochner (Statpower Associates), Robert Lundegard (National Institute of Standards and Technology), Gary McDonald (General Motors Research Labs), Nancy Mann (Quality Enhancement Seminar, Inc.), Vijayan Nair (AT&T Bell Laboratories), Frank Proschan (Florida State University), Paul Shaman (University of Pennsylvania), Nora Smiriga (Lawrence Livermore, National Laboratory), Mahesh Vora (Ford Motor Company), William Woodall (University of Alabama) served as members of the Program Committee. Larry Clark, Dean, College of Arts and Science and Judson Sheridan, Dean, Graduate School, University of Missouri-Columbia, made some opening remarks at the conference. I am grateful to Lee Bain (University of Missouri-Rolla), Dipankar Bandyopadhyay (Bowling Green State University), Nader Ebrahimi (Northern Illinois University), Hamid Fallahi (Eastman Kodak Company), Deborah Flanagan (Oak Ridge National Laboratory), William Griffith (University of Kentucky), James Higgins (Kansas State University), Shriniwas Katti (University of Missouri-Columbia), James Lechner (National Institute of Standards and Technology), Robert Lochner (Statpower

Associates), Steven Rigdon (Southern Illinois University), Ehsan Soofi (University of Wisconsin-Milwaukee), Paul Speckman (University of Missouri-Columbia), Robert Tsutakawa (University of Missouri-Columbia), Tim Wright (University of Missouri-Columbia), and Mahesh Vora (Ford Motor Company) for chairing different sessions. Thanks are also due to the following for serving as referees of various papers in this volume: Lee Bain (University of Missouri-Rolla), Dipankar Bandyopadhyay (Bowling Green State University), Frank Guess (University of Tennessee), Nader Ebrahimi (Northern Illinois University), Mohamed Habibullah (University of Wisconsin-Superior), John Kitchin (Digital Equipment Corporation), John Klein (Ohio State University), Mei-Ling Ting Lee (Harvard University), Wang-Shu Lu (University of Missouri-Columbia), Michael LuValle (AT&T Bell Laboratories), Chiranjit Mukhopadhyay (Ohio State University), Giovanni Parmigiani (Duke University), Prakash Patil (Australian National University), Hakan Polatoglu (University of Wisconsin-Milwaukee), Nick Polson (University of Chicago), Ananda Sen (University of Wisconsin-Madison), Refik Soyer (George Washington University), Kai Sun (University of Missouri-Columbia), Ram Tiwari (University of North Carolina-Charlotte), Tim Wright (University of Missouri-Columbia), and Jyoti Zalkikar (Florida International University).

I am grateful to the contributors to this volume and to Dr. Gerard Wanrooy, Dr. Wim Spaans and other staff members at Elsevier Science Publishers B.V. for their excellent cooperation.

This conference was sponsored by A.B. Chance Company, Digital Equipment Corporation, National Science Foundation and the University of Missouri-Columbia. Special thanks are due to Frank Myers of A.B. Chance Company and John Kitchin of Digital Equipment Corporation for their help. Also I would like to thank Judy Dooley, secretary of Department of Statistics, Kai Sun, a graduate student at the University of Missouri-Columbia, and my wife Sandra Basu for helping me with the editing of this volume.

Asit P. Basu
Columbia, Missouri

CONTENTS

ADVANCES IN RELIABILITY
Edited by Asit P. Basu
1993 Elsevier Science Publishers B.V.

A Comparison Of Various Estimators In Reliability Models Involving Mutual Censorship Of Component Lifelengths[1]

Robin Antoine
University of the West Indies
Hani Doss
Florida State University
Myles Hollander
Florida State University

Abstract
Doss, Freitag and Proschan (Ann. Statist., 1989) considered a model in which coherent systems of independent components are observed continuously until system failure. The failure time of each system, as well as the failure time of each dead component, is noted. They proposed and investigated estimators of the system and component life distributions. This paper considers alternative estimators that are applicable in two special situations. In the first of these it is known that all the components, or some subset of them, have identical life distributions. In the second case, the component life distributions are known to belong to some parametric family. In particular, the case in which they are exponential is considered. The asymptotic distributions of these two estimators are obtained. The asymptotic relative efficiency of each with respect to the estimator proposed by Doss, Freitag and Proschan is calculated.

1 Introduction and Summary

Let S be a coherent system of m independent components with structure function ϕ and reliability function h_ϕ (see Barlow and Proschan, 1981, Chapters 1 and 2, for definitions and basic facts concerning coherent systems). It is assumed that each component, and the system itself, is in either of two states, functioning or failed. Suppose that each of n independent systems identical to S is observed until it fails. For every component in each system, either a failure time or a censoring time is recorded. A censoring time is recorded if the component is still functioning at the time of system failure; otherwise the actual failure time is noted. Let the life distribution of the system be F and that of the jth component be F_j. We wish to estimate the vector $(F_1, F_2, \ldots, F_m)$ and the system distribution, F. This is the model studied by Doss, Freitag and Proschan (1989) (hereafter referred to as DFP). Let T_i be the lifelength

[1]*Key words and phrases:* Kaplan-Meier estimator, reliability function, asymptotic relative efficiency.
Research supported by the Air Force Office of Scientific Research Grants No. 90-0202 and 91-0048.

of system i. F may be estimated by the empirical estimator

$$\hat{F}^{\text{emp}}(t) = \frac{1}{n}\sum_{i=1}^{n} I(T_i \leq t),$$

where $I(A)$ is the indicator of the set A. This, however, does not make use of the information about component failures. DFP proposed estimators, $\hat{F}$ and $\hat{F}_j$, of F and F_j respectively, which use all the information available. They derived the asymptotic distributions of these estimators and calculated the relative efficiency of $\hat{F}(t)$ with respect to $\hat{F}^{\text{emp}}(t)$.

This paper considers two special situations in which more is known about the component life distributions. In the first of these, it is known that the distributions $F_1, \ldots, F_k$ are identical. In the second, the component life distributions are known to belong to some parametric family. In particular, the case in which the distributions are exponential is considered. Estimators of F and F_j are proposed which make use of this extra information, and the asymptotic relative efficiencies of these estimators with respect to the corresponding DFP estimators are calculated.

In order to more fully describe the statistical model and the various estimators being considered, we introduce some notation. Let

$$X_{ij} = \text{lifelength of component } j \text{ in system } i,$$

$$Z_{ij} = \min(X_{ij}, T_i),$$

and

$$\delta_{ij} = I(X_{ij} \leq T_i).$$

Thus $X_{1j}, X_{2j}, \ldots, X_{nj}$ are iid $\sim F_j$.

Define H_j to be the common distribution of $Z_{1j}, Z_{2j}, \ldots, Z_{nj}$. Here, and hereafter, i indexes systems and j indexes components; i ranges from 1 to n and j ranges from 1 to m.

Following Barlow and Proschan (1981), for $y = (y_1, \ldots, y_m) \in [0,1]^m$, $\alpha \in [0,1]$, and $j = 1, \ldots, m$, let

$$(\alpha_j, y) = (y_1, \ldots, y_{j-1}, \alpha, y_{j+1}, \ldots, y_m).$$

For a distribution function G, let $\bar{G}(t)$ denote $1 - G(t)$. It is well known (Barlow and Proschan, 1981, page 23) that

$$\bar{H}_j(t) = \bar{F}_j(t) h_\phi(1_j, \underline{\bar{F}}(t)),$$

where

$$\underline{\bar{F}}(t) = (\bar{F}_1(t), \ldots, \bar{F}_m(t)).$$

Let $Z_{(1)j} \leq Z_{(2)j} \leq \cdots \leq Z_{(n)j}$ be the ordered Z_{ij} and $T_{(1)} \leq T_{(2)} \leq \cdots \leq T_{(n)}$ be the ordered T_n.

Define

$$\delta_{(i)j} = \begin{cases} 1 & \text{if } Z_{(i)j} \text{ corresponds to an uncensored observation,} \\ 0 & \text{otherwise.} \end{cases}$$

Let $\hat{F}_j(t)$ be the Kaplan-Meier estimator of F_j,

$$\hat{F}_j(t) = 1 - \prod_{i:Z_{(i)j} \le t} \left(\frac{n-i}{n-i+1} \right)^{\delta_{(i)j}}. \tag{1}$$

Note that

$$\bar{F}(t) = h_\phi(\bar{F}_1(t), \ldots, \bar{F}_m(t)). \tag{2}$$

DFP estimate $F(t)$ by

$$\hat{F}(t) = \begin{cases} 1 - h_\phi(\hat{\bar{F}}_1(t), \ldots, \hat{\bar{F}}_m(t)) & \text{if } 0 \le t < T_{(n)}, \\ 1 & \text{if } t \ge T_{(n)}. \end{cases} \tag{3}$$

Each component is censored by the system lifelength, which is not independent of the component lifelengths. DFP overcome this difficulty by showing that, corresponding to a generic system with generic random variables X_j, Z_j, δ_j and T, such that the random vector (X_j, Z_j, δ_j, T) has the same distribution as $(X_{ij}, Z_{ij}, \delta_{ij}, T_i)$ for $i = 1, 2, \ldots, n$, and $j = 1, 2, \ldots, m$, there exist random variables Y_j such that

$$(Z_j, \delta_j) = (X_j \wedge Y_j, I(X_j \le Y_j))$$

and X_j and Y_j are independent.
Define the binary function ϕ_j by

$$\phi_j(u_1, \ldots, u_m) = \phi(1_j, u_1, \ldots, u_m), \qquad u_k = 0, 1, \qquad k = 1, \ldots, m,$$

where ϕ is the structure function.
Then

$$Y_{ij} = \sup\{t; \phi_j(I(X_{i1} > t), \ldots, I(X_{im} > t)) = 1\}. \tag{4}$$

Y_j may be thought of as being the lifelength of the system if X_j is replaced by ∞.
Using this construction, DFP prove the following theorems.

Theorem 1 (DFP) *Suppose $F_1, \ldots, F_m$ are continuous and suppose that T is such that $F_j(T) \le 1$ for $j = 1, 2, \ldots, m$. Then, as $n \to \infty$,*

$$n^{1/2}(\hat{F}_1 - F_1, \ldots, \hat{F}_m - F_m) \to (W_1, \ldots, W_m)$$

weakly in $D^m[0, T]$, where $W_1, \ldots, W_m$ are independent *mean 0 Gaussian processes such that*

$$\operatorname{Cov}(W_j(t_1), W_j(t_2)) = \bar{F}_j(t_1)\bar{F}_j(t_2)) \int_0^{t_1} \frac{dF_j(u)}{\bar{H}_j(u)\bar{F}_j(u)} \tag{5}$$

for $0 \le t_1 \le t_2 \le T$.

Theorem 2 (DFP) *Suppose $F_1, F_2, \ldots, F_m$ are continuous, and suppose T is such that $F_j(T) < 1$ for $j = 1, \ldots, m$. Then, as $m \to \infty$,*

$$n^{1/2}(\hat{F} - F) \to W \qquad \textit{weakly in } D[0, T],$$

where W is a mean 0 Gaussian process with covariance structure given by

$$Cov(W(t_1), W(t_2)) = \sum_{j=1}^{m} I_j(t_1) I_j(t_2) \bar{F}_j(t_1) \bar{F}_j(t_2) \int_0^{t_1} \frac{dF_j(u)}{\bar{H}_j(u) \bar{F}_j(u)} \tag{6}$$

for $0 \le t_1 \le t_2 \le T$,
and where

$$I_j(t) = \left. \frac{\partial h_\phi}{\partial u_j}(u_1, \ldots, u_m) \right|_{u_j = F_j(t),\ j=1,\ldots,m.} \tag{7}$$

Suppose that components $1, \ldots, k$ have identical life distributions, i.e.

$$F_1 = F_2 = \cdots F_k = G.$$

Then we may pool these these observations to form a single set and construct a Kaplan-Meier estimator of G based on this set of nk observations. Let $Z_{[i]}$ be the ith largest observation obtained from this (pooled) set of observations $Z_{11}, \ldots, Z_{nk}$. Define the pooled estimator, $\hat{G}$, of G to be

$$\hat{G}(t) = 1 - \prod_{Z_{[i]} \le t} \left(\frac{nk - i}{nk - i + 1} \right)^{\delta_i}. \tag{8}$$

The pooled estimator, $\hat{G}$, may be used to define a second estimator, $\tilde{F}$, of F as follows:

$$\tilde{F}(t) = \begin{cases} 1 - h_\phi(u_1, \ldots, u_k, \ldots, u_m) & \text{if} \quad t \le T_{(n)}, \\ 1 & \text{otherwise}, \end{cases}$$

where

$$u_l = \begin{cases} \hat{\hat{G}}(t) & l \le k, \\ \hat{\hat{F}}_l(t) & l > k. \end{cases}$$

The following example uses hypothetical data to illustrate how the estimators of the component and system life distributions are calculated.

Example 1 Consider the following system:

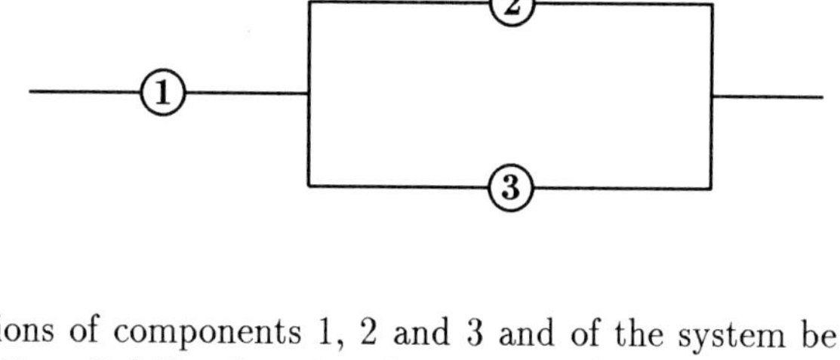

Let the distributions of components 1, 2 and 3 and of the system be F_1, F_2, F_3 and F respectively. The reliability function, h, of the system is

$$h(u_1, u_2, u_3) = u_1(1 - (1 - u_2)(1 - u_3)).$$

Suppose that six observations are made on the system:

$$\big((1.8,1),(1.8,0),(1.8,0)\big),\ \big((2.3,1),(1.7,1),(2.3,0)\big),\ \big((2.4,1),(2.4,0),(2.0,1)\big),$$
$$\big((2.7,0),(2.7,1),(2.1,1)\big),\ \big((1.9,0),(1.9,1),(1.5,1)\big),\ \big((2.5,1),(2.5,0),(2.5,0)\big).$$

The ordered observations corresponding to component 2 are: 1.7, 1.8, 1.9, 2.4, 2.5 and 2.7. The observations 1.7, 1.9 and 2.7 are uncensored. The Kaplan-Meier estimator places mass only on the uncensored observations. Using the redistribute to the right algorithm (Miller, 1981), a mass of 1/6 is placed on 1.7. Zero mass is placed on the censored observation, 1.8. A mass of (1/4)(5/6) is placed on 1.9. Zero mass is put on 2.4 and 2.5. The remaining mass of (3/4)(5/6) is put on 2.7. Thus the Kaplan-Meier estimate of $F_2(t)$ is

$$\hat{F}_2(t) = \begin{cases} 0 & \text{if } 0 < t < 1.7, \\ 1/6 & \text{if } 1.7 \le t < 1.9, \\ 9/24 & \text{if } 1.9 \le t < 2.7, \\ 1 & \text{if } t \ge 2.7. \end{cases}$$

The ordered observations from component 3 are 1.5, 1.8, 2.0, 2.1, 2.3 and 2.5. Masses of 1/6, (1/4)(5/6) and (1/3)(3/4)(5/6) are placed on the uncensored observations 1.5, 2.0 and 2.1 respectively. Thus

$$\hat{F}_3(t) = \begin{cases} 0 & \text{if } 0 < t < 1.5, \\ 1/6 & \text{if } 1.5 \le t < 2.0, \\ 9/24 & \text{if } 2.0 \le t < 2.1, \\ 14/24 & \text{if } t \ge 2.1. \end{cases}$$

Note that, using our definition of the Kaplan-Meier, the estimator of $F(t)$ is not arbitrarily defined to be 1 for values of t exceeding the largest observation, when that observation is censored.

Calculations similar to the above show that the estimator of $F_1(t)$ is

$$\hat{F}_1(t) = \begin{cases} 0 & \text{if } 0 < t < 1.8, \\ 1/6 & \text{if } 1.8 \le t < 2.3, \\ 9/24 & \text{if } 2.3 \le t < 2.4, \\ 14/24 & \text{if } 2.4 \le t < 2.5, \\ 19/24 & \text{if } t \ge 2.5. \end{cases}$$

DFP estimate $\bar{F}$ by substituting the estimates of $\bar{F}_1(t)$, $\bar{F}_2(t)$ and $\bar{F}_3(t)$ into the reliability function. Thus

$$\hat{F}(t) = 1 - \hat{\bar{F}}(t) = \hat{\bar{F}}_1(t)\big(1 - (1 - \hat{\bar{F}}_2(t))(1 - \hat{\bar{F}}_3(t))\big).$$

Suppose now that the distributions of components 2 and 3 are identical with common distribution G. The pooled estimator of $G(t)$ is calculated by constructing the Kaplan-Meier estimator based on the the sample obtained by pooling the observations from components 2 and 3. The ordered observations are therefore (1.5, 1), (1.7, 1),

(1.8, 0), (1.8, 0), (1.9, 1), (2.0, 1), (2.1, 1), (2.3, 0), (2.4, 0), (2.5, 0), (2, 5, 0) and (2.7, 1). Thus the pooled estimator of $G(t)$ is

$$\hat{G}(t) = \begin{cases} 0 & \text{if } 0 \leq t < 1.5, \\ 4/48 & \text{if } 1.5 \leq t < 1.7, \\ 8/48 & \text{if } 1.7 \leq t < 1.9, \\ 13/48 & \text{if } 1.9 \leq t < 2.0, \\ 18/48 & \text{if } 2.0 \leq t < 2.1, \\ 23/48 & \text{if } 2.1 \leq t < 2.7, \\ 1 & \text{if } t \geq 2.7. \end{cases}$$

Substituting $\hat{\bar{F}}_1(t)$, $\hat{\bar{F}}_2(t) = \hat{\bar{F}}_3(t) = \hat{G}(t)$ into the reliability function, we get that the pooled estimator of F is

$$\tilde{F}(t) = 1 - \tilde{\bar{F}}(t) = 1 - \hat{\bar{F}}_1(t)\Big(1 - (1 - \hat{\bar{G}}(t))^2\Big).$$

In Section 2 the asymptotic distributions of $\hat{G}$ and $\tilde{F}$ are derived. In Section 3 it is shown that they are at least as efficient as the corresponding DFP estimators. However, it is somewhat surprising that for certain systems, including the series, parallel and k-out-of-n systems, the pooled estimators are only *equal* in efficiency to the DFP estimators.

In Section 4, the efficiencies of the maximum likelihood estimators of F and F_j with respect to the DFP estimators are calculated. It is assumed that the component life distributions are exponential. It is not surprising that the DFP estimators are less efficient than their parametric counterparts (see Miller, 1983).

2 The Asymptotic Distributions of the Pooled Estimators $\hat{G}$ and $\tilde{F}$.

The asymptotic properties of the estimator, $\hat{G}$, can be obtained using methods almost identical to those used by DFP in obtaining Theorems 1 and 2.

We show, for any $T > 0$ satisfying $\max\{G(T), F_{k+1}(T), \ldots, F_m(T)\} < 1$, that

$$n^{1/2}\Big(\frac{\hat{G} - G}{\bar{G}}, \frac{\hat{F}_{k+1} - F_{k+1}}{\bar{F}_{k+1}}, \ldots, \frac{\hat{F}_m - F_m}{\bar{F}_m}\Big) \xrightarrow{d} \Big(W^*, W^*_{k+1}, \ldots, W^*_m\Big), \tag{9}$$

where $W^*, W^*_{k+1}, \ldots, W^*_m$ are independent mean zero Gaussian processes, and the symbol $\xrightarrow{d}$ signifies weak convergence in $D^{m-k+1}[0, T]$.

This is done by using a method introduced by Aalen (1978) and later refined by Gill (1980). Define the stopped processes

$$F^*_j(t) = F_j(t \wedge Z_{(n)j}) \quad \text{and} \quad G^*(t) = G(t \wedge Z_{[n]}).$$

We show that

$$n^{1/2}\Big(\frac{\hat{G} - G^*}{\bar{G}^*}, \frac{\hat{F}_{k+1} - F^*_{k+1}}{\bar{F}^*_{k+1}}, \ldots, \frac{\hat{F}_m - F^*_m}{\bar{F}^*_m}\Big) \xrightarrow{d} \Big(W^*, W^*_{k+1}, \ldots, W^*_m\Big). \tag{10}$$

This is enough to prove (9) since it is easy to see that the difference between the left side of (9) and the left side of (10) converges to 0 in probability in $D^{m-k+1}[0,T]$.

Let us consider an arbitrary censoring mechanism for component j of system i. Let

$$C_{ij} = I(X_{ij} \text{ is under observation at time t}).$$

We now define the following processes. The processes on the left involve no censoring. For each such process, there is a corresponding process for the case involving censoring, on the right.

Define, for $j = 1, \ldots, m$:

$$N_{ij}(t) = I(X_{ij} \le t), \qquad N_{ij}^c(t) = \int_0^t C_{ij}(s)dN_{ij}(s);$$

$$N_j(t) = \sum_{i=1}^n N_{ij}(t), \qquad N_j^c(t) = \sum_{i=1}^n N_{ij}^c(t);$$

$$V_{ij}(t) = I(X_{ij} \ge t), \qquad V_{ij}^c(t) = C_{ij}(t)V_{ij}(t);$$

$$V_j(t) = \sum_{i=1}^n V_{ij}(t), \qquad V_j^c(t) = \sum_{i=1}^n V_{ij}^c(t);$$

$$A_{ij}(t) = \int_0^t \frac{V_{ij}(s)}{\bar{F}_{ij}(s)} dF_j(s), \qquad A_{ij}^c(t) = \int_0^t \frac{V_{ij}^c(s)}{\bar{F}_{ij}(s)} dF_j(s);$$

$$A_j(t) = \int_0^t \frac{V_j(s)}{\bar{F}_j(s)} dF_j(s), \qquad A_j^c(s) = \int_0^t \frac{V_j^c(s)}{\bar{F}_j(s)} dF_j(s);$$

$$M_{ij}(t) = N_{ij}(t) - A_{ij}(t), \qquad M_{ij}^c(t) = N_{ij}^c(t) - A_{ij}^c(t);$$

$$M_j(t) = N_j(t) - A_j(t), \qquad M_j^c(t) = N_j^c(t) - A_j^c(t);$$

$$N(t) = \sum_{j=1}^k N_j(t), \qquad N^c(t) = \sum_{j=1}^k N_j^c(t);$$

$$V(t) = \sum_{j=1}^k V_j(t), \qquad V^c(t) = \sum_{j=1}^k V_j^c(t);$$

$$A(t) = \int_0^t \frac{V(s)}{\bar{G}(s)} dG(s) = \sum_{j=1}^k A_j(t), \qquad A^c(t) = \int_0^t \frac{V^c(s)}{\bar{G}(s)} = \sum_{j=1}^k A_j^c(t)$$

and

$$M(t) = N(t) - A(t), \qquad M^c(t) = N^c(t) - A^c(t).$$

Define the filtrations:

$$\mathcal{F}_t = \text{completion of } \sigma(N_{ij}(s)) \text{ for } 1 \le i \le n,\ 1 \le j \le m,\ s \le t,$$
$$\mathcal{F}_t^c = \text{completion of } \sigma(N_{ij}^c(s)) \text{ for } 1 \le i \le n,\ 1 \le j \le m,\ s \le t.$$

Define also

$$J_j^c(t) = I(V_j^c(t) > 0) \quad \text{and} \quad J^c(t) = I(V^c(t) > 0).$$

To show (9) we first establish that for each n,

$$\Big\{n^{1/2}\Big(\frac{\hat{F}_j(t) - F_j^*(t)}{\bar{F}_j^*(t)}\Big); t \in [0,T]\Big\}, \qquad j = k+1, \dots, m \tag{11}$$

and

$$\Big\{n^{1/2}\Big(\frac{\hat{G}(t) - G^*(t)}{\bar{G}^*(t)}\Big); t \in [0,T]\Big\} \tag{12}$$

are orthogonal square integrable martingales with respect to an appropriate family of σ-fields. Weak convergence in (9) then follows from a multivariate version of a martingale central limit theorem due to Rebolledo (1980).

DFP show (Proposition 2.2, DFP) the following.

Theorem 3 *Suppose that $F_1, \dots, F_m$ are continuous and $\max_{1 \le j \le m} F_j(t) < 1$ for all $t \ge 0$. Then for each j and for all n,*

$$n^{1/2}\Big(\frac{\hat{F}_j(t) - F_j^*(t)}{\bar{F}_j^*(t)}\Big) = n^{1/2}\int_0^t \frac{J_j^c(s)}{V_j^c(s)} \frac{\hat{\bar{F}}_j(s^-)}{\bar{F}_j(s)} dM_j^c(s). \tag{13}$$

Similarly, it can be easily seen that

$$n^{1/2}\Big(\frac{\hat{G}(t) - G^*(t)}{\bar{G}^*(t)}\Big) = n^{1/2}\int_0^t \frac{J^c(s)}{V^c(s)} \frac{\hat{\bar{G}}(s^-)}{\bar{G}_j(s)} dM^c(s). \tag{14}$$

The plan is to show that $M^c(t), M_{k+1}^c(t), \dots, M_m^c(t)$ are orthogonal martingales with respect to both $\{\mathcal{F}_t; t \in [0,T]\}$ or $\{\mathcal{F}_t^c; t \in [0,T]\}$. Then we will use Theorem 3 to arrive at the same conclusion for the processes in (9).

DFP show that the censoring processes $C_{ij}(t) = I(Y_{ij} \ge t)$, with Y_{ij} defined as in (4), are $\mathcal{F}_t$-predictable. Making use of the fact that $(M_{ij}(t), \mathcal{F}_t)$ is a martingale on $[0,T]$ and that

$$M_{ij}^c(t) = \int_0^t C_{ij}(s) dM_{ij}(s),$$

they proceed to show that $\{(M_{ij}(t), \mathcal{F}_t); t \in [0,T]\}$ and $\{(M_{ij}^c(t), \mathcal{F}_t^c); t \in [0,T]\}$ are both martingales. Since the sum of martingales is a martingale, this gives that $\{(M_j(t), \mathcal{F}_t); t \in [0,T]\}$ and $\{(M_j^c(t), \mathcal{F}_t^c); t \in [0,T]\}$ are square integrable martingales, with

$$\langle M_{j1}^c, M_{j2}^c \rangle(t) = \begin{cases} A_{j1}^c(t) & \text{if} \quad j_1 = j_2 \\ 0 & \text{if} \quad j_1 \ne j_2. \end{cases} \tag{15}$$

Orthogonality results from the fact that the counting processes $N_1^c(t), \dots, N_m^c(t)$ have no jumps in common, with probability one.

Similarly, the processes $\{(M^c(t), \mathcal{F}_t); t \in [0,T]\}$ and $\{(M^c(t), \mathcal{F}_t^c); t \in [0,T]\}$, being the sum of the martingales, $(M_1^c(t)), \ldots, (M_k^c(t))$, are themselves martingales. Moreover, $(M^c(t))$ is orthogonal to $(M_j^c(t))$ for $j = k+1, \ldots, m$, and

$$\langle M^c, M^c \rangle(t) = A^c(t). \tag{16}$$

We would now like to invoke Theorem 3 to conclude that the processes to the left of (9) are square integrable martingales. We must first check that the appropriate conditions are satisfied. The left continuous processes $J^c(s)$, $\hat{\bar{G}}(s-)$, $V^c(s)$ and $J_j^c(s)$, $\hat{\bar{F}}_j(s-)$ and $V_j^c(s)$ $(j = k+1, \ldots, m)$, are adapted to either $\{\mathcal{F}_t\}$ or $\{\mathcal{F}_t^c\}$. Therefore, the integrands on the right sides of (13) and (14) are predictable with respect to either $\{\mathcal{F}_t\}$ or $\{\mathcal{F}_t^c\}$. Since these integrands are also bounded and since the square integrable martingales M^c and M_j^c are of bounded variation, the integrals in (13) and (14) are also square integrable martingales. The covariation process is given by

$$\begin{aligned}
&\Big\langle n^{1/2}\Big(\frac{\hat{F}_{j1} - F_{j1}^*}{\bar{F}_{j1}^*}\Big), n^{1/2}\Big(\frac{\hat{F}_{j2} - F_{j2}^*}{\bar{F}_{j2}^*}\Big)\Big\rangle(t) \\
&\qquad = n\int_0^t \Big[\frac{J_{j1}^c(s)\hat{\bar{F}}_{j1}(s-)}{V_{j1}^c(s)\bar{F}_{j1}(s)}\Big]\Big[\frac{J_{j2}^c(s)\hat{\bar{F}}_{j2}(s-)}{V_{j2}^c(s)\bar{F}_{j2}(s)}\Big] d\langle M_{j1}^c, M_{j2}^c\rangle(s)
\end{aligned} \tag{17}$$

for $j_1 = k+1, \ldots, m$ and $j_2 = k+1, \ldots, m$,

$$\begin{aligned}
&\Big\langle n^{1/2}\Big(\frac{\hat{G} - G^*}{\bar{G}^*}\Big), n^{1/2}\Big(\frac{\hat{F}_j - F_j^*}{\bar{F}_j^*}\Big)\Big\rangle(t) \\
&\qquad = n\int_0^t \Big[\frac{J^c(s)\hat{\bar{G}}(s-)}{V^c(s)\bar{G}(s)}\Big]\Big[\frac{J_j^c(s)\hat{\bar{F}}_j(s-)}{V_j^c(s)\bar{F}_j(s)}\Big] d\langle M^c, M_j^c\rangle(s)
\end{aligned} \tag{18}$$

for $j = k+1, \ldots, m$,
and

$$\begin{aligned}
&\Big\langle n^{1/2}\Big(\frac{\hat{G} - G^*}{\bar{G}^*}\Big), \Big(\frac{\hat{G} - G^*}{\bar{G}^*}\Big)\Big\rangle(t) \\
&\qquad = n\int_0^t \Big[\frac{J^c(s)\hat{\bar{G}}(s-)}{V^c(s)\bar{G}(s)}\Big]^2 d\langle M^c, M^c\rangle(s).
\end{aligned} \tag{19}$$

Using (15) and (16) we see that (17) is 0 for $j_1 \neq j_2$, (18) is 0 for all j and

$$\begin{aligned}
\Big\langle n^{1/2}\Big(\frac{\hat{F}_j - F_j^*}{\bar{F}_j^*}\Big), n^{1/2}\Big(\frac{\hat{F}_j - F_j^*}{\bar{F}_j^*}\Big)\Big\rangle(t) &= n\int_0^t \Big[\frac{J_j^c(s)\hat{\bar{F}}_j(s-)}{V_j^c(s)\bar{F}_j(s)}\Big]^2 dA_j^c(s) \\
&= n\int_0^t \Big[\frac{J_j^c(s)\big(\hat{\bar{F}}_j(s-)\big)^2}{V_j^c(s)\big(\bar{F}_j(s)\big)^3}\Big] dF_j(s),
\end{aligned}$$

and

$$\left\langle n^{1/2}\Big(\frac{\hat{G}-G^*}{\bar{G}^*}\Big), n^{1/2}\Big(\frac{\hat{G}-G^*}{\bar{G}^*}\Big)\right\rangle(t) = n\int_0^t \left[\frac{J^c(s)\hat{\bar{G}}(s-)}{V^c(s)\bar{G}(s)}\right]^2 dA^c(s)$$

$$= n\int_0^t \left[\frac{J^c(s)\Big(\hat{\bar{G}}(s-)\Big)^2}{V^c(s)\Big(\bar{G}(s)\Big)^3}\right] dG(s).$$

We may now proceed as in Theorem 1.1 of Gill (1983) to prove weak covergence of $n^{1/2}(\hat{F}_j - F_j^*)/\bar{F}_j^*$ and $n^{1/2}(\hat{G} - G^*)/\bar{G}^*$. The joint weak convergence in (9) follows from the orthogonality results of (15) and (16) and a multivariate extension of Theorem V.1 of Rebolledo (1980).
These results lead to Theorem 4 below.

Theorem 4 *$n^{1/2}(\hat{G} - G, \hat{F}_{k+1} - F_{k+1}, \ldots, \hat{F}_m - F_m) \to (W, W_{k+1}, \ldots, W_m)$ weakly in $D^{m-k+1}[0,T]$, where $W, \ldots, W_m$ are independent, mean 0 Gaussian processes such that*

$$Cov(W(t_1), W(t_2)) = \bar{G}(t_1)\bar{G}(t_2)\int_0^{t_1} \frac{dG(u)}{\sum_{j-1}^k \bar{H}_j(u)\bar{G}(u)} \tag{20}$$

for $0 \le t_1 \le t_2 \le T$,
and

$$Cov(W_j(t_1), W_j(t_2)) = \bar{F}_j(t_1)\bar{F}_j(t_2)\int_0^{t_1} \frac{dF_j(u)}{\bar{H}_j(u)\bar{F}_j(u)} \tag{21}$$

for $0 \le t_1 \le t_2 \le T$ and $j > k$.

We now show that $\tilde{F}$ is a strong uniform consistent estimator of F and obtain its asymptotic distribution. We will need the following lemma.

Lemma 5 *For any system of m independent components, let ϕ and $h_\phi(u_1, \ldots, u_m)$ be the structure and reliability functions respectively. Then $h_\phi(u_1, \ldots, u_m)\Big|_{u_1=u_2=\cdots=u_k=u}$ is twice continuously differentiable over $[0,\ 1]^{m-k+1}$ and the first and second partial derivatives are bounded in absolute value by k and k^2 respectively over $[0,\ 1]^{m-k+1}$.*

Proof

The reliability function, h_ϕ, is a polynomial in $u_1, \ldots, u_m$, and the power of u_j in each term is at most 1 . In other words, h_ϕ is a linear combination of terms of the form $u_1^{\alpha_1}u_2^{\alpha_2}, \ldots, u_m^{\alpha_m}$, where α_j is 0 or 1. Thus $h_\phi(u_1, \ldots, u_m)|_{u_1=u_2=\cdots=u_k=u}$ is a polynomial in $u, u_{k+1}, \ldots, u_m$ that is of order k in u and 1 in the other variables. Hence

$$\frac{\partial}{\partial u}\Big(h_\phi|_{u_1=u_2=\cdots=u_k=u}\Big) \le k \max_{j\le k}\Big\{\frac{\partial h_\phi}{\partial u_j}\Big|_{u_1=\cdots=u_k=u}\Big\},$$

$$\frac{\partial^2}{\partial u \partial u_{j_2}}\Big(h_\phi|_{u_1=u_2=\cdots=u_k=u}\Big) \le k \max_{j_1\le k}\Big\{\frac{\partial^2 h_\phi}{\partial u_{j_1}\partial u_{j_2}}\Big|_{u_1=\cdots=u_k=u}\Big\} \quad \text{for } j_2 \ge k+1,$$

and

$$\frac{\partial^2}{\partial^2 u}\left(h_\phi|_{u_1=u_2=\cdots=u_k=u}\right) \le k(k-1)\max_{j_1,j_2\le k}\left\{\frac{\partial^2 h_\phi}{\partial u_{j_1}\partial u_{j_2}}\Big|_{u_1=\cdots=u_k=u}\right\}.$$

But

$$h_\phi(u_1,\ldots,u_m) = u_j h_\phi(1_j,u) + (1-u_j)h_\phi(0_j,u).$$

Hence,

$$\frac{\partial h_\phi}{\partial u_j}\Big|_u = h_\phi(1_j,u) - h_\phi(0_j,u), \tag{22}$$

and

$$\frac{\partial^2 h_\phi}{\partial u_{j1}u_{j2}} = \{h_\phi(1_{j1},1_{j2},u) - h_\phi(1_{j1},0_{j2},u)\} - \{h_\phi(0_{j1},1_{j2},u) - h_\phi(0_{j1},0_{j2})\}. \tag{23}$$

By the monotonicity of the structure function, (22) is positive and lies between 0 and 1. Similarly, (23) lies between 0 and 1. The result follows.

We may now apply the convergence results for the individual $\hat{F}_j$'s given in Foldes, Rejto and Winter (1980), together with Lemma 5, to obtain the strong consistency of $\tilde{F}$.

Let

$$I(t) = \frac{\partial}{\partial u}\left\{h_\phi(u_1,\ldots,u_m)\Big|_{u_1=\cdots=u_k=u}\right\}\Big|_{u=\bar{G}(t) \text{ and } u_j=F_j(t) \text{ for } j=k+1,\ldots,m}. \tag{24}$$

Then, applying Taylor's Theorem and Lemma 5,

$$n^{1/2}\left|\tilde{F}(t) - F(t) - \left\{I(t)(\hat{G}(t)-G(t)) + \sum_{j=k+1}^{m} I_j(t)(\hat{F}_j(t)-F_j(t))\right\}\right| \tag{25}$$

$$\le k^2\frac{n^{1/2}}{2}\Big\{\sum_{j_1=k+1}^{m}\sum_{j_2=k+1}^{m}\Big(\sup_{0\le t\le T}|\hat{F}_{j_1}(t)-F_{j_1}(t)|\Big)\Big(\sup_{0\le t\le T}|\hat{F}_{j_2}(t)-F_{j_2}(t)|\Big)$$

$$+\sum_{j=k+1}^{m}\Big(\sup_{0\le t\le T}|\hat{G}(t)-G(t)|\Big)\Big(\sup_{0\le t\le T}|\hat{F}_j(t)-F_j(t)|\Big)$$

$$+\Big(\sup_{0\le t\le T}|\hat{G}(t)-G(t)|\Big)^2\Big\}.$$

The right side of this expression converges to 0 a.s. by the results of Foldes and Rejto (1981). Since convergence in sup norm implies convergence in the Skorohod topology (see page 111 of Billingsley, 1968), we conclude that the process in (25) converges to 0 a.s. in $D[0,T]$. Moreover, by Theorem 5.1 of Billingsley (1968) and Theorem 4

$$I(t)(\hat{G}(t)-G(t)) + \sum_{j=k+1}^{m} I_j(t)(\hat{F}_j(t)-F_j(t)) \xrightarrow{d} W',$$

where W' is a mean 0 Gaussian process with covariance structure

$$\begin{aligned} Cov(W'(t_1), W'(t_2)) = &\ I(t_1)I(t_2)\bar{G}(t_1)\bar{G}(t_2)\int_0^{t_1}\frac{dG(u)}{\sum_{j=1}^{k}\bar{H}_j(u)\bar{G}(u)} \\ &+ \sum_{j=k+1}^{m} I_j(t_1)I_j(t_2)\bar{F}_j(t_1)\bar{F}_j(t_2)\int_0^{t_1}\frac{dF_j(u)}{\bar{H}_j(u)\bar{F}_j(u)}, \end{aligned} \tag{26}$$

for $0 \le t_1 \le t_2 \le T$.

Hence it follows that $n^{1/2}(\tilde{F} - F) \to W'$ weakly in $D[0,T]$.

3 Asymptotic Relative Efficiency of $\tilde{F}$ vs. $\hat{F}$

In this section we assume that all the components have the same distribution, i.e. $F_j = G$ for all $j = 1, \ldots, m$. Define the asymptotic relative efficiency of an estimator, $\hat{\theta}_1$, of θ to that of another estimator, $\hat{\theta}_2$, to be the ratio of the asymptotic variance of $\hat{\theta}_2$ to that of $\hat{\theta}_1$,

$$ARE(\hat{\theta}_1, \hat{\theta}_2) = \frac{AV(\hat{\theta}_2)}{AV(\hat{\theta}_1)}. \tag{27}$$

It is obvious from (20) and (5) that the asymptotic variance of $\hat{G}(t)$ is less than that of $\hat{F}_j(t)$. Thus

$$ARE(\hat{G}(t), \hat{F}_j(t)) \ge 1.$$

For convenience, let $\underline{\bar{G}}(t)$ be the vector $(\bar{G}(t), \bar{G}(t), \ldots, \bar{G}(t))$ and

$$h_\phi(1_j, \bar{G}(u)) = h_{1j}(u) \qquad \text{and} \qquad h_\phi(0_j, \bar{G}(u)) = h_{oj}(u).$$

In certain situations the h_{1j}'s are all equal. This is true, for example, in the series, parallel and k-out-of-n systems. In this case, from (20) and (5) it is immediate that

$$ARE(\hat{G}(u), \hat{F}_j(u)) = m$$

for all $j = 1, \ldots, m$.
By (6), the asymptotic variance of $\hat{F}(t)$ is

$$\begin{aligned} AV(\hat{F}(t)) &= \sum_{j=1}^{m} I_j^2(t)\bar{G}^2(t)\int_0^t \frac{dG(u)}{\bar{H}_j(u)\bar{G}(u)} \\ &= \sum_{j=1}^{m} I_j^2(t)\bar{G}^2(t)\int_0^t \frac{dG(u)}{\bar{G}(u)^2 h_{1j}(u)}. \end{aligned} \tag{28}$$

The asymptotic variance of $\tilde{F}(t)$ is, by (26),

$$AV(\tilde{F}(t)) = I^2(t)\bar{G}^2(t)\int_0^t \frac{dG(u)}{\bar{G}^2(u)\sum_{j=1}^{m} h_{1j}(u)}. \tag{29}$$

Now, since $h_\phi(u_1, \ldots, u_m) = u_j h(1_j, u_1, \ldots, u_m) + (1-u_j)h(0_j, u_1, \ldots, u_m)$ (Barlow and Proschan, 1981, page 21), it follows that the reliability function is a polynomial in $u_1, \ldots, u_m$ and is linear in each u_j. Hence

$$\frac{d}{du}\left\{h(u_1, \ldots, u_m)\Big|_{u_1=u_2=\cdots=u_m=u}\right\} = \sum_{j=1}^{m}\left\{\frac{\partial h}{\partial u_j}\Big|_{u_1=\cdots=u_m}\right\}.$$

Thus

$$I(t) = \sum_{j=1}^{m} I_j(t) = \sum_{j=1}^{m}(h_{1j}(t) - h_{0j}(t)).$$

Hence

$$AV(\tilde{F}(t)) = \Big(\sum_{j=1}^{m}(h_{1j}(t) - h_{0j}(t))\Big)^2 \bar{G}(t)^2 \int_0^t \frac{dG(u)}{\bar{G}^2(u)\sum_{j=1}^{m} h_{1j}(u)}. \tag{30}$$

Equations (28) and (30) lead to the following theorem.

Theorem 6 $ARE(\tilde{F}(t), \hat{F}(t)) \geq 1$.

Proof
By definition,

$$ARE(\tilde{F}(t), \hat{F}(t)) = \frac{AV(\tilde{F}(t))}{AV(\hat{F}(t))}.$$

We show that $AV(\tilde{F}(t)) - AV(\hat{F}(t)) \geq 0$.

By (30) and (28)

$$\begin{aligned} &AV(\tilde{F}(t)) - AV(\hat{F}(t)) = \\ &\quad \bar{G}^2(t)\int_0^t \Big\{\Big\{\Big(\sum_{j=1}^{m} h_{1j}(u)\Big)\Big\{\sum_{j=1}^{m}\Big(h_{1j}(u)\prod_{l\neq j} h_{1l}(u)(h_{1j} - h_{0j})^2(t)\Big)\Big\} \\ &\quad -\Big(\sum_{j=1}^{m}(h_{1j} - h_{0j})(t)\Big)^2 \prod_{j=1}^{m} h_{1j}(u)\Big\}\Big/\Big(\prod_{j=1}^{m} h_{1j}(u)\Big)\sum_{j=1}^{m} h_{1j}(u)\Big\}\frac{dG(u)}{\bar{G}^2(u)}. \end{aligned} \tag{31}$$

The numerator of the expression inside the integral sign is equal to

$$\sum_{j=1}^{m}\Big\{\Big(\sum_{l\neq j} h_{1l}(u)(h_{1j} - h_{0j})^2(t) + \Big(\prod_{l=1}^{m} h_{1l}(u)\Big)(h_{1j} - h_{0j})^2(t)\Big\}$$

$$-\sum_{j=1}^{m}\Big\{\{\prod_{l=1}^{m} h_{1l}(u)(h_{1j} - h_{0j})^2(t)\} + \Big(\prod_{l=1}^{m} h_{1l}(u)\Big)\Big((h_{1j} - h_{0j}(t)\Big)\Big(\sum_{l\neq j}(h_{1l} - h_{0l})(t)\Big)\Big\}$$

$$= \sum_{j=1}^{m}\Big\{\Big(\sum_{l\neq j} h_{1l}(u)\Big)\Big(\prod_{l\neq j} h_{1l}(u)\Big)(h_{1j}-h_{0j})^2(t)\Big\}$$

$$-\sum_{j=1}^{m}\Big\{\Big(\prod_{l=1}^{m} h_{1l}(u)\Big)(h_{1j}-h_{0j}(t)\sum_{l\neq j}(h_{1l}-h_{0l})(t)\Big\}$$

$$= \sum_{j=1}^{m}\Big\{\Big(\prod_{l\neq j} h_{1l}(u)\Big)(h_{1j}-h_{0j})(t)\Big(\Big(\sum_{l\neq j} h_{1l}(u)\Big)(h_{1j}-h_{0j})(t)$$

$$-h_{1j}(u)\textstyle\sum_{l\neq j}(h_{1l}-h_{0l})(t)\Big)\Big\}$$

Let $a_j = h_{1j}(u)$ and $b_j = (h_{1j}-h_{0j})(t)$ for $j = 1, \ldots, n$. Then the expression becomes

$$\sum_{j=1}^{m}\Big(\prod_{l\neq j} a_l\Big)b_j\Big\{\Big(\sum_{l\neq j} a_l\Big)b_j - \Big(\sum_{l\neq j} b_l\Big)a_j\Big\}$$

$$\begin{aligned}
&= \sum_{j=1}^{m}\Big(\prod_{l\neq j} a_l\Big)\Big(\sum_{l\neq j} a_l\Big)b_j^2 - \sum_{j=1}^{m}\Big(\prod_{l\neq j} a_l\Big)\Big(\sum_{l\neq j} b_l\Big)a_jb_j \\
&= \sum_{j=1}^{m}\Big\{\sum_{l\neq j} a_l^2b_j^2\prod_{k\neq l,j} a_k\Big\} - \sum_{j=1}^{m}\Big\{\Big(\prod_{l\neq j} a_l\Big)b_ja_j\sum_{l\neq j} b_l\Big\} \\
&= \sum_{l\leq j}\Big\{(a_l^2b_j^2 + a_j^2b_l^2)\prod_{k\neq j,l} a_k\Big\} - \sum_{j=1}^{m}\Big\{\sum_{l\neq j}\Big\{a_lb_la_jb_j\prod_{k\neq l,j} a_k\Big\}\Big\} \\
&= \sum_{l\leq j}\Big\{(a_l^2b_j^2 + a_j^2b_l^2)\prod_{k\neq j,l} a_k\Big\} - \sum_{l\neq j} 2a_lb_la_jb_j\prod_{k\neq l,j} a_k \\
&= \sum_{l\leq j}\Big\{(a_jb_l - a_lb_j)^2\Big(\prod_{k\neq l,j} a_k\Big)\Big\} \\
&= \sum_{l\leq j}\Big\{\Big(h_{1j}(u)(h_{1l}-h_{0l})(t) - h_{1l}(u)(h_{1j}-h_{0j})(t)\Big)^2\Big(\prod_{k\neq l,j} h_{1k}(u)\Big)\Big\},
\end{aligned}$$

which is always non-negative. Hence $ARE(\tilde{F}(t), \hat{F}(t)) \geq 0$.

Note that if $h_{11}(t) = h_{12}(t) = \cdots = h_{1m}(t)$ and $h_{01}(t) = \cdots = h_{0m}(t)$ then $AV(\tilde{F}(t)) = AV(\hat{F}(t))$, so that $ARE(\tilde{F}(t), \hat{F}(t)) = 1$. This is the case with series, parallel and other k-out-of-n systems. This observation motivates the following definition.

Definition 1 Suppose that a coherent system S of m components is such that, when the component life distributions are all identical, $h_{11}(t) = h_{12}(t) = \cdots = h_{1m}(t)$ and $h_{01}(t) = \cdots = h_{0m}(t)$ for all t and any common distribution G, then S is said to be h-symmetric.

The following is an example of a system that is h-symmetric but not a k-out-of-n system.

Example 2 Consider the system:

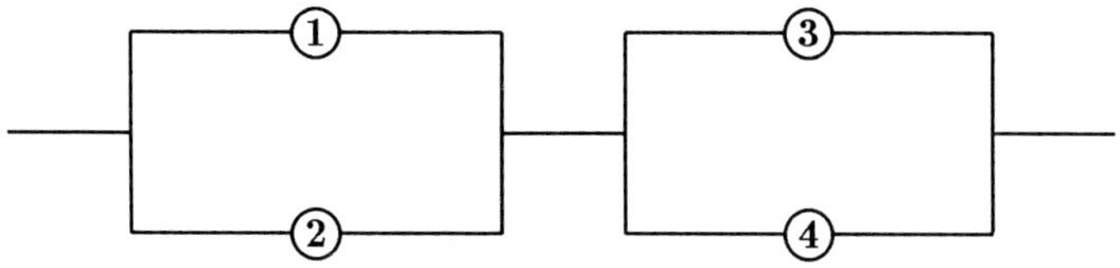

The structure function of this system is

$$\phi(x_1, x_2, x_3, x_4) = \Big(1 - (1 - x_1)(1 - x_2)\Big)\Big(1 - (1 - x_3)(1 - x_4)\Big).$$

The definition leads immediately to the following theorem:

Theorem 7 *If a system is h-symmetric and all the components have identical life distributions, then*

$$ARE(\tilde{F}(t), \hat{F}(t)) = 1.$$

If the system is not h-symmetric then the asymptotic relative efficiency of $\tilde{F}(t)$ versus $\hat{F}(t)$ may be strictly greater than 1. In general, the asymptotic relative efficiency depends on the structure of the system and on the value of t.

Example 3 Consider the system, S, of m components with reliability function

$$h(u_1, u_2, \ldots, u_m) = 1 - (1 - u_1)\Big(1 - \prod_{j=2}^{m} u_j\Big). \tag{32}$$

S consists of component 1 in parallel with the other components, which form a module of $m - 1$ components in series. The following diagram shows the case in which $m = 4$.

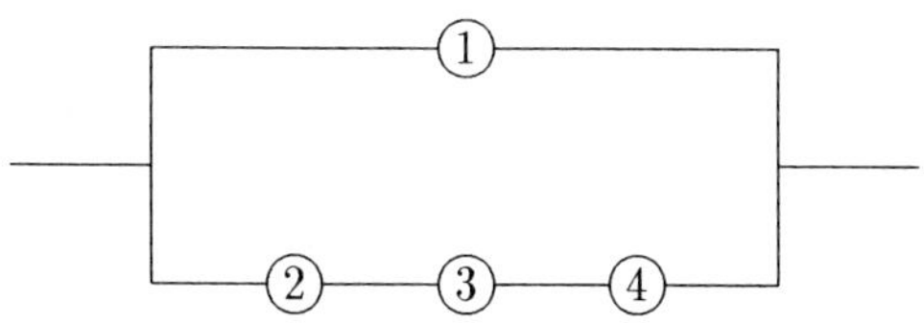

Let the common distribution of the components be G. Then, by (7),

$$I_1(t) = 1 - \Big(\bar{G}(t)\Big)^{m-1} \tag{33}$$

and, for $j > 1$,

$$I_j(t) = G(t)\bar{G}(t)^{m-2}. \tag{34}$$

Figure 1: Plot of ARE versus $G(t)$.

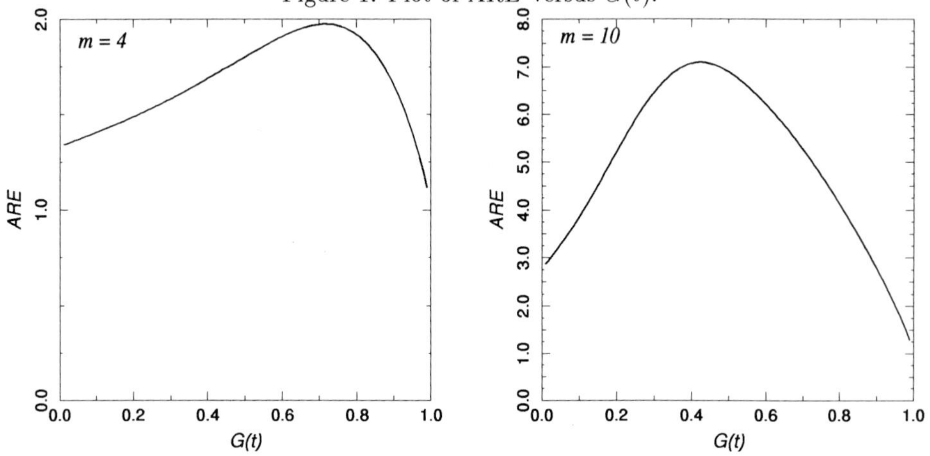

By (28), the asymptotic variance of $\hat{F}(t)$ is

$$AV(\hat{F}(t)) = \sum_{j=1}^{m} I_j^2(t)\bar{G}^2(t) \int_0^t \frac{dG(u)}{\bar{G}(u)^2 h_{1j}(u)}. \tag{35}$$

Now, putting $u_1 = 1$ in (32), we get

$$h_{11}(u) = 1$$

and

$$\begin{aligned} h_{1j}(u) &= 1 - G(u)\left(1 - \bar{G}(u)^{m-2}\right) \\ &= \bar{G}(u)\left\{1 + G(u)\bar{G}(u)^{m-3}\right\}. \end{aligned}$$

Thus

$$\begin{aligned} AV(\hat{F}(t)) &= \left(1 - \bar{G}(t)^{m-1}\right)^2 \bar{G}(t)^2 \int_0^t \frac{dG(u)}{\bar{G}^2(u)} \\ &\quad + (m-1)G^2(t)\bar{G}(t)^{2m-2} \int_0^t \frac{dG(u)}{\bar{G}^3(u)\left\{1 + G(u)\bar{G}^{m-3}(u)\right\}}. \end{aligned}$$

By (30), we have

$$\begin{aligned} AV(\tilde{F}(t) &= \left(1 - \bar{G}(t)^{m-1} + (m-1)G(t)\bar{G}(t)^{m-2}\right)^2 \bar{G}^2(t) \\ &\quad \int_0^t \frac{dG(u)}{\bar{G}^2(u)\left\{1 + (m-1)\bar{G}(u)\left(1 + G(u))\bar{G}(u)^{m-3}\right)\right\}} \end{aligned}$$

The asymptotic variances of $\tilde{F}(t)$ and $\hat{F}(t)$ are functions of t only through $G(t)$. The plots in Figure 1 show the relationship between ARE($\tilde{F}(t), \hat{F}(t)$) and $G(t)$ for $m = 4$ and $m = 10$.

4 A Competing Parametric Estimator

In this section it is assumed that all the components have exponential life distributions,

$$F_j \sim \exp(\beta_j) \quad \text{for} \quad j = 1, \ldots, m.$$

The maximum likelihood estimators for the series and parallel systems are derived, and the asymptotic relative efficiencies of these estimators versus the corresponding DFP estimators are computed.

Suppose that the components of S are in series. Then the DFP estimator of F is the same as the empirical c.d.f.
Thus

$$n\hat{F}(t) \sim \text{Binomial}(n, F(t))$$

and

$$\text{Var}(\hat{F}(t)) = \frac{F(t)(1 - F(t))}{n} = \frac{\exp\{-\beta t\}(1 - \exp\{-\beta t\})}{n},$$

where $\beta = \sum_{j=1}^{m} \beta_j$.

Suppose that the k_i^{th} component failed in the ith system. The likelihood function is then

$$\begin{aligned} L(\beta_1, \ldots, \beta_m) &= \prod_{i=1}^{n} \Big(\frac{f_{k_i}(t_i)}{\bar{F}_{k_i}(t_i)} \prod_{j=1}^{m} \bar{F}_j(t_i) \Big) \\ &= \Big(\prod_{j=1}^{m} \beta_j^{n_j} \Big) \exp\Big\{ (\sum_{j=1}^{m} \beta_j)(\sum_{i=1}^{n} t_i) \Big\}, \end{aligned}$$

where $n_j = \sum_{i=1}^{n} \delta_{ij}$ = number of failures of the jth component.
This implies

$$\hat{\beta}_j = \frac{n_j}{\sum_{i=1}^{n} t_i} \quad \text{and} \quad \hat{\beta} = \sum_{j=1}^{m} \beta_j = \frac{n}{\sum_{i=1}^{n} t_i}.$$

Thus

$$\hat{F}^m(t) = 1 - \exp\left\{ \left(-\frac{n}{\sum_{i=1}^{n} t_i} \right) t \right\},$$

where $\hat{F}^m$ denotes the maximum likelihood estimator of F. Now

$$-\frac{\partial^2 log L}{\partial \beta_j \beta_k} = \begin{cases} \frac{n_j}{\beta_j^2} & \text{if} \quad j = k, \\ 0 & \text{otherwise.} \end{cases}$$

Therefore

$$I_{jk}(\beta_1, \ldots, \beta_m) = \begin{cases} \frac{n}{\beta \beta_j} & \text{if} \quad j = k, \\ 0 & \text{otherwise.} \end{cases}$$

Hence

$$AV(\hat{\beta}) = \sum_{j=1}^{m} AV(\hat{\beta}_j) = \sum_{j=1}^{m} \frac{\beta \beta_j}{n} = \frac{\beta^2}{n}.$$

By Theorem A, page 118 of Serfling, 1980, the asymptotic variance of $\hat{g}(\beta)$, where

$$g(\hat{\beta}) = 1 - \exp\{-\hat{\beta}t\},$$

is

$$g'(\beta)^2 AV(\hat{\beta}) = \frac{\beta^2 t^2 \exp\{-2\beta t\}}{n}.$$

Hence the asymptotic relative efficiency of $\hat{F}^m(t)$ with respect to $\hat{F}(t)$ is

$$\frac{\exp\{\beta t\} - 1}{\beta^2 t^2}.$$

As $t \to 0$ or $t \to \infty$, $ARE(\hat{F}^m(t), \hat{F}(t)) \to \infty$.

The minimum value of the ARE occurs at $t = \dfrac{2}{\beta}$ and is $\dfrac{e^2 - 1}{4}$, or approximately 1.59.

Suppose now that the components of S are in parallel. In this case there is no censoring, so the component lifetimes, x_{ij}, are themselves observed. The MLE of β_j is

$$\hat{\beta}_j = \frac{n}{\sum_{i=1}^n x_{ij}}.$$

Moreover, $\hat{\beta}_j$ is independent of $\hat{\beta}_k$ for $j \neq k$. Thus the MLE of $F(t)$ is

$$\hat{F}^m(t) = \prod_{j=1}^m (1 - \exp\{-\hat{\beta}_j t\}) = \prod_{j=1}^m \left(1 - \exp\left\{-\left(\frac{n}{\sum_{i=1}^n x_{ij}}\right) t\right\}\right).$$

Let

$$K = \log(\hat{F}^m(t)) = \sum_{j=1}^m \log(1 - \exp\{-\hat{\beta}_j(t)\}).$$

Now $\hat{\beta}_j$ is $AN(\beta_j, \beta_j^2/n)$. Therefore by the delta method, K has asymptotic variance

$$AV(K) = \frac{t^2}{n} \sum_{j=1}^m \frac{\beta_j^2 \exp\{-2\beta_j t\}}{(1 - \exp\{-\beta_j t\})^2}.$$

Hence, the asymptotic variance of $\hat{F}^m(t)$ is

$$AV(\hat{F}^m(t)) = \exp\{2\mu_k\} AV(K),$$

where

$$\mu_k = \sum_{j=1}^m \log(1 - \exp\{-\beta_j t\}).$$

Therefore, using the formula for the asymptotic variance of $\hat{F}(t)$ from DFP, we get

$$ARE(\hat{F}^m(t), \hat{F}(t)) = \frac{\sum_{j=1}^m (\exp\{-\beta_j t\}/(1 - \exp\{-\beta_j t\}))}{t^2 \sum_{j=1}^m (\beta_j^2/(1 - \exp\{\beta_j t\})^2)}.$$

As $t \to 0$, $ARE(\hat{F}^m(t), \hat{F}(t)) \to \infty$.

References

Aalen, O. (1978). Nonparametric inference for a family of counting processes. *Ann. Statist.* (6) 701-706.

Barlow, R. and Proschan, F. (1981). *Statistical Theory of Reliability and Life Testing.* To Begin With, Silverspring, Maryland.

Billingsley, P. (1981). *The Convergence of Probability Measures.* John Wiley and Sons, NY.

Doss, H., Freitag, S. and Proschan, F. (1989). Estimating Jointly System and Component Reliabilities Using a Mutual Censorship Approach. *Ann. Statist.* (17) 764-782.

Foldes, A. and Rejto, L. (1981) A LIL type result for the product limit estimator. *Z. Wahrsch. verw. Gebiete* (56) 75-86.

Foldes, A., Rejto, L. and Winter, B. B. (1980) Strong consistency properties of nonparametric estimators for randomly censored data (Part I). *Periodica Math. Hungar.* (11) 233-250.

Gill, R. D. (1980) *Censoring and Stochastic Integrals.* Mathematical Centre Tracts 124. Mathematisch Centrum, Amsterdam.

Gill, R. D. (1980) Large sample behaviour of the product-limit estimator on the whole line. *Ann. Statist.* (11) 49-58.

Miller, R. G. (1981). *Survival Analysis.* John Wiley and Sons, Inc., New York.

Miller, R. G. (1983). What Price Kaplan-Meier? *Biometrics* (39) 1077–1081.

Rebolledo, R. (1980). Central Limit Theorems for Local Martingales. *Z. Wahrsch. verw. Gebiete* (51) 269-289.

Serfling, Robert J. (1980). *Approximation Theorems of Mathematical Statistics.* John Wiley and Sons, NY.

ADVANCES IN RELIABILITY
Edited by Asit P. Basu

Goodness-of-Fit Tests for Power Law Processes

Donna J. Aubrey[a] and Steven E. Rigdon[b]

[a] McDonnell Aircraft Company, St. Louis, Missouri 63145

[b] Department of Mathematics and Statistics, Southern Illinois University at Edwardsville, Edwardsville, IL 62026

Abstract

The power law process, a nonhomogeneous Poisson process with intensity function $(\beta/\theta)(t/\theta)^{\beta-1}$, is a commonly used model for describing the failure times of a repairable system. Here we summarize properties of statistical inference for the power law process, with special attention to goodness-of-fit tests. Two transformations of the failure times can be used to derive goodness-of-fit tests. The problem for testing goodness-of-fit for several repairable systems is also addressed. We discuss a simulation study that was used to assess how the goodness-of-fit tests performed when applied to several nonidentical power law processes.

1. INTRODUCTION TO MODELS FOR REPAIRABLE SYSTEMS

Statistical models for repairable systems must describe the occurrence of events in time. In most cases, such models should be nonstationary, allowing the times between failures to increase (in which case the system is improving), or to decrease (in which case the system is deteriorating). An important quantity for studying models for repairable systems is the intensity function. This is the (limit of the) probability of having a failure in a small time interval divided by the length of the interval; that is,

$$u(t) = \lim_{\Delta t \to \infty} \frac{\mathrm{P}(\text{a failure in } [t,\, t+\Delta t])}{\Delta t}.$$

Here t is global time, that is, the time since the initial start-up of the system. The local time, usually denoted x, is the time since the last failure.

If the intensity function u is a function of t only, then, assuming simultaneous failures cannot occur, the process is a nonhomogeneous Poisson process (NHPP). For an NHPP with continuous intensity function a unit is in exactly the same condition just after a repair as it was just before the failure. For this reason the NHPP is often called a "bad-as-old" or a "same-as-old" model. On the other hand if u is a function of x only, then the times between failure are independent and identically distributed. This leads to the renewal process, or a "good-as-new" or "same-as-new" model.

If the intensity function is constant, $u = \lambda$, then the process is a homogeneous Poisson process and the times between failure are independent and exponentially distributed random variables with mean $1/\lambda$. The homogeneous Poisson process is the only model that is both a Poisson process and a renewal process. The failure times are typically measured in global time t.

2. THE POWER LAW PROCESS

A special case of the NHPP occurs when the intensity function has the form

$$u(t) = \frac{\beta}{\theta}\left(\frac{t}{\theta}\right)^{\beta-1}.$$

This model is referred to as the power law process. Typically, β and θ are unknown parameters and must be estimated from the observed failure data from a repairable system. The intensity function is the same as the hazard function for the Weibull, and for this reason, it is sometimes referred to as the Weibull process. This is misleading because it suggests that the times between failure are independent Weibull random variables. This Weibull renewal process cannot model repairable systems that deteriorate or systems that improve, because the system is assumed to be repaired to a "like-new" state after any failure.

When performing reliability testing, the testing can stop either at a predetermined time T, or after a predetermined number of failures n. If testing stops after n failures, then the test is failure truncated. If testing stops at the predetermined time T, then the test is time truncated. Many of the results on the power law process summarized in this section, and in section 5, were derived by Crow (1974).

The likelihood function for the failure truncated power law process is

$$L(\theta,\beta) = \left(\prod_{i=1}^{n} u(t_i)\right) \exp\{-(t/\theta)^{\beta}\}$$

and the log-likelihood function is

$$\ln L(\theta,\beta) = n \ln \beta - n\beta \ln \theta + \sum_{i=1}^{n} (\beta - 1) \ln t_i - \left(\frac{t_n}{\theta}\right)^{\beta}.$$

The MLE's are then

$$\hat{\beta} = \frac{n}{\sum_{i=1}^{n-1} \ln(t_n/t_i)} \quad \text{and} \quad \hat{\theta} = \frac{t_n}{n^{1/\hat{\beta}}}.$$

The point estimate for the intensity at the time of the last failure is

$$\hat{u} = \frac{\hat{\beta}}{\hat{\theta}}\left(\frac{t_n}{\hat{\theta}}\right)^{\hat{\beta}-1} = \frac{n\hat{\beta}}{t_n}.$$

A useful result for obtaining confidence intervals or tests of hypotheses for β is that $2n\beta/\hat{\beta}$ has a χ^2 distribution with $2(n-1)$ degrees of freedom.

For time truncated tests, testing stops at the predetermined time T, so in this case the number of failures is random and the time T that testing stops is a known constant. Here the number of failures N has the Poisson distribution

$$N \sim \text{POISSON}\left(\int_0^t (\frac{\beta}{\theta})(\frac{y}{\theta})^{\beta-1} dy \right)$$

and the MLE's are

$$\hat{\beta} = \frac{N}{\sum_{i=1}^{N} \log(T/t_i)}, \qquad \text{and} \qquad \hat{\theta} = \frac{T}{N^{1/\hat{\beta}}}.$$

The estimate of the intensity function at the time of truncation is then

$$\hat{u} = \frac{N\hat{\beta}}{T}.$$

3. TRANSFORMATION TO A UNIFORM DISTRIBUTION

To construct goodness-of-fit tests for the power law process, one of two transformations of the failure times can be made. The system is assumed to have been observed from the initial start-up with no gaps in the data.

The ratio-power transformation

$$R_i = (t_i/t_n)^{c_n\hat{\beta}}$$

can be used to transform to order statistics from a uniform distribution. The rationale for this transformation is that if the expected number of failures before time t is $m(t) = (t/\theta)^{\beta}$, as it is for the power law process, then we define

$$R_i = \frac{m(t_i)}{m(t_n)} = \left(\frac{t_i/\theta}{t_n/\theta} \right)^{\beta} = \left(\frac{t_i}{t_n} \right)^{\beta} \quad \text{for } i = 1, 2, \ldots, n-1 .$$

If the power law process is the model describing the failures of the repairable system, then after this transformation the R_i's are distributed as $n-1$ order statistics from a uniform distribution over the interval [0,1]. So if we knew β, we could perform a goodness-of-fit test for the uniform distribution. If we replace β with the estimator $c_n\hat{\beta}$ we get

$$\hat{R}_i = \left(\frac{t_i}{t_n}\right)^{c_n\hat{\beta}} \quad \text{for } i = 1, 2, \ldots, n-1.$$

Crow (1974) chose $c_n = (n-2)/n$ which makes $c_n\hat{\beta} = (n-2)\hat{\beta}/n = \overline{\beta}$, an unbiased estimator for β. Park and Kim (1992) chose $c_n = 1$.

Since the distribution of the $\hat{R}_i$'s is independent of the parameters β and θ the Kolmogorov-Smirnov, Cramér-von Mises or Anderson-Darling goodness of fit tests may be used.

4. TRANSFORMATION TO AN EXPONENTIAL DISTRIBUTION

For failure truncated data, we can apply the transformation

$$u_i = -\log t_i/t_n, \quad i = 1, 2, \ldots, n-1.$$

If the power law process is the true model then the u_i's will be distributed as $n-1$ order statistics from an exponential distribution with unknown mean $1/\beta$. Thus any goodness-of-fit test for the exponential distribution can be applied to the u_i's. One approach for testing exponentiality is to make the further transformation $z_i = F(u_i)$, where F is the c.d.f. of the exponential distribution using $\overline{u}$ as the estimated mean, i.e.

$$z_i = 1 - \exp(u_i/\overline{u})$$

where

$$\overline{u} = \sum_{i=1}^{n-1} u_i/(n-1).$$

Once the z_i's are obtained, the goodness-of-fit tests of Kolmogorov-Smirnov, Cramér-von-Mises, or Anderson-Darling can be applied to test for goodness-of-fit for the power law process. Stephens (1974) studied these procedures for testing exponentiality, and he gave tables of critical values. Rigdon (1989) showed that if $c_n = (n-1)/n$ is used in making the R transformation described in section 3, then exactly the same test statistics are obtained for the three tests mentioned above.

Rigdon (1989) studied the power of these goodness-of-fit tests and found that none of these tests have good power against a renewal process alternative. Consider the further transformation of the u_i's

$$D_i = (n-i+1)(u_i - u_{i-1}), \quad i = 1, 2, \ldots, n-1.$$

If the u_i's are distributed as the order statistics from an exponential distribution, and they are if the power law process describes the failure times, then the D_i's are i.i.d. exponential with mean $1/\beta$. This transformation is called Durbin's modification, and was proposed by Lewis (1965). The above goodness-of-fit tests can then be applied to the D_i's. This further transformation substantially increases the power for a renewal alternative.

5. MULTIPLE SYSTEMS

Suppose now that k identical copies of a system are placed on test, and that the q-th system is observed from time 0 through T_q. The time T_q would be the time of the last failure if data from the q-th system were failure truncated, and the time of the last failure would occur before time T_q if the data were time truncated. Let n_q denote the number of observed failures for the q-th system. Suppose that the same power law process governed the failure times of all k systems. If we denote the failure times of the q-th system by $t_{1q} < t_{2q} < \ldots$, then the MLE's of β and θ are the solutions of the equations

$$\hat{\beta} = \frac{\sum_{q=1}^{k} n_q}{\sum_{q=1}^{k} (T_q/\hat{\theta})^{\hat{\beta}} \ln T_q - \sum_{q=1}^{k} \sum_{i=1}^{n_q} \ln T_{i,q}}$$

and

$$\hat{\theta} = \left(\frac{\sum_{q=1}^{k} T_q^{\hat{\beta}}}{\sum_{q=1}^{k} n_q} \right)^{1/\hat{\beta}}$$

If $k = 1$ then these equations yield the closed form expressions for the MLE's given in section 1. If all k systems are time truncated at the same time T then the MLE's have the closed form expressions

$$\hat{\beta} = \frac{\sum_{q=1}^{k} n_q}{\sum_{q=1}^{k} \sum_{i=1}^{n_q} \ln(T/t_{iq})}$$

and

$$\hat{\theta} = \frac{k^{1/\hat{\beta}}\ T}{\left(\sum_{q=1}^{k} n_q \right)^{1/\hat{\beta}}}.$$

In other cases, iterative methods must be used to solve the likelihood equations.

If we let $m_q = n_q - 1$ when the data from the qth system are failure truncated and $m_q = n_q$ when the data from the qth system are time truncated, and if we let

$$m = \sum_{q=1}^{k} m_q$$

then an unbiased estimator for β has the closed form expression

$$\overline{\beta} = \frac{m - 1}{\sum_{q=1}^{k} \sum_{i=1}^{n_q} \ln(T_q/t_{i,q})}.$$

The fact that $2(m-1)$ has a chi-square distribution with $2m$ degrees of freedom can be used to obtain a confidence interval for β.

Crow (1974) suggested the following procedure for testing goodness-of-fit for several identical systems. Let

$$Y_{i,q} = t_{i,q}/T_q,\ i = 1, 2, \ldots k.$$

The m values $Y_{i,q}$ are then ordered and labeled $Z_1 < Z_2 < \ldots < Z_m$. The Cramér-von Mises statistic is then

$$C^2 = \frac{1}{12m} + \sum_{j=1}^{m} \left[Z_j^{\bar{\beta}} - \frac{2j-1}{2m} \right]^2 .$$

Crow (1974, 1990) gave tables of critical values for C^2. These tables were determined through simulation, and the entries in the latter reference are more precise due to the larger number of simulated samples.

This analysis assumes that the same power law process governed the failure times for each system. It may be the case in practice that the systems are similar, but not identical. Each system may, for example, be governed by a power law process, but the power law processes may be different for different systems. We ran some simulations to study the effect of having different failure processes govern the failure times of *two* systems. Table 1 shows the number of rejections from simulating 1000 times the failures of two repairable systems.

Different θ's seem to have little effect on the nominal level of the goodness-of-fit tests. Even θ's differing by a factor of eight lead to rejection at about the nominal level. Different β's seem to have some effect, but not a large effect unless the β's are very much different. For example, when $\beta_1 = 0.5$ and $\beta_2 = 2.0$, the effect on the goodness-of-fit test will be minimal.

Table 1

Number of rejections out of 1000 simulations of two power law processes for $n = 10$, 20, and 40 failures

	Parameters of the two systems				Significance level α				
n	θ_1	β_1	θ_2	β_2	20%	15%	10%	5%	1%
10	1.0	1.0	1.0	1.0	206	150	94	52	7
	1.0	1.0	2.0	1.0	195	146	91	41	14
	1.0	1.0	4.0	1.0	188	145	97	46	4
	1.0	1.0	8.0	1.0	190	143	93	45	9
	1.0	0.5	2.0	0.5	177	128	82	36	5
	1.0	0.5	4.0	0.5	203	151	97	46	10
	1.0	0.5	8.0	0.5	216	154	96	49	14
	1.0	1.0	0.5	1.5	215	160	115	58	12
	1.0	1.0	2.0	1.5	228	174	106	58	11
	1.0	1.0	0.5	0.5	252	203	149	79	21
	1.0	0.5	1.0	2.0	522	467	395	294	147
	1.0	0.5	1.0	1.5	375	323	256	180	84
	1.0	1.0	1.0	0.5	256	202	142	84	18
	1.0	1.0	1.0	1.5	228	169	118	65	21
	1.0	1.0	1.0	2.0	269	214	168	99	38

Table 1 (continued)
Number of rejections out of 1000 simulations of two power law processes for $n = 10$, 20, and 40 failures

	Parameters of the two systems				Significance level α				
20	1.0	1.0	1.0	1.0	212	134	96	51	9
	1.0	1.0	2.0	1.0	210	165	108	47	6
	1.0	1.0	4.0	1.0	200	153	112	58	20
	1.0	1.0	8.0	1.0	198	149	94	44	8
	1.0	0.5	2.0	0.5	188	145	92	43	10
	1.0	0.5	4.0	0.5	187	140	89	46	9
	1.0	0.5	8.0	0.5	223	164	112	64	13
	1.0	1.0	0.5	1.5	256	188	130	75	15
	1.0	1.0	2.0	1.5	223	171	114	61	15
	1.0	1.0	0.5	0.5	289	232	172	107	32
	1.0	0.5	1.0	2.0	728	695	612	519	309
	1.0	0.5	1.0	1.5	488	438	373	279	133
	1.0	1.0	1.0	0.5	292	231	179	110	25
	1.0	1.0	1.0	1.5	209	165	114	74	24
	1.0	1.0	1.0	2.0	288	220	167	119	45
n	θ_1	β_1	θ_2	β_2	20%	15%	10%	5%	1%
40	1.0	1.0	1.0	1.0	208	155	108	53	11
	1.0	1.0	2.0	1.0	219	168	114	56	11
	1.0	1.0	4.0	1.0	212	158	111	57	16
	1.0	1.0	8.0	1.0	180	125	88	40	11
	1.0	0.5	2.0	0.5	204	141	93	46	9
	1.0	0.5	4.0	0.5	193	151	101	50	15
	1.0	0.5	8.0	0.5	201	152	103	51	13
	1.0	1.0	0.5	1.5	223	169	108	56	12
	1.0	1.0	2.0	1.5	228	173	119	65	24
	1.0	1.0	0.5	0.5	345	293	224	138	49
	1.0	0.5	1.0	2.0	887	866	824	752	581
	1.0	0.5	1.0	1.5	686	624	538	421	239
	1.0	1.0	1.0	0.5	331	274	201	116	46
	1.0	1.0	1.0	1.5	227	175	122	55	13
	1.0	1.0	1.0	2.0	325	267	195	113	45

6. SUMMARY AND CONCLUSIONS

Goodness-of-fit tests for the power law process can be derived by making one of two transformations, the ratio-power transformation

$$\hat{R}_i = (t_i/t_n)^{c_n\hat{\beta}},$$

or the log-ratio transformation $u_i = -\ln(t_i/t_n)$. Durbin's modification applied to the u_i's substantially increases the power for a renewal alternative.

When several identical systems are tested, a goodness-of-fit test can be constructed by combining data from the various systems. This test seems to have nearly the correct nominal level even when the systems have different θ's. If the β's are substantially different, then the goodness-of-fit test is rejected at a higher rate than the nominal level. We agree with Crow (1990) that "the reliability characteristics for each system under study be analyzed before the failure data are combined."

REFERENCES

[1] Crow, L. R. (1974) "Reliability analysis for complex systems", in *Reliability and Biometry*, F. Proschan and R. J. Serfling eds., SIAM, Philadelphia, 379 – 410.

[2] Crow, L. R. (1990) "Evaluating the reliability of repairable systems," *1990 Proceedings of the Annual Reliability and Maintainability Symposium*, 275 – 279.

[3] Lewis, P. A. W. (1965) "Some results on tests for Poisson processes," *Biometrika*, **52**, 67 – 77.

[4] Park, W. and Kim, Y. G. (1992) "Goodness-of-fit tests for the power-law process," *IEEE Transactions on Reliability* **41**, 107 – 111.

[5] Rigdon, S. E. (1989) "Testing goodness-of-fit for the power law process," *Communications in Statistics: Theory and Methods*, **18**, 4665 – 4676.

[6] Stephens, M. A. (1974) "EDF statistics for goodness of fit and some comparisons," *Journal of the American Statistical Association*, **69**, 730 – 737.

ADVANCES IN RELIABILITY
Edited by Asit P. Basu

SOME RECENT APPLICATIONS OF STOCHASTIC INEQUALITIES IN SYSTEM RELIABILITY THEORY

Philip J. Boland[a], Frank Proschan[b] and Y. L. Tong[c]

[a]University College, Dublin, Department of Statistics, Belfield, Dublin 4, Ireland

[b]Department of Statistics, Florida State University, Tallahassee, Florida 32306

[c]School of Mathematics, Georgia Institute of Technology, Atlanta, Georgia 30332

Abstract

Stochastic inequalities have played an important role in modern reliability theory. In this article we review some recent applications of inequalities in system reliability theory. These mainly consist of applications of majorization–related and dependence–related inequalities; the results yield bounds and monotonicity properties for reliability functions of systems consisting of heterogeneous and/or dependent components. They also provide solutions for optimal assembly of systems and optimal arrangement of components, and give optimal redundancy policies. Most of the papers referenced were published within the past ten years.

1. INTRODUCTION

Stochastic inequalities have played an important role in modern reliability theory. Certain notions in reliability theory can be treated under the general framework of inequalities, and on the other hand many results in stochastic inequalities have been applied to obtain useful applications in reliability. A standard example is the notion of a coherent system and the concept of the association of random variables; in this case it is well–known that association–related inequalities can be applied to provide bounds on the reliability functions of coherent systems.

There are many recent applications of inequalities in system reliability theory, especially for parallel and series systems. In this paper we present a survey of some of these applications. The inequalities that yield such applications include (a) majorization–related inequalities (Section 2), (b) mixture–type inequalities (Section 3), (c) dependence–related inequalities (Section 3), and (d) optimal assembly and redundancy inequalities (Section 4).

2. APPLICATIONS OF MAJORIZATION–RELATED INEQUALITIES

The notion of majorization concerns the diversity of the components of a vector. For fixed $n \geq 2$ let $\mathbf{a} = (a_1, \ldots, a_n)$, $\mathbf{b} = (b_1, \ldots, b_n)$ denote two real vectors. Let $a_{[1]} \geq a_{[2]} \geq \cdots \geq a_{[n]}$, $b_{[1]} \geq b_{[2]} \geq \cdots \geq b_{[n]}$ be their ordered components.

Definition 2.1. **a** is said to majorize **b**, in symbols $\mathbf{a} \succ \mathbf{b}$, if $\sum_{i=1}^{m} a_{[i]} \geq \sum_{i=1}^{m} b_{[i]}$ holds for $m = 1, 2, \ldots, n-1$ and $\sum_{i=1}^{n} a_i = \sum_{i=1}^{n} b_i$.

This definition provides a partial ordering, namely, $\mathbf{a} \succ \mathbf{b}$ implies that (for a fixed sum) the a_i's are more diverse than the b_i's. In particular, one immediately has:

Fact 2.2. $\mathbf{a} \succ (\bar{a}, \ldots, \bar{a})$ *holds for all* **a** *where* $\bar{a} = \frac{1}{n}\sum_{i=1}^{n} a_i$.

Fact 2.3. *If* $a_i \geq 0$ $(i = 1, \ldots, n)$, *then* $(\sum_{i=1}^{n} a_i, 0, \ldots, 0) \succ \mathbf{a}$ *holds.*

Fact 2.4. *If* $\mathbf{a} \succ \mathbf{b}$ *holds, then* $\frac{1}{n-1}\sum_{i=1}^{n}(a_i - \bar{a})^2 \geq \frac{1}{n-1}\sum_{i=1}^{n}(b_i - \bar{b})^2$ *holds. Thus majorization is stronger than the variance concept in measuring the diversity of the components of a vector.*

A useful application of the notion of majorization yields inequalities via Schur–concave and Schur–convex functions.

Definition 2.5. A function $f(\mathbf{x}) : \Re^n \to \Re$ is said to be a *Schur–concave (Schur–convex)* function if $\mathbf{x} \succ \mathbf{y}$ implies $f(\mathbf{x}) \leq (\geq) f(\mathbf{y})$ for all $\mathbf{x}, \mathbf{y} \in \Re^n$.

That is, $f(\mathbf{x})$ is a Schur–concave function if the functional value is larger when the components of **x** are less diverse in the sense of majorization.

For a comprehensive treatment of majorization and Schur–concavity (Schur–convexity), see Marshall and Olkin (1979).

2A. Bounds on Reliability Functions of k–out–of–n Systems

Consider a k–out–of–n system that consists of n independent components with component reliabilities $p_1, \ldots, p_n$, respectively. Then the system reliability, as a function of $\mathbf{p} = (p_1, \ldots, p_n)$, is

$$h_k(\mathbf{p}) = \sum_{\mathbf{j} \in B} \prod_{i=1}^{n} p_i^{j_i}(1 - p_i)^{1-j_i}, \tag{2.1}$$

where $\mathbf{j} = (j_1, \ldots, j_n)$ is such that $j_i = 0$ or 1 for $i = 1, \ldots, n$ and

$$B = \{\mathbf{j} : \sum_{i=1}^{n} j_i \geq k\}. \tag{2.2}$$

Earlier results on bounds of $h_k(\mathbf{p})$ include those of Hoeffding (1956), Pledger and Proschan (1971), and Gleser (1975); and a convenient reference is Tong (1980, Section 6.4). Their results state:

Theorem 2.6. *(a)* (Hoeffding). *Let* $\bar{p} = \frac{1}{n}\sum_{i=1}^{n} p_i$ *be the arithmetic mean, and let* $\bar{\mathbf{p}} = (\bar{p}, \ldots, \bar{p})$. *Then*

$$h_k(\mathbf{p}) \geq h_k(\bar{\mathbf{p}}) \quad \text{for } k \leq n\bar{p}, \tag{2.3}$$

$$h_k(\mathbf{p}) \leq h_k(\bar{\mathbf{p}}) \quad \text{for } k \geq n\bar{p} + 1. \tag{2.4}$$

(b) (Gleser). $h_k(\mathbf{p})$ *is a Schur–convex function of* **p** *for* $k \leq [n\bar{p} - 1]$ *and a Schur–concave function of* **p** *for* $k \geq [n\bar{p} + 3]$, *where* $[z]$ *is the largest integer less than or equal to* z.

(c) (Pledger and Proschan). *Let* $\bar{p}^* = (\prod_{i=1}^{n} p_i)^{1/n}$ *denote the geometric mean, and let* $\bar{\mathbf{p}}^* = (\bar{p}^*, \ldots, \bar{p}^*)$. *Then (i)* $h_k(\mathbf{p})$ *is a Schur–convex function of* $(-\ln p_1, \ldots, -\ln p_n)$ *for all* $k \leq n-1$, *and (ii)* $h_n(\mathbf{p}) = h_n(\bar{\mathbf{p}}^*)$.

Note that bounds for parallel and series systems can be obtained by taking $k = 1$ and $k = n$, respectively. Also note that the system reliability with $\mathbf{p} = \bar{\mathbf{p}}$ or $\mathbf{p} = \bar{\mathbf{p}}^*$ is just a binomial probability.

In a 1983 paper, Boland and Proschan obtained a new bound that applies when the component reliabilities are either simultaneously large or simultaneously small.

Theorem 2.7. *For fixed* n *and* k, $h_k(\mathbf{p})$ *is a Schur–convex (a Schur–concave) function of* $\mathbf{p}$ *for* $\mathbf{p} \in \left[\frac{k-1}{n-1}, 1\right]^n$ *(for* $\mathbf{p} \in \left[0, \frac{k-1}{n-1}\right]^n$*).*

A comparison of this result with that in Theorem 2.6 can be found in Boland and Proschan (1983).

2B. Optimal Allocation of Components

Inequalities have become a useful tool for the optimal allocation of components in system reliability theory, and two interesting early papers in this area are those of Derman, Lieberman, and Ross (1972, 1974). Some recent applications of majorization inequalities for parallel–series and series–parallel systems can be found in El–Neweihi, Proschan, and Sethuraman (1986). They considered the problem of the allocation of n independent components to the various positions or slots in these systems, and proved the following theorems:

Theorem 2.8. *Let* $P_1, \ldots, P_k$ *be the disjoint min path sets of a parallel–series system with sizes* $n_1, \ldots, n_k$. *Without loss of generality, assume that* $n_1 \leqq \cdots \leqq n_k$. *Suppose that there are* n $(= n_1 + \cdots + n_k)$ *independent components with reliabilities* $p_1, \ldots, p_n$ *(at time* t_0*) which are to be allocated among these path sets. The reliability of the system (at time* t_0*) is maximized when the* n_1 *most reliable components are allotted to* P_1, n_2 *next most reliable components are allotted to* $P_2, \ldots,$ *and* n_k *least reliable components are allotted to* P_k.

Theorem 2.9. *Consider a series–parallel system with disjoint min cut sets* $C_1, \ldots, C_k$ *of sizes* $n_1, \ldots, n_k$, *respectively. Without loss of generality assume that* $n_1 \leqq \cdots \leqq n_k$. *Suppose that we have* n $(= n_1 + \cdots + n_k)$ *independent components with reliabilities* $p_1, \ldots, p_n$, *which we need to allocate to these cut sets. The reliability of the system is minimized if the* n_k *best components are allocated to* C_k, *the* n_{k-1} *next best components are allocated to* $C_{k-1}, \ldots,$ *and the* n_1 *worst components are allocated to* C_1. *The reliability of the system is maximized at an allocation at which the cut set hazard vector* $\mathbf{Y}$ *is minimized in the sense of majorization, where* $\mathbf{Y} = (Y_1, \ldots, Y_k)$ *is given by*

$$Y_j = \textstyle\sum_{i \in C_j} -\ln(1 - p_i), \quad j = 1, \ldots, k. \tag{2.5}$$

The proof of Theorem 2.8 depends on the fact that the reliability function of such a parallel series system, $h(\mathbf{p}) = 1 - \prod_{j=1}^{k} \left[1 - \exp\left\{-\sum_{i \in P_j}(-\ln p_i)\right\}\right]$, is a Schur–convex function of $\left(\sum_{i \in P_1} -\ln p_i, \ldots, \sum_{i \in P_k} -\ln p_i\right)$. The proof of Theorem 2.9 is similar. Note that the optimal allocation policy in Theorem 2.8 does not depend on the actual values of the p_i's but only on their ordering.

Other interesting recent results include those of Shaked and Shanthikumar (1990), and others.

2C. Schur–Structure Functions and Multistate Coherent Systems

Recently the theory of multistate coherent systems has been developed to describe a system and components which can operate at more than two levels of performance (see, e.g., Barlow and Wu (1978), El–Neweihi, Proschan, and Sethuraman (1978), Ross (1979), Block and Savits (1982), Block, Griffith, and Savits (1989)). A basic notion in this theory is the structure function which relates the level of performance of the system to the levels of performance of its components. In a 1989 paper Abouammah, El–Neweihi, and Proschan studied structure functions that are Schur–concave (Schur–convex). They showed that the Schur properties model the corresponding engineering properties of systems which perform better (worse) when the levels of performances of their components are more (less) homogeneous. They also proved that monotone structure functions possessing the Schur property enjoy a number of basic and interesting structural and probabilistic properties.

2D. Bounds on Software Reliability under the Jelinski–Moranda Model

Perhaps the most widely referenced model in software reliability is that proposed by Jelinski and Moranda (1972). They assume that initially there are an unknown number of faults N in the software system and that at any point in time the failure rate of the system is proportional to the number of faults still resident in the system. They further assume that if X_i is the random variable denoting the time between the discovery of the first $i-1$ and i faults or bugs, then the X_i's are independent exponential random variables with parameters $\tau_i = [N - i + 1]\lambda$ for some proportionality constant $\lambda > 0$, $i = 1, \ldots, N$. Note that in the Jelinski–Moranda model each fault in the system has the same effect on the overall failure rate of the system. Several authors were critical of this model arguing that to assume each fault or bug makes the same contribution to the overall failure rate is not realistic. Thus the question arises: What can be said about the software system if this assumption does not hold in applications?

In a 1987 paper, Boland, Proschan, and Tong applied a known majorization inequality for the multinomial distribution to provide an answer to this question. Let $\boldsymbol{\theta} = (\theta_0, \theta_1, \ldots, \theta_N)$ be an 'initial distribution' for the N bugs or faults in our software system. More specifically we suppose that initially θ_i is the probability of encountering the ith bug in the system as a result of an input ($i = 1, \ldots, N$), while θ_0 is the probability no such bug is encountered. Furthermore, suppose that a bug is eliminated after it is encountered for the kth time, and that when such a bug is eliminated the probability for a no–bug encounter is appropriately modified. Then the following two random variables are of key interest:

(a) $T_{n,\boldsymbol{\theta}}$ = number of inputs (time) until n bugs in the system are eliminated,

(b) $B_{\boldsymbol{\theta}}^m$ = number of bugs eliminated after m inputs to the software system.

It is clear that $T_{n,\boldsymbol{\theta}} \geq m+1 \Leftrightarrow B_{\boldsymbol{\theta}}^m < n$, or in other words that n bugs are eliminated from the system only after $m+1$ inputs, if and only if the number of bugs eliminated from the system after m inputs is less than n. It was shown that

Theorem 2.10. *Let* $\boldsymbol{\theta} = (\theta_0, \theta_1, \ldots, \theta_N)$ *and* $\boldsymbol{\theta}' = (\theta_0', \theta_1', \ldots, \theta_N')$ *be such that*

(i) $\theta_0 \geq \theta_0'$ (2.6)

and

(ii) $\left(\frac{\theta_1}{1-\theta_0}, \ldots, \frac{\theta_N}{1-\theta_0}\right) \succ \left(\frac{\theta_1'}{1-\theta_0'}, \ldots, \frac{\theta_N'}{1-\theta_0'}\right).$ (2.7)

(a) For the standard situation where $k=1$, *it follows that*

$$T_{n,\boldsymbol{\theta}} \geq^{st} T_{n,\boldsymbol{\theta}'} \quad \text{for any } n \tag{2.8}$$

and

$$B_{\boldsymbol{\theta}}^m \leq^{st} B_{\boldsymbol{\theta}'}^m, \quad \text{for any } m. \tag{2.9}$$

(b) For arbitrary k, $T_{N,\boldsymbol{\theta}} \geq^{st} T_{N,\boldsymbol{\theta}'}$.

This theorem yields a stochastic ordering for the random variables $T_{n,\boldsymbol{\theta}}$ and $B_{\boldsymbol{\theta}}^m$, and illustrates how the diversity of the components of $\boldsymbol{\theta}$ (in the sense of majorization) affects the performance of a software package.

Additional recent contributions on the applications of majorization and Schur functions in system reliability theory include the papers by Proschan and Sethuraman (1976) on stochastic comparisons of order statistics and El–Neweihi, Proschan, and Sethuraman (1987) on optimal assembly of systems using Schur functions; Shaked and Shanthikumar (1987, 1988) on the multivariate hazard construction and stochastic convexity; the papers by Shanthikumar (1987) and Liyanage and Shanthikumar (1990) on stochastic majorization and on allocation via Schur convexity.

3. APPLICATIONS OF DEPENDENCE–RELATED INEQUALITIES VIA MIXTURE OF DISTRIBUTIONS

It is well–known that mixtures of distributions play an important role in reliability theory. In certain applications, the system may consist of dependent components or it may contain components that share a common environment; in both cases the system reliability function may depend on a mixture of distributions. Earlier work in this area includes the notions of (i) association of random variables with applications in reliability theory (Esary, Proschan, and Walkup (1967), Barlow and Proschan (1981, Chapter 2)) and (ii) positively–dependent–by–mixture (PDM) random variables (Shaked (1977) and Tong (1977)). Some recent developments on this topic include the following:

3A. Systems of Components Sharing a Common Environment

Lindley and Singpurwalla (1986) and Currit and Singpurwalla (1988) considered a mixture of bivariate exponential distributions of the form

$$F(x_1, x_2) = \int_0^\infty \prod_{i=1}^2 [1 - e^{-\eta\lambda_i x_i}] dG(\eta). \tag{3.1}$$

Here $G(\eta)$ denotes the (prior) distribution of the environment and, for given η, the component life lengths X_1, X_2 are exponentially distributed and are conditionally independent. This formulation of the problem can, of course, be extended to n variables. They showed that for a two–component parallel system when the environment distribution is Gamma, the reliability function of the system may be given explicitly and the marginal distributions are Pareto. Some general bounds and inequalities as well as comparisons with systems of independent components are also given.

3B. Partial Ordering Via Positive Dependence

The earlier results in Shaked (1977) and Tong (1977) concern comparisons between exchangeable random variables and i.i.d. random variables. In Shaked and Tong (1985) partial orderings of positive dependence for exchangeable random variables are given, and inequalities are derived. Let $\mathbf{X} = (X_1, \ldots, X_n)$ and $\mathbf{Y} = (Y_1, \ldots, Y_n)$ be two n–dimensional random vectors such that $X_1, \ldots, X_n$, $Y_1, \ldots, Y_n$ have a common marginal distribution, the X_i's are exchangeable, and the Y_i's are exchangeable. They introduced four partial orderings of positive dependence, and showed that if $\mathbf{X} \overset{d}{>} \mathbf{Y}$ (i.e., if $X_1, \ldots, X_n$ are more positively dependent than $Y_1, \ldots, Y_n$ in a certain fashion), then

$$(E\phi(Y_{(1)}), \ldots, E\phi(Y_{(n)})) \succ (E\phi(X_{(1)}), \ldots, E\phi(X_{(n)})) \tag{3.2}$$

holds for all monotonic functions ϕ such that the expectations exist, and

$$(G_1(t), \ldots, G_n(t)) \succ (F_1(t), \ldots, F_n(t)) \tag{3.3}$$

holds for all t. Here $X_{(1)} \leq \cdots \leq X_{(n)}, Y_{(1)} \leq \cdots \leq Y_{(n)}$ are the order statistics of the X_i's and Y_i's and F_i (G_i) is the distribution function of $X_{(i)}$ ($Y_{(i)}$). In the terminology of reliability theory, consider a parallel or series system that consists of n exchangeable components sharing a common environment. If the environment plays a stronger role, then in the joint distribution the component life lengths are more positively dependent. As a consequence, the parallel (series) system becomes less (more) reliable when the common marginal distribution of the components is fixed.

A partial ordering of positive dependence of random variables with a common marginal distribution (but not necessarily exchangeable) was discussed in Tong (1989). Let $\mathbf{X}, \mathbf{Y}$ be two n–dimensional random variables with a common marginal distribution. $X_1, \ldots, X_n$ are said to be more positively dependent than $Y_1, \ldots, Y_n$ if $E\prod_{i=1}^n \psi(X_i) \geq E\prod_{i=1}^n \psi(Y_i)$ holds for all Borel–measurable functions ψ such that the expectations exist, which implies $P[\bigcap_{i=1}^n \{X_i \in B\}] \geq P[\bigcap_{i=1}^n \{Y_i \in B\}]$ for all Borel sets $B \subset \Re$.

Sufficient conditions under which those inequalities hold were given in terms of the majorization of the dimension vector.

3C. Positive Dependence of a Class of Multivariate Exponential Distributions

Multivariate exponential distributions have many useful applications in reliability theory and shock models, and earlier works in this area include Freund (1961), Marshall and Olkin (1967a, 1967b), Downton (1970), Block and Basu (1974), and others. In a recent paper Olkin and Tong (1990) sudied the positive dependence of a subclass of exponential distributions with applications in shock models and system reliability. The class is characterized by an index vector and a parameter vector, which are used as an ordering to yield degrees of positive dependence. The model represents a modification of that in Marshall and Olkin (1967a); and has a direct implication for the reliability function of a system. A main result states that if the lifelengths of the components (which have a common univariate exponential distribution) are more positive dependent via a majorization ordering of the index vectors, then the corresponding system is less reliable.

3D. Modelling Dependence in Simple and Indirect Majority Systems

A system of n components is said to be a majority system if it functions when and only when a majority of the components function. Let $X_1, \ldots, X_n$ be binary random variables with a common marginal distribution

$$p = P[X_i = 1] = 1 - P[X_i = 0], \quad i = 1, \ldots, n \tag{3.4}$$

If the X_i's are the indicator functions of the components, then the system reliability is defined by

$$h_n(p) = \begin{cases} P_p[S_n \geq (n+1)/2] & \text{for odd } n, \\ P_p[S_n \geq \frac{n}{2} + 1] + \frac{1}{2} P_p[S_n = \frac{n}{2}] & \text{for even } n, \end{cases} \tag{3.5}$$

where $S_n = \sum_{i=1}^{n} X_i$. The problem of interest is how the positive dependence of the random variables $X_1, \ldots, X_n$ affects the reliability of the system. In real world applications, such a problem arises in voting systems, certain juries and committee decisions, and the design of safety systems.

Boland, Proschan, and Tong (1989a) studied this problem under two models of positive dependence of the X_i's. Under one model, it was assumed that there is a common source (or environment) with score Y that affects each of the X_i's such that

$$P[Y = 1] = 1 - P[Y = 0] = p, \tag{3.6}$$

and

$$P[X_i = 1 \mid Y = 1] = 1 - P[X_i = 0 \mid Y = 1] = p + rq, \quad P[X_i = 1 \mid Y = 0] = 1 - P[X_i = 0 \mid Y = 0] = p - rp, \tag{3.7}$$

where $r \in [0,1]$ and $q = 1 - p$. If we rewrite $h_n(p)$ in (3.5) as $h_n(p,r)$ to reflect the fact that it now depends on r, then the X_i's become more positively dependent when r is larger. (It is easy to show that $\mathrm{Corr}(X_i, X_j) = r^2$ for $i \neq j$). Boland, Proschan and Tong (1989a) showed that

Theorem 3.1. *(a) For $p > \frac{1}{2}$, $h_n(p,r)$ is a decreasing function of $r \in [0,1]$.*
(b) For $p < \frac{1}{2}$, $h_n(p,r)$ is an increasing function of $r \in [0,1]$.
(c) For $p = \frac{1}{2}$, $h_n(\frac{1}{2}, r) = \frac{1}{2}$ for all $r \in [0,1]$.

A similar result holds under the other model of positive dependence of the X_i's. In summary, if the component reliability p is large (small), then the system is more reliable when the components are less (more) positively dependent.

3E. Reliability Function of Consecutive k–out–of–$n : F$ Systems with Linearly Dependent Components

In the literature on consecutive k–out–of–$n : F$ systems it is usually assumed that the components function independently, although this assumption is not always satisfied in applications. Boland, Proschan, and Tong (1990) considered a linear dependence model in which $U_1, \ldots, U_n$ are i.i.d. Bernoulli variables with $P[U_i = 1] = 1 - P[U_i = 0] = p$, and the component indicator random variables $X_1, \ldots, X_n$ are such that $X_1 = U_1$,

$$X_i = \begin{cases} X_{i-1} & \text{with probability } r, \\ U_i & \text{with probability } 1 - r \end{cases} \tag{3.8}$$

for $i = 2, \ldots, n$ for some $r \in [0,1]$. It is clear that $X_1, \ldots, X_n$ have a common marginal distribution and they are more positively dependent (in this fashion) when the value of r becomes larger. They proved a MTP_2 property of the joint density of the X_i's, obtained an exact expression for the reliability function of the system for $k \geq n/2$, and showed that

(i) for $k \geq (n+1)/2$ it is a decreasing function of $r \in [0,1]$;
(ii) for $k = n/2$ and even n, it is a decreasing function of r for $r \in [0, 1 - \frac{4}{n^2 p}]$ and an increasing function of r for $r \in [1 - \frac{4}{n^2 p}, 1]$.

For more applications of dependence–related inequalities in system reliability theory, see Boland (1990).

4. SOME OTHER APPLICATIONS

In addition to the contributions on the applications of majorization–related and dependence–related inequalities in system reliability theory, there are certain other results that appeared recently. Many of them concern the optimal arrangement of systems and optimal redundancy policies.

4A. Optimal Assembly of Systems and Optimal Arrangement of Components

It is well-known that inequalities serve as a useful tool for solving the problems of optimal assembly of systems and optimal arrangement of components. A standard example on this topic is the work of Derman, Lieberman, and Ross (1972, 1974). Recent results include Derman, Lieberman, and Ross (1982), Malon (1985), Tong (1985), Du and Hwang (1986, 1990), Griffith (1986), Hwang (1989), and others on consecutive k-out-of-n : F systems; Baxter and Harche (1990a, 1990b) and Malon (1990) on the development of the greedy algorithm for optimal assembly of systems; Natvig (1979, 1985) and Boland, Proschan, and Tong (1989b) for the optimal arrangement of components via partial orderings of their relative importance; Boland, El-Neweihi, and Proschan (1991a, 1991b) on a stochastic order for inspection and repair policies and on redundancy importance and allocation; and other related results.

The main result in Boland, Proschan, and Tong (1989b) deals with an ordering of the relative importance of the components that is stronger than the definition given by Birnbaum (1969). The ordering depends only on the structure function of the system and does not involve the component reliabilities. It is shown that an optimal arrangement policy can be obtained by a pairwise rearrangement whereby a more reliable component is assigned to a more relatively important slot (or position). When applying the main theorem to the consecutive k-out-of-n : F systems, it yields the optimal arrangement policy in Tong (1985) for $k \geq (n+1)/2$ without evaluating the reliability function of the system.

4B. Optimal Redundancy Policies

Another topic that extensively involves the applications of inequalities deals with optimal redundancy policies; and many new results have become available especially for parallel and series systems. For example, Bergman (1985a, 1985b) gave a comprehensive treatment of modern reliability theory with a discussion on redundancy problems; Boland, El-Neweihi, and Proschan (1988, 1991b, 1992) discussed active redundancy allocation and redundancy importance in coherent systems, and presented results on stochastic order for redundancy allocations in series and parallel systems; Boland, Proschan, and Tong (1991a, 1991b) treated the problems of stand-by redundancy in series systems. These results concern a partial ordering of the system reliability function when the component reliability functions are assumed to have the stochastic ordering or likelihood ratio ordering property. Various results have been obtained, and they do not always agree with one's intuition.

FOOTNOTES

The first author's research is partially sponsored by AFOSR Grant No. 91-0048. The second author's research is supported by AFOSR Grant No. 91-0048. The third author's research is sponsored by a grant from AFOSR through NSF grant DMS-9149151.

REFERENCES

[1] Abouammah, A. M., El–Neweihi, E. and Proschan, F. (1989). Schur structure functions. *Prob. Eng. Inf. Sci.* **3**, 581–591.

[2] Barlow, R. E. and Proschan, F. (1981). *Statistical Theory of Reliability and Life Testing: Probability Models.* To Begin With, Silver Spring, MD.

[3] Barlow. R. E. and Wu, A. S. (1978). Coherent systems with multistate components. *Math. Oper. Res.* **3**, 275–281.

[4] Baxter, L. A. and Harche, F. (1990a). On the optimal assembly of series–parallel systems. *Technical Report No. SOR–90–12*, Department of Statistics and Operations Research, New York University, New York, NY.

[5] Baxter, L. A. and Harche, F. (1990b). On the greedy algorithm for optimal assembly. *Technical Report No. SOR–90–14*, Department of Statistics and Operations Research, New York University, New York, NY.

[6] Bergman, B. (1985a). On reliability theory and its applications, with discussion. *Scand. J. Statist.*, **12**, 1–41.

[7] Bergman, B. (1985b). On some new reliability importance measures, *Proc. IFAC SAFECOMP'85,* Como, Italy, W. J. Quirk, ed., pp. 61–64.

[8] Birnbaum, Z. W. (1969). On the importance of different components in a multi–component system. In *Multivariate Analysis II*, P. R. Krishnaiah, ed., pp. 581–592, Academic Press, New York.

[9] Block, H. W. and Basu, A. P. (1974). A continuous bivariate exponential extension. *JASA* **69**, 1031–1037.

[10] Block, H. W., Griffith, W. S. and Savits, T. H. (1989). L–superadditive structure functions. *Adv. Appl. Prob.* **21**, 919–929.

[11] Block, H. W. and Savitis, T. H. (1982). A decomposition for multistate monotone systems. *J. Appl. Prob.* **19**, 391–402.

[12] Boland, P. J. (1990). Modelling dependence in coherent systems. *Technical Report*, Department of Statistics, University College Dublin, Dublin, Ireland.

[13] Boland, P. J., El–Neweihi, E. and Proschan, F. (1988). Active redundancy allocation in coherent systems. *Prob. Eng. Inf. Sci.* **2**, 343–353.

[14] Boland, P. J., El–Neweihi, E. and Proschan, F. (1991a). Stochastic order for inspection and repair policies. *Ann. Appl. Prob.* **I**, 207–218.

[15] Boland, P. J., El–Neweihi, E. and Proschan, F. (1991b). Redundancy importance and allocation of spares in coherent systems. *J. Statist. Plan. Inf.* **29**, 55–66.

[16] Boland, P. J., El–Neweihi, E. and Proschan, F. (1992). Stochastic order for redundancy allocations in series and parallel systems. *Adv. Appl. Prob.* **24**, 161–171.

[17] Boland, P. J. and Proschan, F. (1983). The reliability of k–out–of–n systems. *Ann. Probab.* **11**, 760–764.

[18] Boland, P. J., Proschan, F. and Tong Y. L. (1987). Fault diversity in software reliability. *Prob. Eng. Inf. Sci.* **1**, 175–187.

[19] Boland, P. J., Proschan, F. and Tong Y. L. (1989a). Modelling dependence in simple and indirect majority systems. *J. Appl. Prob.* **26**, 81–88.

[20] Boland, P. J., Proschan, F. and Tong Y. L. (1989b). Optimal arrangement of components via pairwise rearrangements. *Naval Res. Logist.* **36**, 807–815.

[21] Boland, P. J., Proschan, F. and Tong Y. L. (1990). Linear dependence in consecutive k–out–of–n : F systems. *Prob. Eng. Inf. Sci.* **4**, 391–397.
[22] Boland, P. J., Proschan, F. and Tong Y. L. (1991a). Some majorization inequalities for functions of exchangeable random variables. In *Topics in Statistical Dependence,* H. W. Block, A. R. Sampson and T. H. Savits, eds., pp. 85–91. Institute of Mathematical Statistics, Hayward, CA.
[23] Boland, P. J., Proschan, F. and Tong Y. L. (1991b). Standby redundancy policies for series systems. *Technical Report,* Department of Statistics, Florida State University, Tallahassee, FL.
[24] Currit, A. and Singpurwalla, N. D. (1988). On the reliability function of a system of components sharing a common environment. *J. Appl. Prob.* **26**, 763–771.
[25] Derman, C., Lieberman, G. J., and Ross, S. M. (1972). On optimal assembly of systems. *Naval Res. Logist. Quart.* **19**, 569–574.
[26] Derman, C., Lieberman, G. J., and Ross, S. M. (1974). Assembly of systems having maximum reliability. *Naval Res. Logist. Quart.* **21**, 1–12.
[27] Derman, C., Lieberman, G. J., and Ross, S. M. (1982). On the consecutive k–out–of–n : F system. *IEEE Trans. Reliability* **R-31**, 57–63.
[28] Downton, F. (1970). Bivariate exponential distributions in reliability theory. *J. Royal Statist. Soc.* **B32**, 408–417.
[29] Du, D. Z. and Hwang, F. K. (1986). Optimal consecutive 2 out of n systems. *Math. Oper. Res.* **11**, 187–191.
[30] Du, D. Z. and Hwang, F. K. (1990). Optimal assembly of an s–stage k–out–of–n system. *SIAM J. Discrete Math.* **3**, 349–354.
[31] El–Neweihi, E., Proschan, F. and Sethuraman, J. (1978). Multistate coherent systems. *J. Appl. Prob.* **15**, 675–688.
[32] El–Neweihi, E., Proschan, F. and Sethuraman, J. (1986). Optimal allocation of components in parallel–series and series–parallel systems. *J. Appl. Prob.* **23**, 770–777.
[33] El–Neweihi, E., Proschan, F. and Sethuraman, J. (1987). Optimal assembly of systems using Schur functions and majorization, *Naval Res. Logist.* **34**, 705–712.
[34] Esary, J. D., Proschan, F. and Walkup, D. (1967). Association of random variables, with applications. *Ann. Math. Statist.* **38**, 1466–1474.
[35] Freund, J. E. (1961). A bivariate extension of the exponential distribution. *JASA* **56**, 971–977.
[36] Gleser, L. J. (1975). On the distribution of the number of successes in independent trials. *Ann. Probab.* **3**, 182–188.
[37] Griffith, W. S. (1986). On consecutive k–out–of–n : F failure systems and their generalizations. In *Reliability and Quality Control,* A. P. Basu, ed., pp. 157–165. North–Holand, Amsterdam.
[38] Hoeffding, W. (1956). On the distribution of the number of successes in independent trials. *Ann. Math. Statist.* **27**, 713–721.
[39] Hwang, F. K. (1989). Optimal assignment of components to a two–stage k–out–of–n system. *Math. Oper. Res.* **14**, 376–382.

[40] Jelinski, Z. and Moranda, P. B. (1972). Software reliability research. In *Statistical Computer Performance Evaluation*, W. Freiberger, ed., pp. 465–484. Academic Press, New York.

[41] Lindley, D. V. and Singpurwalla, N. D. (1986). Multivariate distributions for the life lengths of components of a system sharing a common environment. *J. Appl. Prob.* **23**, 418–431.

[42] Liyanage, L. and Shanthikumar, J. G. (1990). Allocation through Schur convexity and stochastic arrangement increasingness. *Technical Report*, School of Business, University of California, Berkeley, CA.

[43] Malon, D. M. (1985). Optimal consecutive–k–out–of–n : F component sequencing, *IEEE Trans. Reliability* **R-34**, 46–49.

[44] Malon, D. M. (1990). When is greedy module assembly optimal? *Naval Res. Logist.* **37**, 847–854.

[45] Marshall, A. W. and Olkin, I. (1967a). A multivariate exponential distribution. *JASA* **62**, 30–44.

[46] Marshall, A. W. and Olkin, I. (1967b). A generalized bivariate exponential distribution. *J. Appl. Prob.* **4**, 291–302.

[47] Marshall, A. W. and Olkin, I. (1979). *Inequalities: Theory of Majorization and Its Applications.* Academic Press, New York.

[48] Natvig, B. (1979). A suggestion of a new measure of importance of system components. *Stoch. Proc. Appl.* **9**, 319–330.

[49] Natvig, B. (1985). New light on measures of importance of system components, *Scand. J. Statist.* **12**, 43–54.

[50] Olkin, I. and Tong, Y. L. (1991). Positive dependence of a class of multivariate exponential distributions. *Technical Report,* Department of Statistics, Stanford University, Stanford, CA.

[51] Pledger, G. and Proschan, F. (1971). Comparisons of order statistics and of spacings from heterogeneous distributions. In *Optimizing Methods in Statistics,* J. S. Rustagi, ed., pp. 89–113. Academic Press, New York.

[52] Proschan, F. and Sethuraman, J. (1976). Stochastic comparison of order statistics from heterogeneous populations, with applications in reliability. *J. Multivariate Anal.* **6**, 608–616.

[53] Ross, S. M. (1979). Multivalued state component systems. *Ann. Probab.* **7**, 379–383.

[54] Shaked, M. (1977). A concept of positive dependence for exchangeable random variables. *Ann. Statist.* **5**, 505–515.

[55] Shaked, M. and Shanthikumar, J. G. (1987). The multivariate hazard construction. *Stoch. Proc. Appl.* **24**, 241–258.

[56] Shaked, M. and Shanthikumar, J. G. (1988). Stochastic convexity and its applications. *Adv. Appl. Prob.* **20**, 427–446.

[57] Shaked, M. and Shanthikumar, J. G. (1990). Optimal allocation of resources to nodes of parallel and series systems. *Technical Report,* School of Business, University of California, Berkeley, CA.

[58] Shaked, M. and Tong, Y. L. (1985). Some partial orderings of exchangeable random variables by positive dependence *J. Multivariate Anal.* **17**, 339–349.

[59] Shanthikumar, J. G. (1987). Stochastic majorization of random variables with proportional equilibrium rates. *Adv. Appl. Prob.* **19**, 854–872.

[60] Tong, Y. L. (1977). An ordering theorem for conditionally independent and identically distributed random variables. *Ann. Statist.* **5**, 274–277.

[61] Tong, Y. L. (1980). *Probability Inequalities in Multivariate Distributions.* Academic Press, New York.

[62] Tong, Y. L. (1985). A rearrangement inequality for the longest run, with an application to network reliability. *J. Appl. Prob.* **22**, 386–393.

[63] Tong, Y. L. (1989). Inequalities for a class of positively dependent random variables with a common marginal. *Ann. Statist.* **17**, 429–435.

ADVANCES IN RELIABILITY
Edited by Asit P. Basu

Multivariate survival analysis based on frailty models

Timothy M. Costigan and John P. Klein

Department of Statistics, The Ohio State University, 1958 Neil Avenue, Columbus OH 43210

Abstract

The problem of analyzing multivariate survival data has recently received much attention in the biomedical literature. One approach, which is based on a so called frailty model, assumes that an association is induced between members of sub-populations by individuals sharing a common, unobservable, random covariate that acts multiplicatively on the individual failure rates. In this paper we survey the literature in estimation and modeling based on such models while showing how they may be applied to reliability problems. In a reliability context the frailty may be due to a shared operating environment for some test units or due to some units coming from a common manufacturing lot, for example.

Common models for the frailty include the Gamma, Inverse Gaussian and Positive Stable Distribution. Dependence properties of these models are discussed. We describe for each of these models, parametric techniques for use in estimating random and fixed covariates effects, based on an assumed Weibull model for the baseline failure rate. The problem of semiparametric estimation with an arbitrary baseline failure rate is also discussed. Estimation schemes for all three frailty models based on a modified counting process representation are presented.

1. INTRODUCTION

Traditionally, multivariate failure models commonly employed in reliability analysis have fallen into several broad classes. These include models in which the failure rates of the surviving components of a system change when one of the components fails (Freund (1961), Leurgans et al. (1982)), models based on independent shocks that may simultaneously kill components that are close together (Marshall and Olkin (1967), Klein, Keiding and Kamby (1989)), or models that are mathematically contrived (Gumbel (1960), Downton (1970)), with little if any physical motivation. Recently, a class of models, called *Frailty Models*, has been proposed which induces an association between the various survival times by introducing a random unobserved covariate that acts multiplicatively on the failure rates of all components. This random covariate may represent the effect of a common environment on all components and/or an inherent trait shared by all components (a random genetic or batch effect).

The concept of a frailty was first introduced by Vaupel et al (1979) to account for individual heterogeneity in univariate survival models. Since the appearance of this paper frailty models have been applied to problems in Demography (Heckman and Singer (1982), Manton and Stallard (1981,1988), Manton et al (1981, 1982, 1986) and Vaupel and Yashin (1985)); in economics (Elbers and Ridder (1982), Heckman and Singer (1984a,b), Lancaster (1979) and Lancaster and Nickell (1980)); in biometry (Aalen (1987a,b, 1988, 1990), Clayton (1978), 1988), Clayton and Cuzick (1985), Hougaard (1984, 1986a,b, 1987, 1989), Hougaard et al (1992), Klein (1992), Klein et al (1991), Li (1991), Oakes (1982, 1986a,b, 1989) and Wang (1991)); and in reliability (Lee (1986), Lee and Klein (1988, 1989), Lindley and Singpurwalla (1986), Nayak (1987), and Whitmore and Lee (1991)).

2. PROBABILISTIC PROPERTIES OF FRAILTY MODELS

2.1. Introduction

Suppose that $T_1, ..., T_M$, $M > 1$ are the life times of M components of a system. Let $\Lambda_{oj}(t)$ denote the baseline cumulative hazard rate of T_j, $j = 1, ..., M$ under ideal conditions, such as one may find in the laboratory. We assume that these M components are joined into a system and the system is exposed to an operating environment in which each of the component hazard rates is effected by a common, unobservable, random variable W, called a *frailty*. The resulting conditional marginal hazard rates of the components are $\Lambda_j(t \mid W) = W\,\Lambda_{oj}(t)$. Furthermore, given W, the component life lengths are conditionally independent. The frailty, W, is a representation of the shared common unobservable covariates that jointly effect each of the components of the system. In engineering applications it may represent the effects of the operating environment on the components (Lindley and Singpurwalla (1986)) or an unobserved batch effect (Whitmore and Lee (1991)). In medical applications the frailty may represent the effects of shared genetic or early environmental factors on the life lengths of siblings (Hougaard et al (1992), Klein et al. (1991)) or on the times to different diseases, such as hypertension and heart disease (Wassell (1989)), or it may represent the unmeasured risk factors shared by married couples (Klein (1992)). Note that if W is greater than 1, reflecting a harsher than average environment, all the components in the system tend to be less reliable. Clearly, for this model the lifetimes of the components in the operating environment will be positively associated.

Based on this model the observable system survival function, $S(t) = S(t_1, ..., t_M)$, is

$$S(t) = E_W[\exp\{-w \sum_{j=1}^{M} \Lambda_{oj}(t_j)\}]. \tag{2.1}$$

It follows that $S(t) = LP[\Lambda_{o\cdot}(t)]$ where LP is the Laplace transform of the frailty and $\Lambda_{o\cdot}(t) = \sum_{j=1}^{M} \Lambda_{oj}(t_j)$. Note that marginal hazard rates obtained after adjustment for the random environment are different from those found under ideal conditions.

2.2. Dependence Properties

To study the dependence properties of the frailty models we restrict ourselves to the case $M = 2$. In this case Lee and Klein (1988) show that the joint probability density function corresponding to (2.1) is a TP_2 function (Karlin and Rinott (1980)), regardless of the frailty distribution. This implies that the observable life lengths (X_1, X_2) are associated (Esary et al (1967)) and positive quadrant dependent (Lehmann (1966)). Also all the usual measures of dependence, such as the correlation and Kendall's concordance measure, τ, are non negative. In fact, these measures are strictly positive unless W is degenerate, in which case (X_1, X_2) are independent.

One can also show, using Shaked and Shanthikumar (1990), that (X_1, X_2) are *weakened by failure* (Arjas and Norros (1984)), that is when one of the component fails there is a stochastic decrease in the residual life of the other component; that (X_1, X_2) are *supportive lifetimes* (Norros (1985)), that is upon failure of one component there is an increase in the future accumulation of the failure rate of the remaining component; and (X_1, X_2) are *hazard rate increasing on failure* (Shaked and Shanthikumar (1987)), that is upon failure of one component the hazard rate of the other component is increased.

Lee and Klein (1988) also show that under this model $E[W \mid X_1 > x_1, X_2 > x_2]$ and $Var[W \mid X_1 > x_1, X_2 > x_2]$ are decreasing functions of (x_1, x_2). These results are a consequence of the fact that those systems exposed to harsh environments, where W is large, will tend to fail early. Consequently, systems with long life times tend to have been exposed to less severe conditions.

Lee and Klein (1988) show that the conditional hazard rates of X_1 given X_2, $h(x_1 \mid X_2 = x_2)$ and $h(x_1 \mid X_2 > x_2)$, are decreasing in x_2 for all x_1. Also $h(x_1 \mid X_2 = x_2) > h(x_1 \mid X_2 > x_2)$ for all x_1, x_2. Oakes (1989) defines a "odds ratio" by

$$\theta(x_1,x_2) = P[X_1 > x_1, X_2 > x_2]\, P[X_1 = x_1, X_2 = x_2] / P[X_1 = x_1, X_2 > x_2]\, P[X_1 > x_1, X_2 = x_2], \quad (2.2)$$

Note that $\theta(x_1, x_2) = h(x_1 \mid X_2 = x_2) / h(x_1 \mid X_2 > x_2)$, so this quantity is always greater than one. Oakes shows that this function depends only on (x_1, x_2) through the system reliability, S. Hougaard et al (1992) considers a modification of this function as a dynamic measure of association. Let $\phi(x \mid t) = h(x + t \mid T_2 = t)/h(t + x \mid T_2 > t + x)$. This quantity represents the residual relative risk of failure for a component where the other component has failed at time t as compared to a component in a system where the other component has not yet failed. Note that $\phi(0 \mid t) = \theta(t, t)$ represents the instantaneous increase in the hazard rate of the surviving component when the other component fails at time t.

The functions θ and ϕ are useful in determining the distribution of the frailty. Oakes (1989) shows that $\theta(x_1, x_2) = -S(x_1, x_2)\, q''(S(x_1, x_2))/q'(S(x_1, x_2))$, where $q(\cdot)$ is the inverse of the Laplace transform of the frailty distribution. He also shows that the joint survival function, S, can be expressed as $S(x_1, x_2) = LP\{q[S_1(x_1)] + q[S_2(x_2)]\}$, where S_j is the marginal survival function of X_j obtained from S.

2.3. Gamma Model For Frailty

A common model assumed for the frailty W is the gamma distribution (Clayton (1978), Oakes (1982, 1986a,b), Lindley and Singpurwalla (1986)) with density

$$g(w) = \frac{w^{(1/\alpha - 1)} \exp\{-w/\alpha\}}{\Gamma(1/\alpha)\alpha^{1/\alpha}}, \quad \alpha \geq 0. \quad (2.3)$$

Here the mean frailty is 1 and the variance is α. Therefore $\alpha = 0$ corresponds to independence in which case the frailty is degenerate at 1. As $\alpha \to \infty$ S(x) tends to the upper Fréchet distribution that corresponds to the maximal positive association ($S(x) = \min\{S_j(x_j), j = 1,..., M\}$). The joint survival function is

$$S(x) = \{\sum_{j=1}^{M} S_j(x_j)^{-\alpha} - (M - 1)\}^{-1/\alpha} = \{1 + \alpha \sum_{j=1}^{M} \Lambda_{oj}(x_j)\}^{-1/\alpha} \quad (2.4)$$

where the marginal survival functions are $S_j(x) = \{1 + \alpha\, \Lambda_{oj}(x)\}^{-1/\alpha}$. Note that marginals obtained from the system operating in the random environment are of a different form than the baseline model.

For this model Kendall's τ is $\alpha/(\alpha + 2)$. The cross ratio function, $\theta(x)$, is a constant $(\alpha + 1)$. This model has the property that the strength of the association between the components does not change over time. To see this we note that the conditional value of Kendall's τ, given the system has survived to time t, is $\alpha/(\alpha + 2)$ for all t, and the instantaneous increase in the hazard rates of the surviving components when one of the components fails at time t is also constant over time.

Lindley and Singpurwalla (1986) have studied the bivariate form of this model with exponential baseline distributions. They note that the marginal distributions are Pareto distributions of the second kind or Lomax distributions, yet the bivariate distribution does not have the form of any of the bivariate Pareto distributions studied by Mardia (1962). Nayak (1987) extends the model of Lindley and Singpurwalla to a multicomponent system. He notes that the resulting distribution, which he calls a multivariate Lomax distribution, has a multivariate decreasing failure rate (Brindley and Thompson (1972)). Roy and Mukherjee (1988) study general mixtures of exponential distributions and show that the gamma mixing distribution is the only one that yields linear regressions. They show that the wider class of exponential mixtures is multivariate decreasing failure rate, possesses attractive closure properties, and is characterized by crude hazard rates.

Banydohadhyay and Basu (1990) and Gupta and Gupta (1990) allow the joint life times under ideal conditions to have the bivariate exponential distribution of Marshall and Olkin (1967). They assume a gamma frailty distribution to model the effects of the operating environment. In this model component life times are no longer independent given the frailty. Banydohadhyay and Basu show that the regressions are no longer linear unless conditional independence is assumed.

2.4. Positive Stable Model For Frailty

Hougaard (1986b) suggests that the frailty be modeled by a positive stable distribution with Laplace Transform $LP(u) = \exp\{-u^{\rho}\}$, $0 \leq \rho \leq 1$. This family of distributions, all of which have infinite mean, are centered at one. Positive stable random variables have the property that if Y_j, $j = 1,...n$ are independent identically distributed positive stable random variables then $\sum_{j=1}^{n} Y_j$ is distributed as $n^{1/\rho} Y_1$. Crowder (1989) argues that this fact implies that the positive stable model is more natural than the gamma model since the frailty derives from many small random contributions acting additively on each component. The joint survival function is

$$S(x) = \exp\{ -[\sum_{j=1}^{M} \Lambda_{oj}(x_j)]^{\rho} \}. \tag{2.5}$$

This distribution has marginal survival functions $S_j(x) = \exp[-\Lambda_{oj}(x)^{\rho}]$, so the marginal hazard rates of the components in a random environment are proportional to the hazard rates of the components under laboratory conditions.

For this model the odds' ratio function is $\theta(x) = 1 + (1 - \rho)/(-\rho \log(S(x))$ and Kendall's τ is $1 - \rho$. For the stable frailty model the conditional version of Kendall's tau, given the system has survived to time t, is

$$\tau_t = \{1 - \rho\} [1 - 2^{1/\rho} (-\ln R)^{\rho} \exp\{2(-\ln R)\} \int_{2(-\ln R)}^{\infty} y^{-1/\rho} e^{-y} dy],$$

where R is the probability the system will survive to time t. This quantity is a decreasing function of t. The magnitude of the instantaneous increase in the hazard rates of the surviving components when one of other component fails at time t for the stable model is $1 + ((1 - \rho)/\rho)[-\ln R]^{-1}$, which is also decreasing in t. These two facts imply that the positive stable frailty model is best used to model associations that are initially strong but tend to washout over time.

If one assumes exponential marginals for this model then Hougaard (1986b) shows that one obtains one of the absolutely continuous bivariate exponential distribution proposed by

Gumbel (1960). This results in a new physical interpretation of that model. An attractive feature of this distribution is that the marginals and minimum are exponentially distributed. If one assumes a proportional hazards covariate model for the baseline hazard then the marginal distributions also will follow a proportional hazards assumption.

Hougaard (1986b) shows that a positive stable frailty distribution with baseline Weibull distributions yield a multivariate Weibull distribution with Weibull marginals and Weibull minimum.

2.5. Inverse Gaussian Model For The Frailty

The inverse Gaussian distribution has been used by Houggard (1984) to model heterogeneity in univariate survival models, by Hougaard et al (1992) to model association in the lifetimes of twins and by Whitmore and Lee (1991) to model batch effects in reliability. The one parameter inverse Gaussian model, with mean set to one, has density function

$$f(w; \eta) = (\eta\pi)^{-1/2} \exp(2/\eta)\, w^{-3/2} \exp(-\eta^{-1}w-\eta^{-1}w^{-1}). \tag{2.6}$$

The resulting joint survival function is

$$S(x) = \exp\{2\eta^{-1} - 2[\eta^{-2} + \eta^{-1} \sum_{j=1}^{M} \Lambda_{oj}(x_j)]^{1/2}\}. \tag{2.7}$$

The odds ratio is $\theta(x) = 1 + \eta/(2 - \eta\log(S(x))$ which decreases from $1 + \eta/2$ to 1 as $S(x)$ decreases from 1 to 0. The value of Kendall's τ is

$$.5-2/\eta + (8/\eta^2) \exp(4/\eta) \int_{4/\eta}^{\infty} [(\exp(-u)/u]du.$$

As for the positive stable model, the conditional value of Kendall's τ given survival to time t is a decreasing function of t (see Figure 1), and the size of the instantaneous increase in the hazard rates of the surviving components when one of other component fails at time t is also decreasing in t (see Figure 2).

Whitmore and Lee (1991) consider the multivariate survival distribution derived from an inverse Gaussian mixture of exponential distributions, the IG-exponential distributions. These authors point out the relationship between IG-exponential and IG-Poisson distributions. The later distributions are used to model the number of occurrences of events in an interval when the underlying Poisson intensity process varies from one situation to another. Whitmore and Lee note that if the number of occurrences in a time interval follows an IG-Poisson distribution, then the time between successive occurrences follows an IG-exponential distribution.

2.6. Power Variance Family Of Distributions

Hougaard (1986a) has proposed the use of the family of three parameter positive stable distributions as a general model for the frailty. This model was introduced as an extension of the one parameter stable distribution (see section 2.4) to allow for the closure property that the conditional distribution of the frailty given system survival to time x is in the same family as the unconditional distribution of the frailty. Hougaard et al (1992) applied this model to a study of Danish twins. The density and the Laplace transform of the positive stable distribution, $P(\gamma, \sigma, \rho)$, are given by

$$f(w\,;\gamma,\sigma,\rho) = -\exp\left(-\rho w + \frac{\sigma}{\gamma}\rho^{\gamma}\right)(\pi w)^{-1}\sum_{k=1}^{\infty}\frac{\Gamma(k\gamma+1)}{k!}\left(-\frac{\sigma}{\gamma}w^{-\gamma}\right)^{k}\sin(\gamma k\pi),$$

where $w > 0$, $\gamma \in (0, 1)$, and $LP(s) = \exp\left[-\frac{\sigma}{\gamma}\{(\rho + s)^{\gamma} - \rho^{\gamma}\}\right]$.

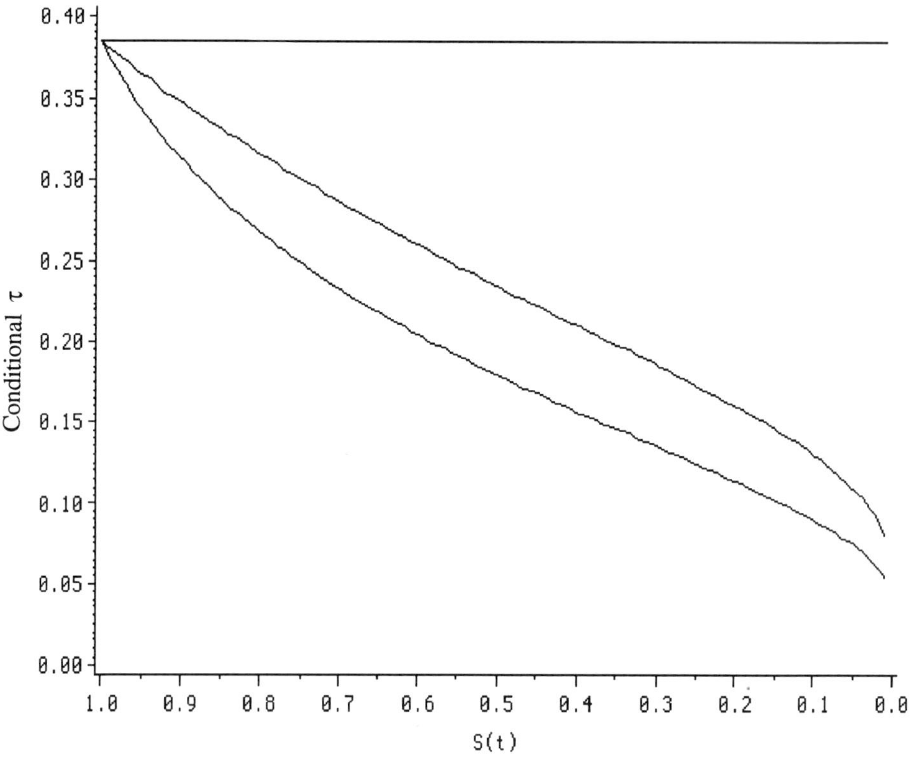

Figure 1. Conditional Value of Kendall's τ Given Survival To Time t For Models with $\tau = .3838$. Curves are Gamma Frailty (Top), Inverse Gaussian Frailty (Middle), Positive Stable Frailty (Lower).

Notice that here σ is a scale parameter. For this class of distributions to have support on the positive real numbers, and thus have W a proper frailty model, γ must be in the range [0, 1]. One can always choose $\sigma = \rho^{1-\gamma}$ so that we have E(W) = 1. It can be shown that, for fixed γ and σ, $P(\gamma, \sigma, \rho)$ is a natural exponential family with power variance $\sigma(1-\gamma)\rho^{v-1}$. For this general model when $\gamma = 1$, W is degenerate, so lifetimes are independent. For $\gamma = 0$, W follows a gamma distribution; when $\rho = 0$ and $\gamma \in [0, 1)$ then this is the positive stable

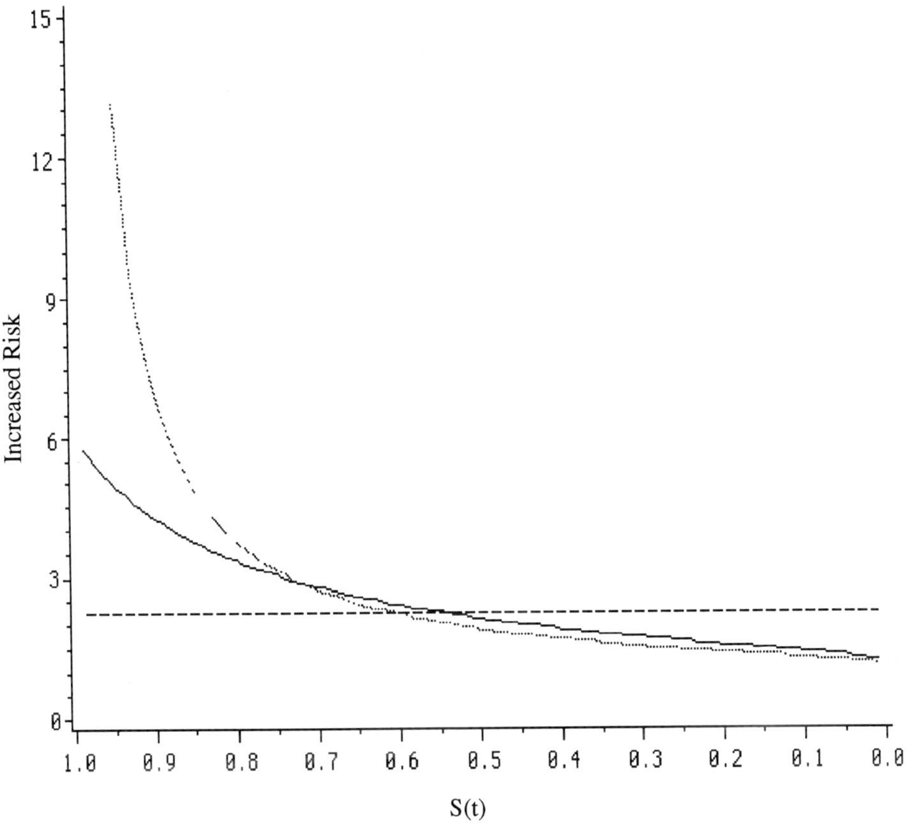

Figure 2. Size Of the Increase in the Hazard Rates Of the Surviving Components When One of Other Component Fails At Time t. Curves are Gamma Frailty (Dashed line), Inverse Gaussian Frailty (Dotted line) Positive Stable Frailty (Solid Line) With $\tau = .3838$.

distribution, and when $\gamma = 1/2$ then this is the inverse Gaussian distribution. For this general family of frailty models the joint survival function is

$$S(x) = \exp\left[-\frac{\sigma}{\gamma} \{ (\rho + \sum_{j=1}^{M} \Lambda_{oj}(x_j))^{\gamma} - \rho^{\gamma} \} \right].$$

Kendall's τ is

$$\tau = (1 - \gamma) - 2\rho + (4\rho^2/\gamma) \exp(2\rho/\gamma) E_{[1/\gamma)-1]}[(2\rho/\gamma)], \text{ where } E_n[y] = \int_y^{\infty} t^{-n} \exp\{-t\, y\} dt.$$

Houggard et al (1992) further shows that for this class of distributions, only the gamma model has the property that the association is constant over time. For the remaining members of this class the strength of the association is a decreasing function of time. This family can be used, with the proper caution about making inferences on the boundary of the parameter space, to differentiate the various models for W.

2.7. Other Frailty Models

Lindley and Singpurwalla (1986) have suggested alternative models for the frailty. In the case when it is reasonable to assume that the operating environment is always more severe than the test environment, they suggest using either a gamma model for W with threshold parameter μ greater than 1 or an exponential distribution with threshold parameter equal to 1. When the operating environment is known to be less severe than the test environment they suggest using a beta model for W.

Lee and Klein (1988) suggest using a uniform model for the frailty, with support either less than 1 (less severe environment), greater than 1 (more severe) or containing one. Detailed graphs of the effects on the system reliability for these models are also given.

Aalen (1988,1990) proposes a class of frailty distributions that extends the three parameter power variance family (see section 2.6). Included in this class of distributions is the power variance family as well as compound Poisson frailty distributions. This compound Poisson frailty distribution has positive mass at 0 so that some components may be immortal.

3. ESTIMATION

3.1. Models And Likelihoods

Frailty models can be applied in two distinct situations. In model I we assume that each system has the same number of components which may have distinct baseline hazard rates. Model II allows for each system to have a different number of similar components each with the same baseline hazard rate. For model I we have the following data: (T_{ij}, I_{ij}, Z_{ij}), $j = 1,..., M$, $i = 1,..., B$, where T_{ij} is the total time on test for the jth component in the ith system; I_{ij} is an indicator of whether this component has failed ($I_{ij} = 1$) or is still functioning ($I_{ij} = 0$); and $Z_{ij} = (Z_{ij1}, \ldots, Z_{ijp})$ is the covariate vector for this component. Under model II we observe (T_{ij}, I_{ij}, Z_{ij}), $j = 1,..., M_i$, $i = 1,..., B$ where $M_i \geq 1$ is the size of the ith group. Some of the covariates associated with the jth component may be specific to the operating conditions of the resulting system (E.g., operating temperature, load or pressure) and some may be specific to the jth component (lot number, manufacturer, etc.).

The first situation (Model I) models a system of M distinct components or sub-systems where, conditional on the frailty, W, the hazard rate for the jth component of a given system is defined by $\lambda_j(t \mid W, Z_j) = W\, \lambda_j(t \mid Z_j)$, $j = 1,..., M$. A common model for $\lambda_j(t \mid Z_j)$ is $\lambda_{oj}(t) \exp[\, Z_j\, \beta_j]$, where $\lambda_{oj}(t)$ is a component specific baseline hazard rate and $\beta_j = (\beta_{1j},..., \beta_{pj})'$ is a parameter vector. This model has been used in medical applications in the competing risks problem of modeling the multivariate distribution of the potential times to competing diseases such as heart disease, hypertension and diabetes. In applications of this model one may have $\lambda_{oj}(\cdot)$ either completely specified up to some unknown parameters or left as an arbitrary baseline intensity function. Note that in this model each test system has the same number of components, $M \geq 2$. The frailty in this formulation may represent the shared effects of the operating environment on all components in the system as well as the effects of unmeasured system characteristics on the components.

The second situation models a collection of systems of possibly different size, M, constructed using components drawn from a common pool. In Model II individual components may have distinct covariates, Z_j, that modify a common baseline hazard rate that is the same for all components, so that, $\lambda_j(t \mid Z_j) = \lambda(t \mid Z_j)$ for all j. Again as in model I, the frailty can be thought of as a random environmental effect shared by all components. An alternative interpretation of the frailty in this model is that it represents a shared random batch effect. When there are no covariates this model then has exchangeable components. Note that for this formulation the value of M may vary from group to group and that "systems" of a single component are allowed. This model has been used by Klein (1992) and Klein et al (1991) to model unobservable risk factors shared by siblings or married couples on either survival or time to heart disease in the Framingham Heart Study.

Under an assumed gamma model for the frailty (2.3) the log likelihood is

$$L_G = \sum_{i=1}^{B} \sum_{j \varepsilon \mathcal{D}_i} \ln[\lambda_j(T_{ij} \mid Z_{ij})] - \sum_{i=1}^{B} (1/\alpha + D_i) \ln[1 + \alpha H_i] + \sum_{i=1}^{B} \sum_{j=1}^{D_i - 1} \ln[\, j\alpha + 1], \tag{3.1}$$

where $\mathcal{D}_i$ is the set of components of the ith system that have failed, D_i is the number of components that have failed in the ith system. Here $H_i = \sum_{j=1}^{M} \int_0^{T_{ij}} \lambda_j(s \mid Z_{ij})ds$ for model I and $H_i = \sum_{j=1}^{M} \int_0^{T_{ij}} \lambda(s \mid Z_{ij})ds$ for model II.

Under the positive stable frailty model (2.5) the log likelihood is:

$$L_{PS} = \sum_{i=1}^{B} \sum_{j \varepsilon \mathcal{D}_i} \ln[\lambda_j(T_{ij} \mid Z_{ij})] - \sum_{i=1}^{B} H_i^{\rho} + \sum_{i:D_i>0} \{\log[\rho] + (\rho - D_i)\ln[H_i]\} + \sum_{i:D_i=2} \ln[1 - \rho + \rho H_i^{\rho}]$$

$$+ \sum_{i:D_i=3} \ln[2 - 3\rho + \rho^2 + 3\rho\,(1-\rho)H_i^{\rho} + \rho^2 H_i^{2\rho}], \tag{3.2}$$

when there are at most three failures in any given system. Hougaard (1986b) gives a recursion relationship for constructing a likelihood with any number of failures per system.

Under the inverse Gaussian model (2.6) the log likelihood is

$$L_{IG} = \sum_{i=1}^{B} \sum_{j \varepsilon \mathcal{D}_i} \ln[\lambda_j(T_{ij} \mid Z_{ij})] + \sum_{i=1}^{B} (2/\eta)\,[1 - \{1 + \eta\, H_i\}^{-1/2}] + \sum_{i:D_i>0} (D_i/2) \ln[\{1 + \eta\, H_i\}]$$

$$+ \sum_{i:D_i=2} \ln[1 + (\eta/2)\,\{1 + \eta\, H_i\}^{-1/2}]$$

$$+ \sum_{i:D_i=3} \ln[1 + (3\eta/2)\{1 + \eta\, H_i\}^{-1/2} + (3\eta^2/4)\,\{1 + \eta\, H_i\}^{-1}], \tag{3.3}$$

when there are at most three failures in any given system.

3.2. Estimation Without Covariates

Gamma frailty model

Several authors have considered estimation of the association parameter as well as the marginal survival distributions under the gamma frailty model (2.3) when there are no covariates. In this case the hazard rate for the jth component in a certain group is $\lambda_j(t) = W\,\lambda_o(t)$. Clayton (1978) considers estimation for the bivariate case (M = 2) when the baseline hazard function for a given individual has a Weibull form ($\lambda_o(t) = \lambda\mu t^{\mu-1}$).

Oakes (1982) considers estimation of the gamma frailty dependence parameter for bivariate systems when the marginal survival functions, S_1 and S_2, are known up to Lehmann alternatives. That is $S_1(t) = G^{\phi}(t)$ and $S_2(t) = H^{\varphi}(t)$ where G and H are known survival functions. The transformations U –> $-\log G(S_1)$ and V -> $-\log H(S_2)$ makes the marginal densities exponential with parameters ϕ and φ, respectively. Oakes (1986b) contains a detailed discussion of parametric inference for this model based on (3.1).

Clayton (1978) obtains the following nonparametric estimator of the gamma dependence parameter. He forms an augmented sample space of paired survival times (Y_{1k}, Y_{2k}) obtained by matching with each first component failure time the failure times of all the second components that failed. Note that there are $D = D_1 \cdot D_2$ points in this augmented sample space, where D_i is the number of failures of component i observed. Assume that there are no ties among the death times of the Y_{1k}'s or the Y_{2k}'s. For (t_1, t_2) in the augmented space define the risk set to be all pairs in that space such that $Y_1 \geq t_1$ and $Y_2 \geq t_2$ and let $N(t_1, t_2)$ be the size of the risk set. Let $\Delta(t_1, t_2) = 1$ if the deaths at (t_1, t_2) were from the same system and 0 otherwise. Clayton shows that $P(\Delta(t_1, t_2) = 1 \mid N(t_1, t_2)) = \alpha/\{N(t_1, t_2) - 1 + \alpha\}$. Clayton uses this fact to argue that

$$L = \prod_{k=1}^{D} [\alpha/\{N(t_1, t_2) - 1 + \alpha\}]^{\Delta(t_{1k}, t_{2k})} [1 - \alpha/\{N(t_1, t_2) - 1 + \alpha\}]^{1-\Delta(t_{1k}, t_{2k})} \qquad (3.4)$$

can be treated as a conditional likelihood. He estimates α by finding the value that maximizes this conditional likelihood. The estimator of the variance of this estimator is the inverse of the second derivative of the log likelihood.

Oakes (1982) criticizes Clayton's estimator of the variance. The product in (3.4), which Oakes (1986b) calls a psuedo-likelihood, are not independent so that the observed information underestimates the variance of Clayton's estimator. Oakes shows that (3.4) is not a partial likelihood. He finds the correct marginal likelihood for n = 3 pairs and shows that it does not equal the psuedo-likelihood. He also claims that the marginal likelihood is not tractable for larger n, and consequently proposes a simpler method of moments type estimator based on Kendall's measure of concordance, $\tau = \alpha/(2 + \alpha)$, and he develops an estimator of the variance of this statistic. In a subsequent paper Oakes (1986a) obtains a valid variance estimator of Clayton's maximum likelihood estimator obtained from (3.4). In an article in which he reviews parametric and nonparametric estimation for the gamma frailty model, Oakes (1986b) suggests that one should use Clayton's maximum likelihood estimator with its valid variance estimator rather than his concordance estimator, because the former is more efficient. The concordance estimator is valuable for diagnostic checks.

Positive stable frailty model

Hougaard (1986b) derives maximum likelihood estimators under the positive stable model using (3.2) when the baseline distributions are Weibull and there are no covariates. He also develops a two stage estimation procedure for the nonparametric model. In the first stage one ignores dependence and estimates the integrated hazards of the marginal distributions by the Nelson-Aalen estimator (Nelson 1972, Aalen 1978). In the second stage one uses these estimated marginal distributions in (3.2) to find the maximum profile likelihood estimator of the stable dependence parameter, $\tilde{\rho}$. A conservative test for independence is obtained by comparing the log likelihood at $\rho = \tilde{\rho}$ with the log likelihood under independence ($\rho = 1$). An advantage of this procedure is that the same estimation technique can be applied to a variety of frailty distributions. Consequently, this technique is valuable during the initial search for a frailty distribution. However, the statistical properties of this procedure have not been studied. This technique is the first step of an EM algorithm approach that we shall discuss in the next section.

Inverse Gaussian frailty models

Whitmore and Lee (1991) consider the estimation problem for Inverse Gaussian frailties and exponential baseline distributions ($\lambda_{oj}(t) = \lambda_j$) in detail. In these applications it is assumed that the frailty represents a batch effect and that batch sizes are sufficiently large to allow accurate estimation of batch specific quantities. Empirical Bayes techniques are used to estimate the group hazard rate. Maximum likelihood techniques are used to estimate the inverse Gaussian parameters. The posterior predictive survival function is derived.

Whitmore and Lee also consider the assessment of the adequacy of the fit of this model. They claim that one should first test the assumption that times within each group are exponentially distributed by standard methods and pool tests across groups. This is feasible when the group is a manufacturing batch of several items, but not when groups consist of two components within a system. The second assumption that should be checked is the form of the mixing distribution. This is done by comparing the observed and expected sums of the survival times within groups, calculating the corresponding p-value for each group, and constructing a PP plot. It would be useful if these techniques for checking model assumptions could be extended to other distributions.

3.3. Estimation Of Covariate Effects

Recent work on frailty models has focused on the problem of estimating the effects of covariates on the distribution of the time to event after an adjustment for the unobservable frailty. Such adjustments need to be made since one can show that if the association between or within individuals is ignored the effects of covariates will be underestimated. In the sequel we assume that associated with each event time is a vector of potential explanatory variables, z. Under Model I, where the frailty represents the unmeasured characteristics shared by each of the failure modalities of a particular system, some of these covariates could represent the measured potential risk factors for the system (eg. load, operating temperature, etc.) and some could be component specific risk factors (eg. component manufacturer). Under model II, where the frailty represents a batch effect or the effect of a shared environment, the covariates would be specific to each component.

Several authors have studied the problem of estimating covariate effects under an assumed Weibull model for the baseline hazard rate Λ_{oj}. Estimation in this case involves a straight forward maximization of the appropriate likelihood (see 3.1, 3.2, 3.3). Details of this maximization are found in Klein et. al (1991) for the gamma frailty, Houggard (1986b, 1989) for the positive stable frailty, and Li (1991) for the Inverse Gaussian frailty.

When the baseline hazard rate is left unspecified, several semiparametric approaches to estimation of the covariate effects have been suggested in the literature. To explain these techniques we shall consider the bivariate case ($M = 2$) where, given the frailty, the component hazard rates are $\lambda(x \mid Z_{ij}, W_i) = \lambda_{0j}(x) \exp(\beta_j z_{ij}) W_i$, $j = 1, 2$, $i = 1,..., B$. For simplicity we assume that the frailties are independent identically distributed gamma random variables with density (2.3). Let $(t_{ij}, \delta_{ij}, z_{ij})$ denote the observed life times, failure indicator and covariate values for the $j = 1, 2$; $i = 1,..., B$.

Klein (1992) has developed an estimation procedure based on the profile likelihood construction (Johansen, 1983). In this approach, the EM algorithm (Dempster et. al, (1977) is used. His analysis is summarized as follows.

Given the observed data set $(t_{ij}, \delta_{ij}, Z_{ij})$ $(i = 1,..., B, j = 1, 2)$ and the unobserved frailty W_i's, the augmented log likelihood is given by $L[\alpha, \beta_1, \beta_2, \Lambda_{01}, \Lambda_{02} \mid W, (t_{ij}, \delta_{ij}, Z_{ij})] = L_1(\beta_1, \beta_2, \Lambda_{01}, \Lambda_{02}) + L_2(\alpha)$, where

$$L_1(\beta_1,\beta_2, \Lambda_{01}, \Lambda_{02}) = \sum_{i=1}^{B} \sum_{j=1}^{2} \delta_{ij}[\log \lambda_{0j}(t_{ij}) + \beta_j Z_{ij}] - W_i\Lambda_{0j}(t_{ij})\exp(\beta_j Z_{ij}) \quad (3.5)$$

and

$$L_2(\alpha) = -B [\log\Gamma(\alpha^{-1}) + \alpha^{-1}\log\alpha] + (\alpha^{-1} + D_i - 1) \sum_{i=1}^{B} \log W_i - \alpha^{-1} \sum_{i=1}^{B} W_i . \quad (3.6)$$

One can show that given the data the W_i's are independent gamma random variables with shape parameters $A_i = [1/\alpha + \delta_{i1} + \delta_{i2}]$ and scale parameters $C_i = [1/\alpha + \sum_{j=1}^{2} \Lambda_{0j}(t_{ij})\exp\{\beta Z_{ij}\}]$. In the E-step of the EM algorithm we replace the unobserved W_i's in (3.5, 3.6) by $\hat{w}_i = E[W_i \mid (t_{ij}, \delta_{ij})] = A_i / C_i$, and $\log W_i$ by $[\psi[A_i] - \ln(C_i)]$, where Ψ is the digamma function, based on the current values of α, β_1 and β_2. The M-step involves maximization of the resulting likelihoods. Maximization of (3.6) follows directly, while maximization of (3.5) involves the unspecified nuisance functions Λ_{0j}. A profile likelihood construction is then used to obtain an appropriate partial likelihood for the β_j's. That is, we maximize $L_1(\beta_1, \beta_2, \Lambda_{01}, \Lambda_{02})$ with respect to Λ_{0j} for fixed β_j, $j = 1, 2$. Hence, we have a nonparametric estimate of Λ_{0j} given by

$$\hat{\Lambda}_{0j}(t) = \sum_{t_{ij} \le t} \frac{\delta_{ij}}{\sum_{t_{kl} \ge t_{ij}} \hat{w}_k \exp(\beta_j Z_{kl})} \quad (3.7)$$

Substitution of $\hat{\Lambda}_0$ into $L_1(\beta_1, \beta_2, \Lambda_{01}, \Lambda_{02})$ yields a profile likelihood given by

$$L_3(\beta_1, \beta_2) = \sum_{i=1}^{B} \sum_{j=1}^{2} \delta_{ij} [\beta_j Z_{ij} - \log\{ \sum_{t_{kl} \ge t_{ij}} \hat{w}_k \exp(\beta_j Z_{kl}) \}]. \quad (3.8)$$

Thus, the updated estimates of $(\alpha, \beta_1, \beta_2)$ can be derived by solving the score equations based on (3.6) and (3.8). Note that the partial likelihood used in the M-step is a standard Cox likelihood with the inclusion of an additional fixed covariate, $\ln[\hat{w}_k]$, with known coefficient. The estimation algorithm now iterates between updating the estimate of $\hat{w}_k$ based on the current parameter values and using these values to update the estimates of $(\alpha, \beta_1, \beta_2)$. Detailed variance formulas for the resulting estimator can be found in Klein (1992). Hougaard(1989) considers estimation of α only based the first step of this algorithm.

To summarize the estimation routine proceeds as follows:

Step 1: Using a standard Cox regression program obtain initial estimates of $\beta_1, \beta_2, \Lambda_{o1}$ and Λ_{o2} from (3.8) and (3.7), respectively with $\hat{\omega}_k = 1$ (i.e., $\alpha = 0$).

Step 2: Using the current values of $\beta_1, \beta_2, \Lambda_{01}$, and Λ_{02} compute A_i, C_i, and $\hat{\omega}_k$.

Step 3: Update the estimate of α using (3.6). Update the estimate of β_1, β_2 (and $\Lambda_{o1}, \Lambda_{o2}$) using (3.8) and (3.7)

Step 4: Iterate between steps 2 and 3 until convergence.

This methodology has been applied to the positive stable frailty model (Wang (1991)) and to the inverse Gaussian model (Li (1991)). The extension of this procedure to these and other frailty models is straight-forward. This procedure can also be used when a parametric model is suggested for the baseline hazard rates. In that case the E-step of the algorithm involves direct maximization of (3.5).

Several alternate approaches to estimation for the semi-parametric gamma frailty model have been suggested. Clayton and Cuzick (1985) suggest an iterative approach based several approximations to the likelihood generated by the bivariate generalized ranks. Wang (1991) has shown that this technique yields identical estimates to those obtained by the above procedure. Self and Prentice (1986) provide an estimation scheme based on a counting process formulation of the problem in which the distribution of the frailty depends on the observed history process. Their approach, then treats the frailty as an additional time dependent covariate in a Cox regression model. This technique is extremely computer intensive since it requires updates of the expected value of the frailty at each event time for each iteration of an EM-like algorithm. There has been little work on obtaining a variance estimator for this technique.

ACKNOWLEDGMENT

This research was supported by a grant from the International Life Sciences Institute.

REFERENCES

1. Aalen, O. (1978). Nonparametric inference for a family of counting processes. *The Annals of Statistics*, 6, 701-726.
2. Aalen, O. (1987a). Two examples of modelling heterogeneity in survival analysis. *Scandinavian Journal of Statistics*, 6, 701-726.
3. Aalen, O. (1987b). Mixing distributions on a Markov chain. *Scandinavian Journal of Statistics*, 14, 281-289.
4. Aalen, O. (1988). Heterogeneity in survival analysis. *Statistics in Medicine* 14, 19-25.
5. Aalen, O. (1990). Modelling heterogeneity in survival analysis by the compound Poisson distribution. Unpublished report, University of Oslo.

6. Arjas, E. and Norros, I. (1984). Life lengths and association: a dynamic approach. *Mathematics of Operations Research*, 9, 151-158.
7. Bandyopadhyay, D. and Basu, A. P. (1990). On a generalization of a model by Lindley and Singpurwalla. *Annals of Applied Probability*, 22, 498-500.
8. Brindley, E. C. and Thompson, W. A. (1972). Dependence and aging aspects of multivariate survival. *Journal of the American Statistical Association*, 67, 822- 830.
9. Clayton, D. G. (1978). A model for association in bivariate life tables and its application in epidemiological studies of familial tendency toward chronic disease. *Biometrika*, 65, 141-151.
10. Clayton, D. G. (1988). The analysis of event history data: a review of progress and outstanding problems. *Statistics in Medicine*, 7, 819-841.
11. Clayton, D. G. and Cuzick, J. (1985). Multivariate generalizations of the proportional hazards model (with Discussion). *Journal of the Royal Statistical Society*, Series A 148, 82-117.
12. Crowder, M. (1989). A multivariate distribution with Weibull connections. *Journal of the Royal Statistical Society*, Series B 51, 93-107.
13. Dempster, A. P., Laird, N. M. and Rubin, D. R. (1977). Maximum likelihood estimation for incomplete data via the EM algorithm (with Discussion). *Journal of the Royal Statistical Society*, Series B 39, 1-38.
14. Downton, F. (1970). Bivariate exponential distributions in reliability theory. *Journal of the Royal Statistical Society*, Series B 32, 408-417.
15. Elbers, C., and Ridder, G. (1982). True and spurious duration dependence: the identifiability of the proportional hazards model. *Review of Economic Studies*, XLIX, 403-409.
16. Esary, J. D., Proschan, F. and Walkup, D. (1967). Association of random variables with applications. *Annals of Mathematical Statistics*, 38, 1466-1474.
17. Freund, J. E. (1961). A bivariate extension of the exponential distribution. *Journal of the American Statistical Association*, 56, 971-977.
18. Gumbel, E. J. (1960). Bivariate exponential distributions. *Journal of the American Statistical Association*, 56, 698-707.
19. Gupta, P. L. and Gupta, R. D. (1990). A bivariate environmental stress model. *Advances in Applied Probability*, 22, 501-503.
20. Heckman, J. J. and Singer, B. (1982). Population heterogeneity in demographic models. *Multidimensional Mathematical Demography*, 567-595.
21. Heckman, J. J. and Singer, B. (1984a). A method of minimizing the impact of distributional assumptions in econometric models for duration data. *Econometrica*, 52, 271-320.
22. Heckman, J. J. and Singer, B. (1984b). The identifiability of the proportional hazards model. *Review of Economic Studies*, 60, 231-243.
23. Hougaard, P. (1984). Life table methods for heterogeneous populations: Distributions describing the heterogeneity. *Biometrika*, 71, 75-83.
24. Hougaard, P. (1986a). Survival models for heterogeneous populations derived from positive stable distributions. *Biometrika*, 73, 387-396.
25. Hougaard, P. (1986b). A class of multivariate failure time distributions. *Biometrika*, 73, 671-687.
26. Hougaard, P. (1987). Modelling multivariate survival. *Scandinavian Journal of Statistics*, 14, 291-304.
27. Hougaard, P. (1989). Fitting a multivariate failure time distribution. *IEEE Transactions on Reliability*, 38, 444-448.
28. Hougaard, P., Harvald, B. and Holm N. (1992). Measuring the similarities between the lifetimes of adult Danish twins born between 1881-1930. *Journal of the American Statistical Association*, 87, 17-24.
29. Johansen, S. (1983). An extension of Cox's regression model. *International Statistical Review*, 51, 258-262.

30. Karlin, S. and Rinott, Y. (1980). Classes of orderings of measures and related correlation inequalities-I: Multivariate totally positive distributions. *Journal of Multivariate Analysis*, 10, 476-498.
31. Klein, J. P. (1992). Semiparametric estimation of random effects using the Cox model based on the EM algorithm. *Biometrics*, To Appear
32. Klein, J. P., Costigan, T. M. and Moeschberger, M. L. (1991). Assessment of risk factors and the strength of dependence when event times are associated within groups. Ohio State University Technical Report.
33. Klein, J. P., Keiding, N. and Kamby, C. (1989). Semiparametric Marshall-Olkin type models applied to the occurrence of metastases at multiple sites. *Biometrics*, 45, 1073-1086.
34. Lancaster, T. (1979). Econometric methods for duration of unemployment. *Econometrika*, 47, 939-956.
35. Lancaster, T. and Nickell, S. J. (1980). The analysis of re-employment probabilities for the unemployed (with discussion). *Journal of the Royal Statistical Society*, Series A 143, 141-165.
36. Lee, S. (1986). Inference for a bivariate survival function induced through the environment. Ph.D. Dissertation, The Ohio State University.
37. Lee, S. and Klein, J. P. (1988). Bivariate models with a random environmental factor. *Indian Journal of Productivity, Reliability and Quality Control,* 13, 1-18.
38. Lee, S. and Klein, J. P. (1989). Statistical Methods for Combining Laboratory and Field Data Based On a Random Environmental Stress Model. In *Recent Developments in Statistics and Their Applications,* Klein and Lee, Eds., 87-116.
39. Lehmann, E. L. (1966), Some concepts of dependence. *Annals of Mathematical Statistics*, 26, 399-419.
40. Leurgans, S., Tsai, W. Y. and Crowley, J. (1982). Freund's bivariate distribution and censoring. *Survival Analysis*, Crowley and Johnson, Eds., 216-229, IMS, 1982.
41. Li, Y. H. (1991). Regression analysis of failure time data. Ph.D. Dissertation, Graduate Program in Biostatistics, The Ohio State University.
42. Lindley, D. V. and Singpurwalla, N. A. (1986). Multivariate distributions for the reliability of a system of components sharing a common environment. *Journal of Applied Probability*, 23, 418-431.
43. Manton, K. G. and Stallard, E. (1981). Methods for evaluating the heterogeneity of aging processes in human populations using vital statistics data: Explaining black/white mortality crossover by a model of mortality selection. *Human Biology,* 53, 47-61.
44. Manton, K. G. and Stallard, E. (1988). *Chronic Disease Modelling*. Griffin, London.
45. Manton, K. G., Stallard, E. and Riggan, W. (1982). Strategies for analyzing ecological health data: Models of the biological risk in individuals. *Statistics in Medicine*, 1, 163-181.
46. Manton, K. G., Stallard, E., and Vaupel, J. W. (1981). Methods of comparing the mortality experience of heterogeneous populations. *Demography*, 18, 389-410.
47. Manton, K. G., Stallard, E., and Vaupel, J. W. (1986). Alternative models for the heterogeneity of mortality risks among the aged. *Journal of the American Statistical Association*, 81, 635-644.
48. Mardia, K. V. (1962). Multivariate Pareto distribution. *Annals of Mathematical Statistics*, 35, 1815-1818.
49. Marshall, A. W. and Olkin, I. (1967). A multivariate exponential distribution. *Journal of the American Statistical Association,* 66, 30-40.
50. Nayak, T. K. (1987). Multivariate Lomax distribution: properties and usefulness in reliability theory. *Journal of Applied Probability*, 24, 170-177.
51. Nelson, W. (1972). Theory and applications of hazard plotting for censored failure time data. *Technometrics*, 14, 945-965.
52. Norros, I. (1985). Systems weakened by failure. *Stochastic Processes Applications*, 20, 181-196.

53. Oakes, D. (1982). A model for association in bivariate survival data. *Journal of the Royal Statistical Society*, Series B 44, 414-422.
54. Oakes, D. (1986a). Semiparametric inference in a model for association in bivariate survival data. *Biometrika*, 73, 353-361.
55. Oakes, D. (1986b). A model for bivariate survival data. In *Modern Statistical Methods in Chronic Disease Epidemiology* , Moolgavkar, and Prentice Eds., 151-166, John Wiley and Sons, New York.
56. Oakes, D. (1989). Bivariate survival models induced by frailties. *Journal of the American Statistical Association*, 84, 487-493.
57. Roy, D. and Mukherjee, S. P. (1988). Generalized mixtures of exponential distributions. J*ournal of Applied Probability*, 25, 510-518.
58. Self, S. G. and Prentice, R. L. (1986). Incorporating random effects into multivariate relative risk regression models. In *Modern Statistical Methods in Chronic Disease Epidemiology,* Moolgavkar and Prentice, Eds., 167-177, John Wiley and Sons, New York.
59. Shaked, M. and Shanthikumar, J. G. (1987). Multivariate hazard rates and stochastic ordering. *Advances in Applied Probability*, 19, 123-137.
60. Shaked, M. and Shanthikumar, J. G. (1990). Multivariate stochastic ordering and positive dependence in reliability theory. *Mathematics of Operations Research*, 15, 545-552.
61. Vaupel, J. W., Manton, K. G. and Stallard, E. (1979). The impact of heterogeneity in individual frailty and the dynamics of mortality. *Demography*, *16*, 439-454.
62. Vaupel, J. W. and Yashin, A. I. (1985). The deviant dynamics of death in heterogeneous populations. In *Sociological Methodology* , Turma Ed.. Jossey-Bass, San Francisco.
63. Wang, S. T. (1991). Estimation for the gamma and positive stable frailty models. Ph.D. Dissertation, Department of Statistics, The Ohio State University.
64. Wassell, J. T. (1989). Bivariate survival methods for epidemiology: An application to the Framingham heart study of risk factors for cardiovascular disease. Ph.D. Dissertation, Department of Preventive Medicine, The Ohio State University.
65. Whitmore, G. A. and Lee, M. L. T. (1991). A multivariate survival distribution generated by an inverse Gaussian mixture of exponentials. *Technometrics,* 33, 39-50.

ADVANCES IN RELIABILITY
Edited by Asit P. Basu
1993 Elsevier Science Publishers B.V.

Lifetime and reliability predictions of advanced ceramics based on creep and creep fracture mechanisms

David C. Cranmer

Ceramics Division, National Institute of Standards and Technology,
Gaithersburg, MD 20899, USA

Abstract

Advanced structural ceramics have reached a stage of development where long term reliability and lifetime are critical issues. High temperature deformation and failure due to creep and creep fracture, is one of the limiting phenomena to use of these materials. Creep behavior is controlled by a variety of factors including microstructure, temperature, and stress. Understanding the effects of these parameters on damage evolution and development of appropriate models of how these parameters affect creep and damage evolution is crucial to the development of an adequate predictive capability for determining lifetime and reliability. Experimentally, the creep behavior of a material can be characterized using simple, inexpensive, uniaxial tensile and compressive tests to measure deformation behavior and fracture lifetime, and this information can be used, in turn, to predict the behavior of more complicated stress states such as exist in flexure and C-ring specimens, and to predict the lifetime of advanced ceramics such as silicon carbide and silicon nitride.

Empirically, creep fracture, or time-to-failure due to creep, for these materials seems to follow a power law function of the strain rate, generally independent of applied stress or temperature. This means the temperature and stress dependencies of the failure time are determined solely by the temperature and stress dependencies of the creep rate. The mechanisms and microstructural features of the creep and creep fracture process will be discussed, and related to models of creep and creep fracture.

The data developed above are used to establish lifetime prediction curves of the maximum stress allowed for a given temperature. In addition, this type of data can be combined with similar data on fast fracture, fatigue failure, and failure via a slow crack growth mechanism to create a fracture mechanism map. This map can provide the design or systems engineer with guidance as to the stress-temperature regimes to work in or avoid for a particular application.

1. INTRODUCTION

Advanced structural ceramics such as aluminum oxide (Al_2O_3), silicon carbide (SiC), and silicon nitride (Si_3N_4) have reached a stage of development where long term reliability and lifetime are critical issues. The temperature range at which these materials are expected to be used generally lies between about 1000°C to 1500°C, but cyclic excursions from room temperature to the service temperature must also be considered as must cyclic stresses which occur at temperature. At the low end of this use-temperature range are turbochargers used in some current models of automobiles; at the high end are gas turbines being developed for cruise missiles and automobile engines. To successfully use these materials over the range from room temperature to the maximum use temperature, the designer must be able to predict a variety of properties (strength, creep and creep fracture, fatigue, etc.) over a range of applied stresses and temperatures. The applied stress state for the overall component may be very complex as a result of a combination of applied stresses which might have tensile, compressive, and shear stress components.

Strength is probably the most studied and well understood of the mechanical properties. For this property, the concept of the weakest link has proved extremely useful. This concept assumes that the strength of the material is controlled by pre-existing flaws or defects that can propagate unstably if the applied stress exceeds the fracture stress. Alternatively, at lower stresses, stable crack extension can cause crack growth until one reaches a critical size, at which point the component fails by brittle fracture. In many ceramic systems, it has been demonstrated that crack growth can be enhanced by the chemistry of the service environment.[1]

For design and use purposes, it is extremely important to know that crack extension is not the only mechanism which can control the fracture behavior of a material, particularly at elevated temperatures. Long time exposure to stresses at elevated temperatures can lead to creation of localized damage, and failure can then occur by the accumulation of such damage. The long term deformation behavior, or creep, then is one of the key properties which determine the usefulness of a material in many applications for advanced ceramics. This deformation can be characterized under a number of different stress states (tension, compression, flexure, C-ring), and must be conducted under a variety of temperature and applied stress conditions. Understanding the effects of microstructure, temperature, and stress on damage evolution and being able to model these effects is crucial to development of an adequate predictive capability for determining lifetime and reliability.

In this paper, the current concepts and models of creep and creep fracture will be reviewed. The use of appropriate experimental data and models of varying stress states to predict the creep behavior and damage zones of a C-ring and flexure bar of a siliconized silicon carbide will be shown. A lifetime prediction methodology for a silicon nitride will also be demonstrated, as will the integration of creep data and models into a fracture mechanism map which can provide design guidance.

2. CREEP MECHANISMS

The physical process of creep involves the permanent deformation over time of the material in response to an applied stress. Specific mechanisms depend on atomic motion in the form of diffusion of material along grain boundaries[2] (Coble creep) or through the lattice[3] (Nabarro-Herring creep); on a dissolution-reprecipitation process[4]; or on viscous flow[5]. Each of these processes can be described mathematically in terms of material constants and properties. For the case of Coble creep, where diffusion of material along a grain boundary is the rate-limiting process, the relevant equation is:

$$\dot{\epsilon}=\frac{47\Omega\delta D_b\sigma}{kTd^3} \qquad (1)$$

where $\dot{\epsilon}$ is the strain rate, σ is the applied stress, Ω is the vacancy volume, δ is the grain boundary width, D_b is the diffusion coefficient in the grain boundary, k is Boltzman's constant, T is the absolute temperature, and d is the grain size. The ability to use these equations to predict the behavior of a material is limited by our knowledge of the material's properties such as diffusivities as a function of time and composition, and adequate understanding of the effects of complex microstructures. The model developed for Coble creep relies on a number of assumptions which are not directly applicable to real materials. When the microstructure does not consist of ideal geometries and compositions, the model predictions can be inaccurate. In addition, it is frequently unclear when the mechanism of creep changes, based solely upon knowledge of diffusivities. In many instances, it is easier and quicker to measure the creep properties themselves than to measure the diffusivities.

It has also been noted, particularly for two-phase materials that there are differences in creep behavior when measured in tension and compression. These differences are related to differences in the physical processes which control the motion of individual grains and the formation of cavities. These concepts are discussed in more detail elsewhere.[6]

Of particular importance to the advancement of our understanding of the creep and creep fracture process is the role of microstructure. Failure to understand what happens to the microstructure and how those changes relate to changes in the creep and creep fracture behavior will lead to inaccurate and ultimately useless predictions.

3. CREEP TESTING METHODOLOGY

Initially, bend, or flexure, tests were used[7] because of their simplicity of specimen design and preparation, but the flexure geometry results in a complex stress state which is only now beginning to be properly understood.[8-9] One particularly important feature to note about this geometry is the dramatic shift of the neutral axis of stress in the flexure geometry when the creep behavior under tension and compression differ as discussed in references 8 and 9. The models which account for this behavior have only recently been developed[10] and are still being refined. Details of the specific models for creep will be discussed in a later section.

Uniaxial tensile and compressive tests[11-12] have also been used to gain a better understanding of creep mechanisms and models of behavior. These tests can be used to predict the behavior of more complex geometries such as bend bars and C-rings. The advantages of the uniaxial tests are that the stress state is simple, and the data can be readily combined to predict the behavior of materials subjected to more complicated stress states. Care must be exercised in all of the tests to ensure that temperature, applied stress, and specimen and system alignments are properly controlled.

One convenient apparatus for measuring uniaxial tensile creep uses an inexpensive, dogbone-shaped specimen[13] in combination with a laser extensometer. Other systems using different specimen sizes and geometries can also be used[14] but have similar components. These other devices may also be used for additional tests such as fatigue or high temperature strength measurements.

The C-ring test is substantially similar to the flexure and tensile tests, and the test specimens can be prepared either from plates or tubes of material. The experimental set-up differs from the tensile creep set-up in that two high temperature push rods and platens are used to compress the C-ring.[15] Alignment of the load train and the specimen is achieved by proper machining of the radius of the ring. The creep rate is determined from measurements of the load-point displacement as function of time. The load-point displacement is determined in this case through the use of a travelling optical telescope with a resolution of 1 micrometer. The data developed from these C-ring tests have been used to verify the predictions of models which use uniaxial tensile and compressive data. These models are discussed below.

4. CREEP MODELS

Models of the creep behavior of a particular geometry are typically empirical. An initial model[16] of creep in the flexure bar was developed based on simple beam theory. The model assumed that the material behaves identically in both tension and compression, and gave reasonable predictions for some but not all ceramics. A more recent model[10] allows for a difference in creep rates under tension and compression. In the case where the creep rates are different, the relation between the steady-state creep rate and the applied stress takes the form:

$$\dot{\epsilon}_s = A_c \left(\frac{\sigma}{\sigma_0}\right)^{n_c} \tag{2}$$

for compression, and:

$$\dot{\epsilon}_s = A_t \left(\frac{\sigma}{\sigma_0}\right)^{n_t} \tag{3}$$

for tension, and where A and n are creep constants and σ_0 is a normalization constant which has the same units as stress. The respective constants can be determined from uniaxial testing, or can be estimated from analysis of creep data from more complex loadings such as flexure. In principal, if a sufficiently wide range of curvature rate can be accessed, it is possible to extract the power law parameters for both tension and compression from flexure data. For most materials, this appears not to be the case, but a combination of any two (flexure, tension, or compression) can be used to extract the parameters for the third.[17]

In the case of the C-rings discussed above, the model can be based on either simple curved beam theory to analyze stress state[15] or a finite element method.[18] The creep rate can be related to measured load-point displacement either geometrically or by an energy method. The shape of the damage zone which develops due to the creep process is predicted numerically. Important features are migration of the neutral axis, the moment-curvature rate relationship, the load-point displacement rate, and damage zone shape. Using simple beam theory, fair agreement between predictions and experimental observations has been achieved.

5. LIFETIME AND RELIABILITY

Given the empirical models and adequate experimental data, it is possible to predict the lifetime of a material when it is subjected to creep conditions. The creep fracture, or time-to-failure data for many ceramic systems have been found to fit a power-law description of the form:[19]

$$t_f = A\dot{\epsilon}_s^m \tag{4}$$

The creep rate for the material can generally be expressed as a function of applied stress and temperature of testing as:

$$\dot{\epsilon}_s = B\sigma^n \exp(\frac{-\Delta H}{RT}) \tag{5}$$

When these two expressions are combined, a stress-fracture formulation containing the Monkman-Grant exponent (m), the creep stress exponent (n), and the temperature dependence for creep is created,

$$t_f = AB^m \sigma^{mn} \exp(\frac{-m\Delta H}{RT}) \tag{6}$$

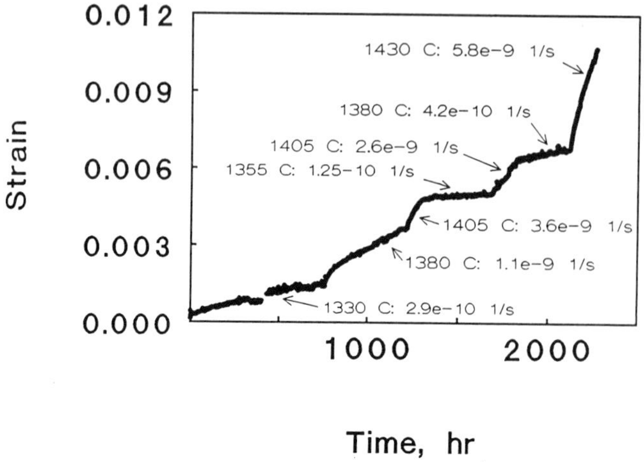

Figure 1. Strain versus time for a silicon nitride as a function of temperature. All data were taken at an applied stress of 100 MPa.

An example of the use of this expression in conjunction with measured creep and creep fracture data for a silicon nitride material is given below.[20] The uniaxial tensile creep curve from a specimen exposed to constant stress at several temperatures is shown in Figure 1. The measured creep rates are between about 10^{-10} s^{-1} at 1325°C, and 6 x 10^{-9} s^{-1} at 1400°C. A plot of the creep rate versus inverse temperature developed from these data is shown in Figure 2, and is used to determine the activation energy for creep (1258 KJ/mol). Using additional data obtained between 1350 and 1385°C over the range of applied stresses from 75 to 150 MPa, shown in Figure 3, the tensile creep stress exponent (n) of the material is obtained, and has a value of 6.9. A typical creep fracture curve obtained is shown in Figure 4. The observed failure times range from 4 to about 250 h. The data were plotted using the Monkman-Grant formulation for log t_f versus log $\dot{\epsilon}$ (Figure 5). The measured slope (m) is the Monkman-Grant exponent and has a value of about -1. If all of the data fall on a single curve, the lifetime of a material can be determined solely by the creep behavior of the solid, and improvements in lifetime can be achieved by modifying the microstructure to improve creep resistance. For the present data, there appears to be a threshold below which the creep behavior is different, thus the lifetime estimates for this range are conservative. Typical structural modifications which can be used to enhance creep resistance include increasing the effective viscosity of the intergranular phase, decreasing the amount of intergranular phase, and increasing the grain size of the solid. Using the expression above, preliminary lifetime predictions for this material have been made (Figure 6). The predictions indicate that the material can withstand a combined stress and temperature of 70 MPa and 1370°C for a period of about 1 year.

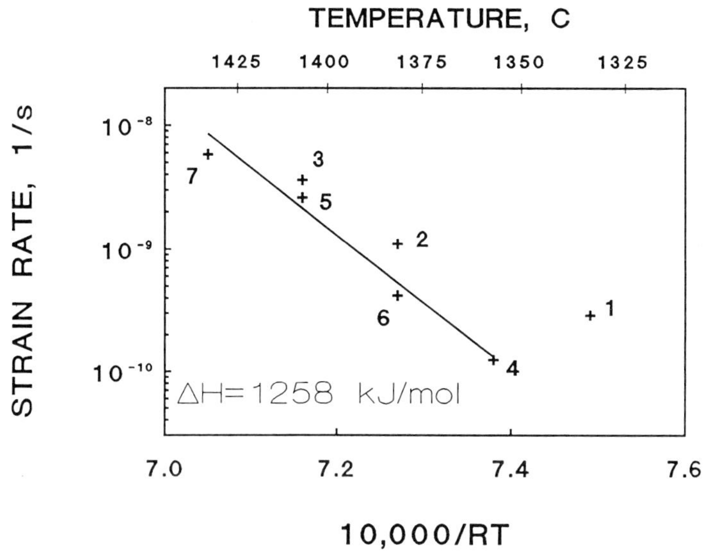

Figure 2. Creep rate versus inverse temperature for a hot-isostatically pressed silicon nitride. Activation energy = 1258 kJ/mol.

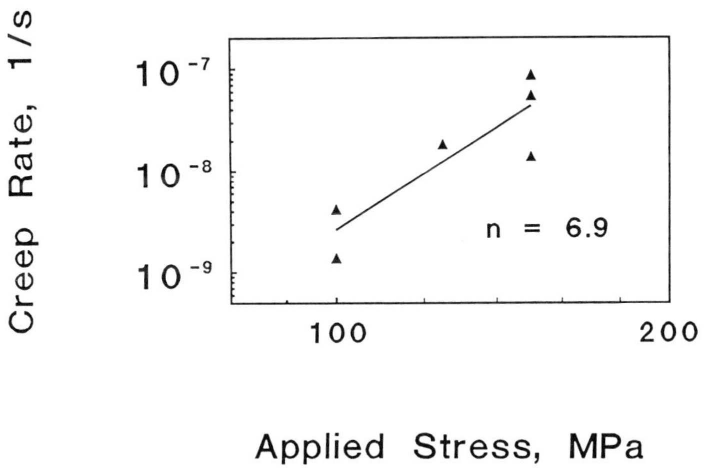

Figure 3. Creep rate versus applied stress at 1350 and 1375°C for a hot-isostatically pressed silicon nitride. Stress exponent = 6.9.

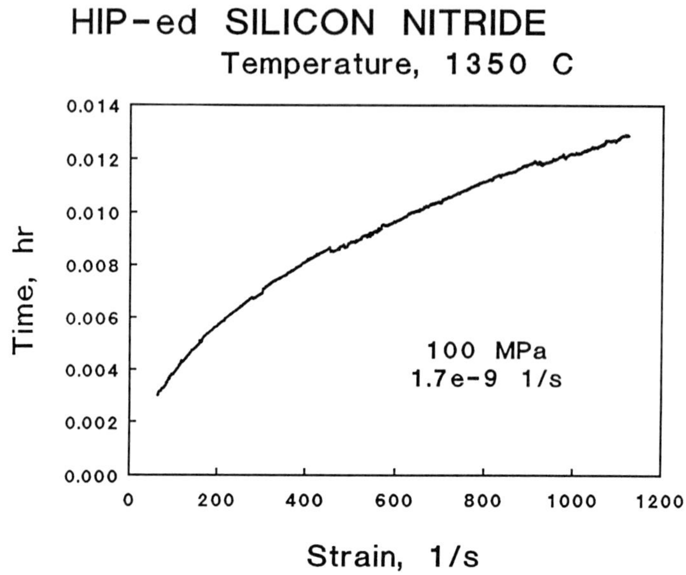

Figure 4. Typical creep curve for a hot-isostatically pressed silicon nitride.

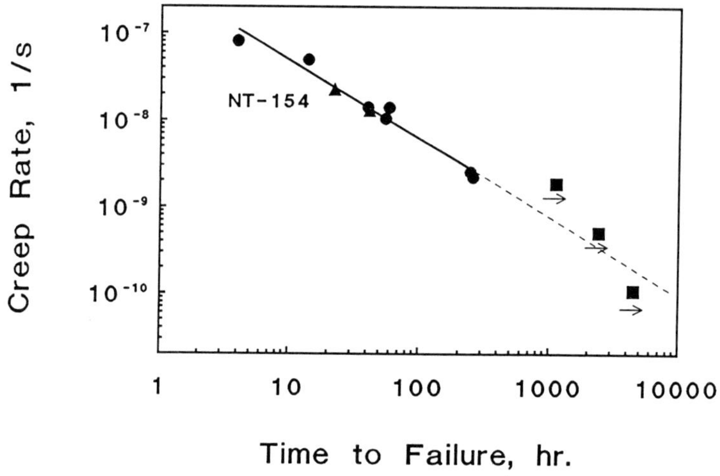

Figure 5. Monkman-Grant representation of creep fracture data for a hot-isostatically pressed silicon nitride. Arrows indicate specimens which did not fail.

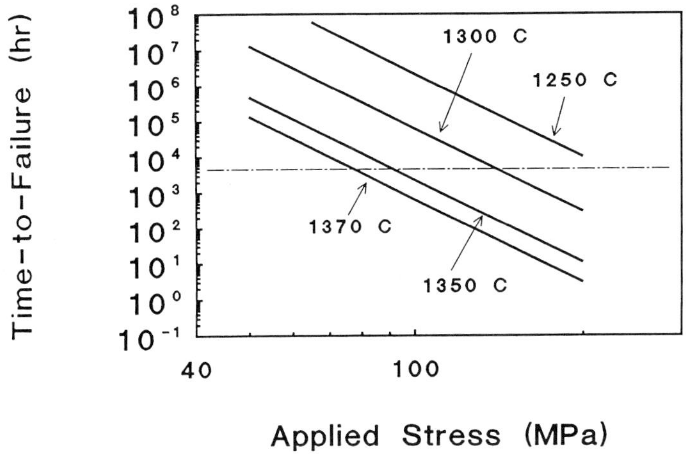

Figure 6. Lifetime predictions for a hot-isostatically pressed silicon nitride based on information about Monkman-Grant representation of creep rate versus time-to-failure, activation energy for creep, and stress exponent for creep.

6. MICROSTRUCTURAL EFFECTS

As noted earlier, measurements of the creep behavior without understanding the changes in and effects of the microstructure on subsequent creep behavior is ultimately accomplishes little more than accumulation of an empirical database. In many silicon nitride materials, differences in creep behavior from one specific material to another is the result of relatively small changes in composition which cause significant changes in grain boundary chemistry and structure. Most of the silicon nitrides available commercially contain some glassy phase at grain boundaries and at triple points in the microstructure. These features are revealed via the use of techniques such as optical, scanning electron (SEM), and transmission electron microscope (TEM). For the specific material above for which the lifetime predictions were discussed, both optical and analytical TEM (ATEM) have been used to characterize the microstructure. Optical examination of polished sections from a series of samples that failed at 1375°C under different loads, together with two other samples that survived for periods of 1100 and 2000 h, revealed no evidence of distributed creep cracks. From this result, it appears that fracture occurs by unstable crack growth, presumably once a critical density of cavities is developed. The microstructure of the as-received material is shown in Figure 7. It can be seen from the figure that the material is fully dense and contains both equiaxed and elongated grains, primarily of β-Si_3N_4. There is some glass at the grain boundaries, and at triple points in the microstructure.

ATEM of specimens subjected to tensile creep/creep fracture at temperatures of 1300 to 1400°C shows that the relatively high resistance to creep of this grade of silicon nitride can be related to two important aspects of its microstructure. The first is related to the "composite-like" microstructure produced by limited grain growth during processing. β-Si_3N_4 grain sizes and morphologies range from equiaxed sub-micrometer

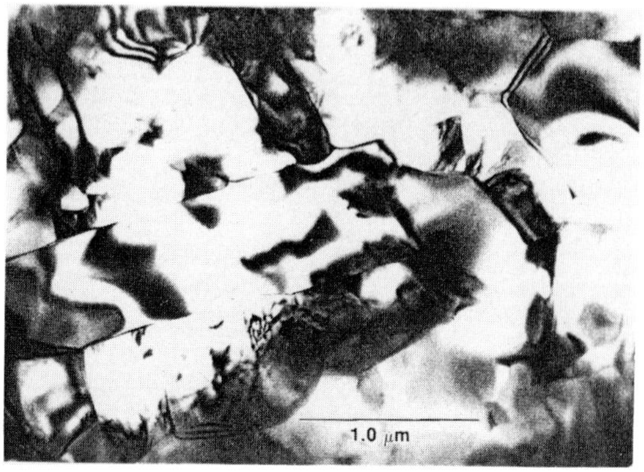

Figure 7. As-received microstructure of a hot-isostatically pressed silicon nitride.

sized grains, to "blocky" 1-3 micrometer size grains, to whisker-like grains, 1-6 micrometers in length with an aspect ratio of ≅3. The latter relates to both the volume fraction and composition of intergranular glass that binds the structure together. This glass is originally a relatively pure yttrium-silicate (only yttria is added as a sintering aid), and although continuous throughout the entire structure, is largely contained within multigrain junctions of nearly micrometer-size dimensions. Within these large intergranular regions, devitrification occurs rapidly at high temperature, forming, in nearly equal quantities, both α-$Y_2Si_2O_7$ (a triclinic yttrium-silicate crystalline phase) and N-Apatite (a hexagonal, yttrium-silicon-oxide nitride crystal phase). While residual glass always exists as a thin interfacial film (1-10 nm in thickness) between all crystalline phases, the composition of this film must necessarily become silica-rich after devitrification. With this change in residual glass composition, the viscosity of the glass will increase loading to a higher resistance to viscous flow and intergranular diffusivity. Moreover, since grain boundary sliding is the principal mechanism of creep deformation in multi-phase ceramics and involves both viscous flow and diffusional transport, the change in residual glass composition is directly related to the relative creep resistance of this material.

The increased resistance to grain boundary sliding associated with both the residual glass composition and the interlocked composite-like grain structure also brings about a change in the process of cavitation, which inevitably occurs and results in creep fracture. Unlike other grades of silicon nitride where rapid cavity growth occurs along interfaces to form creep crack nuclei, the cavitation in this Y_2O_3-Si_3N_4 results in isolated, ellipsoidal cavities. Regardless of test conditions, these cavities have major axis dimensions of 100 to 200 nm along the interface and minor axis dimensions of 50 to 100 nm. Since the interphase interfacial widths are typically ~1 to 10 nm, cavity growth clearly involves diffusive growth into the silicon nitride grains. Although the details of cavity development with strain are still under investigation, it appears that creep fracture occurs with the development of a critical density of these cavities along tensile oriented interfaces (Figure 8). In this material, the cavities are preferentially

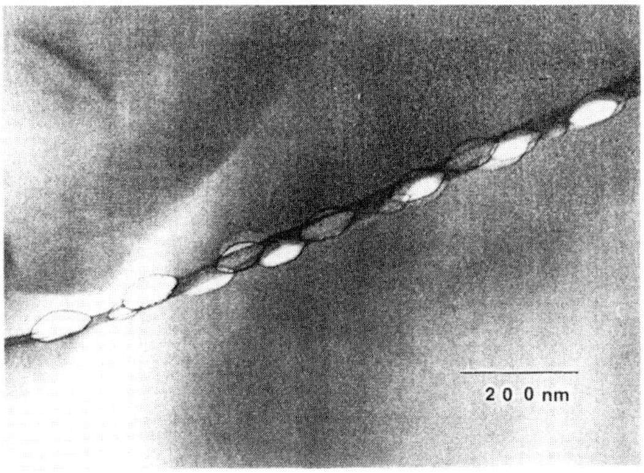

Figure 8. Lenticular shaped cavities formed along grain boundaries in a hot-isostatically pressed silicon nitride during creep.

distributed on grain boundaries that are normal to the tensile stress axis, and are generally in the form of isolated, lens-shaped cavities, showing apparent diffusive growth into the adjacent silicon nitride grains. So far, evidence for full-facet cavitation has only rarely been found.

Whisker reinforcement of silicon nitride is one possible method for improving the high temperature properties, however, in at least one instance, such additions did not improve its creep resistance.[21] Strains to failure of the unannealed whisker-free specimens were approximately twice those of the whisker-reinforced specimens, $\approx 2\%$, indicating a greater tolerance to deformation in the former. Annealing at the test temperature for extended periods improved the creep resistance of both materials, but the creep resistance of the whisker-free material was improved more than that of the whisker-reinforced material. The differences in creep behavior depend on the composition and volume fraction of the intergranular phase in the two materials. In particular, the presence of calcium and other impurities from the whiskers in the retained glass of whisker-reinforced material probably reduced its resistance to creep.

Both the whisker-reinforced and whisker-free materials failed by the nucleation and growth of cavities from the interfaces of large grains or whiskers. In the whisker-reinforced material, the whiskers act as nucleating sites for cavities that grow along the whisker interface to form micrometer- sized, crack-like cavities. These develop further into macroscopic size creep crack segments that eventually link together to cause failure.

7. FRACTURE MECHANISM MAPS

Creep is only one part of predicting reliability. Knowledge and models of it can be combined with similar models of fast fracture, fracture controlled by slow crack growth, and fracture controlled by fatigue to develop design guidance in the form of fracture mechanism maps. The fracture map delineates the temperature-stress regions where each of these mechanisms is dominant, thus indicating to the designer or engineer which regimes may need to be avoided for a particular application or set of applications. A map developed for one particular silicon nitride is shown in Figure 9.

The advantage of the fracture mechanism map is that it presents the design guidance in a simple, visual form. Results from a number of different investigators can be combined, thus making it easier to develop and compare results between investigators, if all of the experimental results are available. The disadvantage is that it requires a large number of individual tests to develop the information which goes into it. In the absence of suitable predictive models, and due to the statistical nature of ceramic failure modes, in excess of 500 specimens may be required.

Using known relations between flexure and tension, a map developed from one type of fracture data (e.g., from flexure) can be converted to a map for a different stress state such as tension.[22]

8. SUMMARY

As can be seen from the above results and discussion, there are a number of physical phenomena which control the long term deformation and fracture behavior of advanced structural ceramics. Our ability to predict lifetime and reliability of these materials is limited by our knowledge of the fundamental phenomena which control creep and damage evolution, and our ability to model these phenomena. A number of models are being used to describe the creep processes but there is still much work, both experimental and theoretical, to be done.

9. ACKNOWLEDGEMENTS

This work was supported by the U.S. Department of Energy, Ceramic Technology for Advanced Heat Engines Program under Interagency Agreement DE-AI05-850R21569. Many helpful discussions were also held with Dr. Sheldon Wiederhorn and Mr. George Quinn.

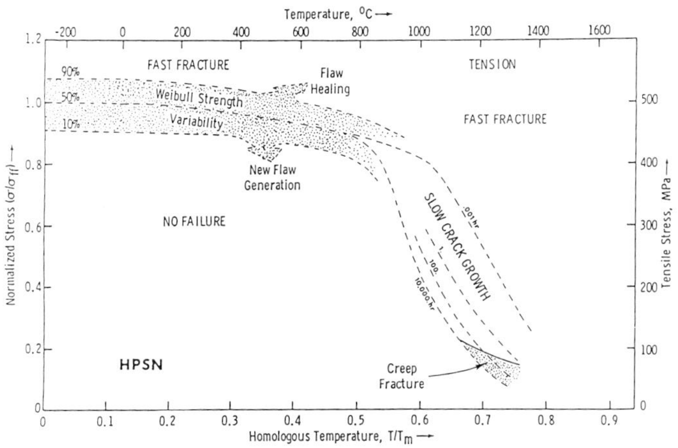

Figure 9. Fracture mechanism map for a hot-pressed silicon nitride. After Quinn (22).

10. REFERENCES

1. S. W. Freiman, "Environmentally Enhanced Fracture of Ceramics", in Materials Stability and Environmental Degradation, A. Barkaat, E. D. Verink, Jr., and L. R. Smith, eds., MRS Symp. Proc. Vol. 125, Materials Research Society, Pittsburgh, 1988, pp. 205-215

2. R. L. Coble, "A Model for Boundary Diffusion Controlled Creep in Ceramic Materials", J. Appl. Phys., 34 1679-1682 (1963).

3. C. Herring, "Diffusional Viscosity of a Polycrystalline Solid", J. Appl. Phys., 21 437-445 (1950).

4. R. Raj, "Creep in Polycrystalline Aggregates by Matter Transport Through a Liquid Phase", J. Geophys. Res., 87 4731-4739 (1982).

5. F. F. Lange, "Non-Elastic Deformation of Polycrystals with a Liquid Boundary Phase" in Deformation of Ceramic Materials, R. C. Bradt and R. E. Tressler, eds., Plenum Press, New York, 1972, pp. 361-381.

6. S. M. Wiederhorn, B. J. Hockey, and D. F. Carroll, "Creep of Two Phase Ceramics", in Sintering of Advanced Ceramics, Ceramic Transactions Vol. 7, C. A. Handwerker, J. E. Blendell, and W. A. Kaysser, eds., Amer. Ceram. Soc., Westerville, 1990, pp. 492-501.

7. G. W. Hollenberg, G. R. Terwilliger, and R. S. Gordon, "Calculation of Stresses and Strains in Four-Point Bending Creep Tests", J. Amer. Ceram. Soc., 54 [4] 196-199 (1971).

8. T.-J. Chuang, S. M. Wiederhorn, and C. F. Chen, "Transient Behavior of Structural Ceramics under Flexural Creep" in Creep and Fracture of Engineering Materials and Structures, B. Wilshire and R. W. Evans, eds., The Institute of Metals, London, 1987, pp. 957-973.

9. C.-F. Chen and T.-J. Chuang, "Improved Analysis for Flexural Creep with Application to Sialon Ceramics", J. Amer. Ceram. Soc., 73 [8] 2366-2373 (1990).

10. T.-J. Chuang, "Estimation of Power-Law Creep Parameters from Bend Test Data", J. Mater. Sci., 21 165-175 (1986).

11. D. F. Carroll and S. M. Wiederhorn, "High Temperature Creep Testing of Ceramics", Int. J. High Tech. Cer., 4 (1988) 227-241.

12. J. E. Lane, C. H. Carter, Jr., and R. F. Davis, "Kinetics and Mechanisms of High Temperature Creep in Silicon Carbide: III, Sintered α-Silicon Carbide", J. Amer. Ceram. Soc., 71 [4] 281-295 (1988).

13. D. F. Carroll, S. M. Wiederhorn, and D. E. Roberts, "A Technique for Tensile Creep Testing of Ceramics", J. Amer. Ceram. Soc., 72 [9] 1610-1614 (1989).

14. K. C. Liu, H. Pih, C. O. Stevens, and C. R. Brinkman, "Tensile Creep Behavior and Cyclic Fatigue/Creep Interaction of Hot-Isostatically-Pressed Si_3N_4," Annual Automotive Technology Development Contractors' Coordination Meeting, October 22-25, 1990, to be published.

15. T.-J. Chuang, W.-J. Liu, and S. M. Wiederhorn, "Steady-State Creep Behavior of Si-SiC C-Rings", J. Amer. Ceram. Soc., 74 [10] 2531-2537 (1991).

16. I. Finnie, "Method for Predicting Creep in Tension and Compression from Bending Tests", J. Amer. Ceram. Soc., 49 [4] 218-220 (1966).

17. R. F. Krause, Jr. and T.-J. Chuang, "A Test Method for Tensile Creep of Structural Ceramics Using Flexure Beams", in Ceramics Today - Tomorrow's Ceramics, P. Vincenzi, ed., Elsevier Science Publishers B.V., Amsterdam, 1991, pp. 1865-1874.

18. T.-J. Chuang, Z.-D. Wang, and D. Wu, "Analysis of Creep in a Si/SiC C-Ring by Finite-Element Method", J. Eng. Mater. Techn., 1991.

19. F. C. Monkman and N. J. Grant, "An Empirical Relationship between Rupture Life and Minimum Creep Rate in Creep-Rupture Tests," Proc. ASTM 56, 593-620 (1956).

20. D. C. Cranmer, B. J. Hockey, S. M. Wiederhorn, and R. Yeckley, "Creep and Creep-Rupture of HIP-ed Si_3N_4", Cer. Eng. Sci. Proc., 12 [9-10] 1862-1872 (1991).

21. B. J. Hockey, S. M. Wiederhorn, W. Liu, J. G. Baldoni and S.-T. Buljan, "Tensile Creep of Whisker-Reinforced Silicon Nitride," J. Mater. Sci., 26 3931-3939 (1991).

22. G. D. Quinn, "Fracture Mechanism Maps for Advanced Structural Ceramics, Part 1, Methodology and Hot-Pressed Silicon Nitride Results", J. Mat. Sci., 25 (1990) 4361-4376.

ADVANCES IN RELIABILITY
Edited by Asit P. Basu

Estimating an optimal block replacement policy using stochastic approximation

John I. Crowell[a] and Pranab K. Sen[b]

[a]Department of Mathematics and Statistics, Western Michigan University, Kalamazoo, MI 49008-5152, USA
[b]Department of Biostatistics, University of North Carolina, Chapel Hill, NC 27599 USA

Abstract

Stochastic approximation methodology is applied to the estimation of optimal block replacement policies. In the simplest block replacement policy (BRP) one carries out planned replacements of a functioning unit or part at times $0, T, 2T, \ldots$ for some fixed $T > 0$, as well as at any intervening failures. The motivation for such a policy is that stochastic part failure and replacement may be more disruptive and costly than planned part replacement. The spirit of the proposed procedure is to constantly adjust T so as to ultimately arrive at the optimal value ϕ^* which minimizes the long-run expected cost per unit time. Under certain conditions the algorithm does provide a strongly consistent and asymptotically normal estimate of ϕ^*. Various advantages and disadvantages of such a procedure are also considered.

1. INTRODUCTION

Consider a system consisting of a single component or part. It is assumed that this part can be replaced by a like part at any time and is not repaired following failure. Suppose that all parts used have independent and identically distributed lifetimes with continuous distribution F. Further assume that the cost C_1 of having to unexpectedly replace a failed part is greater than the cost C_2 of making a replacement at a planned time. If F has increasing failure rate, then it may be less expensive to substitute new parts for aging parts near failure than to only make replacements following failure. Two maintenance plans motivated by this consideration are the block replacement policy (BRP) and the age replacement policy (ARP). Both of these policies are treated in detail by Barlow and Proschan (1965, 1975), where some historical development is also given. Suppose a new part is installed at time 0. In the simplest BRP the part in use is replaced by a new part at times $T, 2T, 3T, \ldots$ and replaced following failure at any other time. (It is assumed for both the BRP and the ARP that all replacements are made instantaneously.) Here T is some positive constant. For this BRP, the expected long-run cost per unit time as a function of T is

$$B(T) = \frac{C_1 H(T) + C_2}{T}.$$

Here H is the renewal function associated with the lifetime distribution F. Thus $H(T)$ is the expected number of failures for each of the intervals $(0, T], (T, 2T], \ldots$.

Under an ARP, if a part has not failed and been replaced prior to reaching some fixed age T, it is preventively replaced at age T. The expected long-run cost per unit time can be shown to be

$$A(T) = \frac{C_1 F(T) + C_2[1 - F(T)]}{\int_0^T [1 - F(t)]dt}.$$

For both the age and block replacement policies an important practical question is what value of T, if any, minimizes expected cost per unit time. Barlow and Proschan (1965) and Bergman (1979) give sufficient conditions for the existence of an optimal ARP T_o such that T_o minimizes $A(T)$. Frees and Ruppert (1985) summarize some approaches that have been taken to the problem of estimating T_o for the ARP when the distribution F is unknown, contrasting fixed-sample procedures with sequential procedures. Of relevance here is a sequential scheme proposed by Bather (1977) that uses estimates of T_o to make planned replacements with the objective of reducing costs due to part failures. Bather's procedure has the desirable feature that asymptotically the incurred cost per unit time is minimized, as though the optimal ARP were used all along. With similar motivation, Frees and Ruppert (1985) successfully apply stochastic approximation methodology to the problem of finding the optimal ARP when the distribution F is unknown.

Now suppose there exists a finite positive value ϕ^* which minimizes the BRP cost $B(T)$ in T. The authors are unaware of any nonparametric procedures for estimating ϕ^* when the distribution F is unknown. Here we investigate using stochastic approximation to estimate an optimal BRP in a nonparametric setting. The idea is similar to that of Frees and Ruppert (1985). To suggest why sequential methods like stochastic approximation may be useful for estimating an optimal block replacement policy, consider those sequential methods for estimating ϕ^* which can be carried out concurrently with a random BRP. A random BRP implies that if the n-th planned replacement takes place at time t, the next planned replacement will be made at time $t + T_n$, where T_n is a random variable observable at time t. In practice, T_n could be an estimate of ϕ^*. For such a procedure let $N_0(t)$ and $N(t)$ denote the number of failure and planned replacements, respectively, through time t. Define $R(t) = C_1 N_0(t) + C_2 N(t)$. Then $R(t)/t$ is the incurred cost per unit time of using the procedure through time t. One desirable property that such a procedure might possess is that $\lim_{t\to\infty} R(t)/t = B(\phi^*)$, the minimum possible expected long-run cost per unit time. Theorem 1.1 below states that this is indeed the case (under some regularity conditions) when $T_n \to \phi^*$ $a.s.$ The result and proof are similar to Theorem 2.1 of Frees and Ruppert (1985), which dealt with random age replacement policies.

For the remainder of this article let $\{X_{ij}\}_{i,j=1}^{\infty}$ be independent and identically distributed random variables (part lifetimes) with lifetime distribution F which are defined on a probability space $(\Omega, \mathcal{F}, P)$. Take $N_n = \{N_n(t), t \geq 0\}$ to be the renewal process associated with the sequence $\{X_{nj}\}_{j=1}^{\infty}$ with renewals occurring at times $S_{n1} = X_{n1}, S_{n2} = X_{n1} + X_{n2}, ..., S_{nj} = X_{n1} + X_{n2} + ... + X_{nj}, ...$ for $n = 1, 2, ..., j = 1, 2, ...$ Define the σ-fields $\mathcal{F}_n = \sigma(X_{ij}, i = 1, 2, ..., n-1, j = 1, 2, ...)$.

The symbol $\circ$ will be used to denote the composition of functions. A superscript of the form $^{(p)}$ where p is a nonnegative integer will indicate the p-th derivative of a function, with composition taking precedence to differentiation. For example, $f \circ g^{(p)} = (f \circ g)^{(p)}$. Wherever used, $K_1, K_2, \ldots$ will represent appropriate constants. **E, Var**, and **P** will be used for expectation, variance, and probability, respectively. Convergence in distribution will be denoted by $\xrightarrow{D}$. Finally, $\chi[\quad]$ will be the indicator function of the event within the brackets.

In carrying out a random BRP as described above, planned replacements would be made at times $T_1, T_1 + T_2, T_1 + T_2 + T_3, \ldots$. During the time interval between the $(n-1)$st and n-th planned replacements, one would observe the truncated renewal process $\{N_n(t); 0 \leq t \leq T_n\}$ corresponding to any part failures during this time interval. At the time of the n-th planned replacement, observation of the truncated renewal process $\{N_{n+1}(t); 0 \leq t \leq T_{n+1}\}$ would begin. Thus in practice one would never observe in entirety the renewal process N_n defined above.

Theorem 1.1: Suppose $\{T_n\}$ is a sequence of random variables being used to carry out a random BRP as previously defined. Assume T_n is $\mathcal{F}_n$- measurable, $T_n \to \phi^*$ *a.s.* as $n \to \infty$, and H is continuous at ϕ^*. Define $R_n = C_1 \sum_{j=1}^{n} N_j(T_j) + C_2 n$, the incurred cost through the time of the n-th planned replacement. Then

$$\lim_{t\to\infty} \frac{R_{N(t)}}{t} = B(\phi^*).$$

The proof of Theorem 1.1 appears in Section 4. Theorem 1.1 says that it may be possible for the incurred cost per unit time to approach the minimum possible expected cost per unit time by carrying out a random BRP as suggested above. With this motivation, a stochastic approximation algorithm for the estimation of ϕ^* is introduced in Section 2; under certain circumstances it can be carried out in conjunction with a random BRP to achieve this goal. Asymptotic properties are given in Theorem 2.1. Section 3 contains some general discussion of the benefits and limitations of the proposed method. Section 4 contains proofs.

2. THE ESTIMATION PROCEDURE AND PROPERTIES

Recall that with previously defined notation the cost function for the BRP is

$$B(t) = \frac{C_1 H(t) + C_2}{t}$$

where H is the renewal function associated with F and $C_1 > C_2$. The objective is to minimize B in t. Assume that F is absolutely continuous so that the renewal density $h = H^{(1)}$ exists. Define

$$D(t) = [tC_1 h(t) - C_1 H(t) - C_2]/t^2,$$

so that $D(t) = B^{(1)}(t)$. Suppose that $B(t)$ has a unique minimum at $t = \phi^*$ with $D(\phi^*) = 0$. Let $\{a_n\}$ be a sequence of positive constants decreasing to zero. Given an initial estimate ϕ_1^* of ϕ^*, the Kiefer-Wolfowitz stochastic approximation algorithm suggests defining successive estimates of ϕ^* by the recursive formula

$$\phi_{n+1}^* = \phi_n^* - a_n D_n(\phi_n^*)$$

where $D_n(\phi_n^*)$ is an estimate of $D(\phi_n^*)$. This procedure would have the disadvantage of possibly generating negative estimates when necessarily ϕ^* is greater than zero. If it is given that ϕ^* is greater than some known $\delta > 0$, then one might use the algorithm

$$\phi_{n+1}^* = \max(\delta, \phi_n^* - a_n D_n(\phi_n^*)).$$

Rather than restrict the procedure in this fashion, we consider the following unconstrained stochastic approximation procedure which is relatively simple to analyze.

Assume there exists a known $\delta, 0 < \delta < \phi^*$. Let $g{:}\mathbb{R} \to (\delta, \infty)$ be a known, continuous, strictly increasing function which will be required to be differentiable a certain number of times. Suppose ϕ minimizes $B \circ g(t)$ in t so that $\phi^* = g(\phi)$. Define

$$Dg(t) = \{g(t)C_1 H \circ g^{(1)}(t) - g^{(1)}(t)[C_1 H \circ g(t) + C_2]\} \Big/ [g(t)]^2.$$

Since $Dg(t) = B \circ g^{(1)}(t)$ and $Dg(\phi) = 0$, we now consider using a stochastic approximation procedure to estimate the zero of Dg. Given an initial estimate ϕ_1 such that $\mathbf{E}\phi_1^2 < \infty$, define successive estimates of ϕ by

$$\phi_{n+1} = \phi_n - a_n Dg_n(\phi_n), \tag{2.1}$$

where $Dg_n(\phi_n)$ is an estimate of $Dg(\phi_n)$. Given ϕ_n, take $g(\phi_n)$ as an estimate of ϕ^*.

The next step is to define an estimator $Dg_n(t)$ of $Dg(t)$. Take $N_n = \{N_n(t), t \geq 0\}, n = 1, 2, \cdots$, to be the renewal processes defined in the introduction. We first consider defining an estimate of $H \circ g^{(1)}(t)$ assuming that for some integer $r \geq 1$ the function $H \circ g^{(r+1)}$ exists and is bounded. Denote by M the class of all Borel-measurable bounded functions K that vanish outside the interval [−1,1]. Take

$$\mathbb{K} = \Big\{K \in M \ \text{ such that } \ \frac{1}{j!}\int_{-1}^{+1} u^j K(u)du = 1, \text{ if } j = 0; \ 0, \text{if } j = 1, \cdots, r-1\Big\}.$$

For $K \in \mathbb{K}$ define

$$\begin{aligned} hg_n(t) &= \frac{1}{c_n}\int_\delta^\infty K\left(\frac{g^{-1}(u) - t}{c_n}\right) dN_n(u) \\ &= \frac{1}{c_n}\sum_{i=1}^\infty K\left(\frac{g^{-1}(S_{ni}) - t}{c_n}\right) \end{aligned} \tag{2.2}$$

where $\{c_n\}$ is a sequence of positive constants and $g^{-1}(x) = -\infty$ for $x \leq \delta$. The simplest and perhaps most intuitive choice for K would be $K(u) = \chi[-1 \leq u \leq 1]/2$. In this case we would have

$$hg_n(t) = \frac{N_n(g(t + c_n)) - N_n(g(t - c_n))}{2c_n}, \tag{2.3}$$

as an estimate of $H \circ g^{(1)}(t)$. Finally, let

$$Dg_n(t) = \frac{g(t)C_1hg_n(t) - g^{(1)}(t)[C_1N_n(g(t)) + C_2]}{[g(t)]^2}, \tag{2.4}$$

with estimates of ϕ defined recursively by (2.1). In practice one would only observe N_n through time $g(\phi_n + c_n)$; the estimate ϕ_n could then be updated according to (2.1). At this time, a new part would be put in place and observation of N_{n+1} would begin. Thus to implement the random BRP discussed in Section 1, one would take $T_n = g(\phi_n + c_n)$. The following assumptions will be employed.
A1: F is an absolutely continuous lifetime distribution with continuous density f.
A2: There exists known δ such that $0 < \delta < \phi^*$. For $t \in (\delta, \infty), D(t)(t - \phi^*) > 0$ for $t \neq \phi^*$.
A3: For some integer $r \geq 1$, $g : \mathbb{R} \to (\delta, \infty)$ is a known, continuous, strictly increasing function with $r+1$ bounded derivatives. Furthermore, $H \circ g^{(1)}$ and $H \circ g^{(r+1)}$ exist, are bounded on the entire real line, and are continuous in a neighborhood of ϕ. If $r \geq 3$, then either $H \circ g^{(2)}$ or $H \circ g^{(3)}$ is bounded on the entire real line.
A4: $\{a_n\}$ and $\{c_n\}$ are sequences of positive constants such that $\lim_{n\to\infty} a_n = \lim_{n\to\infty} c_n = 0, \sum_{n=1}^{\infty} a_n = \infty, \sum_{n=1}^{\infty} a_n c_n < \infty$, and $\sum_{n=1}^{\infty} a_n^2/c_n < \infty$.
A5: The function $Dg^{(1)}(t)$ exists and is continuous in a neighborhood of ϕ. For constants $A, C > 0$, let $a_n = An^{-1}$ and $c_n = Cn^{-\gamma}$, where $\gamma = (2r+1)^{-1}$ for the r of assumption A3. Further assume that $1 - \gamma < 2\Gamma$, where $\Gamma = A \cdot Dg^{(1)}(\phi)$.

Theorem 2.1 gives sufficient conditions for $a.s.$ convergence and asymptotic normality of the estimates generated by the procedure.

<u>**Theorem 2.1**</u>: (a) Assume A1-A3 and either A4 or A5 holds. Then $\phi_n \to \phi$ $a.s.$

(b) Assume A1-A3 and A5 hold. Then $n^{(1-\gamma)/2}(\phi_n - \phi) \xrightarrow{D} N(\mu_1, \sigma_1^2)$ where

$$\mu_1 = -\frac{C_1AC^rH \circ g^{(r+1)}(\phi)\beta_1}{g(\phi)[\Gamma - (1-\gamma)/2]}$$

and

$$\sigma_1^2 = \frac{C_1^2A^2C^{-1}H \circ g^{(1)}(\phi)\beta_2}{[g(\phi)]^2[2\Gamma - (1-\gamma)]}$$

with $\beta_1 = \int_{-1}^{+1} u^rK(u)du/r!$ and $\beta_2 = \int_{-1}^{+1} [K(u)]^2du$.

Of course our real interest is in estimating $g(\phi) = \phi^*$. The following corollary states asymptotic properties of $\phi_n^* = g(\phi_n)$.

<u>**Corollary 2.2**</u>: (a) Assume A1-A3 and either A4 or A5 holds. Then $\phi_n^* \to \phi^*$ a.s.
(b) Assume A1-A3 and A5 hold. Then $n^{(1-\gamma)/2}(\phi_n^* - \phi^*) \xrightarrow{D} N(g^{(1)}(\phi)\mu_1, [g^{(1)}(\phi)]^2\sigma_1^2)$.

3. SOME CONSIDERATIONS

Here we list some possible benefits and limitations of using a stochastic approximation procedure to estimate an optimal BRP. Perhaps the most troublesome assumption is A2, whose validity depends on both the unknown distribution F and the relative magnitudes of the costs C_1 and C_2. To give a simple example, suppose that F has density $f(t) = te^{-t}$ for $t \geq 0$, that is, F is the gamma distribution with shape parameter 2. Then $H(t) = t/2 - (1/4) + (1/4)e^{-2t}$ and $t \cdot h(t) - H(t)$ is a strictly increasing function of t. In this case a unique finite optimal BRP exists and A2 is satisfied if and only if $C_1 > 4C_2$. (See Barlow and Proschan (1965)). Crowell (1990) gives plots of the function $t \cdot h(t) - H(t)$ for the gamma distributions with shape parameters 2, 4, and 6 to illustrate how both the existence of an optimal BRP and A2 depend on the costs C_1 and C_2. A sufficient condition for both the existence of a unique ϕ^* and A2 is that $h^{(1)}(t)$ be nonnegative for all t, there exists a finite ϕ^* such that $D(\phi^*) = 0$, and $h^{(1)}$ be strictly positive in a neighborhood of ϕ^*. Unfortunately, this condition is rarely met and like A2 itself would be impossible to verify in practice.

A second major concern suggested by Theorem 2.1 and Corollary 2.2 is the slow rate of convergence of the estimator to the optimal value. Using the uniform kernel to estimate the renewal density as in (2.3) results in a rate of convergence $n^{2/5}$ at best. (This is assuming $r = 2$ in assumption A3.) See Crowell (1990) for a more complete description of the procedure's properties when the uniform kernel is used. If H is differentiable a sufficient number of times, then the rate of convergence can be quickened by choosing a higher order kernel that reduces the bias of the renewal density estimator. However, this will be achieved at the cost of some simplicity. We mention one possibility for partially overcoming the effect of a slow rate of convergence. Frees and Ruppert (1985) were able to significantly improve finite sample properties of their stochastic approximation procedure by using the tuning constants $a_n = A(n + K_a)^{-1}$ and $c_n = C(n + K_c)^{-1}$ for some positive constants K_a and K_c, rather than taking $K_a = 0$ and $K_c = 0$ as in A5. Hopefully this same modification could be employed with similar effect in the present setting.

The most important benefit of using a sequential procedure of the type proposed here has already been emphasized in Theorem 1.1: by simultaneously estimating an optimal BRP and carrying out a program of planned replacements, significant cost savings may be possible. Another benefit of the stochastic approximation procedure is that the current estimate is updated using only information about the last parts used. This provides some protection against changes in the lifetime distribution of the parts over time. If a block replacement policy is to be used over a long period of time, there might be some benefit to using a sequential control procedure to periodically adjust the replacement interval to reflect the stochastic behaviour of the parts currently in use.

4. PROOFS

Before proceeding with the proofs, some simple consequences of the assumptions will be noted. For any sequence of real numbers $\{x_n\}$, let $Y_n = N_n(g(x_n+c_n)) - N_n(g(x_n-c_n))$. Assuming A3 holds and $c_n = o(1)$, by an appropriate Taylor expansion,

$$\mathbf{E}Y_n = H \circ g^{(1)}(x_n)2c_n + o(c_n).$$

Also, by A1 and A3, for $k \geq 2$,

$$\mathbf{P}\{Y_n = k\}$$
$$\leq \sum_{j=1}^{\infty} [F^{(j)}(g(x_n + c_n)) - F^{(j)}(g(x_n - c_n))][F(g(x_n + c_n) - g(x_n - c_n))]^{k-1}$$
$$\leq K_1 c_n [F(g(x_n + c_n) - g(x_n - c_n))]^{k-1}. \tag{4.1}$$

Since $c_n \downarrow 0$, ultimately $F(g(x_n + c_n) - g(x_n - c_n)) < 1$, and thus for any real number $p \geq 0$,

$$\mathbf{E}\{Y_n^p \chi[Y_n \geq 2]\} \leq K_1 c_n \sum_{k=2}^{\infty} k^p [F(g(x_n + c_n) - g(x_n - c_n))]^{k-1}$$
$$= o(c_n). \tag{4.2}$$

The above implies that $\mathbf{P}\{Y_n = 1\} = H \circ g^{(1)}(x_n) 2c_n + o(c_n)$ and that for any real number $p > 0$

$$\mathbf{E}Y_n^p = H \circ g^{(1)}(x_n) 2c_n + o(c_n). \tag{4.3}$$

Now let F be any continuous lifetime distribution and $\{N(t), t \geq 0\}$ a renewal process with underlying distribution F. If F does not have finite moments of all orders, truncate F at some positive real number T such that $0 < F(T) < 1$ and define the distribution

$$F_*(x) = \begin{cases} F(x)/F(T), & x \leq T; \\ 1, & x > T. \end{cases}$$

Next let $\{N_*(t), t \geq 0\}$ be a renewal process with underlying distribution F_*. Then by Theorem 1 of Smith (1959), for any fixed integer $k \geq 1$,

$$\mathbf{E}[N(t)]^k \leq \mathbf{E}[N_*(t)]^k \leq K_1 + K_2 t^k \tag{4.4}$$

and

$$\mathbf{Var}\ \{N(t)\} \leq \mathbf{E}[N_*(t)]^2 \leq K_3 + K_4 t^2 \tag{4.5}$$

for appropriate positive constants K_1, K_2, K_3, and K_4.

Proof of Theorem 1.1: Let $U_n = \sum_{j=1}^{n} [N_j(T_j) - H(T_j)]$, so that $\{U_n, \mathcal{F}_{n+1}\}$ is a martingale. By the assumptions and (4.5),

$$\sum_{n=1}^{\infty} \mathbf{E}_{\mathcal{F}_n} [U_{n+1} - U_n]^2 / n^2 \leq \sum_{n=1}^{\infty} \frac{K_1 + K_2 T_n^2}{n^2} < \infty \ a.s.$$

This implies that $U_n/n \to 0$ *a.s.* (See Theorem 5 of Chow (1965).) By the continuity of H at ϕ^* and since $T_n \to \phi^*$ *a.s.*, it follows that $n^{-1}\sum_{j=1}^n N_j(T_j) \to H(\phi^*)$ *a.s.* and $R_n/n \to C_1 H(\phi^*) + C_2$ *a.s.*

Define $L_n = \sum_{j=1}^n T_j$, the time of the n-th planned replacement. Necessarily, $L_{N(t)} \le t < L_{N(t)+1}$, so that

$$\frac{N(t)}{N(t)+1}\cdot\frac{R_{N(t)}/N(t)}{L_{N(t)+1}/[N(t)+1]} \le \frac{R_{N(t)}}{t} \le \frac{R_{N(t)}/N(t)}{L_{N(t)}/N(t)}.$$

holds. (This is identical in form to (1.8) of Frees (1983).) Thus with probability one $T_n \to \phi^*$ and $L_n/n \to \phi^*$. The theorem follows by letting $t \to \infty$. ∎

The next two lemmas concern properties of the renewal density estimator hg_n. Analogous results for a kernel estimator of a probability density function were given by Singh (1977) and Frees and Ruppert (1985).

Lemma 4.1: Assume A3 holds. Let $K \in \mathbb{K}$ and $hg_n(x)$ be defined by (2.2). Assume $c_n = o(1)$. Then
(a) $\sup_x |\mathbf{E}hg_n(x) - H\circ g^{(1)}(x)| = O(c_n^r)$.
(b) Also suppose that $\{x_n\}$ is a sequence such that $\lim_{n\to\infty} x_n = x_0$, where $H\circ g^{(r+1)}(x)$ is continuous in a neighborhood of x_0. Then

$$\lim_{n\to\infty} c_n^{-r}[\mathbf{E}hg_n(x_n) - H\circ g^{(1)}(x_n)] = H\circ g^{(r+1)}(x_o)\frac{1}{r!}\int_{-1}^{+1} u^r K(u)du.$$

Proof: By the substitution $t = (g^{-1}(u) - x)/c_n$

$$\begin{aligned}\mathbf{E}hg_n(x) &= \frac{1}{c_n}\int_\delta^\infty K\Big(\frac{g^{-1}(u)-x}{c_n}\Big)h(u)du\\ &= \int_{-1}^{+1} K(t)h\circ g(x+c_n t)g^{(1)}(x+c_n t)dt\\ &= \int_{-1}^{+1} K(t)H\circ g^{(1)}(x+c_n t)dt.\end{aligned}$$

Expanding $H\circ g^{(1)}(x+c_n t)$ in a Taylor series about x, where $|\nu(t)-x| \le |c_n t|$,

$$\begin{aligned}\mathbf{E}hg_n(x) &= \sum_{j=0}^{r-1}\frac{H\circ g^{(j+1)}(x)}{j!}c_n^j\int_{-1}^{+1}K(t)t^j dt + \int_{-1}^{+1}K(t)\frac{H\circ g^{(r+1)}(\nu(t))}{r!}(c_n t)^r dt\\ &= H\circ g^{(1)}(x) + O(c_n^r)\end{aligned}$$

since $K \in \mathbb{K}$ and $H\circ g^{(r+1)}$ is bounded. This establishes (a). By part (a) it follows that

$$\begin{aligned}\lim_{n\to\infty} c_n^{-r}[\mathbf{E}hg_n(x_n) - H\circ g^{(1)}(x_n)] &= \lim_{n\to\infty}\int_{-1}^{+1}K(t)\frac{H\circ g^{(r+1)}(\nu(t))}{r!}t^r dt\\ &= H\circ g^{(r+1)}(x_o)\frac{1}{r!}\int_{-1}^{+1}K(t)t^r dt\end{aligned}$$

as $n \to \infty$, by the bounded convergence theorem and the continuity of $H \circ g^{(r+1)}$ at x_o. Thus (b) holds. ∎

Lemma 4.2: Assume A1 and A3 hold. Let $K \in \mathbb{K}$ and $hg_n(x)$ be defined by (2.2). Assume $\{x_n\}$ is a sequence such that $\lim_{n\to\infty} x_n = x_0$. Further assume that $c_n = o(1)$.
(a) If $t > -1$, then $\sup_x \mathbf{E}|hg_n(x)|^{t+1} = O(c_n^{-t})$.
(b) If $H \circ g^{(1)}$ is continuous at x_o, then

$\lim_{n\to\infty} c_n \mathbf{E}[hg_n(x_n)]^2 = H \circ g^{(1)}(x_o) \int_{-1}^{+1} [K(u)]^2 du.$

Proof: By the definition of hg_n, (4.3), and the boundedness of K and $H \circ g^{(1)}$

$$\begin{aligned} c_n^t \mathbf{E}|hg_n(x)|^{t+1} &= \frac{1}{c_n}\mathbf{E}| \int_\delta^\infty K\Big(\frac{g^{-1}(u)-x}{c_n}\Big) dN_n(u)\ |^{t+1} \\ &\le K_1 \frac{1}{c_n}\ \mathbf{E}[N_n(g(x+c_n)) - N_n(g(x-c_n))]^{t+1} \\ &\le K_2 \end{aligned}$$

This establishes (a).

As before, let $Y_n = N_n(g(x_n+c_n)) - N_n(g(x_n-c_n))$. Using the notation of Section 2,

$$\begin{aligned} c_n[hg_n(x_n)]^2 &= \frac{1}{c_n}\sum_{i=1}^\infty \left[K\left(\frac{g^{-1}(S_{ni})-x_n}{c_n}\right)\right]^2 \\ &+ \frac{1}{c_n}\sum_{\substack{i,j=1\\ i\ne j}}^\infty K\left(\frac{g^{-1}(S_{ni})-x_n}{c_n}\right) K\left(\frac{g^{-1}(S_{nj})-x_n}{c_n}\right). \end{aligned}$$

By the substitution $t = [g^{-1}(u) - x_n]/c_n$,

$$\begin{aligned} \mathbf{E}\frac{1}{c_n}\sum_{i=1}^\infty \left[K\left(\frac{g^{-1}(S_{ni})-x_n}{c_n}\right)\right]^2 &= \mathbf{E}\frac{1}{c_n}\int_\delta^\infty \left[K\left(\frac{g^{-1}(u)-x_n}{c_n}\right)\right]^2 dN_n(u) \\ &= \int_{-1}^{+1} [K(t)]^2 H \circ g^{(1)}(x_n + c_n t) dt \\ &\to H \circ g^{(1)}(x_o) \int_{-1}^{+1} [K(v)]^2 dv, \end{aligned}$$

as $n \to \infty$ by the bounded convergence theorem. Using (4.2) and letting $||\cdot||$ denote the supremum norm,

$$\begin{aligned} \mathbf{E}\,|\,\frac{1}{c_n}\sum_{\substack{i,j=1\\ i\ne j}}^\infty K\left(\frac{g^{-1}(S_{ni})-x_n}{c_n}\right) K\left(\frac{g^{-1}(S_{nj})-x_n}{c_n}\right)| &\le \frac{1}{c_n}||K||^2 \mathbf{E}\{Y_n^2 \chi[Y_n \ge 2]\} \\ &\to 0 \end{aligned}$$

as $n \to \infty$, which completes the proof of the lemma. ▮

The following theorem is the main tool used in establishing Theorem 2.1 (a).

Theorem (Robbins and Siegmund (1971)): Let $\mathcal{F}_n$ be a nondecreasing sequence of sub σ-fields of $\mathcal{F}$. Suppose that X_n, β_n, ν_n, and γ_n are nonnegative $\mathcal{F}_n$-measurable random variables such that

$$\mathbf{E}_{\mathcal{F}_n} X_{n+1} \leq X_n(1+\beta_n) + \nu_n - \gamma_n \quad \text{for} \quad n = 1, 2, \ldots.$$

Then, $\lim_{n\to\infty} X_n$ exists and $\sum_{n=1}^{\infty} \gamma_n < \infty$ on the set such that $\sum_{n=1}^{\infty} \beta_n < \infty$ and $\sum_{n=1}^{\infty} \nu_n < \infty$.

Proof of Theorem 2.1 (a): Letting $U_n = \phi_n - \phi$, we have by (2.1) that

$$\mathbf{E}_{\mathcal{F}_n} U_{n+1}^2 = U_n^2 + a_n^2 \mathbf{E}_{\mathcal{F}_n}[Dg_n(\phi_n)]^2 - 2a_n U_n \mathbf{E}_{\mathcal{F}_n}[Dg_n(\phi_n)] \tag{4.6}$$

since ϕ_n is $\mathcal{F}_n$-measurable. Define the bias term $\Delta_n = \mathbf{E}_{\mathcal{F}_n}[Dg_n(\phi_n) - Dg(\phi_n)]$. By Lemma 4.1 (a),

$$|\Delta_n| = C_1 |\mathbf{E}_{\mathcal{F}_n} hg_n(\phi_n) - H \circ g^{(1)}(\phi_n)\,|\,/g(\phi_n) \leq K_1 c_n^r.$$

By Lemma 4.2 (a) with $t = 1, \mathbf{E}_{\mathcal{F}_n}|hg_n(\phi_n)|^2 \leq K_1 c_n^{-1}$ so that by the moment property (4.4)

$$\begin{aligned}\mathbf{E}_{\mathcal{F}_n}[Dg_n(\phi_n)]^2 &\leq K_1 \frac{\mathbf{E}_{\mathcal{F}_n}|hg_n(\phi_n)|^2}{[g(\phi_n)]^2} + \frac{K_2 + K_3 \mathbf{E}_{\mathcal{F}_n}[N_n(g(\phi_n))]^2}{[g(\phi_n)]^4} \\ &\leq K_4 c_n^{-1} + K_5.\end{aligned}$$

Thus, since $\mathbf{E}_{\mathcal{F}_n} Dg_n(\phi_n) = Dg(\phi_n) + \Delta_n$, by (4.6)

$$\begin{aligned}\mathbf{E}_{\mathcal{F}_n} U_{n+1}^2 &\leq U_n^2 + a_n^2 \mathbf{E}_{\mathcal{F}_n}[Dg_n(\phi_n)]^2 + 2a_n|U_n||\Delta_n| - 2a_n(\phi_n - \phi)Dg(\phi_n) \\ &\leq U_n^2 + K_6 a_n^2 c_n^{-1} + 2a_n[K_7 + K_8 U_n^2]c_n^r - 2a_n(\phi_n - \phi)Dg(\phi_n) \\ &\leq U_n^2\{1 + 2K_8 a_n c_n^r\} + \{K_6 a_n^2 c_n^{-1} + 2K_7 a_n c_n^r\} - 2a_n(\phi_n - \phi)Dg(\phi_n).\end{aligned}$$

Since the required summations are finite by A4 or A5, $\phi_n \to \phi$ *a.s.* by an application of the Robbins-Siegmund result. ▮

Define $V_n = c_n^{1/2}[Dg_n(\phi_n) - Dg(\phi_n) - \Delta_n]$, where Δ_n was defined in the proof of Theorem 2.1 (a). The proof of Theorem 2.1 (b) will depend on the following lemma which describes the asymptotic behaviour of V_n.

Lemma 4.3: Assume A1, A3, and A5 hold. Then

$$\lim_{n\to\infty} \mathbf{E}_{\mathcal{F}_n} V_n^2 = C_1^2 H \circ g^{(1)}(\phi) \int_{-1}^{+1} [K(u)]^2 du/[g(\phi)]^2.$$

Proof: By definition $V_n^2 = c_n C_1^2 [X_n - Y_n - Z_n]^2$ where $X_n = [g(\phi_n)]^{-1} h g_n(\phi_n), Y_n = [g(\phi_n)]^{-1} \mathbf{E}_{\mathcal{F}_n} h g_n(\phi_n)$, and $Z_n = [g(\phi_n)]^{-2} g^{(1)}(\phi_n)[N_n(g(\phi_n)) - H \circ g(\phi_n)]$. We consider separately the conditional expectations $\mathbf{E}_{\mathcal{F}_n} X_n^2, \mathbf{E}_{\mathcal{F}_n} X_n Y_n$, etc.
(i) Since $\phi_n \to \phi$ *a.s.*, by Lemma 4.2 (b)

$$c_n \mathbf{E}_{\mathcal{F}_n} X_n^2 \to H \circ g^{(1)}(\phi) \int_{-1}^{+1} [K(u)]^2 du / [g(\phi)]^2.$$

(ii) By A3 and Lemma 4.2 (a) with $t = 0$, $c_n \mathbf{E}_{\mathcal{F}_n} X_n Y_n \le c_n O(1) \to 0$.
(iii) By (4.5), Lemma 4.2 (a) with $t = 1$, and the conditional version of the Cauchy-Schwarz inequality,

$$\begin{aligned} c_n \mathbf{E}_{\mathcal{F}_n} |X_n Z_n| &\le c_n [g(\phi_n)]^{-3} g^{(1)}(\phi_n) \mathbf{E}_{\mathcal{F}_n} \{h g_n(\phi_n) |N_n(g(\phi_n)) - H \circ g(\phi_n)|\} \\ &\le c_n K_1 \{\mathbf{E}_{\mathcal{F}_n} [h g_n(\phi_n)]^2\}^{1/2} \{\mathbf{E}_{\mathcal{F}_n} [N_n(g(\phi_n)) - H \circ g(\phi_n)]^2 / [g(\phi_n)]^2\}^{1/2} \\ &\le c_n K_2 c_n^{-1/2} \to 0. \end{aligned}$$

(iv) Also by Lemma 4.2 (a) with $t = 0$, $c_n \mathbf{E}_{\mathcal{F}_n} Y_n^2 \le c_n K_1 \to 0$.
(v) For all $n, \mathbf{E}_{\mathcal{F}_n} Y_n Z_n = 0$.
(vi) It follows from (4.5) that $c_n \mathbf{E}_{\mathcal{F}_n} Z_n^2 \le K_2 c_n \to 0$.
The result follows from (i) - (vi). ∎

We quote part of a theorem on asymptotic normality of stochastic approximation procedures as adapted by Frees and Ruppert (1985).

Theorem (Fabian (1968)): Suppose $\mathcal{F}_n$ is a nondecreasing sequence of sub σ-fields of $\mathcal{F}$. Suppose U_n, V_n, T_n, Γ_n, and Φ_n are random variables such that Γ_n, Φ_{n-1}, and V_{n-1} are $\mathcal{F}_n$-measurable. Let β, T, Σ, Γ, and Φ be real constants with $\Gamma > 0$ such that

$$\Gamma_n \to \Gamma, \Phi_n \to \Phi, \ T_n \to T \text{ or } \mathbf{E}|T_n - T| \to 0, \mathbf{E}_{\mathcal{F}_n} V_n = 0,$$

and there exists a constant C' such that $C' > |\mathbf{E}_{\mathcal{F}_n} V_n^2 - \Sigma| \to 0$.
Suppose, with $\sigma_{j,r}^2 = \mathbf{E}\chi[\ V_j^2 \ge rj\ ] V_j^2$, that

$$\lim_{n \to \infty} n^{-1} \sum_{j=1}^{n} \sigma_{j,r}^2 = 0 \ \text{ for all } \ r.$$

Let $0 \le \beta < 2\Gamma$, and $U_{n+1} = U_n[1 - n^{-1}\Gamma_n] - n^{-(1+\beta)/2} \Phi_n V_n + n^{-1-\beta/2} T_n$. Then

$$n^{\beta/2} U_n \xrightarrow{D} N\left(\frac{T}{(\Gamma - \beta/2)}, \frac{\Sigma \Phi^2}{(2\Gamma - \beta)}\right).$$

Proof of Theorem 2.1 (b): By the definition of $V_n, c_n^{-1/2} V_n = D g_n(\phi_n) - D g(\phi_n) - \Delta_n$. Expanding $Dg(\phi_n)$ about $Dg(\phi) = 0$ one has that $Dg(\phi_n) = (\phi_n - \phi) D g^{(1)}(\nu_n)$, where $|\nu_n - \phi| \le |\phi_n - \phi|$. As before, let $U_n = \phi_n - \phi$. Then

$$\begin{aligned} U_{n+1} &= U_n - a_n [c_n^{-1/2} V_n + Dg(\phi_n) + \Delta_n] \\ &= U_n[1 - a_n D g^{(1)}(\nu_n)] - A n^{-1} C^{-1/2} \ n^{\gamma/2} V_n - a_n \Delta_n, \end{aligned}$$

where $\gamma = (2r+1)^{-1}$ as in assumption A5. Set $\Gamma_n = ADg^{(1)}(\nu_n), \Phi_n = \Phi = -AC^{-1/2}, T_n = -An^{(1-\gamma)/2}\Delta_n$, and $\beta = 1-\gamma$. Then

$$\begin{aligned} U_{n+1} &= U_n[1 - n^{-1}\Gamma_n] + n^{\gamma/2-1}\Phi_n V_n + n^{-1}n^{-(1-\gamma)/2}T_n \\ &= U_n[1 - n^{-1}\Gamma_n] + n^{-(1+\beta)/2}\Phi_n V_n + n^{-1-(\beta/2)}T_n. \end{aligned}$$

Let $\Gamma = ADg^{(1)}(\phi)$. Since $Dg^{(1)}(t)$ is continuous at ϕ and $\phi_n \to \phi$ *a.s.* by Theorem 2.1 (a), it follows that $\Gamma_n \to \Gamma$ as $n \to \infty$. The asymptotic behaviour of T_n now depends on how many derivatives of $H \circ g$ exist. Since $\phi_n \to \phi$ *a.s.*, using Lemma 4.1 (b) we have that

$$\begin{aligned} T &= \lim_{n\to\infty} T_n \\ &= -AC_1 \lim_{n\to\infty} [g(\phi_n)]^{-1} C^r c_n^{-r} \left(\mathbf{E}_{\mathcal{F}_n}[hg_n(\phi_n)] - H \circ g^{(1)}(\phi_n)\right) \\ &= -\frac{AC^r C_1 H \circ g^{(r+1)}(\phi) \int_{-1}^{+1} u^r K(u)du}{g(\phi) r!}. \end{aligned}$$

By definition

$$V_n = c_n^{1/2} C_1 \left\{ \frac{[hg_n(\phi_n) - \mathbf{E}_{\mathcal{F}_n} hg_n(\phi_n)]}{g(\phi_n)} - \frac{g^{(1)}(\phi_n)[N_n(g(\phi_n)) - H(g(\phi_n))]}{[g(\phi_n)]^2} \right\}.$$

Using the inequality $(a+b+c)^t \le 3^t(a^t + b^t + c^t)$ for any nonnegative constants a, b, c, and t, we have that for any $t \ge 0$

$$\begin{aligned} |V_n|^t \le c_n^{t/2} K_1 \Big\{ &|hg_n(\phi_n)|^t [g(\phi_n)]^{-t} + |\mathbf{E}_{\mathcal{F}_n} hg_n(\phi_n)|^t [g(\phi_n)]^{-t} \\ &+ |N_n(g(\phi_n)) - H(g(\phi_n))|^t [g(\phi_n)]^{-2t} \Big\}. \end{aligned} \tag{4.7}$$

By definition, $\mathbf{E}_{\mathcal{F}_n} V_n = 0$. Furthermore, by Lemma 4.2 (a), (4.5), and (4.7), $\mathbf{E}_{\mathcal{F}_n} V_n^2$ is bounded. Thus using Lemma 4.3, there exists a positive constant C' such that $C' > |\mathbf{E}_{\mathcal{F}_n} V_n^2 - \Sigma| \to 0$, where Σ is the limit indicated by Lemma 4.3.

We show that $\sigma_{n,r}^2 = \mathbf{E}\{V_n^2 \chi[V_n^2 \ge rn]\} \to 0$ as $n \to \infty$ for $r = 1, 2, \dots$. Take $p > 2$. By Lemma 4.2 (a), $c_n^p \mathbf{E}_{\mathcal{F}_n} |hg_n(\phi_n)|^p \to 0$ and $c_n^p |\mathbf{E}_{\mathcal{F}_n} hg_n(\phi_n)|^p \to 0$ as $n \to \infty$. Thus by (4.4) and (4.7), for any $p > 2, \mathbf{E}_{\mathcal{F}_n} |c_n^{1/2} V_n|^p \to 0$ as $n \to \infty$. For the $\gamma \in (0,1)$ of assumption A5, take $q > 1$ such that $\gamma \le 1/q$. Then take p such that $2/p + 1/q = 1$. The fact that $\sigma_{n,r}^2 \to 0$ for $r = 1, 2, \dots$ follows from the Holder and Markov inequalities in the manner indicated on page 661 of Frees and Ruppert (1985). Thus $\lim_{n\to\infty} n^{-1} \sum_{j=1}^n \sigma_{j,r}^2 = 0$ for all r. Part (b) of Theorem 2.1 now follows from Fabian's theorem. ∎

<u>Proof of Corollary 2.2</u>: Part (a) is immediate from part (a) of Theorem 2.1 by the continuity of g. The "δ-method" can be used to show that part (b) follows from part (b) of Theorem 2.1. ∎

Acknowledgements

The authors thank the referee for his or her suggestions.

REFERENCES

Bather, J. A. (1977), "On the sequential construction of an optimal age replacement policy," *Bull. of the Intern. Statis. Inst. 47*, 253-266.

Barlow, R. E. and Proschan, F. (1965), *Mathematical Theory of Reliability,* Wiley, New York.

Barlow, R. E. and Proschan, F. (1975), *Statistical Theory of Reliability and Life Testing*, Holt Reinhart and Winston, Inc., New York.

Bergman, B. (1979), "On age replacement and the total time on test concept," *Scand. J. of Statistics 6*, 161-168.

Chow, Y. S. (1965), "Local convergence of martingales and the law of large numbers," *Ann. Math. Statist. 36*, 552-558.

Crowell, J. (1990), "On sequential estimation of the renewal function, optimal block replacement policies and fixed width confidence bands," Institute of Statistics Mimeo Series #2025, the consolidated University of North Carolina.

Fabian, V. (1968), "On asymptotic normality in stochastic approximation," *Ann. Math. Statis. 39*, 1327-1332.

Frees, E. W. (1983), "On construction of sequential age replacement policies via stochastic approximation," Institute of Statistics Mimeo Series #1525, the consoli-dated University of North Carolina.

Frees, E. W. and Ruppert, D. (1985), "Sequential nonparametric age replacement policies," *Ann. Statist. 13*, 650-662.

Robbins, H. and Siegmund, D. (1971), "A convergence theorem for nonnegative almost supermartingales and some applications," *Optimizing Methods in Statistics* (J. S. Rustagi, Ed.), 233-257. Academic Press, New York.

Singh, R. S. (1977), "Improvement on some known nonparametric uniformly consistent estimators of derivatives of a density," *Ann. Statist. 5*, 394-399.

Smith, W. L. (1959), "On the cumulants of renewal processes," *Biometrika 46*, 1-29.

ADVANCES IN RELIABILITY
Edited by Asit P. Basu

The role of information theory in reliability analysis

N. Ebrahimi[a], M. Habibullah[b], and E. S. Soofi[c]

[a]Division of Statistics, Northern Illinois University, DeKalb, IL 60115-2888, USA

[b]Division of Sciences and Mathematics, University of Wisconsin-Superior, Superior, WI 54880, USA

[c]School of Business Administration, University of Wisconsin-Milwaukee, P.O. Box 742, Milwaukee, WI 53201, USA

Abstract

A brief review of reliability literature indicates that while classical and Bayesian methods have been well developed and fairly broadly used in the field, the information-theoretic methods are just beginning to draw more than isolated attention. This paper reviews some aspects of information-theoretic statistics for reliability analysts. Basic information-theoretic quantities, measures of information loss due to censoring, and information diagnostics for reliability modelling are reviewed.

1. INTRODUCTION

The past 40 years have seen numerous advances in statistical theory and methods relevant to reliability. One of the major activities in the 1960's and 1970's was in the area of estimation and hypothesis testing procedures of various parametric life distributions for both censored and uncensored data. Bain [1], Lawless [2], Mann, Schafer and Singpurwalla [3], and Nelson [4] review much of the work in this area.

Nonparametric methods have been considered for various problems in reliability theory and advances are made in the following directions. First, major accomplishment was the development of Kaplan-Meier estimator of the survival function [5]. Closely related to this are nonparametric estimation of the cumulative hazard function [6] and nonparametric estimation of the mean residual life function [7-11]. Second, nonparametric tests for the equality of two or more survival functions under different sampling schemes have been developed [12, 13].

Third, "Aging" is a natural phenomenon, and change in the hazard function with the "age" is also quite common. Nonparametric classes of life distributions based on the notion of aging [increasing (decreasing) failure rate (IFR(DFR)), increasing (decreasing) failure rate in average (IFRA(DFRA)), increasing (decreasing) mean residual life (IMRL(DMRL)), new better (worse) than used (NBU(NWU)), new better (worse) than used in expectation (NBUE(NWUE)), harmonic new better (worse) than used in expectation (HNBUE(HNWUE)), and new better (worse) than used at specified aged t_0 (NBU-t_0 (NWU-t_0))] have been defined. Statistical methods for studying these broad classes of life distributions have been developed [14-17].

Emphasis on increased productivity has heightened interest in developing reliability techniques that use all available information in a cost effective manner. Because of that, among statisticians working in reliability theory perhaps the most significant trend is the growing recognition of the usefulness of the Bayesian approach. The past three decades witnessed the growth and implementation of a large body of Bayesian reliability analysis. Extensive references are given in [3, 18- 21].

The issue that concerns us here, and in our opinion has not received the attention that it deserves, is the role of "information theory" in the field of reliability. Scientists from diverse fields have been drawn to the field of information theory "because of the apparent impossibility of capturing the intangible concept of information' [22]. Statisticians have long endeavored to develop precise notion of information. In the Fisherian tradition, the amount of information provided by data for estimating a parameter is inversely related to the variance of the sampling distribution of the estimator. Consequently, the common measure of information utilized in statistics is the inverse of variance. In parametric estimation, particularly in the case of normal distribution, Fisher's notion of information turns out to be reasonable. However, the information measures based on the entropy are shown to be more general than Fisher's definition [23, 24]. The information-theoretic measures are very useful for applications in statistical analysis.

Ever since the fundamental works of Shannon [25] and Wiener [26] many researchers have devoted their attention to various areas of the information theory. In particular, a significant portion of statistics literature is devoted to studying the properties of the basic information functions and establishing relationships between statistical concepts (estimation, hypothesis testing, and decision theory) and the information theory [22, 27-32]. Soofi [33] reviews this line of research. We believe that information theory provides a convenient framework for many usual problems we face in the field of reliability and our intention is to review some recent applications of this theory.

To our knowledge, information-theoretic measures in the reliability analysis was first introduced by Tribus [34]. He showed that basic reliability models such as exponential, truncated normal, and Poisson are maximum entropy models (see (5) below) under various engineering constraints. Evans [35] discussed usefulness of entropy as a measure of information at a conceptual level, however, he believed that at the time, "the principle cannot be evoked to solve practical problems in reliability". El-Sayyad [36] used entropy to quantify amount of information contained in exponential samples about various functions of the exponential parameter. This work illustrated that information-theoretic approach can provide useful solutions to practical problems faced by reliability engineers. This line of research was later continued by a number of researchers [37-42]. In section 3 of this paper we report a summary of these works.

Another information-theoretic line of research which has wide applications in reliability modeling is development of entropy-based diagnostics and tests of distributional hypotheses [43-52]. A summary of this line of work will be presented in section 4.

2. MEASURES OF INFORMATION

In the information theory, the uncertainty associated with the distribution of a random variable X is given by

$$H(X) \equiv H[f(x)] = -\sum f(x) \ln f(x) \qquad \text{if X is discrete,}$$

$$= -\int f(x) \ln f(x)\, dx \qquad \text{if X is continuous,} \tag{1}$$

where f is the probability density (mass) function of X [25]. H(X) is called the entropy of the distribution and gives a measure of expected uncertainty about the outcome of a random draw from X.

For discrete distributions, H(X) is only a function of the probabilities f(x) and it does not depend on the values of X. For discrete distributions with finite support $H(X) \geq 0$. For continuous distributions, H(X) is not scale invariant, $H(cX) = \ln c + H(X)$, but it is translation invariant so, $H(c+X) = H(X)$. The differential entropy may be negative and infinite. However, boundedness of f(x) implies $H(X) > -\infty$, and $Var(X) < \infty$ implies $H(X) < \infty$. For some distributions such as Cauchy for which the variance does not exist, the entropy is finite. Some researchers have shown concerns about the negativeness of differential entropy. Hollander, Proschan and Sconing [39] suggested use of entropy for quantifying loss of information in the randomly right censored discrete data. However, for the continuous case on the ground that the entropy could be negative, they reverted from the entropy-based measures and used the theory of majorization to study the information loss. In many applications, entropy-based measures are used for quantifying uncertainty changes due to changes in the state of nature. Hence the magnitudes of the information-theoretic measures of entropy changes are of interest rather than the magnitude of H(X).

The most basic information-theoretic quantity for measuring a distributional change is the entropy difference

$$\delta H(f_1,f_2) = H[f_1(x)] - H[f_2(x)]. \tag{2}$$

Note that $\delta H(f_1,f_2)$ is scale invariant. A positive (negative) $\delta H(f_1,f_2)$ indicates an information increase (decrease) due to the change in the process, so negativeness of the individual entropies would be of no concern. For example, if due to a change in a process the distribution of X changes from an exponential distribution with parameters θ_1 to another exponential distribution with parameter θ_2, then the change in the uncertainty may be measured by the entropy difference $\delta H(f_1,f_2) = \ln(\theta_2/\theta_1)$. So the uncertainty in the process decreases if $\theta_2 > \theta_1$; e.g. when $\theta_2 = e\theta_1$, ($e = 2.71$) the uncertainty reduction is 1 nits of information.

A number of indices based on $\delta H(f_1,f_2)$ have been defined for comparison of distributions. Shannon [25] used the normal distribution as the basis for comparing uncertainty in continuous distributions. Among the distribution with a given variance σ^2, the normal distribution has the maximum entropy of $\frac{1}{2}\ln(2\pi e\sigma^2)$. Shannon defined the *entropy power* of a distribution f(x) as the variance of any normal density with the same entropy as H[f(x)]. That is,

$$N(f) \equiv \frac{\exp\{2H[f(x)]\}}{2\pi e}\,. \tag{3}$$

Note that $N(f) = \exp[2\delta(f, \phi)]$, where $\phi = N(\mu, 1)$. Thus N(f) is a measure of "closeness" of f to the normal density. Dudewicz and Van der Meulen [44] defined the *entropy power variance ratio (EPVR)* of distributions by the ratio $EPVR(f) \equiv \exp\{H[f(x)]\}/\sigma(f) = [2\pi e N(f)]^{1/2}/\sigma(f)$. They introduced the concept of *EPVR-distinguishability* $[EPVR(f_1) \neq EPVR(f_2)]$ and *EPVR-uniqueness* $[EPVR(f_1) = EPVR(f_2)$ when $f_1 \neq f_2]$.

Gokhale [47] formalized these ideas as follows. Consider the class of distributions

$$\Omega_\eta = \left\{ f(x|\theta)\colon\ E_f[T_j(X)] = \eta_j(\theta) = \eta_j\,,\ j=0,1,\ldots,m \right\}, \tag{4}$$

where $\eta = [1,\eta_1(\theta),...,\eta_m(\theta)]$, θ is a vector of parameters, $T_0(x) = 1$, and T_j, $j=1,...,m$ are absolutely integrable functions with respect to f. The maximum entropy model $f^*(x|\theta)$ in Ω_η is in the form of

$$f^*(x|\theta) = C(\theta)\exp\left[u_1(\theta)T_1(x) + ... + u_m(\theta)T_m(x)\right], \tag{5}$$

where $C(\theta)$ is the normalizing constant. The maximum entropy (ME) principle [53, 54] is used for modeling and inference by researchers in many fields [55]. In a class of probability models the ME model is preferred to all others because, it is "maximally uncommital" to information that are not included by the researcher in Ω_η. Many lifetime distributions such as exponential, Gamma, Weibull, Log-normal,... are ME distributions for various $T_j(X)$; see [34, 47, 55] for details.

Gokhale [47] defined the *Entropy Power Fraction (EPF)* of f(x) in Ω_η as

$$\mathrm{EPF}(f) \equiv \frac{\exp\{H[f(x|\theta)]\}}{\exp\{H[f^*(x|\theta)]\}} = \exp[\delta H(f,f^*)], \tag{6}$$

where $f^*(x|\theta)$ is the ME model in Ω_η. EPF is an index for comparing distributions in the class and $0 < \mathrm{EPF}(f) \leq 1$. Two distributions $f_1 \neq f_2$ in Ω_η are said to be *EPF-unique* if $\mathrm{EPF}(f_1) = \mathrm{EPF}(f_2)$. Otherwise, f_1 and f_2 are *EPF-distinguishable.*

The best known information function for measuring discrepancy between distributions is the cross-entropy between two distributions defined by

$$\begin{aligned} I(f_1:f_2) &= \sum f_1(x) \ln \frac{f_1(x)}{f_2(x)} && \text{if X is discrete,} \\ &= \int f_1(x) \ln \frac{f_1(x)}{f_2(x)} dx, && \text{if X is continuous,} \end{aligned} \tag{7}$$

provided that f_1 is absolutely continuous with respect to f_2. $I(f_1:f_2)$ is called Kullback-Leibler discrimination information function (number). $I(f_1:f_2) \geq 0$ and the equal sign holds if and only if $f_1(x) = f_2(x)$ almost everywhere. Although $I(f_1:f_2)$ is not a distance function in the usual sense, it is often thought of as a pseudo-distance or a "closeness" measure between f_1 and f_2. $I(f_1:f_2)$ is not symmetric in its two arguments; f_2 is called the *reference distribution.*

The expected amount of information in an observation of a random variable X for distinguishing between two potential probability distributions f_1 and f_2 for X is measured by $I(f_1:f_2)$. When f_i is the distribution of X under a hypothesis H_i, then the quantity $\ln[f_1(x)/f_2(x)]$ is interpreted as the amount of information contained in an observation x for the discrimination in favor of H_1. The discrimination function (7) quantifies the expected amount of information contained in an observation for discrimination between H_1 and H_2 in terms of the expected log-odds in favor of H_1. An application of the Bayes Theorem gives

$$I(f_1:f_2) = E_1\left\{\ln\frac{P(H_1|X)}{P(H_2|X)} - \ln\frac{P(H_1)}{P(H_2)}\right\}$$

$$= E_1\left\{\ln\frac{P(H_1|X)}{P(H_1)} - \ln\frac{P(H_2|X)}{P(H_2)}\right\}. \tag{8}$$

Here, $P(H_i)$ and $P(H_i|x)$ denote the prior and the posterior probabilities of H_i, respectively. That is, $I(f_1:f_2)$ is the expected difference between the posterior and prior log-odds in favor of f_1, under H_1; see [29]. Thus, large values of $I(f_1:f_2)$ indicate that X favors H_1 over H_2.

Let $X_1,\ldots,X_n$ be samples from the family of distributions $\{f(x|\theta), \theta\in\Theta\}$ and suppose $Y = T(X)$ is a statistic. If T is a measurable function and $g_i(y) = f_i[T^{-1}(y)]$, then $I[f_1(x):f_2(x)] \geq I[g_1(y):g_2(y)]$, with equality if and only if Y is a sufficient statistic for θ. Thus information in a sample for discrimination "cannot be increased by any statistical operations and is invariant (not decreased) if and only if sufficient statistics are employed" [23].

Soofi, Ebrahimi, and Habibullah [52] defined the *information discrimination (ID) index* of distributions in Ω_η as

$$ID(f:f^*) \equiv 1 - \exp[-I(f:f^*)]\,. \tag{9}$$

Note that $0 \leq ID(f:f^*) < 1$. Two distributions $f_1 \neq f_2$ in Ω_η are said to be *ID-unique* if $ID(f_1:f^*) = ID(f_2:f^*)$. Otherwise the two distributions are *ID-distinguishable*. They showed that if the reference distribution in the cross-entropy (7) is the ME model in Ω_η, then for all f in Ω_η,

$$I(f:f^*) = H[f^*(x|\boldsymbol{\theta})] - H[f(x|\boldsymbol{\theta})]. \tag{10}$$

Thus $ID(f:f^*) = 1 - EPF(f)$.

Let the F(x) with support $[0, \infty)$ be a lifetime distribution, and let

$$r(x) = \frac{f(x)}{1- F(x)}, \qquad \text{for } x \geq 0, \tag{11}$$

denote the hazard rate function of F. Using $\ln[r(x)] = \ln[f(x)] - \ln[1-F(x)]$, and noting that $E\{\ln[1-F(x)]\} = -1$, the entropy expression (1) can be written as

$$H(X) = 1 - E\{\ln[r(X)]\}. \tag{12}$$

Teitler, Rajagopal, and Ngai [56] discussed some implications of this relationship for the reliability analysis. Soofi, Ebrahimi, and Habibullah [52] give further implications of the relationship (12) both for reliability analysis and the information-theoretic statistics.

The discrimination function (7) may be related to the hazard rates functions of F_1 and F_2. Let $r_i(x)$, $i = 1,2$, denote respective hazard rate functions. Then

$$I(f_1:f_2) = \int_0^\infty f_1(x) \ln \frac{r_1(x)}{r_2(x)}\, dx, \tag{13}$$

whenever $E_1\{\ln[1 - F_2(X)]\} = -1$. For example, (13) holds when $f_1(x)$ is in a class of distributions with mean μ, and f_2 is the exponential density with the same mean. From (13) we observe that, on the average, the two densities would be closer to each other than the corresponding hazard rate functions whenever $E_1\{\ln[1 - F_2(X)]\} > -1$.

3. INFORMATION LOSS UNDER CENSORING

An important factor that distinguishes survival analysis and life testing from other fields of statistics is censoring. Vaguely speaking, a censored experiment contains only partial information about the quantities of interest. A censored experiment benefits from cost reduction at the expense of some information loss. Frequently, the amount of reduction in the cost of an experiment is computable, but the notation of information loss due to censoring remains intuitive. A number of authors have used information-theoretic measures for quantifying information loss due to censoring.

The most common measure used for quantifying the amount of information in the sample X about a parameter θ in the Bayesian framework is

$$\vartheta(\Theta|X) = E_X\{\delta H[p(\theta),p(\theta|x)]\} = I[f(x,\theta):f(x)p(\theta)] \geq 0. \tag{14}$$

Here, $p(\theta)$ and $p(\theta|x)$ are the prior and posterior distribution over the parameter space Θ, and $f(x)$ is the marginal distribution of X [28]. Let $X_1,...,X_n$ denote the lifetimes of n objects that are under a reliability test. In a censored experiment there will be only r, $r \leq n$ observations taken. Let T_s $s = r, n$, denote a (sufficient) statistic of interest. A comparison of the information quantities $\vartheta(\Theta|T_r)$ of the censored sample and $\vartheta(\Theta|T_n)$ of the complete sample gives an indicator of the information loss due to censoring.

Brooks [37] used $\vartheta(\Theta|X)$ to study information loss in censored samples under the following assumptions:

$$f(x|\theta) = \theta e^{-\theta x}, \qquad p(\theta) = \frac{\beta^{\alpha}}{\Gamma(\alpha)} \theta^{\alpha-1} e^{-\beta\theta}. \qquad \alpha > 0,\ \beta > 0. \tag{15}$$

He basically examined information loss in type I censoring, however, he also briefly discussed type II censoring and categorized data. For Type I censored data the calculation of $\vartheta(\Theta|X)$ requires numerical integration. Brooks used a large sample approximation under weak prior to study the information loss.

Turrero [40] used $\vartheta(\Theta|X)$ for quantifying information loss in random and fixed right censoring in a more general set up. He considered random partitions of type $\{(x_{i-1}, x_i], i = 1,...,k+1, x_{k+1} = \infty\}$, with probability of censoring in x_i equals to c_i. The parameters of interest are θ_i, the probability of death in $(x_{i-1}, x_i]$. He studied effects of varying k and c_i on $\vartheta(\Theta|X)$ under Dirichlet prior for θ_i.

Ebrahimi and Soofi [41] used $\vartheta(\Theta|X)$ to study information loss in type II censored samples under the model (15). Let $y_1 < y_2 <...< y_r$, be the r order statistics from a type II censored experiment and $t_r = y_1 + ... + y_r + (n-r)y_r$. Then the following index measures the relative change in information content of the data for posterior analysis when only r observations has been taken instead of n:

$$\xi(r,n,\alpha) = \frac{\vartheta_\alpha(\Theta|t_n) - \vartheta_\alpha(\Theta|t_r)}{\vartheta_\alpha(\Theta|t_n) - \vartheta_\alpha(\Theta|t_1)}\ 100\% . \tag{16}$$

Here,

$$\vartheta_\alpha(\Theta|t_s) = \alpha[\psi(\alpha+s) - \psi(\alpha)] + s[\psi(\alpha+s) - 1] + \ln[\Gamma(\alpha)/\Gamma(\alpha+s), \tag{17}$$

where $\psi(u) = d\ln[\Gamma(u)]/du$ is the digamma function. The information index $\xi(r,n,\alpha)$

monotonically decreases in r with $\xi(n,n) = 0\%$ and $\xi(1,n) = 100\%$ It monotonically increases in the prior shape parameter α. $\xi(r,n,\alpha)$ provides a diagnostic for determining the relative information loss of type II censored experiment data for Bayesian analysis. It can also be used to examine the incremental contribution of additional prior precision in reduction of relative information loss in type II censoring.

Ebrahimi and Soofi [41] also proposed the following index for measuring the relative change in information content of the maximum likelihood estimate when only r observations has been taken instead of n:

$$\xi(r,n) = \frac{H(W_n) - H(W_r)}{H(W_n) - H(W_1)} 100\% . \tag{18}$$

Here $W_r=(t_r/r)$ is the maximum likelihood estimator of $\mu = 1/\theta$, and

$$H(W_s) = -\ln(\theta) + \ln[\Gamma(s)/s] + (1 - s)\psi(s) + s , \tag{19}$$

is the expected amount of uncertainty in using W_s for inference about μ. This index is also monotone decreasing function of r with $\xi(n,n) = 0\%$ and $\xi(1,n) = 100\%$. $\xi(r,n)$ provides a diagnostic for determining the relative information loss in the maximum likelihood analysis of Type II censored exponential data.

The above indices are useful for determining the trade off between cost of sampling and cost of prior versus the amount of information loss due to censoring. Suppose the fixed cost of running an experiment is C_0, the cost of collecting one observation is C_1, and the cost of acquiring prior precision is C_α. Then take the first r + 1 observations if

$$\frac{C_0 + (r + 1)C_1 + C_\alpha}{C_0 + r\, C_1 + C_\alpha} < \frac{\xi(r,n,\alpha)}{\xi(r + 1,n,\alpha)} \tag{20}$$

otherwise take the first r observations. For the maximum likelihood analysis let $C_\alpha = 0$, and use $\xi(r,n)$ in (20).

Finally, we conclude this section with a brief description of an interesting problem studied by Abel and Singpurwalla [42]. Suppose n identical components are under testing for estimating (a) the mean lifetime, and (b) the failure rate of a component. The question is that which components provide more information about the mean lifetime and about the failure rate of the component? Those that do fail at time t, or those that do not fail? They show that when the lifetimes follow an exponential distribution, then the components that fail are more informative about the mean life and those that survive are more informative about the failure rate. That is, for information about the failure rate observing survivals are preferred, however, for information about the mean observing failures are preferable.

4. INFORMATION DIAGNOSTICS FOR RELIABILITY MODELING

Entropy-based tests and diagnostics for reliability modelling are basically those that are proposed for measuring departure of data distribution from an ME distribution. These procedures consists of computing a nonparametric measure of sample entropy and comparing it with the entropy of the ME distribution under consideration. A number of nonparametric entropy estimators are available, see [49, 57]. Mack [49] studied tests based on a number of nonparametric entropy estimates. Thus far an estimator proposed by Vasicek [43] has gained prominence in testing hypotheses of ME distributions.

In the class of distributions with a given variance, $\Omega_\sigma = \{ f(x|\sigma): E_f(X^2) = \sigma^2 \}$, with support (∞, ∞), the normal distribution is the ME model [25]. When the support is $[0, \infty)$, then the ME model is the truncated normal [34]. Vasicek [43] proposed a test of normality based on the comparison of the sample entropy computed using his procedure and the entropy of normal distribution $H[f^*(x|s^2)] = \frac{1}{2}\ln(2\pi e s^2)$, s^2 being the maximum likelihood estimate of the variance. He actually used the transformation (6) of the estimated entropies. Arizono and Ohta [50] derived Vasicek's test of normality based on the discrimination information function using the equation (10).

When there is only one constraint in (4), $T_0(X) = 1$, and the support of distributions in Ω_1 is the interval [a, b], then the ME distribution is the uniform distribution over [a, b]. Dudewicz and Van der Meulen [44] proposed a test of uniformity over [0, 1] based on the comparison of the sample entropy estimated by Vasicek's procedure and the maximum entropy $H[f^*(x)]=0$.

More generally, using a parametric entropy estimate of the ME in Ω_η denoted by $H[f^*(x|\theta_n)]$ and a nonparametric entropy estimate denoted by H_n in the expression (6) we obtain an information statistic,

$$EPF_n(f) = \frac{\exp\{H_n\}}{\exp\{H[f^*(x|\theta_n)]\}} . \tag{21}$$

This provides diagnostics for examining the distribution of the data generating process.

An important case in the reliability analysis is when the class of distributions (4) is defined by $\Omega_\mu = \{ f(x|\mu): E_f(X)=\mu \}$. In this class the ME distribution is the exponential distribution $f^*(x|\mu) = (1/\mu)\exp(-x/\mu)$, with $H[f^*(x|\mu) = \ln \mu + 1$. Moreover, this is also the ME model in the larger class $\Omega_{\leq\mu} = \{ f(x|\mu): E_f(X) \leq \mu \}$. In terms of the failure rate, from (12) we see that in the class $\Omega_{\geq\theta} = \{ f(x|\theta): r(x) \geq \theta \}$ the ME model is the exponential distribution $f^*(x|\theta) = \theta\exp(-\theta x)$. For $\mu = 1/\theta$, $\Omega_{\leq\mu}$ and $\Omega_{\geq\theta}$ are equivalent; see [58] for detail.

Gokhale [47] proposed a test of exponentiality based on $EPF_n(f)$ using Vasicek's estimator in the numerator of (21) and $H[f^*(x|\overline{x})] = \ln \overline{x} + 1$ as the maximum likelihood estimate of the exponential entropy in the denominator. Ebrahimi, Habibullah, and Soofi [51] derived Gokhale's test of exponentiality based on the discrimination information function using the equation (10). They showed (by simulations) that the discrimination test performs very well in comparison with all existing tests.

Soofi, Ebrahimi, and Habibullah [52] proposed entropy estimation based on the hazard rate function formula (12). They developed a number of information diagnostics by estimating the ID index defined in (9) and used them for assessing departure of data generating distribution from exponentiality in the class of IFR distributions. Simulation results reported there indicate that hazard rate-based entropy estimates perform superior to the corresponding density based entropy estimates. These results also indicate that discrimination information statistics perform very well as tests of exponentiality. In this line of research, there are still an enormous scope for further developments. For example, information diagnostics based on the ID index may be developed in other classes of life distributions, say IFRA, NBU, and so on.

Some authors have developed other types of entropy-based tests of exponentiality. Chandra and Singpurwalla [45] developed a test of exponentiality using an entropy-based "redundancy" measure. Chandra, De Wet, and Singpurwalla [46] studied sampling properties of this test. Parametric entropy-based procedures useful for reliability analysis have also been developed. Arizono and Ohta [48] used cross-entropy for testing hypothesis about the exponential parameter in the context of acceptance sampling.

5. REFERENCES

1 L.J. Bain, Statistical Analysis of Reliability and Life Testing Models, Marcel Dekker, New York, 1978.

2 J.F. Lawless, Statistical Models and Methods for Lifetime Data, New York: John Wiley, 1982.

3 N.R. Mann, R.E. Schafer and N.D. Singpurwalla, Methods for Statistical Analysis of Reliability and Life Data, New York: John Wiley, 1982.

4 W.B. Nelson, Applied Life Data Analysis, New York: John Wiley, 1982.

5 E.L. Kaplan and P. Meier, Nonparametric Estimation From Incomplete Observations, JASA, 53, (1958) 457-481.

6 W.B. Nelson, Theory and Applications of Hazard Plotting for Censored Failure Data, Technometrics, 14, (1972) 945-965.

7 G.L. Yang, Estimates of Biometric Function, The Annals of Statistics, Vol. 6 (1978) 112-117.

8 R.C. Johnson and N.L. Johnson, Survival Models and Data Analysis, New York: John Wiley, 1980.

9 R. Dykstra and C. Feltz, Maximum Likelihood Estimation of the Survival Function of N Stochastically Ordered Random Variables, Journal of the American Statistical Association, 80 (1984), 1014-1019.

10 N. Ebrahimi, On Estimating Change Point in a Mean Residual Life Function, Sankhya, Series A, Vol. 53 (1991).

11 N. Ebrahimi, Estimation of Two Ordered Mean Residual Lifetime Functions, due to appear in Biometrics (1991).

12 R.L. Prentice, Linear Rank Tests With Right Censored Data, Biometrika, 65 (1978), 167-179.

13 W.E. Franck, A Likelihood Ratio Test for Stochastic Ordering, JASA, 79 (1984), 686-691.

14 M. Hollander and F. Proschan, Nonparametric Concepts and Methods in Reliability, in Handbook of Statistics, Vol. 4: Nonparametric methods (1984) (eds., P. R. Krishniah and P. K. Sen).

15 R.E. Barlow and F. Proschan, Statistical Theory of Reliability and Life Testing (1981), to begin with.

16 A.P. Basu and N. Ebrahimi, HNBUE and HNWUE distributions (survey), On Reliability and Quality Control, edited by A.P. Basu, 33-46, North-Holland, 1986.

17 N. Ebrahimi and M. Habibullah, Testing Whether Survival Function is New Better than used at Specified Time t_0, Biometrika, 77 (1990), 212-215.

18 H.F. Martz and R.A. Waller, Bayesian Reliability Analysis, New York: John Wiley, 1982.

19 R.E. Barlow and J. Wu, Prepostrior Analysis of Bayes Estimators of Mean Life, Biometrika, Vol. 68 (1981), 403-410.

20 A.P. Basu and N. Ebrahimi, Bayesian Approach to Life Testing and Reliability Estimation Using Asymmetric Loss Function, JSPI, 29 (1991), 21-33.

21 A.P. Basu and N. Ebrahimi, Bayesian Approach to Life Testing and Reliability Estimation, Technical Report (1991), Department of Statistics, University of Missouri-Columbia.

22 M.T. Cover and J.A. Thomas, Elements of Information Theory, New York: John Wiley, 1991.

23 S. Kullback and R.A. Leibler, On Information and Sufficiency, Ann. of Math Statist., 22 (1951), 79-86.

24 S. Kullback, Certain Inequalities in Information Theory and the Cramer-Rao Inequality, Ann. Math. Statist., 25 (1954), 745-751.

25 C.E. Shannon, A Mathematical Theory of Communication, Bell Sys. Tech. J. (1948), 279-423 and 623-656.

26 N. Wiener, Cybernetics, New York: John Wiley, 1948.

27 A.I. Khinchin, Mathematical Foundations of Information Theory, New York: Dover, 1957.

28 D.V. Lindley, On a Measure of Information Provided by an Experiment, Annals of Mathematical Statistics, 27 (1956), 986-1005.

29 S. Kullback, Information Theory and Statistics, New York: Wiley, 1959 (reprinted in 1969 by Dover).

30 H. Akaike, Information Theory and an Extension of the Maximum Likelihood Principle, in B. N. Petrov and F. Csaki, 267-81 (eds.), 2nd International Symposium on Information Theory, Akademia Kiado, Budapest, 1973.

31 I. Csiszar, I-divergence Geometry of Probability Distributions and Minimization Problems. Ann. Probability, 3 (1975), 146-158.

32 J.M. Bernardo, Expected Information as Expected Utility, Ann. Statist., 7 (1979), 686-690.

33 E.S. Soofi, On the Information Theory and Statistics, paper presented at the First Iranian Statistics Conference, Isfahan, Iran (1992).

34 M. Tribus, The Use of Maximum Entropy Estimate in the Estimation of Reliability, Recent Development in Information and Decision Process, Edited by Machol, R.E. and Gray, P., New York: Macmillan, 1962.

35 R.A. Evans, The Principle of Minimum Information, IEEE Transactions on Reliability, R-18 (1969), 87-90.

36 G.M. El-sayyad, Information and Sampling from the Exponential Distribution, Technometrics, 11 (1969), 41-45.

37 R.J. Brooks, On the Loss of Information Through Censoring, Biometrika, 69 (1982), 137-144.

38 R.E. Barlow and Jaw Huan Hsiung, Expected Information From a Life-Test Experiment,

Statistician, Vol. 32 (1983), 35-45.

39 M. Hollander, F. Proschan and J. Sconing, Measuring Information in Right Censored Models, Naval Research Logistics, 34 (1987), 669-681.

40 A. Turrero, On the Relative Efficiency of Grouped and Censored Survival Data, Biometrika 76 (1989), 125-131.

41 N. Ebrahimi and E.S. Soofi, Relative Information Loss Under Type II Censored Exponential Data. Biometrika, 77 (1990), 429-435.

42 P. Abel and N. Singpurwalla, To Survive or to Fail: That is the Question, Paper presented at the 4th International Research Conference on Reliability, Columbia, MO (1991).

43 O. Vasicek, A Test for Normality Based on Sample Entropy, Journal of Royal Statistical Society, Ser. B, 38 (1976), 54-59.

44 E.J. Dudewicz and E.C. Van der Meulen, Entropy-Based Tests of Uniformity, Journal of the American Statistical Association, 76 (1981), 967-974.

45 M. Chandra and N.D. Singpurwalla, Relationships Between Some Notions Which are Common to Reliability Theory and Economics, Mathematics of Operations Research, 6 (1981), 113-121.

46 M. Chandra, T. De Wet, and N.D. Singpurwalla, On the Sample Redundancy and a Test for Exponentiality, Commun. Statist. Theor. Meth., 11(4) (1982), 429-438.

47 D.V. Gokhale, On entropy-based goodness-of-fit tests, Computational Statist. and Data Analysis, 1 (1983), 157-165.

48 H. Ohta and I. Arizono, A Simplified Design Procedure for Life Tests Based on Kullback-Leibler Information, IEEE Transactions on Reliability, R-34, 4 (1985), 363-365.

49 S. Mack, A Comparative Study of Entropy Estimators and Entropy-Based Goodness-of-fit Tests, Ph.D. Dissertation, Univ. of California, Riverside, 1988.

50 I. Arizono, and H. Ohta, A Test of Normality Based on Kullback-Leibler Information, The American Statistician, 34 (1989), 20-23.

51 N. Ebrahimi, M. Habibullah, and E.S. Soofi, Testing Exponentiality Based on Kullback-Leibler Information, J. Roy. Statist. Soc., B, 54 (1992), 739-749.

52 E.S. Soofi, N. Ebrahimi and M. Habibullah, Information Diagnostics for Analysis of Failure Data, submitted for publication (1992).

53 J.E. Shore, and R.W. Johnson, Axiomatic Derivation of the Principle of Maximum Entropy and the Principle of Minimum Cross-Entropy, IEEE Transaction on Information Theory, IT-26 (1980), 26-37.

54 E.T. Jaynes, On the Rationale of Maximum-Entropy Methods, Proceedings of IEEE, 70 (1982), 939-952.

55 J.N. Kapur, Maximum Entropy Models in Science and Engineering, New York: John Wiley, 1989.

56 S. Teitler, A.K. Rajagopal, and K.L. Ngai, Maximum Entropy and Reliability Distributions, IEEE Transactions on Reliability, R-35 (1986), 391-395.

57 N. Ebrahimi, K. Pflughoeft, and E.S. Soofi, Two Measures of Sample Entropy, submitted for publication, 1992.

58 N. Ebrahimi, G.G. Hamadani, and E.S. Soofi, Maximum Entropy Modeling with Partial Information on Failure Rate, Working Paper, Management Research Center, University of Wisconsin-Milwaukee, 1991.

ADVANCES IN RELIABILITY
Edited by Asit P. Basu
1993 Elsevier Science Publishers B.V.

Statistical analysis of a power-law process with left-truncated data

Max Engelhardt[a], David H. Williams[b] and Lee J. Bain[c]

[a]Idaho National Engineering Laboratory, EG&G Idaho, Inc., P. O. Box 1625, Idaho Falls, ID 83415, USA

[b]Department of Statistics, University College Dublin, Dublin 4, Ireland

[c]Department of Mathematics and Statistics, University of Missouri-Rolla, Rolla, MO 65401, USA

Abstract

Much of the recent work on modeling and analysis of reliability for repairable systems assumes that failures occur according to a nonhomogeneous Poisson process known as a Power-Law process, also known as a Weibull process. There are situations in which some early failure times are not observed, and the data are either left-truncated or left-censored, depending on whether the number of missing observations is known. However, most of the existing statistical methods for this model assume that all of the data are available.

1. INTRODUCTION

Much of the theory of reliability deals with nonrepairable systems or devices, and it emphasizes the study of lifetime models. A *nonrepairable system* can fail only once, and a lifetime model such as the Weibull distribution provides the distribution of the time at which such a system fails. On the other hand, a *repairable system* can be repaired and placed back in service. Thus, a model for repairable systems must allow for a whole sequence of repeated failures, and it should be capable of reflecting changes in the reliability of the system as it ages.

A repairable system is often modeled by means of a *counting process*. Let $N(t)$ represent the number of failures of a repairable system in the time interval $[0, t]$. It follows that $N(t)$ is nonnegative and integer-valued, and if $t > s$, the difference $N(t) - N(s)$ is the number of failures in the interval $(s, t]$. Another characterization can be given in terms of successive failure times $T_1, T_2, \ldots, T_n, \ldots$

An often used approach to the analysis of data from repairable systems involves parametric assumptions which reflect important aspects of the system being modeled. For example, if a system is repaired to "like new" condition following each failure, then it might be reasonable to assume that the times between failures are independent and identically distributed. This would correspond to assuming that the system is modeled by a renewal process.

A different kind of situation, commonly encountered with repairable systems, involves changes in the reliability of the system as it ages. For example, when a complex system is in the development stage, early prototypes will often contain design flaws. During the early testing phase, design changes are made to correct such problems. If the development program is succeeding, one would expect a tendency toward longer times between failures. When this occurs, such systems are said to be undergoing reliability growth. On the other hand, if a system is deteriorating and it is given only minimal repairs when it fails, one would expect a tendency toward shorter times between failures as the system ages.

Poisson processes are often used in the modeling of repairable systems. A common approach in characterizing such a process is to specify a set of axioms or properties which describe the probabilistic behavior of $N(t)$. For example, the following is a typical set of axioms which characterize a Poisson process:

- $N(0) = 0$
- Independent Increments
- $P[N(t+h) - N(t) \geq 2] = o(h)$
- $P[N(t+h) - N(t) = 1] = \nu(t)h + o(h)$

where $o(h)$ has the usual meaning that $o(h)/h \to 0$ as $h \to 0$. The quantity $\nu(t)$ is called the *intensity function*, or *rate of occurrence of failures*, and it can be shown that the number of failures in the interval $[0, t]$ is Poisson distributed with mean $M(t) = E[N(t)] = \int_0^t \nu(s)ds$. More complete information on Poisson processes can be found in the literature (see e.g. Ross, 1983).

The best known case of a Poisson process is a *homogeneous Poisson process* (HPP), in which case the intensity function is constant, say $\nu(t) = \lambda$. However, much of the recent work on repairable systems involves Poisson processes with nonconstant intensity functions. Such a process is usually called a *nonhomogeneous Poisson process* (NHPP). Of course, an NHPP would be capable of modeling systems which are undergoing either reliability growth or deterioration. In particular, if the intensity function $\nu(t)$ is decreasing, the times between failures tend to be longer, and if it is increasing they tend to be shorter.

Much of the recent work on modeling and analysis of repairable systems is based on the assumption of a special type of NHPP known as a *power-law process*. This model is also known in the literature as a *Weibull process*. The name Weibull process derives primarily from the resemblance of the intensity function of the process to the hazard function of a Weibull distribution. In particular the intensity function has the form $\nu(t) = \lambda\beta t^{\beta-1}$ for parameters $\lambda > 0$ and $\beta > 0$. However, to avoid confusion between the present model, which is a NHPP, and a Weibull distribution, which is a model for nonrepairable systems, the name power-law process is often used. Also, the notions of intensity and hazard rate should not be confused with one another. The latter is a relative rate of failure for nonrepairable systems, whereas the former is an absolute rate of failure for repairable systems. Further discussion on this point is provided by Ascher and Feingold (1984, p.33). The mean value function of a power-law process has the form $M(t) = \lambda t^{\beta}$. The parameter β is called a shape parameter, and if $\beta = 1$ the process is an HPP, which is often used to model *renewal*. Otherwise, a power-law process provides a model for a system whose reliability changes as it ages. If $\beta > 1$, it can model a deteriorating system, and when $\beta < 1$, it can model reliability growth.

In order to obtain data it is necessary to cease taking further observations at some point. In general, the process is said to be *failure truncated* if it is observed until a fixed number of failures have occurred, and it is said to be *time truncated* if it is observed for a fixed length of time. With failure truncation, the data consists simply of the set of observed failure times, whereas with time truncation the number of occurrences in the interval of observation is also part of the data set. Specifically, if the system is observed until a fixed number, say n, failures have occurred, and if all the failure times are observed, then the data consists simply of the ordered times. On the other hand, if the process is observed for a fixed length of time, say τ, then the number of failures, $N(\tau) = n$, is random, and there are two cases. If no failures are observed in the interval $[0, \tau]$, then the dataset includes only the value $n = 0$, but if some failures are observed, then the dataset consists of the ordered times, as well as n. We will derive some results involving time truncated data. Strictly speaking, both types of truncation should be referred to as *right truncation*, since the unobserved failures are to the right of the truncation point. In this paper, we will also consider the situation in which there is *left truncation*, and in some situations, *left censoring* of data. Papers dealing with statistical analyses of time truncated power-law processes include Bain and Engelhardt (1980), Crow (1974, 1982), Engelhardt (1988) and Møller (1976). In the papers of Bain and Engelhardt (1980) and Crow (1982), Uniformly Most Powerful Unbiased (UMPU) tests are derived for the value of the intensity function at time τ with β an unknown nuisance parameter. Such tests are useful, for example, in evaluating the reliability growth of a developmental system, and for testing the current system reliability at time τ. Tests which are UMPU for the shape parameter β are also discussed by Bain and Engelhardt (1980) and Crow (1974).

All of the tests mentioned above assume that all of the occurrence times from the start of the experiment (time zero) until truncation time τ are available. However, there are also situations in which some of the early occurrence times are missing. For example, such a problem was encountered by Yeoman (1987), in a study with the objective to forecast building maintenance requirements at a military base. It was found that many of the records of past maintenance actions were available, but in some cases the early data had been lost. In this paper, we will consider such a problem, in which some of the early failure times may be missing. The problem of estimating parameters for a power-law process in the case of left truncation, in which an unknown amount of early failure data might be unobserved, was considered by Crow (1974), but properties of the estimators were not investigated, and inference procedures such as tests and confidence intervals were not considered for this case. In another case, the number of missing failure times would be known, but the exact times not known. This is similar to the situation of left censoring which is sometimes encountered in standardlife testing.

2. MAXIMUM LIKELIHOOD ESTIMATION WITH LEFT-TRUNCATED DATA

Let $0 < \tau_1 < \tau_2 < \infty$, and suppose r failures are observed in the interval $(\tau_1, \tau_2]$, but nothing is known about possible failures in the interval $[0, \tau_1]$. Define a random variable $R = N(\tau_2) - N(\tau_1)$, and denote by r the observed value of R. We say that a process is *truncated from the left* at time τ_1 and *truncated from the right* at time τ_2 if the only available data are the occurrence times in the interval $[\tau_1, \tau_2]$, say $t_1 < \cdots < t_r$. In other words, the right truncation time, τ_2, is selected for the purpose of stopping observation and analyzing the data, while the left truncation time, τ_1, is a point below which failure times cannot be observed. In this section we consider the situation in which the number of failures below τ_1 is unknown, which is typical in the case of truncation. However, in a later section we will also consider what happens if the number of failures less than τ_1 is known.

THEOREM 1. If a power-law process has truncation from the left at time τ_1 and time truncation from the right at time τ_2, with $R = r$ observed failure times, $t_1 < \cdots < t_r$, in the interval $[\tau_1, \tau_2]$, then the likelihood function is given by

$$L(\lambda, \beta) = (\lambda\beta)^r \left[\prod_{i=1}^{r} t_i\right]^{\beta-1} exp\left[-\lambda\left(\tau_2^\beta - \tau_1^\beta\right)\right]$$

if $r \geq 1$ and $\tau_1 < t_1 < \cdots < t_r < \tau_2$, and

$$L(\lambda, \beta) = exp[-\lambda(\tau_2^\beta - \tau_1^\beta)]$$

if r = 0.

Proof: In order to obtain the joint distribution of the number of failures $R = r$ and the failure times, $T_1, \ldots, T_r$, we consider first the original power-law process and define a related process, say, $N_1(t)$, which only counts the failures if they occur after time τ_1. This is a left truncated process, and it can be expressed in terms of the original process as $N_1(t) = 0$ if $0 \leq t < \tau_1$ and $N_1(t) = N(t) - N(\tau_1)$ if $\tau_1 \leq t$. In other words, $N_1(t)$ is a counting process which ignores the failures times which occur in the interval $[0, \tau_1)$, but counts all failures in the interval $[\tau_1, \infty)$. The process $N_1(t)$ is not a power-law process as defined previously, but it is a Poisson process with mean function $M_1(t) = 0$ if $0 \leq t < \tau_1$ and $M_1(t) = \lambda(t^\beta - \tau_1^\beta)$ if $\tau_1 \leq t$. Furthermore, by differentiation we obtain the intensity function which has the form $\nu_1(t) = 0$ if $0 \leq t < \tau_1$ and $\nu_1(t) = \lambda\beta t^{\beta-1}$ if $\tau_1 \leq t$. Now $R = N_1(\tau_2)$ is the number of failure times counted in the interval $[0, \tau_2]$ by the left truncated process, and it follows from well-known results about Poisson processes that conditional on $R = r$, the failure times, $T_1 < \cdots < T_r$, are distributed as order statistics of a random sample of size r from a distribution with density function

$$
\begin{aligned}
f_1(t) &= \frac{\nu_1(t)}{M_1(\tau_2)} \\
&= \frac{\lambda\beta t^{\beta-1}}{\lambda\tau_2^\beta - \lambda\tau_1^\beta} \\
&= \frac{\beta t^{\beta-1}}{\tau_2^\beta - \tau_1^\beta} \qquad (1)
\end{aligned}
$$

if $\tau_1 \leq t \leq \tau_2$, and zero otherwise. It follows that conditional on $R = r$, the conditional density function of the observed failure times $T_1, \ldots, T_r$ is

$$
\begin{aligned}
f(t_1, \ldots, t_r \mid r) &= r! \prod_{i=1}^{r} \left[\frac{\beta t_i^{\beta-1}}{\tau_2^\beta - \tau_1^\beta}\right] \\
&= r! \, \frac{\beta^r \left[\prod_{i=1}^{r} t_i\right]^{\beta-1}}{\left(\tau_2^\beta - \tau_1^\beta\right)^r} \qquad (2)
\end{aligned}
$$

if $\tau_1 \leq t_1 < \cdots < t_r \leq \tau_2$, and zero otherwise. Since $R \sim \mathrm{POI}\left[\lambda\left(\tau_2^\beta - \tau_1^\beta\right)\right]$, it follows that the joint density of $T_1, \ldots, T_r$ and R is

$$
\begin{aligned}
f(t_1, \ldots, t_r, r) &= f(t_1, \ldots, t_r \mid r) P[R = r] \\
&= r! \, \frac{\beta^r \left[\prod_{i=1}^{r} t_i\right]^{\beta-1}}{\left(\tau_2^\beta - \tau_1^\beta\right)^r} \times \frac{\left[\lambda\left(\tau_2^\beta - \tau_1^\beta\right)\right]^r exp\left[-\lambda\left(\tau_2^\beta - \tau_1^\beta\right)\right]}{r!} \\
&= (\lambda\beta)^r \left[\prod_{i=1}^{r} t_i\right]^{\beta-1} exp\left[-\lambda\left(\tau_2^\beta - \tau_1^\beta\right)\right] \qquad (3)
\end{aligned}
$$

which agrees with the statement of the theorem when $r \geq 1$. Of course, it is possible to not observe any failures in the interval $[\tau_1, \tau_2]$, and this happens with probability $P[R=0] = exp[-\lambda(\tau_2^\beta - \tau_1^\beta)]$, which is the likelihood when $r=0$. This concludes the proof.

It is possible, based on the results of the theorem, to compute the MLE's under certain conditions. In order to state the conditions conveniently, we adopt the notation $\tilde{t}$ for the *geometric mean* of the observed failure times. In other words,

$$\tilde{t} = \left[\prod_{i=1}^{r} t_i\right]^{1/r} \tag{4}$$

Similarly, we will denote the geometric mean of the truncation times as $\tilde{\tau} = \sqrt{\tau_1 \tau_2}$. We can now state the following corollary:

COROLLARY 1. If a power-law process is truncated from the left at τ_1 and truncated from the right at τ_2, the MLE's are solutions the following equations:

$$\lambda = r\Big/\left(\tau_2^\beta - \tau_1^\beta\right) \tag{5}$$

$$\frac{\tau_2^\beta \ln \tau_2 - \tau_1^\beta \ln \tau_1}{\tau_2^\beta - \tau_1^\beta} - \frac{1}{\beta} = \ln \tilde{t} \tag{6}$$

and the MLE's, say $\beta = \hat{\beta}$ and $\lambda = \hat{\lambda}$, exist if and only if $\tilde{t} > \tilde{\tau}$.

Proof: For $r \geq 1$, the log-likelihood is

$$\begin{aligned} \ln L &= r(\ln\lambda + \ln\beta) + (\beta-1)\ln\left[\prod_{i=1}^{r} t_i\right] - \lambda\left(\tau_2^\beta - \tau_1^\beta\right) \\ &= r(\ln\lambda + \ln\beta) + r(\beta-1)\ln\tilde{t} - \lambda\left(\tau_2^\beta - \tau_1^\beta\right) \end{aligned} \tag{7}$$

In the case where $r = 0$, there are no observed failure times and the MLE's of λ and β do not exist. The partial derivatives are as follows:

$$\frac{\partial}{\partial\lambda}\ln L = \frac{r}{\lambda} - \left(\tau_2^\beta - \tau_1^\beta\right) \tag{8}$$

$$\frac{\partial}{\partial\beta}\ln L = \frac{r}{\beta} + r\ln\tilde{t} - \lambda\left(\tau_2^\beta \ln\tau_2 - \tau_1^\beta \ln\tau_1\right) \tag{9}$$

The MLE's are obtained by equating (8) and (9) to zero and solving the resulting system of equations. In order to establish conditions under which the solutions exist, we note first that the MLE of β as given by (6) is the same as the conditional MLE of β given $R = r$, is obtained by finding the value of β which maximizes (2). It follows from this that the failure times in the interval $[\tau_1, \tau_2]$ can be transformed into variables which are distributed conditionally as an ordered random sample from a truncated exponential distribution. In particular, we know that $T_1 < \cdots < T_r$ are distributed, conditional on $R = r$, as order statistics for a random sample of size r from a distribution with density function given by (1). We now consider the transformation

$$x_i = \ln(\tau_2 / t_{r-i+1}) \tag{10}$$

for $i = 1, \ldots, r$, and note that this reverses the order of the variables and that the range of the transformed variables $X_1, \ldots, X_r$ is from $0 = ln(\tau_2/\tau_2)$ to $\psi \equiv ln(\tau_2/\tau_1)$. Furthermore, the absolute value of the Jacobian is of the following form: $|J| = \prod_{i=1}^{r}[\tau_2 exp(-x_i)] = \tau_2^r exp(-\sum_{i=1}^{r} x_i)$, yielding the conditional density

$$g(x_1, \ldots, x_r \,|\, r) = r! \frac{\beta^r \left\{ \prod_{i=1}^{r} [\tau_2 exp(-x_i)] \right\}^{\beta-1}}{\left(\tau_2^\beta - \tau_1^\beta\right)^r} |J|$$

$$= r! \beta^r \left[\frac{\tau_2^\beta}{\tau_2^\beta - \tau_1^\beta} \right]^r exp\left(-\beta \sum_{1=1}^{r} x_i \right)$$

$$= r! \frac{\beta^r exp\left(-\beta \sum_{1=1}^{r} x_i \right)}{[1 - exp(-\beta\psi)]^r} \tag{11}$$

if $0 \le x_1 < \cdots < x_r \le \psi$, and zero otherwise. This is the same as the joint density of the order statistics for a sample of size r from a *truncated exponential* distribution with parameter $\theta = 1/\beta$ and right truncation point ψ. The problem of maximum likelihood estimation of θ for this distribution was studied by Deemer and Votaw (1955). In particular, they found the MLE of θ to be the solution $\hat{\theta}$ of the equation

$$\frac{\hat{\theta}}{\psi} - \frac{1}{exp(\psi/\hat{\theta}) - 1} = \frac{\bar{x}}{\psi} \tag{12}$$

if $0 < \bar{x}/\psi < 1/2$, with no solution if $1/2 \le \bar{x}/\psi$. If we relate this to the original problem of estimating β for a power-law process, the result is equivalent to (6). If we substitute $\hat{\theta} = 1/\hat{\beta}$ and $\bar{x} = ln\,\tau_2 - ln\,\tilde{t}$ in (12), then multiply by ψ, we obtain

$$\frac{1}{\hat{\beta}} - \frac{\psi}{exp(\hat{\beta}\psi) - 1} = ln\,\tau_2 - ln\,\tilde{t}$$

Since $\psi = ln(\tau_2/\tau_1)$, it follows that

$$ln\,\tilde{t} = ln\,\tau_2 + \frac{ln\,\tau_2 - ln\,\tau_1}{(\tau_2/\tau_1)^{\hat{\beta}} - 1} - \frac{1}{\hat{\beta}}$$

$$= ln\,\tau_2 + \frac{\tau_1^{\hat{\beta}} ln\,\tau_2 - \tau_1^{\hat{\beta}} ln\,\tau_1}{\tau_2^{\hat{\beta}} - \tau_1^{\hat{\beta}}} - \frac{1}{\hat{\beta}}$$

$$= \frac{\tau_2^{\hat{\beta}} ln\,\tau_2 - \tau_1^{\hat{\beta}} ln\,\tau_2 + \tau_1^{\hat{\beta}} ln\,\tau_2 - \tau_1^{\hat{\beta}} ln\,\tau_1}{\tau_2^{\hat{\beta}} - \tau_1^{\hat{\beta}}} - \frac{1}{\hat{\beta}}$$

$$= \frac{\tau_2^{\hat{\beta}} ln\,\tau_2 - \tau_1^{\hat{\beta}} ln\,\tau_1}{\tau_2^{\hat{\beta}} - \tau_1^{\hat{\beta}}} - \frac{1}{\hat{\beta}}$$

which agrees with (6). It is also easily verified that $0 < \bar{x}/\psi < 1/2$ if and only if $\tilde{\tau} = \sqrt{\tau_1 \tau_2} < \tilde{t}$, which completes the proof of the corollary. Figure 1 shows the left side of (6) as a function of β.

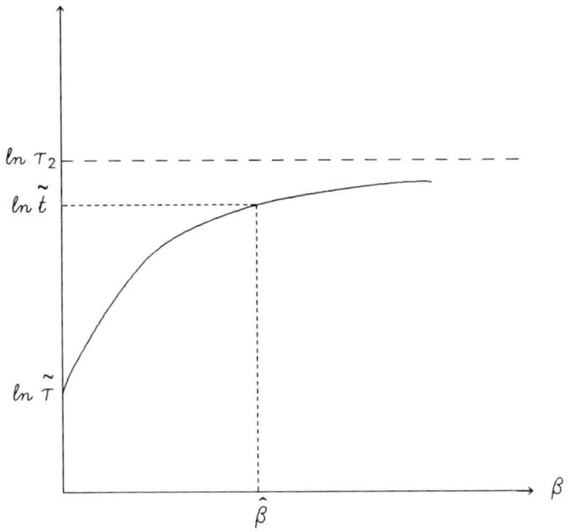

Figure 1. Maximum likelihood estimation of β from left truncated data.

3. INFERENCES WITH LEFT-TRUNCATED DATA

In Bain and Engelhardt (1980), and also in Crow (1974, 1982), UMPU tests were derived the in the case of right time truncation for a reparametrized model. The parameters, in this case, were β and $\nu = \nu(\tau_2)$. We will attempt to extend those results to the case of left truncation. The reparametrization is accomplished by substituting $\lambda(\tau_2^\beta - \tau_1^\beta) = (\tau_2/\beta)\nu[1 - (\tau_1/\tau_2)^\beta]$.

It follows from (3) that the density of $T_1, \ldots, T_r$ and R given $R \geq 1$ is

$$f_c(t_1, \ldots, t_r, r) = \frac{f(t_1, \ldots, t_r, r)}{P[R \geq 1]}$$

$$= \frac{(\lambda\beta)^r \left[\prod_{i=1}^{r} t_i\right]^{\beta-1} exp\left[-\lambda\left(\tau_2^\beta - \tau_1^\beta\right)\right]}{1 - exp\left[-\lambda\left(\tau_2^\beta - \tau_1^\beta\right)\right]}$$

$$= \left(\lambda\beta\tau_2^{\beta-1}\right)^r \left[\prod_{i=1}^{r} (t_i/\tau_2)\right]^{\beta-1} \times C(\nu, \beta)$$

$$= \nu^r \left[\prod_{i=1}^{r} (t_i/\tau_2)\right]^{\beta-1} \times C(\nu, \beta) \tag{13}$$

if $1 \le r$ and $\tau_1 \le t_1 < \cdots < t_r \le \tau_2$, and zero otherwise, where

$$C(\nu,\ \beta) = \frac{exp\left\{-(\tau_2/\beta)\nu\left[1-(\tau_1/\tau_2)^{\beta}\right]\right\}}{1-exp\left\{-(\tau_2/\beta)\nu\left[1-(\tau_1/\tau_2)^{\beta}\right]\right\}}$$

Notice that (13) can be expressed in the regular exponential form of Lehmann (1959, p. 136), namely

$$f_c(t_1,\ldots,t_r,\ r) \ = \ C(\nu,\ \beta)h(t_1,\ldots,t_r,\ r)exp[r\,ln(\nu) + w\beta] \tag{14}$$

where $W = ln[\ \prod_{i=1}^{r}(T_i/\tau_2)] \ = \ \sum_{i=1}^{r} ln(T_i/\tau_2)$. Thus, the pair $(R,\ W)$ is complete and sufficient for $(\nu,\ \beta)$. Furthermore, it is possible to derive UMPU tests for each parameter with the other parameter an unknown nuisance parameter. Each test is expressed as a conditional test of one of the statistics given the other. We consider first the procedure for β, which is based on the conditional distribution of W given $R = r$. For convenient, we will base the test on the related statistic $S = -W = -\sum_{i=1}^{r} ln(T_i/\tau_2) = \sum_{i=1}^{r} ln(\tau_2/T_i) = \sum_{i=1}^{r} X_i$, where the X_i's are defined by the transformation (10). The advantage is that S is distributed as the sum of order statistics for a random sample of size r from a truncated exponential distribution with parameter $\theta = 1/\beta$. Of course, this is also the distribution of the sum of the random sample, since it involves all of the variables. The distribution of this sum was derived by Bain and Weeks (1964), and in the present notation we obtain the conditional density of S given $R = r$,

$$f_{S\,|\,r}(s;\beta) = \frac{e^{-\beta s}}{[(1-e^{-\beta\psi})/\beta]^{r}(r-1)!}\sum_{i=0}^{m}\binom{r}{i}(-1)^{i}(s-i\psi)^{r-1} \tag{15}$$

if $0 < s \le r\psi$ with $m = m(s) = [s/\psi]$ where $[\cdot]$ is the greatest integer function.

Any test for the parameter of the truncated exponential distribution can be easily adapted, using the fact that small values of θ correspond to large values of β, and visa versa. These results are summarized in the following theorem:

THEOREM 2. If a power-law process has truncation from the left at time τ_1 and time truncation from the right at time τ_2, with $R = r$ observed failure times, $t_1 < \cdots < t_r$, in the interval $[\tau_1,\ \tau_2]$, then

1. A size α UMPU test of $H_0{:}\,\beta \le \beta_0$ versus $H_a{:}\,\beta > \beta_0$ is to reject H_0 if $s \le s_{\alpha}(r;\beta_0)$.
2. A size α UMPU test of $H_0{:}\,\beta \ge \beta_0$ versus $H_a{:}\,\beta < \beta_0$ is to reject H_0 if $s \ge s_{1-\alpha}(r;\beta_0)$.
3. A size α test of $H_0{:}\,\beta = \beta_0$ versus $H_a{:}\,\beta \ne \beta_0$ is to reject H_0 if either $s \le s_{\alpha/2}(r;\beta_0)$ or $s \ge s_{1-\alpha/2}(r;\beta_0)$.

where $s_{\gamma}(r;\beta)$ is the $\gamma \times 100$th percentile of (15).

Notice that it is not claimed that the two-tailed test of part 3 is UMPU. In order to achieve unbiasedness, an additional constraint is needed. This is usually rather difficult to achieve, and in practice equal-tail tests are often used instead. We will not pursue this point further. Also, confidence limits can also be obtained by inverting tests of hypotheses, but this will not be a trivial matter in this instance since there is no pivotal quantity for β, and (15) is a rather complicated function of β.

It is also possible to derive UMPU tests of the parameter $\nu = \nu(\tau_2)$ with β an unknown nuisance parameter. According to Lehmann's theory, such tests are conditional tests, based on the conditional distribution of R given $S=s$. We will now consider the problem of deriving this conditional distribution. The following theorem provides the conditional density of R given $S=s$ and $R\geq 1$.

THEOREM 3. If a power-law process is left truncated at time τ_1 and right truncated at time τ_2, with $R=r$ failure times $T_1,\ldots,T_r$ in the interval $[\tau_1, \tau_2]$, and if $S=\sum_{i=1}^{r} ln(\tau_2/T_i)$, then the conditional probability mass function of R given $S=s$ and $R\geq 1$ is

$$f_{R\mid s}(r;\nu)=\frac{\dfrac{(\tau_2\nu)^r}{(r-1)!}\displaystyle\sum_{i=0}^{m}\frac{(-1)^i}{i!(r-i)!}(s-i\psi)^{r-1}}{\displaystyle\sum_{j=m+1}^{\infty}\frac{(\tau_2\nu)^j}{(j-1)!}\sum_{i=0}^{m}\frac{(-1)^i}{i!(j-i)!}(s-i\psi)^{j-1}} \tag{16}$$

if $r>m=[s/\psi]$ and $s>0$.

Proof: The conditional density of R conditional on $S=s$ and $R\geq 1$ is

$$f_{R\mid s}(r;\nu) = \frac{f_{S\mid r}(s;\beta)P[R=r\mid R\geq 1]}{\sum_{j=1}^{\infty}f_{S\mid j}(s;\beta)P[R=j\mid R\geq 1]}$$

$$= \frac{f_{S\mid r}(s;\beta)P[R=r\mid R\geq 1]}{\sum_{j=m+1}^{\infty}f_{S\mid j}(s;\beta)P[R=j\mid R\geq 1]} \tag{17}$$

if $r>m=[s/\psi]$. Note that the sum in the denominator is over terms such that $j\geq m+1$. This is because the density of S given j is positive if and only if $0<s\leq j\psi$, which is equivalent to $j>[s/\psi]$, or $j\geq m+1$. Similarly, the numerator is nonzero if and only if $r\geq m$. Next we note that for $j\geq 1$

$$P[R=j\mid R\geq 1] = \frac{\left\{(\tau_2/\beta)\nu\left[1-(\tau_1/\tau_2)^{\beta}\right]\right\}^j C(\nu,\,\beta)}{j!}$$

$$= \frac{(\tau_2\nu)^j[(1-e^{-\beta\psi})/\beta]^j C(\nu,\,\beta)}{j!}$$

so that

$$f_{S\mid j}(s;\beta)P[R=j\mid R\geq 1] = \frac{e^{-\beta s}}{[(1-e^{-\beta\psi})/\beta]^j(j-1)!}\sum_{i=0}^{m}\binom{j}{i}(-1)^i(s-i\psi)^{j-1}\times$$

$$\times\frac{(\tau_2\nu)^j[(1-e^{-\beta\psi})/\beta]^j C(\nu,\,\beta)}{j!}$$

$$= C(\nu,\,\beta)e^{-\beta s}\frac{(\tau_2\nu)^j}{(j-1)!j!}\sum_{i=0}^{m}\binom{j}{i}(-1)^i(s-i\psi)^{j-1}$$

Substituting this in (17) yields

$$f_{R|s}(r;\nu) = \frac{C(\nu,\beta)e^{-\beta s}\dfrac{(\tau_2\nu)^r}{(r-1)!r!}\sum_{i=0}^{m}\binom{r}{i}(-1)^i(s-i\psi)^{r-1}}{C(\nu,\beta)e^{-\beta s}\sum_{j=m+1}^{\infty}\dfrac{(\tau_2\nu)^j}{(j-1)!j!}\sum_{i=0}^{m}\binom{j}{i}(-1)^i(s-i\psi)^{j-1}}$$

$$= \frac{\dfrac{(\tau_2\nu)^r}{(r-1)!}\sum_{i=0}^{m}\dfrac{(-1)^i}{i!(r-i)!}(s-i\psi)^{r-1}}{\sum_{j=m+1}^{\infty}\dfrac{(\tau_2\nu)^j}{(j-1)!}\sum_{i=0}^{m}\dfrac{(-1)^i}{i!(j-i)!}(s-i\psi)^{j-1}}$$

if $r > m$ and $s > 0$.

From (16) it is possible to construct tests for $\nu = \nu(\tau_2)$ as follows:

THEOREM 4. For a power-law process with truncation from the left at time τ_1 and truncation from the right at time τ_2,

1. A size α UMPU test of H_0: $\nu \le \nu_0$ versus H_a: $\nu > \nu_0$ is to reject H_0 if

   $$\sum_{j=r}^{\infty} f_{R|s}(j;\nu_0) \le \alpha.$$

2. A size α UMPU test of H_0: $\nu \ge \nu_0$ versus H_a: $\nu < \nu_0$ is to reject H_0 if

   $$\sum_{j=1}^{r} f_{R|s}(j;\nu_0) \le \alpha.$$

3. A size α test of H_0: $\nu = \nu_0$ versus H_a: $\nu \ne \nu_0$ is to reject H_0 if $\sum_{j=1}^{r} f_{R|s}(j;\nu_0) \le \alpha_2$ or $\sum_{j=r}^{\infty} f_{R|s}(j;\nu_0) \le \alpha_2$ with $\alpha_1 > 0$, $\alpha_2 > 0$ and $\alpha = \alpha_1 + \alpha_2$.

Proof: By the theorem of Lehmann, a UMPU test for an upper one-sided alternative is a conditional test based on R given $S = s$, with rejection of H_0 for large r.

Strictly speaking the two-tailed test of part 3 is not unbiased unless an additional constraint is also satisfied, but a test with is free of the nuisance parameter β is obtained regardless of this constraint. Since R is discrete, it will usually not be possible to construct an exact size α test unless randomization is employed. In general, rather than randomizing, it is recommended to construct a test such as that of part 3 which is simpler, and is also conservative. Although (16) is rather complicated, it is not too difficult to evaluate with the aid of a computer. We now consider an approximate test for the shape parameter.

A Beta-approximation was derived by Bain, Engelhardt and Wright (1977) which should be useful in constructing approximate tests for β. Specifically, the distribution of the variable $S/(r\psi) = \overline{X}/\psi$ can be approximated by a Beta distribution with parameters

$$a = \frac{r\mu\left[\mu(1-\mu) - \sigma^2/r\right]}{\sigma^2} \qquad b = \frac{r(1-\mu)\left[\mu(1-\mu) - \sigma^2/r\right]}{\sigma^2} \tag{18}$$

where μ and σ^2 are the mean and variance of the distribution of X/ψ, namely

$$\mu = \eta\left\{1 - \frac{exp(-1/\eta)}{\eta[1 - exp(-1/\eta)]}\right\} \qquad \sigma^2 = \eta^2\left\{1 - \frac{exp(-1/\eta)}{\eta^2[1 - exp(-1/\eta)]^2}\right\} \tag{19}$$

with $\eta = 1/(\beta\psi)$. A convenient variation on this is to base a test on an approximate F-statistic,

$$Y = \left(\frac{a}{b}\right)\left(\frac{r\psi}{S} - 1\right) \sim \mathrm{F}(2b,\ 2a) \tag{20}$$

Tests for the parameter β can be expressed in terms of (20). For example, an approximate size α test of $\mathrm{H}_0\!:\beta \le \beta_0$ versus $\mathrm{H}_a\!:\beta > \beta_0$ is obtained by rejecting H_0 if $y \ge f_\alpha(2b,\ 2a)$.

4. MAXIMUM LIKELIHOOD ESTIMATION WITH LEFT-CENSORED DATA

It is also possible to develop methods of analysis for the case in which it is known how many failures have occurred before τ_1, but the exact values of the failure times are only known after τ_1. In other words, we will consider Type I left-censoring.

Suppose now there are k failures in the interval $[0,\ \tau_1]$, and r failures in the interval $(\tau_1,\ \tau_2]$. The numbers k and r are observed values of random variables $K = N(\tau_1)$ and $R = N(\tau_2) - N(\tau_1)$. For convenience we will define $N = N(\tau_2)$, and note that $N = K + R$. We say that a process is *censored from the left* at time τ_1 and *truncated from the right* at time τ_2 if the only available data are k, the number of failures in the interval $[0,\ \tau_1)$, and the occurrence times in the interval $[\tau_1,\ \tau_2]$, $t_1 < \cdots < t_r$. In other words, τ_1 is a point below which it is not possible to observe the exact failure times, but the number of unobserved failures k is known. We state the following theorem:

THEOREM 5. If a Weibull process has Type I censoring from the left at time τ_1 and time truncation from the right at time τ_2, with $K = k$ failures in the interval $[0,\ \tau_1)$, and $R = r$ observed failure times, $t_1 < \cdots < t_r$, in the interval $[\tau_1,\ \tau_2]$, then the likelihood function is given by

$$L(\lambda,\ \beta) = \frac{(\beta/\tau_2)^r\left[(\tau_1/\tau_2)^k \prod_{i=1}^{r}(t_i/\tau_2)\right]^{\beta - 1}(\lambda\tau_2^\beta)^{r+k}(\tau_1/\tau_2)^k exp(-\lambda\tau_2^\beta)}{k!} \tag{21}$$

for $k \ge 0$, $r \ge 1$ and $\tau_1 < t_1 < \cdots < t_r < \tau_2$, and

$$L(\lambda,\ \beta) = exp[-\lambda(\tau_2^\beta - \tau_1^\beta)]$$

if r = 0.

Proof: Recall that $f(t_1, \ldots, t_r, r)$, the joint density of $T_1, \ldots, T_r$ and R, is given by (3).

Furthermore, the K failures in the interval $[0,\ \tau_1)$ occur independently of failures in the time interval $[\tau_1,\ \tau_2]$, and $K \sim \mathrm{POI}(\lambda\tau_1{}^\beta)$. Thus, the joint density of $T_1, \ldots, T_r$, R and K is

$$
\begin{aligned}
f(t_1,\ldots,t_r,r,\ k) &= f(t_1,\ldots,t_r,r)P[K=k]\\
&= (\beta/\tau_2)^r\tau_2^{r\beta}\left[\prod_{i=1}^{r}(t_i/\tau_2)\right]^{\beta-1}\lambda^r exp\left[-\lambda\left(\tau_2^\beta-\tau_1^\beta\right)\right]\times\\
&\quad\times\frac{(\lambda\tau_1^\beta)^k\ exp\left(-\lambda\tau_1^\beta\right)}{k!}\\
&= \frac{(\beta/\tau_2)^r\left[\prod_{i=1}^{r}(t_i/\tau_2)\right]^{\beta-1}(\lambda\tau_2^\beta)^r(\lambda\tau_1^\beta)^k exp\left(-\lambda\tau_2^\beta\right)}{k!}\\
&= \frac{(\beta/\tau_2)^r\left[\prod_{i=1}^{r}(t_i/\tau_2)\right]^{\beta-1}(\lambda\tau_2^\beta)^r(\lambda\tau_2^\beta)^k(\tau_1/\tau_2)^{\beta k} exp\left(-\lambda\tau_2^\beta\right)}{k!}\\
&= \frac{(\beta/\tau_2)^r\left[(\tau_1/\tau_2)^k\prod_{i=1}^{r}(t_i/\tau_2)\right]^{\beta-1}(\lambda\tau_2^\beta)^{r+k}(\tau_1/\tau_2)^k exp\left(-\lambda\tau_2^\beta\right)}{k!}
\end{aligned}
$$

which agrees with (21), concluding the proof.

The joint maximum likelihood estimates (MLE's) can be obtained as follows:

COROLLARY 2. If a Weibull process is censored from the left at τ_1 and truncated from the right at τ_2, the MLE's of λ and β are

$$\hat{\lambda} = (r+k)\Big/\tau_2^{\hat{\beta}} \tag{22}$$

$$\hat{\beta} = r\Big/\left[\sum_{i=1}^{r} ln(\tau_2/t_i)+k\,ln(\tau_2/\tau_1)\right] \tag{23}$$

Proof: As in the case of Corollary1, the MLE's are obtained by taking the partial derivatives of the logarithm of the likelihood function, which for $r \geq 1$ is

$$lnL = \text{constant} + rln(\beta/\tau_2) + (\beta-1)\left[kln(\tau_1/\tau_2) + \sum_{i=1}^{r} ln(t_i/\tau_2)\right] + nln\lambda + n\beta ln\tau_2 - \lambda\tau_2^\beta$$

As in the case of left truncation, when $r = 0$, there are no observed failure times and the MLE's of λ and β do not exist. The details are straightforward and they will be omitted.

5. INFERENCES WITH LEFT-CENSORED DATA

It is also possible to derive the joint density of $T_1,\ldots,T_r$, R and K given $R \geq 1$, as given by the following theorem:

THEOREM 6. The joint density of $T_1, \ldots, T_r$, R and K given $R \geq 1$ is

$$f_c(t_1, \ldots, t_r, r, k) = \frac{\nu^r \left[(\tau_1/\tau_2)^k \prod_{i=1}^{r} (t_i/\tau_2)\right]^{\beta - 1} (\tau_1/\tau_2)^k [(\tau_2/\beta)\nu]^k exp[-(\tau_2/\beta)\nu]}{k! q_1(\nu, \beta)}$$

if $0 \leq t_1 < \cdots < t_r \leq \tau$, $1 \leq r$ and $0 \leq k$ where

$$q_1(\nu, \beta) = 1 - exp\left\{-(\tau_2/\beta)\nu\left[1 - (\tau_1/\tau_2)^\beta\right]\right\} \tag{24}$$

Proof: This is similar to the derivation of (21), except we multiply by $P[R = r \mid R \geq 1]$ rather than $P[R = r]$ to obtain

$$f_c(t_1, \ldots, t_r, r, k) = (\beta/\tau_2)^r \left[(\tau_1/\tau_2)^k \prod_{i=1}^{r} (t_i/\tau_2)\right]^{\beta - 1} (\lambda \tau_2^\beta)^{r+k} \times$$

$$\times \frac{(\tau_1/\tau_2)^k exp(-\lambda \tau_2^\beta)}{k! \left\{1 - exp\left[-\lambda\left(\tau_2^\beta - \tau_1^\beta\right)\right]\right\}} \tag{25}$$

for $k \geq 0$, $r \geq 1$ and $\tau_1 < t_1 < \cdots < t_r < \tau_2$. We now reparametrize in terms of β and $\nu = \nu(\tau_2)$. This can be done by substituting $\lambda \tau_2^\beta = (\tau_2/\beta)\nu$ and $\lambda \tau_1^\beta = (\tau_1/\tau_2)^\beta (\tau_2/\beta)\nu$ in Equation (25), yielding the following

$$f_c(t_1, \ldots, t_r, r, k) = (\beta/\tau_2)^r \left[(\tau_1/\tau_2)^k \prod_{i=1}^{r} (t_i/\tau_2)\right]^{\beta - 1} [(\tau_2/\beta)\nu]^{r+k} \times$$

$$\times \frac{(\tau_1/\tau_2)^k exp[-(\tau_2/\beta)\nu]}{k! q_1(\nu, \beta)}$$

$$= \nu^r \left[(\tau_1/\tau_2)^k \prod_{i=1}^{r} (t_i/\tau_2)\right]^{\beta - 1} [(\tau_2/\beta)\nu]^k \times \frac{(\tau_1/\tau_2)^k exp[-(\tau_2/\beta)\nu]}{k! q_1(\nu, \beta)} \tag{26}$$

for $k \geq 0$, $r \geq 1$ and $\tau_1 < t_1 < \cdots < t_r < \tau_2$. This conclude the proof.

This density can be written in an exponential form, but not a regular form. Specifically, (26) can be written in a form similar to (14), namely

$$f_c(t_1, \ldots, t_r, r, k) = C_1(\nu, \beta) h_1(t_1, \ldots, t_r, r, k) exp[r ln(\nu) + w_1 \beta + k ln(\nu/\beta)]$$

where the variables w_1, r and k correspond to the following three statistics: $W_1 = \sum_{i=1}^{r} ln(T_i/\tau_2) + k ln(\tau_1/\tau_2)$, $R = N(\tau_2) - N(\tau_1)$ and $K = N(\tau_1)$. As a consequence, it follows that the triple (R, K, W_1) is sufficient for the pair (ν, β). This is not ideal, since the dimension of the sufficient statistic is higher than the dimension of the parameter space, and thus, the statistic is not complete. As a consequence, Lehmann's theory does not apply, and whatever inference procedures are obtained may not be optimal.

A modest amount of algebraic manipulation with this function yields an alternate form in which the exponent is $(r + k) ln(\nu) + w_1 \beta + k ln(1/\beta) = n ln(\nu) + w_1 \beta + k ln(1/\beta)$. This still leaves three statistics to construct inference procedures for two parameters, but it suggests a more natural grouping where N is associated with the parameter ν, and the pair (K, W_1) is associated with β. We will consider some inference procedures based on this grouping.

Recall, in the proof of Corollary 1, equation (10) provided a way of transforming the failure times above τ_1 into the first r ordered observations from an exponential distribution. We will adapt this idea for left censored data. Thus, conditional on $N=n$, the transformed data $x_1 < \cdots < x_r$, obtained from (10), are distributed as a Type I censored random sample with from an exponential distribution with mean $\theta = 1/\beta$, with $k = n-r$ a known number of censored values. This suggests the possibility of adapting methods of inference for the mean θ of an exponential distribution with Type I censoring such as those discussed by Bain and Engelhardt (1991, p. 136). For example, since $\beta = 1/\theta$, the reciprocal of a lower confidence limit for θ would correspond to an upper confidence limit for β, and tests of hypotheses can be similarly modified.

For inferences on β, in the case of a left censored process, there are essentially two possible approaches, one based on the MLE $\hat{\beta}$, and the other based on the number of observed failures, $R=r$, but not on the exact failure times. Although R is Poisson distributed, its conditional distribution given $N=n$ is binomial with parameters n and $p = 1-(\tau_1/\tau_2)^{\beta}$. Consequently, a number of common inference procedures for the binomial parameter p are available for analyzing β. Note that a small value of β corresponds to a small value of p, in which case r, the observed value of R, will tend to be small. Thus, for example, with Type I censored data, a size α test of H_0: $\beta \le \beta_0$ versus H_a: $\beta > \beta_0$ is to reject H_0 if $\mathrm{B}(r, n, p_0) \le \alpha$ with $p_0 = 1-(\tau_1/\tau_2)^{\beta_0}$. Of course, for large n asymptotic normal results for p (or $q = 1-p$) can be used. For example, approximate $(1-\alpha)\times 100$ percent confidence limits for β are of the form

$$ln\left(\hat{q} \pm z_{\alpha/2}\sqrt{\hat{p}\hat{q}/n}\right)\Big/\psi \tag{27}$$

where $\hat{q} = k/n$, $\hat{p} = 1-\hat{q}$ and $\psi = ln(\tau_2/\tau_1)$. Although this is relatively simple and convenient to apply, an obvious matter of concern is the amount of efficiency lost by not using the actual failure times. This question, for the corresponding exponential problem was studied by Bartholomew (1963). By comparing asymptotic variances, an asymptotic efficiency of about 96 percent was found when $p \le 0.5$, which in the present problem corresponds to $\beta \le ln2/\psi$. For large β, there is an approximate normal test, also due to Bartholomew, for the exponential problem which uses all of the data. When adapted to the problem of testing β for a Weibull process, the test statistic is

$$Z = \frac{V\sqrt{np}}{\sqrt{1-2(qlnq)V/p + qV^2}} \tag{28}$$

where $V = \beta/\hat{\beta} - 1$ and $p = 1-(\tau_1/\tau_2)^{\beta} = 1-e^{-\beta\psi}$. Conditional on both $N=n$ and $R>0$, this statistic has an approximate standard normal distribution, with the best approximation occurring when $p \ge 0.5$ (or when $\beta \ge ln2/\psi$), which is the range where the test based only on r is least efficient. Thus, it seems reasonable to use (28) for testing large values of β, and R for small values of β. For example, if $\beta_0 \ge (ln2)/\psi$, and z_0 is an observed value of (28), an approximate size α test of H_0: $\beta \le \beta_0$ versus the alternative H_a: $\beta > \beta_0$ is to reject H_0 if $z_0 \le z_\alpha$.

The best way to proceed in constructing a test for $\nu = \nu(\tau_2)$ is not as clear. The remarks following Theorem 6 suggest using a conditional test given $K=k$ and $W_1 = w_1$. Note also that W_1 is a function of K and $S = \sum_{i=1}^{r} ln(\tau_2/T_i)$. In particular, we can write $W_1 = Kln(\tau_1/\tau_2) - S$, so that conditioning on K and W_1 is equivalent to conditioning on K and S.

One possible approach is suggested by a result of Engelhardt and Bain (1991), involving Type II left censored data from a Power Law process. In that work, the number of left-censored observations was a fixed integer k, and an optimal test for ν

was derived. The test involved the variable R conditional on $U = u$, where $U = R/\hat{\beta}_2$ with $\hat{\beta}_2$ the MLE of β under Type II left-censoring. A convenient approximation was also derived by Engelhardt and Bain (1991). In particular, it was found in that work that if $h = u\tau_2\nu$, then for large h,

$$E(R \mid u) \quad \cong \sqrt{h} + \frac{1-2k}{4} \tag{29}$$

and

$$Var(R \mid u) \quad \cong \frac{\sqrt{h}}{2} \tag{30}$$

Also, with the notation $\mu_0 = \sqrt{h} + (1-2k)/4$, $\sigma_0^2 = \sqrt{h}/2$, then conditional on $U = u$,

$$Z = \frac{R - \mu_0}{\sigma_0} \xrightarrow{d} \mathrm{N}(0, 1) \tag{31}$$

as $h \to \infty$. Consequently, limits for an approximate $(1-\alpha) \times 100$ percent confidence interval for ν can be expressed as solutions to the equations

$$\begin{aligned} \pm z_{\alpha/2} &= \frac{r + \frac{1}{2} - \mu_0}{\sigma_0} \\ &= \frac{r + \frac{1}{2} - [\sqrt{u\tau_2\nu} + (1-2k)/4]}{\sqrt{\sqrt{u\tau_2\nu}/2}} \end{aligned} \tag{32}$$

The solutions to (32) are of the form

$$\left[\left(r + \frac{1}{4} + \frac{k}{2}\right) + \frac{z_{\alpha/2}^2}{4} \pm \frac{z_{\alpha/2}}{2}\sqrt{2\left(r + \frac{1}{4} + \frac{k}{2}\right) + \frac{z_{\alpha/2}^2}{4}}\,\right]^2 / (u\tau_2) \tag{33}$$

Of course, our focus here is on inference procedures for ν in the case of Type I left-censored data. The basic idea is that even though a procedure derived for use with Type II censoring is not theoretically correct when applied with Type I censored data, if the amount of censoring is not too large, a useful approximation might result. Our proposed method for the case of Type I censoring is based on using (31) for constructing tests, and (33) for computing confidence bounds, except that now we define the statistic $U = R/\hat{\beta}_1$ where $\hat{\beta}_1 = \hat{\beta}$ is the MLE under Type I censoring given by Corollary 2. This can also be written as $U = S + K\psi$. This procedure was studied by simulating 1000 sets of Type I left-censored data, and computing the relative frequency of coverage with intervals based on (33) and with the modifications mentioned above. The result for a 95 percent confidence level and various values of τ_1, τ_2 and $\nu = \nu(\tau_2)$ are shown in Table 1. It appears from this study that the approximation works fairly well provided that τ_1 is fairly small relative to τ_2, say $\tau_1/\tau_2 \leq 1/3$.

One other idea, which was suggested earlier, is to construct a conditional test based on the statistic N given $K = k$ and $S = s$. Consider, for example, a test with critical region of the form $N \leq c_1$, with c_1 chosen to give a test of size α. In other words, $\alpha = P[N \leq c_1 \mid K = k, S = s, R \geq 1]$. Since $N = R + K$, this would be equivalent to a test with critical region of the form $R \leq c_2$, where c_2 is chosen to give a size α test, $\alpha = P[R \leq c_2 \mid K = k, S = s, R \geq 1] = P[R \leq c_2 \mid S = s, R \geq 1]$. This procedure is equivalent to the test for ν which was derived in Section 3 for a truncated process, in which the number of failures before time τ_1 is not known. Thus, although such a test succeeds in eliminating the nuisance parameter β, it apparently makes no use of the information that k failures occurred in the interval $[0, \tau_1)$.

Table 1. Simulated confidence levels for approximate limits for $\nu = \nu(\tau_2)$. (Nominal level 95 percent)

τ_1	τ_2	$\nu = 1$	$\nu = 2$	$\nu = 5$
5	10	.94	.94	.93
	15	.94	.95	.96
	20	.95	.95	.96
	25	.95	.95	.95
10	15	.90	.86	.73
	20	.94	.93	.90
	25	.95	.96	.93
	30	.95	.95	.95
15	20	.81	.65	.26
	25	.91	.89	.75
	30	.93	.92	.86
	35	.95	.94	.90
	40	.95	.95	.92
	45	.95	.95	.95

6. ACKNOWLEDGMENTS

Dr. Engelhardt's work was supported by the U.S. Department of Energy under Contract DE-AC07-76ID01570. Dr. William's support to participate in this research was provided by a Jefferson Smurfit Fellowship.

7. REFERENCES

1 Ascher, H. and Feingold, H., *Repairable Systems Reliability.* Marcel Dekker: New York, 1984.

2 Bain, L. J. and Engelhardt, M., "Inferences on the Parameters and Current System Reliability for a Time Truncated Weibull Process." *Technometrics*, 22, (1980), 421.

3 Bain, L. J. and Engelhardt, M., *Statistical Analysis of Reliability and Life-Testing Models.* Marcel Dekker: New York, 1991.

4 Bain L. J. and Weeks, D. L., "A Note on the Truncated Exponential Distribution." *Annals of Mathematical Statistics*, 35, (1964), 1366.

5 Bartholomew, D. J., "The Sampling Distribution of an Estimate Arising in Life-Testing." *Technometrics,* 5, (1963), 361.

6 Crow, L., "Reliability Analysis of Complex, Repairable Systems." *Reliability and Biometry.* ed. by F. Proschan and R. J. Serfling, (1974), 379.

7 Crow, L., "Confidence Interval Procedures for the Weibull Process with Applications to Reliability Growth." *Technometrics*, 24, (1982), 67.

8 Engelhardt, M., *Encyclopedia of Statistical Sciences.* ed. by N. L. Johnson and S. Kotz. John Wiley and Sons: New York, 1988.

9 Engelhardt, M. and Bain, L. J., "Statistical Analysis of a Weibull Process with Left-Censored Data." *Survival Analysis: State of the Art,* ed. by J. P. Klein and P. Goel, NATO ASI Series, Series E: Applied Sciences-Vol. 211, (1992), 173.

10 Lehmann, E. L., *Testing of Statistical Hypotheses.* John Wiley and Sons: New York, 1959.

11 Møller, S. K., "The Rasch-Weibull Process." *Scand. J. Statist.*, 3, (1976), 107.
12 Ross, S. M., *Stochastic Processes.* John Wiley and Sons: New York, 1983.
13 Yeoman, A., *Forecasting Building Maintenance Using the Weibull Process.* M. S. Thesis, University of Missouri-Rolla, (1987).

ADVANCES IN RELIABILITY
Edited by Asit P. Basu
1993 Elsevier Science Publishers B.V.

Nonparametric empirical Bayes reliability qualification testing in the Poisson distribution

Robin C. Fisher,[a] Harry F. Martz,[b] and William J. Zimmer[c]

[a]Statistical Methods Division, Current Population Survey Branch, US Bureau of the Census, Washington, DC 20233

[b]Group A-1, MS F600, Los Alamos National Laboratory, Los Alamos, NM 87545

[c]Department of Mathematics and Statistics, University of New Mexico, Albuquerque, NM 87131

Abstract

A new nonparametric empirical Bayes estimator is presented for the failure rate λ in the Poisson distribution for the case of unequal test times. Both large and small sample performance properties of the estimator are discussed, and some Monte Carlo comparisons with an existing well-known empirical Bayes estimator are presented. A reliability qualification test procedure based on the new estimator is developed. The procedure is applied to a real-world problem in which an IBM engineer was asked to develop an accelerated reliability qualification test for a new modem which is similar to a previously qualified and satisfactory modem. The objective of the test is to show that the new modem has a 90% upper failure probability limit after testing which does not exceed that of the original modem. Suitably adjusted test results on both the earlier and yet another modem are used to determine the testing requirements for the new product. Test plans are presented which meet this objective, and which require significantly less testing than the original plan.

1. INTRODUCTION

Consider the common reliability problem in which a random variable (rv) X, the number of failures of some device of interest, follows a Poisson distribution with mean λT, where λ is the unknown failure rate and T is the known total time-on-test (or test time). Further assume that λ is the (unknown) realization of an rv Λ with a completely unknown and unspecified cumulative distribution function (cdf) $G(\lambda)$ and corresponding density $g(\lambda)$.

We now assume that a total of m "past" experiments have occurred which have yielded the data (x_1,T_1), ..., (x_m,T_m). Suppose that we are in the current (or present) experiment in which x_{m+1} failures have been observed in test time T_{m+1}. For notational convenience, we will frequently drop the subscripts m+1 when referring to the current experiment. It is further assumed that all experiments are intrinsically related in the sense that the experiments, conditional on λ_j, are all mutually independent and that the underlying λ_j values are all independent realizations from the common cdf $G(\lambda)$. The general problem considered here is: How to use the past and current data to estimate λ in such a way that the chosen estimator has good small sample (i.e., for small values of m) performance regardless of the form of $G(\lambda)$? This is the usual nonparametric empirical Bayes (EB) small-sample problem. This EB small-sample problem was partially motivated by the following real-world reliability qualification testing problem. A computer manufacturer was in the process of developing two new modems similar to a modem that had been released some months earlier. The earlier modem was currently in production and demonstrating satisfactory reliability results. The production line that was building the successful modem would produce the new modems and most components would be the same for all three models. Components that were different would be taken from the same families: chip resistors, capacitors, SAW filter, and crystals. Thus, it is reasonable to assume that the reliability qualification tests constitute a set of three related experiments in the EB sense. The idea is to use the test results from the prior qualification of the earlier modem and one of the new modems to reduce the testing costs when qualifying the remaining modem; thus, $m = 2$ here. It was required to determine a test plan which would show that, after testing and with high confidence, the two new modems are no less reliable than the original modem.

An estimation procedure for $g(\lambda)$ is presented in Section 2 along with the corresponding EB estimator for λ. We also summarize the large-sample behavior of the estimator for g and show that, while it performs better for large values of m than a similar existing estimator, it is generally not consistent for estimating g.

In Section 3 we examine the Bayes risk of the estimator for small m. The performance of this new EB estimator is also compared to the usual parametric EB estimator obtained by assuming a gamma prior distribution.The reliability qualification test problem described above is again considered in Section 4, and some conclusions are finally presented in Section 5.

2. A NONPARAMETRIC EMPIRICAL BAYES ESTIMATOR

Fröhner (1985A, 1985B) suggested an estimator for the prior density g when the test times vary. In each of the past experiments an artificial prior in the gamma family,

$$p(\lambda_j) = \frac{\beta^\alpha}{\Gamma(\alpha)} \lambda_j^{\alpha-1} e^{-\lambda_j\beta} \quad \lambda_j>0,\ \alpha,\beta>0\,, \tag{1}$$

which will be denoted as $\Gamma(\alpha,\beta)$, is placed on λ_j. Bayes' theorem is then used to combine this prior with the Poisson conditional distribution

$$f_{T_j}(x|\lambda_j) = \frac{(\lambda_j T_j)^x e^{-\lambda_j T_j}}{x!} \quad x=0,\ 1,\ .\ .\ .\ ,\ \lambda_j,\ T_j>0, \tag{2}$$

denoted by $P(\lambda_j T_j)$, to produce an artificial posterior density for the jth past experiment. This density is

$$\hat{g}_j(\lambda|x_j, T_j) = \frac{f_{T_j}(x|\lambda)p(\lambda)}{\int f_{T_j}(x|\lambda)p(\lambda)d\lambda}$$

$$= \frac{(\beta+T_j)^{\alpha+x_j}}{\Gamma(\alpha+x_j)} \lambda^{\alpha+x_j-1} e^{-(\beta+T_j)} \quad \alpha,\ \beta,\ \lambda, T_j>0,\ x_j=1,\ .\ .\ .\ . \tag{3}$$

The simple mixture of these distributions is then used as the estimated prior distribution for λ; that is,

$$\hat{g}(\lambda) = \sum_{j=1}^{m} \frac{1}{m}\hat{g}_j(\lambda|x_j, T_j)\ . \tag{4}$$

Bayes' theorem is again used to obtain the estimated posterior distribution for λ,

$$\hat{g}(\lambda|x,T) = \frac{\hat{g}(\lambda)f_T(x|\lambda)}{\int \hat{g}(\lambda)f_T(x|\lambda)d\lambda}\ . \tag{5}$$

Martz and Zimmer (1990), (1992) consider a similar approach for the case of binomial sampling.

This procedure has an appealing Bayesian interpretation. The statistician may believe that prior knowledge about λ_j in the jth previous experiment enables one to put a prior p on λ_j and obtain the posterior $\hat{g}_j(\lambda|x_j,T_j)$. Because λ_j is random in each of the experiments, the more familiar approach

$$g(\lambda|\{x_j T_j\}_{j=1}^{m}) = \frac{g(\lambda)\Pi_{j=1}^{m} f_{T_j}(x_j|\lambda)}{\int g(\lambda)\Pi_{j=1}^{m} f_{T_j}(x_j|\lambda)d\lambda} \tag{6}$$

is not appropriate.

Unfortunately, the estimator based on the estimated prior in (4) is not asymptotically optimal according to Robbin's (1955) definition. In fact, Fisher (1990) shows that the estimated prior in (4) does not, in general, converge to the true prior nor to any particular distribution.

As it is our purpose to investigate the performance of the estimator for small values of m, the asymptotic performance of the estimator is not of primary concern. However, the convergence of Fröhner's estimate can be improved by weighting the past data portion of the parameters of the component gamma distribution in (4). We merely point out that the EB estimator based on the prior mixture

$$\hat{g}_{MF}(\lambda) = \sum_{j=1}^{m} \frac{1}{m} \hat{g}_{j,MF}(\lambda|x_j, T_j), \tag{7}$$

where $\hat{g}_{j,MF}(\lambda|x_j, T_j) = \Gamma(\alpha + hw_j x_j, \beta + hw_j T_j)$, and where $h = h(m) \to \infty$ as $m \to \infty$, is asymptotically optimal if $T_j \to \infty$ for all j [Fisher (1990)]. Notice that the weight w_j may be interpreted as the relevance of the jth past experiment to the current experiment, while $h(m)$ ensures that the component gamma variances all vanish in the limit.

Given the current data, the corresponding posterior density of λ is

$$\hat{g}_{MF}(\lambda|x,T) = \frac{\sum_{j=1}^{m} \frac{(\beta + hw_j T_j)^{\alpha + hw_j x_j}}{\Gamma(\alpha + hw_j x_j)} \lambda^{\alpha + hw_j x_j + x - 1} e^{-\lambda(\beta + hw_j T_j + T)}}{\sum_{j=1}^{m} \frac{(\beta + hw_j T_j)^{\alpha + hw_j x_j}}{\Gamma(\alpha + hw_j x_j)} \frac{\Gamma(\alpha + hw_j x_j + x)}{(\beta + hw_j T_j + T)^{\alpha + hw_j x_j + x}}}. \tag{8}$$

For squared-error loss, the EB estimator of λ becomes

$$E_{\hat{g}_{MF}}(\lambda|x,T) = \frac{\sum_{j=1}^{m} \frac{(\beta + hw_j T_j)^{\alpha + hw_j x_j}}{\Gamma(\alpha + hw_j x_j)} \frac{\Gamma(\alpha + hw_j x_j + x)}{(\beta + hw_j T_j + T)^{\alpha + hw_j x_j + x}} \left(\frac{\alpha + hw_j x_j + x}{\beta + hw_j T_j + T} \right)}{\sum_{j=1}^{m} \frac{(\beta + hw_j T_j)^{\alpha + hw_j x_j}}{\Gamma(\alpha + hw_j x_j)} \frac{\Gamma(\alpha + hw_j x_j + x)}{(\beta + hw_j T_j + T)^{\alpha + hw_j x_j + x}}}. \tag{9}$$

This estimator, along with the corresponding prior and posterior distributions in (7) and (8), will be referred to later simply as the "MF" (for Modified Fröhner) estimator and distributions, respectively. Fisher (1990) considers successive iterations of (9) to obtain additional EB estimators for λ. Although he shows that such iterated EB estimators do indeed have smaller estimated Bayes risks than (9), the improvements are slight and have diminishing returns. Thus, they are not considered here.

If m and T_j are not large, the prior (7) may still be highly influenced by the few specified gammas in the mixture and it is possible that the procedure may not have sufficient flexibility to provide robust estimates. This topic, along with the small sample performance study, is the subject of the next section.

3. SMALL-SAMPLE PERFORMANCE

In many practical cases the value of m is quite small and, consequently, the large sample performance properties are irrelevant. Analytic evaluation of the small sample behavior of the EB estimator in (9) is difficult because of the complexity (e.g., the nonlinearity) of the estimator. Monte Carlo simulation has been traditionally used in such instances to examine the small-sample Bayes risk of EB estimators. Monte Carlo simulation is likewise used here to examine the Bayes risk (for squared-error loss) of the estimator in (9) and to compare its performance with perhaps the most commonly used parametric EB (PEB) estimator. The common PEB estimator is the posterior mean obtained by assuming a gamma prior distribution and using the past data to estimate the gamma hyperparameters by either the method of moments or maximum likelihood (ML) [see Martz and Waller (1991) and Morris et al (1989)]. Although only the Bayes risk is considered here, Fisher (1990) also considers the Pitman's distance [Pitman (1937)] of each estimator relative to the usual ML estimator.

Our main interest is the performance of the estimators with respect to a variety of priors when the test times in the previous experiments vary. There are clearly some reliability problems for which the prior distribution may be multimodal. A multimodal prior could arise, for example, in a situation in which there exists only a few (but unknown) randomly occurring distinct causes of failure. Consequently, it is important to consider the performance of a nonparametric EB estimator for a multimodal prior distribution.

The estimators considered here are appropriate for the situation in which the test times vary among experiments. Although Fisher (1990) considers several cases, only two sets of test times are considered here:

- $T_j = 4{,}000$ hr. for $j=1, \ldots, m+1$ (the equal test times case), and
- $T_j = t_j$ for $j=1, \ldots, m+1$;

in these simulations the times t_j were randomly selected from a $U(0, 10^4$ hr.) distribution. The selection was made only once and the resulting set of times was used for all the simulations in this category.

Two different-shaped priors were used:

- The bimodal piecewise linear prior shown in Figure 1; and
- $\Gamma(2, 10^4$ hr.).

This choice of test times and priors is somewhat motivated by practical reliability considerations coupled with our own interest. In practice, it is often the case that, although the test times for highly reliable devices are measured in the thousands of hours, the failures are rather sparse.

Because the purpose of the Monte Carlo simulation is to study the performance of the estimators when the number of previous experiments is small, the values of m used were 2,3,4,5,10,25, and 50.

Other variables we need to consider include the choice of h, the values of α and β, and the weights w_j. No attempt is made here to determine optimal choices for

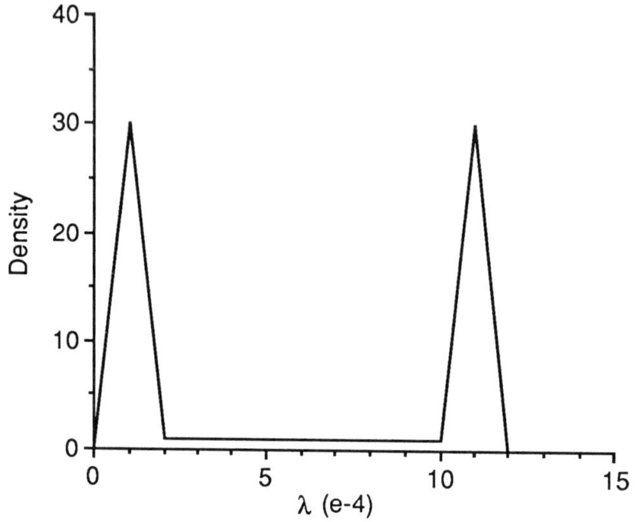

Figure 1. Bimodal prior distribution for λ.

any of these. Instead, the following values are used in the simulation: $hw_j = 1$, $\alpha = .1$, $\beta = 0$. This choice for α and β ensures that the EB estimator always exists, while that for hw_j corresponds to the unweighted case.

There has been some debate in EB circles over whether it is best to use all m+1 experiences to estimate the prior or just use the m previous experiences, which are independent of the current experience. In our simulation each EB estimate corresponds to an estimated prior distribution based only on the observed data from the m previous experiments.

The simulation proceeded in the usual way. For each combination of prior distribution, set of test times, and value of m, 10,000 Monte Carlo repetitions were performed. In each repetition, m+1 values of Λ, $\lambda_1, \ldots, \lambda_{m+1}$, were independently drawn from the prior distribution. The values of X, $x_1, \ldots, x_{m+1}$, were then conditionally generated from the Poisson distribution with corresponding mean $\lambda_j T_j$. Both the PEB and EB estimates of λ_{m+1} were then computed. The Bayes risk for squared-error loss was then estimated for each estimator by averaging the corresponding squared-errors over the 10,000 repetitions.

First consider the case where all of the test times are equal. The estimated Bayes risk results for both priors are shown in Figure 2. The estimated Bayes risk of the usual ML estimator of λ is also shown, along with the minimum Bayes risk associated with the Bayes estimator. Note that the ML Bayes risk is theoretically constant over m and the observed sampling variation is due to the simulation error. We could, of course, also calculate the ML Bayes risk analytically.

The MF estimator has consistently lower estimated Bayes risk than the PEB estimator for $m = 2$ or 3. In the case of a gamma prior, the PEB estimator has much smaller Bayes risk than the MF estimator for $m > 5$. In the case of the

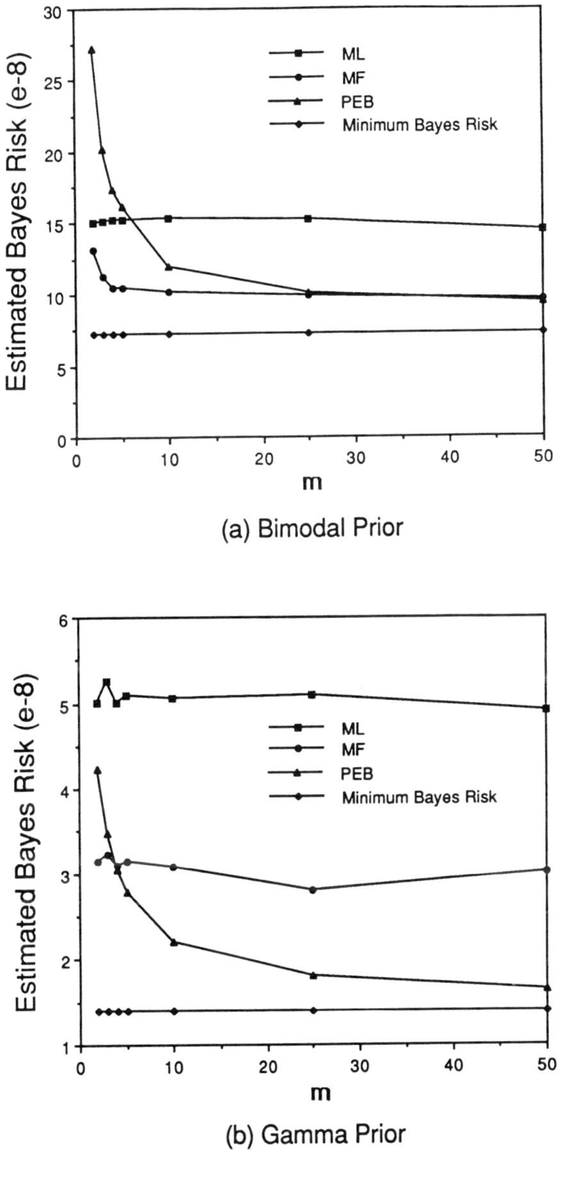

(a) Bimodal Prior

(b) Gamma Prior

Figure 2. Estimated Bayes risk for two prior distributions when the test times are all 4,000 hours.

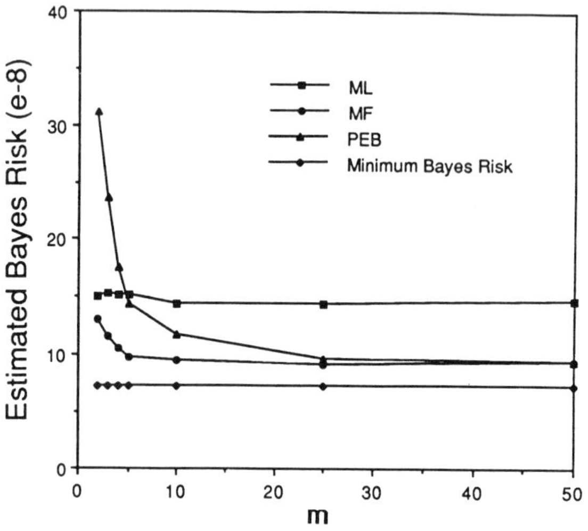

(a) Bimodal Prior

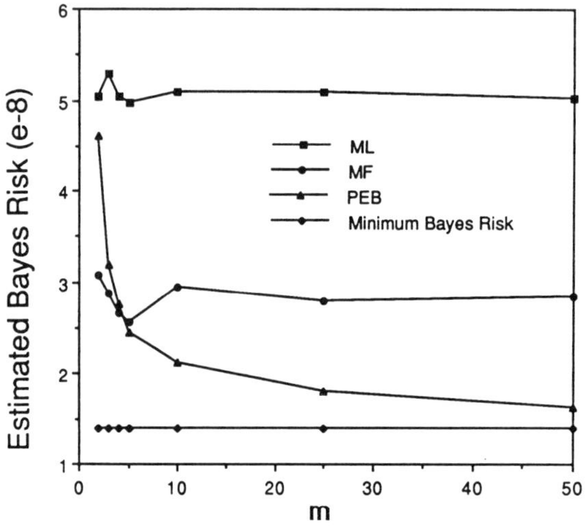

(b) Gamma Prior

Figure 3. Estimated Bayes risk for two prior distributions when the test times are all distributed U(0, 10^4 hours).

bimodal prior, it is not surprising that the estimated Bayes risk of the MF estimator is lower than that of the PEB estimator for all values of m.

When the test times are distributed uniformly, the estimated Bayes risks follow a similar pattern.The results are shown in Figure 3 along with the results for the ML estimator. For both priors the MF estimator performs well for $m = 1$ or 2. When $m = 25$ or 50, the PEB estimator always performs well. In the case of the bimodal prior, the MF and PEB estimators perform equally well for $m = 25$ or 50, while the MF estimator is clearly superior for $m \leq 10$.

These and other simulation results described in Fisher (1990) suggest that the MF estimator has excellent Bayes risk for extremely small values of m (such as $m = 2$ or 3) regardless of the shape of the prior distribution. On the other hand, the PEB estimator performs well when m is at least 10, particularly when the prior is a gamma distribution.

4. RELIABILITY QUALIFICATION TEST EXAMPLE

In this section we illustrate the application of the MF estimator to the real-world reliability qualification testing problem introduced in Section 1 and first considered by Hart (1990), with subsequent commentary by Ganter et al (1990). Hart, an IBM engineer, participated in a project whose goal was to produce two new modems similar to a modem that had been released some months earlier. The major difference was that the new modems would operate at frequencies different from the original. The earlier modem was currently in production and demonstrating satisfactory reliability results. In the following discussion, "A" denotes the original modem, while "B1" and "B2" denote the two new modems.

Recall that the production line that built the A modems would produce B1 and B2 and that most components would be the same for all three models. Because the three designs were quite similar, Hart proposed and felt justified in considering a Bayesian reliability qualification test procedure for the two newer modems. Qualification tests of the A modem involved binomial (success/failure) testing of 150 sample units in which 6 failures were observed. Consistent with these test results and a Bayesian approach, Hart considered a beta distribution for the failure probability p of the A modems given by

$$f(p;6,144) = \frac{\Gamma(150)}{\Gamma(6)\Gamma(144)} p^{6-1} (1-p)^{144-1}, \ 0 \leq p \leq 1, \tag{10}$$

which we denote as $\beta(6,144)$.

When the production qualification of the B1 and B2 modems began, the developers asked Hart how many units he needed for testing. In addressing this question, Hart's intent was to use the test results from the prior qualification of A for that of B1 and B2. Also, the results of the concurrent qualification of B1 would serve as additional prior information for B2 and conversely. Hart's qualification test objective was to show that B1 and B2 have a 90% upper failure probability limit after testing which does not exceed that of A.

Consider the required qualification test plan for modem B2. In addition to the observed test results on A, Hart also used the observed test results on B1, in which 1 failure was observed in 36 tests, to determine the required test plan for B2. Hart's idea was that the A and B1 test results should each receive some weight in forming the beta prior for B2. Hart weighted the A test results 0.6 and the B1 test results 0.8, the assumption being that each A unit was worth 60% of a B2 unit on test and each B1 was worth 80% of a B2 unit on test.

Because of the natural conjugate property of the beta distribution, Hart formed the combined $\beta[.6(6) + .8(1) = 4.4, .6(144) + .8(35) = 114.4]$ prior distribution for B2. Assuming X2 failures in N2 tests on B2, the corresponding B2 posterior is the $\beta(X2 + 4.4, N2 - X2 + 114.4)$ distribution.

Hart chose to satisfy the test objective by matching this posterior beta distribution and the $\beta(6,144)$ distribution for A. This match guarantees that the 0.9 quantile of the posterior B2 distribution equals the 0.9 quantile of the beta distribution assigned to A. However, Hart failed to recognize that matching distributions is an approach that is too conservative when only a single quantile match is really required. It is generally true that the required quantile match can be obtained with fewer test units than found by completely matching the distributions.

Following the Hart procedure yields the pair of equations given by $X2 + 4.4 = 6$ and $N2 - X2 + 114.4 = 144$, whose solution gives $X2 = 1.6$ and $N2 = 31.2$ as the desired test plan. The practical (conservative) implementation of this plan requires testing 32 B2 units and accepting (passing) B2 if no more than 1 unit fails the test. The 32 units represent a significant decrease in tests units over the 150 A units originally tested, a result of the use of strong prior information about B2.

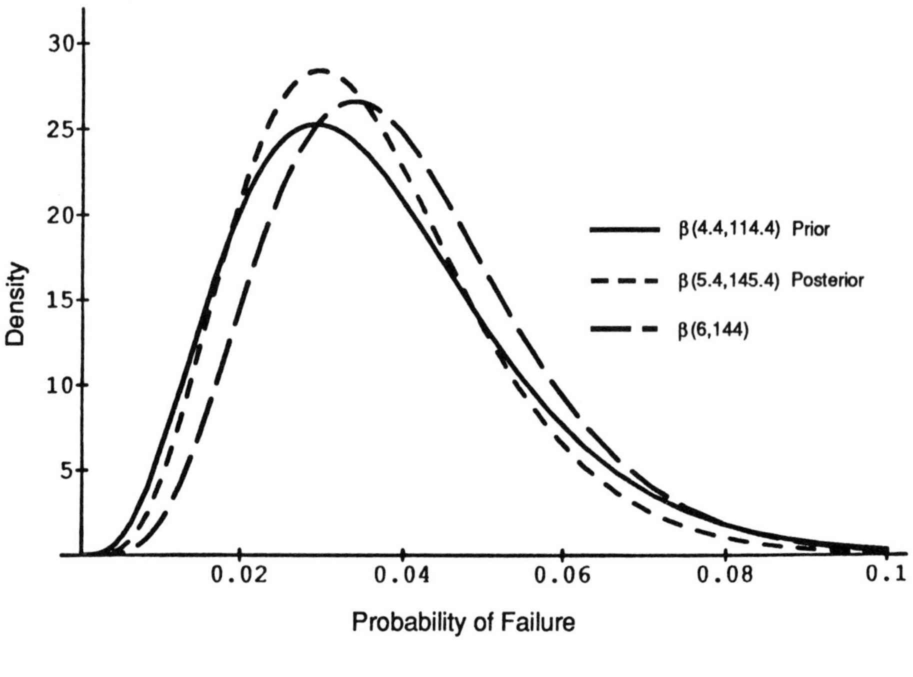

Figure 4. Results for the Hart (1990) Bayesian test plan (X2 = 1, N2 = 32).

The prior and posterior beta distributions corresponding to this plan are plotted in Figure 4 along with the $\beta(6, 144)$ A-modem reference distribution. The 0.9 quantile of the A-modem $\beta(6,144)$ distribution is 0.062. Recall that a test plan for B2 is sought whose corresponding posterior quantile does not exceed this value. Given 1 failure in 32 tests on B2, the posterior $\beta(5.4, 145.4)$ distribution has a 0.9 quantile of 0.056, which clearly meets the requirement.

However, Hart fails to recognize that the $\beta(4.4, 114.4)$ prior distribution itself also meets the requirement! The 0.9 quantile of $\beta(4.4, 114.4)$ is $0.060 < 0.062$; thus, although it may not be a satisfactory overall solution, any further testing of B2 is quite unnecessary for satisfying the single quantile test objective. This clearly illustrates the lack of optimality (in the sense of minimized testing) produced by unknowingly using a more severe test criterion than is actually required. In the EB procedure that follows, the designated single quantile test criterion is used to determine the required test plan.

Because $n = 150$ and $n = 36$ are rather large and p is small, we consider the Poisson approximation to the binomial in developing the EB test plan. The estimated Bayes risk performance of the MF estimator was shown in Section 3 to be excellent when $m = 2$ or 3; thus, we will use the MF procedure here.

With $m = 2$, $\alpha = \beta = 0$, $h = 1$, and the data given above in (7), the MF prior distribution for the probability of failure p of the B2 modems becomes

$$\hat{g}_{MF}(p) = \frac{(150w_A)^{6w_A}}{2\Gamma(6w_A)} p^{6w_A-1}e^{-150w_Ap} + \frac{(36w_{B1})^{w_{B1}}}{2\Gamma(w_{B1})} p^{w_{B1}-1}e^{-36w_{B1}p}, \; p > 0. \tag{11}$$

The insignificant density beyond $p = 1$ can be ignored. Recall from Section 2 that the weight w_A rescales the A-modem data set according to the perceived worth of an A unit relative to a B2 unit on test. The weight w_{B1} behaves in the same way. A weight less than one increases the variance of the corresponding component in the mixture without changing its mean. Because this interpretation corresponds to that of Hart, we also consider $w_A = 0.6$ and $w_{B1} = 0.8$. However, we will examine the sensitivity of the EB test plan to this choice later.

Given X2 failures in N2 tests on B2, the corresponding mixed gamma posterior cdf corresponding to (8) can be expressed in terms of the standard incomplete gamma function $\gamma(a,z) = \int_0^z y^{a-1}\exp(-y)dy$, $a>0$, as

$$\hat{G}_{MF}(p|X2,N2) = \left\{\frac{(150w_A)^{6w_A}}{\Gamma(6w_A)} (150w_A+N2)^{-6w_A-X2}\, \gamma[6w_A+X2,\, p(150w_A+N2)] + \frac{(36w_{B1})^{w_{B1}}}{\Gamma(w_{B1})} (36w_{B1}+N2)^{-w_{B1}-X2}\, \gamma[w_{B1}+X2,\, p(36w_{B1}+N2)]\right\} / K(X2,N2), \; p>0, \tag{12}$$

where K(X2,N2) is defined by

$$K(X2,N2) = \frac{\Gamma(6w_A+X2)}{\Gamma(6w_A)}\left(\frac{150w_A}{150w_A+N2}\right)^{6w_A}(150w_A+N2)^{-X2}$$
$$+\ \frac{\Gamma(w_{B1}+X2)}{\Gamma(w_{B1})}\left(\frac{36w_{B1}}{36w_{B1}+N2}\right)^{w_{B1}}(36w_{B1}+N2)^{-X2}, \tag{13}$$

and where $w_A = 0.6$ and $w_{B1} = 0.8$.

For fixed values of X2 and p, $\widehat{G}_{MF}(\cdot)$ is an increasing function of N2. Thus, for a given value of X2 = 0,1,2,..., the desired EB test plan may be found by simply solving

$$\widehat{G}_{MF}(0.062|X2,N2) \geq 0.9 \tag{14}$$

for the smallest integer value of N2 that satisfies this inequality. Unless there is a compelling reason to do otherwise, the analyst will usually choose X2 to be either 0 or 1 in order to minimize the testing costs. For the chosen value of X2, the EB plan is implemented as follows: if no more than X2 failures occur in a binomial test of N2 B2-modems, then the qualification test is passed.

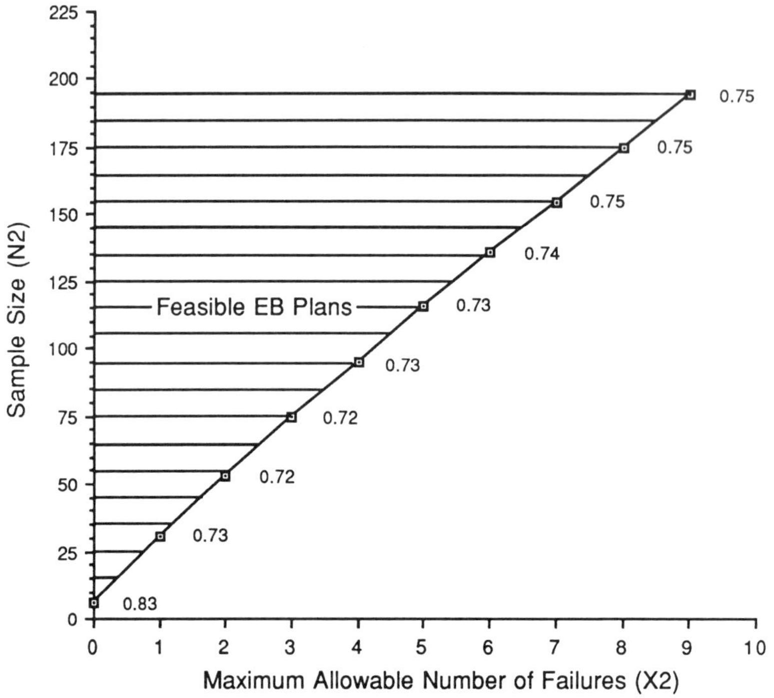

Figure 5. Empirical Bayes test plans and corresponding unconditional probabilities of passing the test.

If the test is passed, then the claim can be made that the 90% upper failure probability limit of B2 does not exceed that of A, which is the desired objective.

The unconditional (marginal) probability of passing the EB test is also of interest and is easily computed as

$$P(X \leq X2; N2) = 0.5 \sum_{x=0}^{X2} (N2)^x K(x,N2)/x! \tag{15}$$

where K(·) is defined in (13). For a given candidate EB test plan, this probability should be sufficiently large to provide satisfactory assurance that the proposed test will be passed.

Figure 5 illustrates the EB test plan region found using (14) for X2 = 0,1,...,9. In addition, for each optimal EB test plan, Figure 5 gives the associated unconditional probability of passing the test. For example, for the EB plan (X2 = 0, N2 = 6), $\hat{G}_{MF}(0.062|0,6) = 0.902 > 0.9$, and the probability of passing this test is P(X=0;6) = 0.83.

The EB prior, EB posterior, and the β(6,144) A-modem reference distribution for the EB plan (X2 = 0, N2 = 6) are shown in Figure 6. When X2 = 0 the posterior has the same general shape as the prior because both components in the MF posterior have the shape parameter values as in the prior. However, each component is now weighted differently in the mixture and the scale of both components is changed.

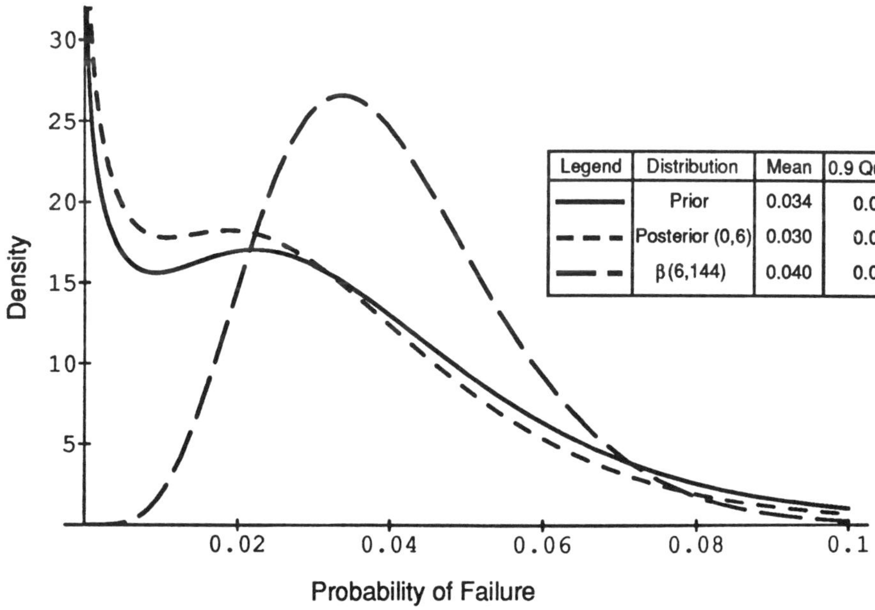

Legend	Distribution	Mean	0.9 Quantile
———	Prior	0.034	0.068
- - -	Posterior (0,6)	0.030	0.061
— —	β(6,144)	0.040	0.062

Figure 6. Results for the empirical Bayes test plan (X2 = 0, N2 = 6).

Figure 6 also gives the mean and 0.9 quantile for all three distributions. The EB posterior 0.9 quantile is 0.061 < 0.062, as required. As a matter of interest we note that, unlike Hart's prior, the EB prior 0.9 quantile does not automatically satisfy the criterion without the need for additional testing. The reason for this is that the EB prior is more diffuse than a prior consisting of only a single gamma (or beta) component.

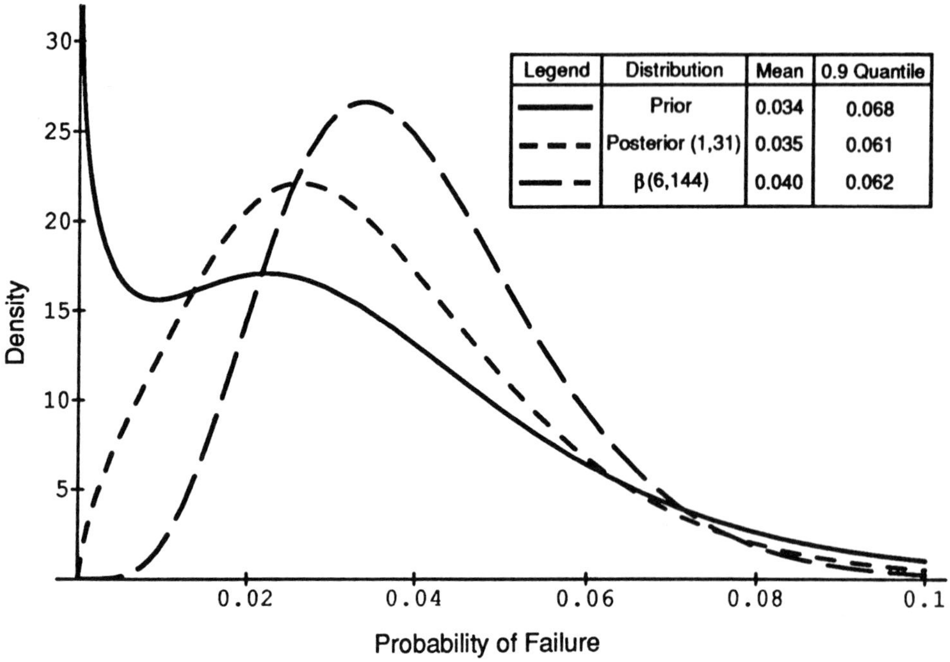

Figure 7. Results for the empirical Bayes test plan (X2 = 1, N2 = 31).

Table 1
Sensitivity of N2 to the choice of weights wA and wB1 for X2 = 0 and X2 = 1 failures

	X2 = 0			X2 = 1		
wB1	wA			wA		
	0.2	0.6	1.0	0.2	0.6	1.0
0.6	9	7	3	36	32	28
0.8	9	6	3	35	31	26
1.0	8	6	2	34	30	25

While not directly considered, note also that both the prior and posterior means are less than that of the $\beta(6,144)$ distribution.

The same three distributions for the EB plan ($X2 = 1$, $N2 = 31$) are displayed in Figure 7. From Figure 5, the probability of passing this test is 0.73. Note also that the prior and posterior no longer have the same shape, which they did for the preceding EB plan. The quantile test criterion is again satisfied and, although not required, the posterior mean is again less than 0.04.

It is interesting to examine the sensitivity of these two EB plans to changes in the prior weights w_A and w_{B1}. Table 1 gives the minimum required EB sample size N2 as a function of a selected range of values for w_A and w_{B1} for $X2 = 0$ and $X2 = 1$ failures. First, consider $X2 = 0$. As w_A changes by 400% (an effective A-modem equivalent sample size ranging between 30 and 150), N2 decreases roughly 67-75%, depending on the value of w_{B1}. As w_{B1} changes by 67% (an effective B1-modem equivalent sample size ranging between 22 and 36), N2 decreases roughly 11-22%, depending on w_A. Thus, for $X2 = 0$, N2 is quite sensitive to the choices of w_A and w_{B1}.

Now consider $X2 = 1$. As w_A increases by 400%, N2 decreases roughly 22-27%, depending on w_{B1}. As w_{B1} again increases by 67%, N2 decreases roughly 6-10%. Thus, when $X2 = 1$, N2 is significantly less sensitive to the choices of w_A and w_{B1} than when $X2 = 0$. Sensitivity to the choices of other EB parameters, such as α, β, and h, are not considered for this example. If there were multiple samples from different tests on the modems, it would be possible to make better choices of the values of w_A, w_{B1}, α, β, and h. One could use the marginal distribution and standard statistical estimation procedures to obtain more objective choices.

The nonparametric EB test plan ($X2 = 1$, $N2 = 31$) compares favorably with Hart's Bayesian test plan ($X2 = 1$, $N2 = 32$) based on the assumption of a beta prior distribution. For the added restriction of a beta prior, Hart's plan ensures that the corresponding beta posterior distribution closely matches the $\beta(6,144)$ reference prior. Matching both distributions occurs at the expense of overlooking plans which indeed satisfy the desired quantile criterion and require less testing. On the other hand, the EB procedure considered here identifies a set of alternate plans and, indeed, the plan ($X2 = 0$, $N2 = 6$) satisfies the desired quantile criterion with significantly less testing than Hart's plan. However, if a Bayesian single quantile criterion is used to determine a plan, less testing than that of Hart's procedure would also be required (recall that Hart's beta prior satisfies the quantile criterion with no additional testing).

It is interesting to compare the EB test plans for the B2-modem to the corresponding classical test plans which, of course, ignore the A- and B1-modem test data. The results provide some idea of the value of these related data in the EB test plan procedure. First, consider $X2 = 0$. One possible corresponding classical test plan would be to calculate the minimum number of binomial tests such that, if 0 failures occurred, the upper 90% confidence limit on p would be less than or equal to 0.062. It is easy to show that $N2 = 36$ is the minimum sample size required; thus, the corresponding classical plan is ($X2 = 0$, $N2 = 36$). We see that 4 times as much testing with 0 failures is required when using this classical approach as opposed to an EB approach. This clearly illustrates one important economic advantage in EB reliability qualification testing - decreased testing costs. Similarly, the classical plan for X2 = 1 is (X2 = 1, N2 = 62), which requires twice as much testing as the corresponding EB plan.

There is an even better way to assess the contribution of the past data to the EB test plan results. Because of the philosophical differences in classical and Bayesian statistics, it is more appropriate to compute the corresponding Bayesian test plans using a noninformative prior distribution and to compare the results. Using Jeffreys' noninformative prior, the corresponding Bayesian test plans are (X2 = 0, N2 = 21) and (X2 = 1,N2 = 50). As in the case of the classical test plans, the use of prior data in the EB procedure clearly results in significantly decreased testing requirements. Comparing the noninformative Bayes plan (X2 = 0, N2 = 21) with the EB plan (X2 = 0, N2 = 6), we see that the prior distribution in the EB analysis is roughly "worth" the equivalent of 0 failures in 15 B2 tests for the case in which X2 = 0.

Thirty-six B2-modems were ultimately tested and no failures occurred. Figure 8 illustrates the corresponding Hart $\beta(4.4, 150.4)$ and EB mixed gamma posteriors given these test results. The 0.9 quantile of the posterior $\beta(4.4, 150.4)$ distribution is 0.046, while the 0.9 quantile of the EB mixed gamma posterior is 0.041. These are also shown in Figure 8. Both quantiles are less than the required limit of 0.062, and the B2-modems were subsequently qualified for production.

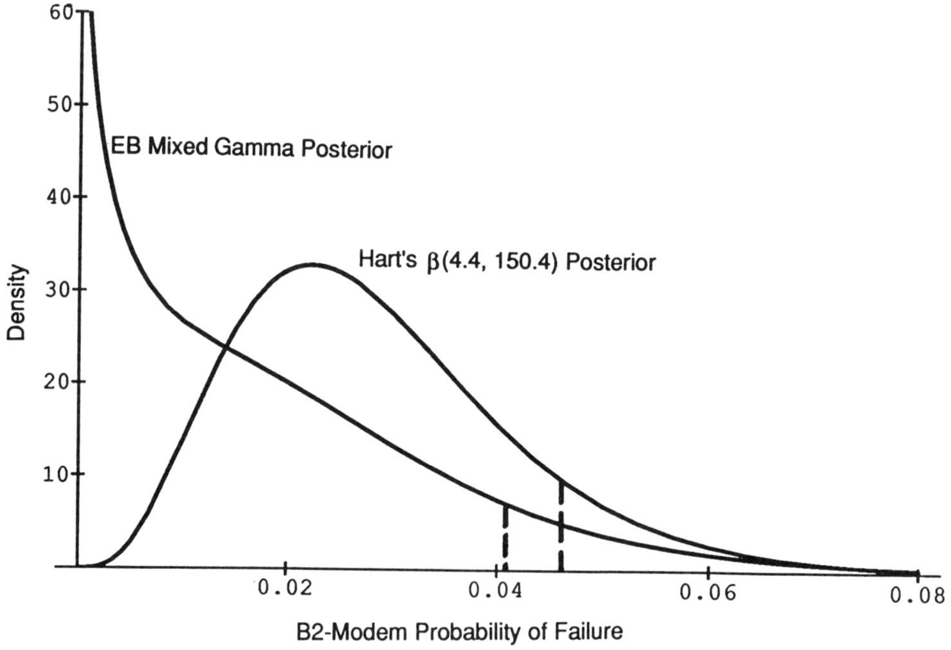

Figure 8. Posterior results given 36 B2-modem tests with 0 failures.

5. CONCLUSIONS

A nonparametric EB procedure was developed and used to determine a required reliability qualification test plan in a real-world example. The EB plan requires less testing than both an existing Bayesian plan (due to the incorrect consideration of the test objective) and a comparable plan based on the use of classical statistical methods. Although the method is not asymptotically optimal, it has been shown to perform quite well when m, the number of past experiments, is extremely small. Recall that $m = 2$ in the example considered here.

The main advantage in using nonparametric EB estimation methods in determining reliability test plans is the lack of necessity to specify a prior distributional family. This usually produces a test plan that is more robust than PEB test plan procedures which require such as assumption. This fact was illustrated here. This feature becomes important in practice, particularly when there are few past data sets (such as for the example considered here) which can be used to examine and establish the validity of such a parametric prior assumption.

Although only a single quantile test criterion was considered here, other single or multiple criteria could be considered in developing EB test plans. Martz and Waller (1991), Chapter 10, discuss some alternate criteria which could be used. Future research will consider dual test criteria (such as specified posterior producer's and consumer's risks) in developing the required EB test plan. Future research will also consider using parametric EB notions to estimate w_A, w_{B1}, α, β, and h.

6. REFERENCES

Fisher, R. C. (1990), "Nonparametric Empirical Bayes Estimation of the Poisson Failure Rate," Ph.D. Dissertation, University of New Mexico, Albuquerque, NM.

Fröhner, F. H. (1985A), "Analytic Bayesian Solution of the Two-Stage Poisson-Type Problem in Probabilistic Risk Analysis," Risk Analysis, 5, 217-225.

Fröhner, F. H. (1985B), "The Two-Stage Poisson-Type Problem in Probabilistic Risk Analysis," Risk Analysis, 5, 231-234.

Ganter, W. A., Hart, L., Martz, H. F., and Fisher, R. (1990), "Comment on: Reliability of Modified Designs," IEEE Transactions on Reliability, 39, 520-522.

Hart, L. (1990), "Reliability of Modified Designs: A Bayes Analysis of An Accelerated Test of Electronic Assemblies," IEEE Transactions on Reliability, 39, 140-144.

Martz, H. F., and Waller, R. A. (1991). Bayesian Reliability Analysis. Malabar, Florida: Krieger Publishing Co.

Martz, H. F., and Zimmer, W. J. (1990), "A Nonparametric Bayes Empirical Bayes Procedure for Estimating the Percent Nonconforming in Accepted Lots," Journal of Quality Technology, 22, 95-104.

Martz, H. F., and Zimmer, W. J. (1992), "The Risk of Catastrophic Failure of the Solid Rocket Boosters on the Space Shuttle," The American Statistician, 46, 42-47.

Morris, C. et al (1989), "Empirical Bayes Methodology in Traffic Accident Analyses," Technical Report, Center for Statistical Sciences, University of Texas at Austin, Austin, TX.

Pitman, E. J. G. (1937), "The Closest Estimates of Statistical Parameters," Proceedings of the Cambridge Philosophical Society, 33, 212-222.

Robbins, H. (1955), "An Empirical Bayes Approach to Statistics," Proc. Third Berkeley Symposium on Math. Statistics and Probability, 1, 157-164, Univ. of California Press.

ADVANCES IN RELIABILITY
Edited by Asit P. Basu

A Bayesian Approach to the Estimation of Change-Point in a Hazard Rate

Jayanta K. Ghosh[a], Shrikant N. Joshi[b], and Chiranjit Mukhopadhyay[c]

[a] *Indian Statistical Institute, 203 B. T. Road, Calcutta 700035, India* and *Department of Statistics, Purdue University, West Lafayette, IN 47907*

[b] *Indian Statistical Institute, 8th Mile Mysore Road, Bangalore 560059, India* and *Department of Statistics, Ohio Statte University, Columbus, OH 43210*

[c] *Department of Statistics, University of Missouri, Columbia, MO 65211*

Abstract

The hazard rate h(t) is assumed to be a constant α till time τ and a constant β after time τ. A Bayesian approach is taken to estimate the change-point τ in the hazard rate. A diffuse prior on the parameters has been utilized to accomplish this goal. Analytical properties of the marginal posterior density function of τ has been studied. Asymptotic expansion of the marginal posterior density of τ has been obtained to establish the consistency of the posterior mode. The behavior of the posterior estimates under different parametric bounds has been studied through simulation. The simulation also compares the performances of the classical and Bayesian estimates of τ.

1. Introduction

Consider the following model

$$h(t) = \begin{cases} \alpha & t \le \tau \\ \beta & t > \tau \end{cases} \quad \alpha > \beta \qquad \ldots \qquad (1.1)$$

where h(t) is the hazard rate of a non-negative random variable T'. A typical h(t) as in (1.1) is plotted in Figure 1. The corresponding probability density function f(t) is given in equation (1.2) and a typical density function as in (1.2) is plotted in Figure 2. Observe that f(t) has a jump at the point τ, which we wish to estimate.

$$f(t) = \begin{cases} \alpha\ e^{-\alpha t} & t \le \tau \\ \beta e^{-(\alpha-\beta)\tau}\ e^{-\beta t} & t > \tau \end{cases} \qquad \ldots \qquad (1.2)$$

Chernoff and Rubin (1956) first considered densities of above type having a jump at an unknown change-point τ. They considered arbitrary probability densities parametrized by θ and a change-point τ and having both left and right hand limits at τ. For such densities under certain regularity conditions and under the assumption of the consistency of the MLE's of (θ,τ) they derived the asymptotic distribution of $n(\hat{\tau}-\tau)$, $\hat{\tau}$ being the MLE of τ. More recently, Ibragimov and Has'minskii (1981, chapter V) have considered densities $f(x;\theta)$ having finitely many jumps. They also assumed the existence of the left and right hand limits of $f(x;\theta)$ at the points of discontinuities and have studied asymptotic behaviors of both MLE and Bayesian estimates of θ. They however have not considered the problem of estimating the change-points for such densities unlike Chernoff and Rubin.

The model in (1.1) (or 1.2) was introduced by Matthews and Farewell (1982) to model the relapse rate of leukemia patients. They were interested in testing whether a new therapy produces a departure from the usual experience of a constant relapse rate after induction of remission. A departure of interest is that for which a constant relapse rate is evident for some period after induction, but subsequently a substantial reduction in relapse rate occurs. In that case, if T' denotes the time of relapse, then (1.1) can be used to model such a departure. (1.1) can also be viewed as a special case of the two-step regression model introduced by Anderson and Senthilselvan (1982), where they used their model for cancer mortality studies.

There are however other motivations for studying the above model. A common problem in reliability engineering is that of 'burn-in'. Certain items produced by a manufacturing process are inherently weak and these weak items are screened out by burning in all the manufactured items. When the hazard rate of the items produced are known, different cost-optimal burn-in schemes have been proposed in the literature. Jensen and Petersen (1978) suggested different graphical methods for the burn-in procedure. Kuo (1984) discusses the optimal burn-in time for series systems, with reliability enhancement, when the truncated bath-tub shaped hazard-rates for each component are known. Bergman and Klefsjo (1984), Spizzichino (1991), Clarotti and Spizzichino (1990), all have taken a non-parametric approach (that is, they do not assume any parametric form for the hazard rate) and discuss the optimal burn-in strategy. However, they also assumed that the hazard-rate is known. While these approaches to burn-in are feasible for products for which the hazard-rates are well-known, they do not address the statistical problem of modelling and estimating the hazard rate of a freak contaminated population of manufactured items.

Towards this end, we propose the above model (1.1) to model the hazard-rate of newly manufactured items with the following heuristic argument. We assume that at the beginning, the population of manufactured items has a high failure-rate α. If the items are burnt-in till the threshold time τ, the failure-rate might be expected to fall down to β due to the failure and hence removal of some of the weak items from the population of manufactured items. Thus when τ is known, items could be tested up to time τ and only survivors sold. In practise, usually screening is done by subjecting items to shocks, thermal or electrical, and then selling only the survivors. Screening based on inference about τ would be an alternative. Also after the hazard rate has been estimated, one might employ one of the above-mentioned cost-optimal procedures to obtain the optimal burn-in time.

This paper is organized as follows. In §2 we briefly discuss the available classical estimates of τ. In §3 we study the properties of the likelihood function of τ by assuming that α and β are known. In §4 we study the small sample properties of the marginal posterior of τ under two different kinds of prior. In §.5 we obtain asymptotic approximations of the marginal posterior of τ for any arbitrary prior $\pi(\alpha,\beta)$ on (α,β). Finally in §6 we compare the classical and Bayesian estimates of τ through simulation study.

2. Classical Estimates of τ

We have three different classical estimates of τ available to us.

Let $T_1, \dots, T_n$ be a random sample from (1.1). From this point onwards, without loss of generality, we will assume that $T_1 \le \dots \le T_n$, that is after making suitable rearrangement we will always observe the order statistics $T_1, \dots, T_n$.

Nguyen, Rogers and Walker (1984) (abbreviated to NRW from now on) gets a consistent estimator of τ in the model (1.1) as follows :

Define,

$$R(t) = \sum_{i=1}^{n} I_{[T_i \le t]}, \quad A(t) = \frac{1}{(n-R(t))} \sum_{i=R(t)+1}^{n} T_i, \quad S^2(t) = \frac{1}{(n-R(t))} \sum_{i=R(t)+1}^{n} T_i^2 - A(t)^2 \text{and}$$

$$X_n(t) = \frac{1}{n} S(t) \left[(n-R(t)) \log\left(\frac{n}{(n-R(t))} \right) - R(t) \right] + \frac{R(t)}{n} A(t) - A(0) \log\left(\frac{n}{(n-R(t))} \right).$$

Now if one defines $Y_n(T_i) = X_n(T_i)$ and the remaining points of the Y process are defined by linear interpolation, then NRW show that the 0 of this process is a consistent estimator of τ, provided the solution is nearest to a $n^{1/2}$ consistent estimate of τ. Thus define,

$$\hat{\tau}_{NRW} = 0 \text{ of the process } Y_n \quad \dots \quad (2.1)$$

Basu, Ghosh and Joshi (1988) (abbreviated to BGJ from now on) have introduced two semi-parametric estimates of τ for the following model :

$$h(t) = \begin{cases} \lambda(t) & t \le \tau \\ \lambda_0 & t > \tau \end{cases}$$

where $\lambda(t)$ is a non-increasing function of t and $\lambda(\tau) \ge \lambda_0$, under the reasonable conservative assumption that $F(\tau) \le p_0 < 1$, a known constant. From the last assumption it is possible to estimate λ_0 by the method of least-squares with t and $\log(\overline{F}_n(t))$ as the independent and dependent variables respectively. Let $\hat{\lambda}_0$ and $\hat{\xi}_{p_0}$ be the estimates of λ_0 and ξ_{p_0}, the p_0-th population quantile, respectively. Also let $y_n(t) = -\log(\overline{F}_n(t))$, $\varepsilon_n = c(\log n)n^{-1/2}$ and $h_n = n^{-1/4}$. Then BGJ define two estimates for τ as follows :

$$\hat{\tau}_{BGJ1} = \text{Inf}\ \{\ t>0 : y_n(t+h_n) - y_n(t) \le h_n \hat{\lambda}_0 + \varepsilon_n\ \} \qquad \ldots \qquad (2.2)$$

$$\hat{\tau}_{BGJ2} = \text{Inf}\ \{\ t>0 : - y_n(t) - \log(1-p_0) \le \hat{\lambda}_0(\hat{\xi}_{p_0} - t) + \varepsilon_n\ \} \qquad \ldots \qquad (2.3)$$

The motivation for these estimates are as follows. Consider the hypothesis testing problem:

$$H_0 : h(t) = \lambda_0$$
$$H_1 : h(t) > \lambda_0$$

If $\hat{h}_n(t)$ is a consistent estimator of h(t) then one can construct tests by accepting H_0 if $\hat{h}_n(t) \le \hat{\lambda}_0 + \varepsilon_n$. Then the smallest t for which H_0 is accepted would be a candidate for an estimate of τ. For $\hat{\tau}_{BGJ1}$, take $\hat{h}_n(t) = \dfrac{y_n(t+h_n) - y_n(t)}{h_n}$ and for $\hat{\tau}_{BGJ2}$, take $\hat{h}_n(t) = \dfrac{-y_n(t) - \log(1-p_0)}{(\hat{\xi}_{p_0} - t)}$.

BGJ establish the consistency of $\hat{\tau}_{BGJ1}$ and $\hat{\tau}_{BGJ2}$ and in a follow-up paper Ghosh and Joshi (1989) also establish the asymptotic normality of $\hat{\tau}_{BGJ1}$ and $\hat{\tau}_{BGJ2}$. BGJ (1988) also proved the asymptotic normality of $\hat{\tau}_{NRW}$. Note that Loader (1991) seems to be unaware that the limiting distribution of $\hat{\tau}_{NRW}$ is known now.

NRW observed that for their model, the likelihood is unbounded as the MLE $\hat{\tau} \to T_n$. Also as $\hat{\tau} \to T_n$, $\hat{\beta} \to \infty$. However BGJ showed that the MLE of τ exists and is $O_p(n^{-1})$ by imposing some identifiability conditions on the parameter space of (α, β, τ) and restricting the parameters to a compact subset of the parameter space.

The above three estimates are classical in flavor, in the sense, that they rely on the asymptotic properties (consistency and asymptotic normality) of the estimates. Achcar and Bolfarine (1989) considered a Bayesian approach for the above model. But their main interest was focused at the estimation of α and β for known τ. They also considered the problem for a finite discrete prior for τ. In particular they considered a discrete prior on τ, where the support is the set of the order statistics $\{ T_1, \ldots, T_n \}$. This approach, though intuitively reasonable for this particular model, has the serious drawback of the prior being data dependent. The support of the prior would be completely unknown before the data are observed. A reasonable approach would be to place a diffuse prior on τ. In §4 we approach the problem from this point of view.

3. Properties of the Likelihood of τ

In this section we first examine the properties of the likelihood function of τ for fixed α and β. We will denote our data $\{ T_1, \dots, T_n \}$ by **D**, $T_0 = 0$ and $T_{n+1} = \infty$. Also let,

$$\left.\begin{aligned} R(\tau) &= \sum_{i=1}^{n} I_{[T_i \le \tau]} \\ M(\tau) &= \sum_{i=1}^{n} T_i \; I_{[T_i \le \tau]} \\ Q(\tau) &= M(\tau) + (n - R(\tau))\tau \\ T &= \sum_{i=1}^{n} T_i \end{aligned}\right\} \quad \dots \quad (3.1)$$

Then the likelihood L (α, β, τ | **D**) is given by :

$$L(\alpha,\beta,\tau \mid \mathbf{D}) = \alpha^{R(\tau)} e^{-\alpha Q(\tau)} \beta^{(n-R(\tau))} e^{-\beta(T-Q(\tau))} \quad \dots \quad (3.2)$$

It is easier to deal with the log likelihood. Also in this section we are considering α and β to be fixed. So we suppress the dependence of the log likelihood on α and β and denote it by $G(\tau)$. Let $S_i = \sum_{j=1}^{i} T_j$, $i = 0, \dots, n$. Then for $T_i \le \tau < T_{i+1}$, $i = 0, \dots, n$,

$$G(\tau) = [\, n \log\beta \; - \; \beta T \; + \; i\,(\log\alpha - \log\beta\,) \; - \; S_i\,(\alpha - \beta\,)\,] \; - \; (\,n - i\,)\,(\,\alpha - \beta\,)\,\tau \quad \dots \quad (3.3)$$

The following properties of G can be established by simple algebra.

Property 1 : $G(\tau)$ is right continuous but $G(T_i) - G(T_i-) > 0 \; \forall \; i = 1, \dots, n$, so that G is discontinuous at every observed data point and at every discontinuity point $G(\tau)$ gets a jump upwards.

Property 2 : $G(\tau)$ is linear in every continuous strip ($T_i \le \tau < T_{i+1}$, $i = 0, \dots, n$) and its slope is negative, so that $G(\tau)$ is decreasing in each continuous strip.

Property 3 : The slope of $G(\tau)$ is $-(n-i)(\alpha-\beta)$ for $T_i \le \tau < T_{i+1}$, $i = 0, \dots, n$. In particular for $\tau \ge T_n$, $G(\tau)$ is constant.

Property 4 : $G(T_i) - G(T_{i+1}) > 0 \Leftrightarrow (n-i)\,(T_{i+1} - T_i) > \dfrac{\log\alpha \; - \; \log\beta}{\alpha \; - \; \beta}$.

Property 3 is important for we know immediately that if we assign a flat prior for τ, the support of τ has to be bounded above in order to make the posterior a proper density. A typical log likelihood as defined in (3.3) above is plotted in Figure 3. We end this section by noting an interesting fact about the density and proving a proposition about the likelihood function.

Property 5 : If τ_0 is the true unknown value of τ and m_0 is such that $T_{m_0} \le \tau_0 < T_{m_0+1}$, then given m_0, $T_{m_0+1}-\tau_0, \dots \dots, T_n-\tau_0$, can be regarded as an ordered sample of size $(n-m_0)$ from an exponential distribution with failure rate β.

Proposition 3.1 : Let τ_0 be the true unknown value of τ. Let m_0 be such that $T_{m_0} \le \tau_0 < T_{m_0+1}$. Let K be such that τ_{m_0+K} is the <u>first</u> local maximum of G to the right of (and including) τ_0, $0 \le K \le n-m_0$, where $\tau_{m_0} = \tau_0$, and $\tau_{m_0+K} = T_{m_0+K}$, $K \ge 1$. Then $E(K \mid m_0) = q\dfrac{1-q^{n-m_0}}{1-q}$, where $q = 1 - \left(\dfrac{\beta}{\alpha}\right)^{\frac{\beta}{\alpha-\beta}}$. For large n, $E(K \mid m_0)$ can be approximated by $\dfrac{q}{1-q}$.

Proof : Since τ_{m_0+K} is the <u>first</u> local maxima of $G(\tau)$ to the right of τ_0 (inclusive),

$$K = \inf\left\{ k : 0 \le k \le n-m_0-1 \text{ and } G\left(\tau_{m_0+k+1}\right) < G\left(\tau_{m_0+k}\right) \right\} \text{and,}$$
$$K = n-m_0, \text{ if } G(\tau_{m_0+k+1}) \ge G(\tau_{m_0+k}) \ \forall\ 0 \le k \le n-m_0-1 .$$

Thus,

$$\{ K = k \} = \{ G(\tau_{m_0+1}) \ge G(\tau_{m_0}),\ G(\tau_{m_0+2}) \ge G(\tau_{m_0+1}), \dots \dots$$
$$\dots \dots, G(\tau_{m_0+k}) \ge G(\tau_{m_0+k-1}),\ G(\tau_{m_0+k+1}) < G(\tau_{m_0+k})\} \text{for } 0 \le k \le n-m_0-1, \text{ and,}$$
$$\{ K = n-m_0 \} = \left\{ G(\tau_{m_0+1}) \ge G(\tau_{m_0}), \dots \dots, G(\tau_n) \ge G(\tau_{n-1}) \right\} \qquad \dots \qquad (*)$$

For $m > m_0$, it follows from property 4 above that,

$$G(\tau_{m+1}) \ge G(\tau_m) \Leftrightarrow (n-m)\,(\tau_{m+1} - \tau_m) \le \frac{1}{\alpha-\beta}\log\left(\frac{\alpha}{\beta}\right) = c \text{ (say)} \qquad \dots \qquad (**)$$

Similarly one can show that, $G(\tau_{m_0+1}) \ge G(\tau_0) \Leftrightarrow (n-m_0)\,(\tau_{m_0+1} - \tau_0) \le c$.

Define,

$$Y_i = (n-m_0-(i-1))\,(\tau_{m_0+i} - \tau_{m_0+i-1}),\ 1 \le i \le n-m_0.$$

Thus using $(**)$ above,

$$G(\tau_{m_0+i}) \ge G(\tau_{m_0+i-1}) \Leftrightarrow Y_i \le c,\ 1 \le i \le n-m_0.$$

and (∗) reduces to,

$$\{ K = k \} = \{ Y_1 \le c, \dots\dots, Y_k \le c, Y_{k+1} > c \}, 0 \le k \le n\text{-}m_0\text{-}1, \text{ and}$$

$$\{ K = n - m_0 \} = \left\{ Y_1 \le c, \dots\dots, Y_{n-m_0} \le c \right\}.$$

Thus by property 5 above the scaled differences $Y_1, Y_2, \dots, Y_{n\text{-}m_0}$ are i.i.d. exponential (β). Let $p = e^{-\beta c}$ and $q = 1\text{-}p$. Thus $P(K{=}k \mid m_0) = q^k p$, $0 \le k \le n\text{-}m_0\text{-}1$, and $P(K{=}n\text{-}m_0 \mid m_0) = q^{(n\text{-}m_0)}$. $E(K \mid m_0) = q\dfrac{1-q^{n-m_0}}{1-q}$. The result follows by substituting the value of c. For the limiting result observe that $(n\text{-}m_0) \to \infty$ as $n \to \infty$, because m_0/n has a finite limit < 1. •

Among other things Proposition 5.3.1 shows that for q even moderately greater than half the first local maxima after τ_0 on an average quite steps away from τ_0. For α=1.0, β=0.5, τ=2.0 and n=25, the probability that there is a local maximum at T_n or T_{n-1} is greater than 0.3. The probability of a local maximum in a neighbourhood of T_n is larger for smaller sample size and is high for certrain other combinations of parameter values also. These calculations indicate that even when α and β are known the posterior of τ under the uniform prior has the unpleasant property of having a local maximum far from τ_0, the true value of τ. Simulations show the same phenomenon when α and β are known and α=3.0, β=2.0, τ=0.15. In fact his forces us to recomend truncating the range of t in a data dependent manner.

4. Posterior Analysis

In this section we propose a diffuse prior for (α, β, τ) and then obtain the marginal posterior of τ by integrating out the nuisance parameters α and β. We start with the natural parameter space and then gradually put the necessary restrictions on the domain of (α, β, τ) so that the required integrals exist in order to make the marginal posterior of τ a proper density function. α and β being scale parameters, we start with the following prior :

$$\pi(\alpha, \beta, \tau) = \frac{1}{\alpha\beta} \quad \dots \quad (4.1)$$

Natural restrictions on the parameters are $0 < \beta < \alpha < \infty$, $0 < \tau < \infty$.

In order to get the joint posterior $\pi(\alpha, \beta, \tau \mid \mathbf{D})$ we multiply the likelihood (3.2) by the prior (4.1) and obtain the following :

$$\pi(\alpha, \beta, \tau \mid \mathbf{D}) \propto \alpha^{R(\tau)-1} e^{-\alpha Q(\tau)} \beta^{(n-R(\tau)-1)} e^{-\beta(T-Q(\tau))} \quad \dots \quad (4.2)$$

In order to get the marginal posterior of τ we first obtain the joint posterior of (β, τ) by integrating α out of (4.2) on its natural range (β, ∞). Two different cases arise :

Case 1 : $0 < \tau < T_1$

$$\pi(\beta,\tau \mid \mathbf{D}) \propto \beta^{n-1} e^{-\beta(T-n\tau)} g(\beta) \qquad \dots \qquad (4.3)$$

where $g(\beta) = \int_{\beta}^{\infty} \alpha^{-1} e^{-n\tau\alpha} d\alpha$

Case 2 : $T_i \le \tau < T_{i+1}, \; i = 1, \dots, n.$

$$\pi(\beta,\tau \mid \mathbf{D}) \propto \frac{(i-1)!}{(Q(\tau))^i} \sum_{j=0}^{i-1} \frac{(Q(\tau))^j}{j!} \beta^{n-i+j-1} e^{-T\beta} \qquad \dots \qquad (4.4)$$

The marginal posterior of τ is obtained by integrating out β from (4.3) and (4.4). The natural range of β is $(0, \infty)$. But for $T_n \le \tau < \infty$

$$\pi(\beta,\tau \mid \mathbf{D}) \propto \frac{(n-1)!}{T^n} \sum_{j=0}^{n-1} \frac{T^j}{j!} \beta^{j-1} e^{-T\beta} \qquad \dots \qquad (4.5)$$

(see (4.4) and (3.1) above)

Now it is easy to see that when $j = 0$ in (4.5) above, $\int_0^{\infty} \beta^{-1} e^{-T\beta} d\beta$ diverges. Thus we cannot let β vary over its natural range $(0, \infty)$. This leads us to our first domain restriction on the natural parameter space namely,. $0 < \beta_0 \le \beta$.

The above restriction can be justified from a heuristic viewpoint too. $\beta \to 0$ implies that all the failures are due to the weak components and none of the strong components has failed and thus the lifetimes of the manufactured items do not have a change-point τ in their failure-rate. For most practical problems there should not be any difficulty in specifying a lower bound for the failure-rate of the strong components. By specifying a lower bound for the failure rate of the strong components the manufacturer is not at any risk, rather the producer is protected by saying that the average life-span of the burnt-in strong components is at most $1/\beta_0$.

In the following development we will ignore the proportionality constant of $\pi(\beta, \tau \mid \mathbf{D})$, and will treat the right hand side of (4.3) and (4.4) above as $\pi(\beta, \tau \mid \mathbf{D})$. The proportionality constant of $\pi(\tau \mid \mathbf{D})$ will also be ignored following proposition 4.1. Actually ignoring the proportionality constant of $\pi(\tau \mid \mathbf{D})$ can be legitimized only after proving proposition 4.5. The marginal posterior of τ is obtained by integrating out β from (4.3) and (4.4) over the range $[\beta_0, \infty)$. Three different cases arise.

Case 1 : $0 < \tau < T_1$

$\pi(\tau|\mathbf{D})$ cannot be obtained in a closed form. So we first check the legitimacy of $\pi(\tau|\mathbf{D})$ by checking that $\pi(\tau|\mathbf{D})$ is finite $\forall\ 0 < \tau < T_1$. Then we study some of its properties.

Proposition 4.1 : $\pi(\tau \mid \mathbf{D}) \propto \int_{\beta_0}^{\infty} \pi(\beta,\tau \mid \mathbf{T})\, d\beta < \infty,\ \forall\ 0 < \tau < T_1.$

Proof : $g(\beta) = \int_{\beta}^{\infty} \alpha^{-1} e^{-n\tau\alpha}\, d\alpha < \beta^{-1} \int_{\beta}^{\infty} e^{-n\tau\alpha}\, d\alpha = \frac{1}{n\tau\beta} e^{-n\tau\beta}$. Thus, $\int_{\beta_0}^{\infty} \pi(\beta,\tau \mid \mathbf{D})\, d\beta < \int_{0}^{\infty} \beta^{n-1} e^{-\beta(T-n\tau)} g(\beta)\, d\beta < \frac{1}{n\tau} \int_{0}^{\infty} \beta^{n-2} e^{-T\beta}\, d\beta = \frac{1}{n\tau} \frac{\Gamma(n-1)}{T^{n-1}}$. Thus $\pi(\tau \mid \mathbf{D})$ is finite $\forall\ 0 < \tau < T_1$. •

Proposition 4.2 : For $0 < \tau < T_1$, $\pi(\tau \mid \mathbf{D})$ is a decreasing function of τ.

Proof : Can be shown by taking the differences of the integrals and by using $\alpha > \beta$. •

Proposition 4.3 : $\lim_{\tau \to 0+} \pi(\tau \mid \mathbf{D}) = \infty$.

Proof : Take any sequence $\tau_m \downarrow 0$. Let $h_m(\alpha, \beta) = \pi(\alpha, \beta, \tau_m \mid \mathbf{D})$. By proposition 4.2 above, $h_m(\alpha, \beta) \uparrow \pi(\alpha, \beta, 0 \mid \mathbf{D}) = \alpha^{-1} \beta^{n-1} e^{-\beta T} = h(\alpha, \beta)$ (say) $\forall\ \alpha, \beta$. Thus by Monotone Convergence Theorem, $\int_{\beta_0}^{\infty}\int_{\beta}^{\infty} h_m(\alpha,\beta)\, d\alpha\, d\beta \uparrow \int_{\beta_0}^{\infty}\int_{\beta}^{\infty} h(\alpha,\beta)\, d\alpha\, d\beta = \infty$. •

Proposition 4.4 : $\lim_{\tau \to T_1-} \pi(\tau \mid \mathbf{D}) < \frac{1}{nT_1} \frac{(n-2)!}{T^{n-1}} e^{-T\beta_0} \sum_{k=0}^{n-2} \frac{(T\beta_0)^k}{k!} = \pi(T_1 \mid \mathbf{D})$.

Proof : Fix $t_0 < T_1$. By proposition 4.2, for $t_0 < \tau < T_1$, $\pi(\alpha, \beta, \tau \mid \mathbf{D}) < \alpha^{-1} \beta^{n-1} e^{-\beta T} e^{-n(\alpha-\beta)t_0} = h(\alpha, \beta)$ (say). Now following the same argument as in the proof of proposition 4.1, one can show that $\int_{\beta_0}^{\infty}\int_{\beta}^{\infty} h(\alpha, \beta)\, d\alpha\, d\beta < \infty$. Take any sequence $\tau_m \uparrow T_1$, and let $g_m(\alpha, \beta) = \pi(\alpha, \beta, \tau_m \mid \mathbf{D})$, then $g_m(\alpha, \beta) < h(\alpha, \beta)\ \forall\ \alpha, \beta$ and for infinitely many m. Thus by using the Dominated Convergence Theorem we can pass to the limit within the integral sign to obtain $\lim_{\tau \to T_1-} \pi(\tau \mid \mathbf{D}) = \int_{\beta_0}^{\infty}\int_{\beta}^{\infty} \alpha^{-1} \beta^{n-1} e^{-nT_1\alpha} e^{-(T-nT_1)\beta}\, d\alpha\, d\beta$.

From the proof of proposition 4.1, $\int_{\beta}^{\infty} \alpha^{-1}\, e^{-nT_1\alpha}\, d\alpha < \frac{1}{nT_1}\, \beta^{-1}\, e^{-nT_1\beta}$. Hence,

$\lim_{\tau \to T_1-} \pi(\tau \mid \mathbf{D}) < \int_{\beta_0}^{\infty} \frac{1}{nT_1}\, \beta^{n-2}\, e^{-T\beta}\, d\beta = \frac{1}{nT_1} \frac{(n-2)!}{T^{n-1}} e^{-T\beta_0} \sum_{k=0}^{n-2} \frac{(T\beta_0)^k}{k!}$. The last equality can be obtained by integrating by parts. It follows from (4.6) below that the above quantity equals $\pi(T_1 \mid \mathbf{D})$. ●

Proposition 4.5 : $\int_0^{T_1} \pi(\tau \mid \mathbf{D})\, d\tau < \infty$.

Proof : Fix any $t_0 < T_1$.

$$\begin{aligned}
&\int_0^{t_0} \pi(\tau \mid \mathbf{D})\, d\tau \\
&= \int_{\beta_0}^{\infty}\int_{\beta}^{\infty} \alpha^{-1}\, \beta^{n-1}\, e^{-T\beta} \left(\int_0^{t_0} e^{-n(\alpha-\beta)\tau}\, d\tau \right) d\alpha\, d\beta \\
&= \int_{\beta_0}^{\infty}\int_{\beta}^{\infty} \alpha^{-1}\, \beta^{n-1}\, e^{-T\beta}\, \frac{1}{n(\alpha-\beta)} \left[1-e^{-nt_0(\alpha-\beta)}\right] d\alpha\, d\beta
\end{aligned}$$

Consider the last integral first for a fixed β. For α bounded away from β the integrand is of order $(1/\alpha^2)$ and thus is convergent, while for α in a negihbourhood of β $\left[1-e^{-nt_0(\alpha-\beta)}\right]$ dominates $\frac{1}{n(\alpha-\beta)}$ and thus the integral is convergent. Integrating over β does not pose any additional problems. ●

Although by proposition 4.5 the marginal posterior of τ is integrable on $(0, T_1]$, by proposition 4.3 $\lim_{\tau \to 0+} \pi(\tau \mid \mathbf{D}) = \infty$ and thus the posterior mode would always be unbounded if restricted to the natural parameter space $(0, \infty)$ of τ. If $\tau=0$ is included the model is non-identifiable. This point was also noted by BGJ, where they imposed compactness and identifiability restriction in order to ensure the consistency of the MLE of τ. Following them we will impose the second domain restriction of $\tau \geq a$ in the following development.

Case 2 : $T_i \leq \tau < T_{i+1}$, $i = 1, \ldots, n-1$.

$\pi(\tau \mid \mathbf{D})$ can be obtained by integrating the RHS of (4.4) w.r.t. β over the range $[\beta_0, \infty)$ yielding :

$$\pi(\tau \mid \mathbf{D}) \propto \frac{(i-1)!}{((Q(\tau))^i} e^{-T\beta_0} \sum_{j=0}^{i-1} \left[\left\{ \sum_{k=0}^{n-i+j-1} \frac{(T\beta_0)^k}{k!} \right\} \frac{(n-i+j-1)!}{j!} \frac{((Q(\tau))^j}{T^{n-i+j}} \right] \quad \ldots (4.6)$$

The following two properties of $\pi(\tau \mid \mathbf{D})$ are useful, at least for computational purposes. They are also immediate after numerically graphing the posterior. These properties can be established by straight forward algebra.

Property 1 : $\pi(\tau \mid \mathbf{D})$ is a decreasing function of τ on $[T_i, T_{i+1})$.

Property 2 : $\lim_{\tau \to T_i -} \pi(\tau \mid \mathbf{D}) \leq \pi(T_i \mid \mathbf{D}) \ \forall \ i = 1, \ldots, n.$

Case 3 : $\tau \geq T_n$.

By integrating the RHS of (4.4) with i = n, one obtains,

$$\pi(\tau \mid \mathbf{D}) \propto \frac{(n-1)!}{T^n}\left[h(\beta_0) + \sum_{j=1}^{n-1}\left(\sum_{k=0}^{j-1}\frac{(T\beta_0)^k}{k!}\right)\frac{1}{j!}e^{-T\beta_0}\right] \quad \ldots \quad (4.7)$$

where $h(\beta_0) = \int_{\beta_0}^{\infty} \beta^{-1} e^{-T\beta} \, d\beta$.

It is obvious from (4.7) that $\pi(\tau \mid \mathbf{D})$ is constant in this range of τ, $[T_n, \infty)$. It can also be shown that $\lim_{\tau \to T_n -} \pi(\tau \mid \mathbf{D}) \leq \pi(T_n \mid \mathbf{D})$. Due to the constancy of $\pi(\tau \mid \mathbf{D})$ on $[T_n, \infty)$ obviously we have to impose the third restriction of $\tau \leq b < \infty$, on the natural parameter space $(0, \infty)$ of τ. This is the final domain restriction on the parameter space. Thus finally our prior is :

$$\pi(\alpha, \beta, \tau) = \frac{1}{\alpha\beta}, \quad 0 < \beta_0 \leq \beta < \alpha < \infty, 0 < a \leq \tau \leq b < \infty \quad \ldots \quad (4.8)$$

The properties of the posterior arising out of this prior has been summarized below and such a typical posterior has been plotted in Figure 4.

Property 1 : $\pi(\tau \mid \mathbf{D})$ is finite for all values of τ.

Property 2 : $\pi(\tau \mid \mathbf{D})$ is non-increasing on $[T_i, T_{i+1}) \ \forall \ i = 0, \ldots, n$. In particular $\pi(\tau \mid \mathbf{D})$ is constant on the last range $[T_n, b)$.

Property 3 : $\lim_{\tau \to T_i -} \pi(\tau \mid \mathbf{D}) < \pi(T_i \mid \mathbf{D}) \ \forall \ i = 1, \ldots, n.$

So at every T_i, $\pi(\tau \mid \mathbf{D})$ gets an upwards jump and thus starts at a higher value for the next interval. Thus possibly the only interesting values of the posterior are $\pi(T_i \mid \mathbf{D})$, $i = 1, \ldots, n$.

Estimation of τ : Point estimates of τ can be obtained by computing the posterior mean or the posterior mode. The computing procedure can be summarized as follows. Let T_r and T_s be such that $T_{r-1} < a \leq T_r \leq \ldots \leq T_s \leq b < T_{s+1}$. Let k =s-r+1 and $U_0 = a$, $U_1 = T_r$, ..., $U_k =$

T_s, $U_{k+1} = b$. By property 2 and property 3 of the posterior it is clear that in order to compute the posterior mode one needs only to evaluate $\pi(U_i \mid \mathbf{D})$, $i = 0, \ldots, k$ and choose the U_i corresponding to the maximum value of the posterior as the posterior mode. The problem of evaluating the posterior mean is computationally equivalent to evaluating any posterior moment or finding the normalizing constant of $\pi(\tau|\mathbf{D})$. The integrals cannot be expressed in closed form and thus we evaluate them by using numerical quadrature. The range of integration being $[a,b]$ and $\pi(\tau|\mathbf{D})$ having jump discontinuities at every U_i, $i = 0, \ldots, k+1$, the integrals are first computed piecewise on each interval $[U_i, U_{i+1})$, $i = 0, \ldots, k$, and finally the total integral is obtained by adding the pieces. Bayesian interval estimates of τ can also be obtained by using the highest posterior density credible sets and the algorithm to compute such credible sets for this posterior is interesting in its own right and will be discussed elsewhere.

When the true values of α and β are close it has been observed that the classical estimates of τ normally gets worse. To investigate this point further, we bound α above by a constant $\alpha_0 < \infty$ in the prior and then try to estimate τ by the mean or the mode of the posterior arising henceforth. Thus, consider the following prior :

$$\pi'(\alpha, \beta, \tau) = \frac{1}{\alpha\beta}, \quad 0 < \beta_0 \leq \beta < \alpha \leq \alpha_0 < \infty, \ 0 < a \leq \tau \leq b < \infty \qquad \ldots \qquad (4.9)$$

Proceeding as before, we obtain the marginal posterior $\pi'(\tau|\mathbf{D})$ by first integrating α over $(\beta, \alpha_0]$ and then β over $[\beta_0, \alpha_0]$. $\pi'(\tau|\mathbf{D})$ have exactly the same properties as $\pi(\tau|\mathbf{D})$. The form of $\pi'(\tau|\mathbf{D})$ has been summarized below and a typical $\pi'(\tau|\mathbf{D})$ is plotted in Figure 5.

$$\pi'(\tau|\mathbf{D}) \propto \int_{\beta_0}^{\alpha_0}\int_{\beta}^{\alpha_0} \alpha^{-1} e^{-n\tau\alpha} \ \beta^{n-1} e^{-(T-n\tau)\beta} \ d\alpha d\beta, \text{ for } 0 < \tau < T_1$$

$$\pi'(\tau|\mathbf{D}) \propto \frac{(i-1)!}{(Q(\tau))^i} \sum_{j=0}^{i-1} \frac{(Q(\tau))^j}{j!} \left[\left\{ \frac{(n-i+j-1)!}{T^{(n-i+j)}} \sum_{l=0}^{n-i+j-1} \frac{T^l}{l!} (\beta_0^l e^{-T\beta_0} - \alpha_0^l e^{-T\alpha_0}) \right\} - \left\{ \frac{(n-i-1)!}{(P(\tau))^{(n-i)}} \sum_{k=0}^{n-i-1} \frac{(P(\tau))^k}{k!} (\beta_0^k e^{-P(\tau)\beta_0} - \alpha_0^k e^{-P(\tau)\alpha_0}) \right\} \right] \text{ for } T_i \leq \tau < T_{i+1}$$

$$(i = 1, \ldots, n-1, \text{ where } P(\tau) = T - Q(\tau))$$

$$\pi'(\tau|\mathbf{D}) \propto \frac{(n-1)!}{T^n} \left[\left\{ \int_{\beta_0}^{\alpha_0} \beta^{-1} e^{-T\beta} d\beta \right\} + \left\{ \sum_{j=1}^{n-1} \frac{T^j}{j!} \left[\frac{(j-1)!}{T^j} \sum_{k=0}^{j-1} \left(\frac{T^k}{k!} (\beta_0^k e^{-T\beta_0} - \alpha_0^k e^{-T\alpha_0}) \right) \right] \right\} - \left\{ \log\frac{\alpha_0}{\beta_0} e^{-T\alpha_0} \sum_{j=0}^{n-1} \frac{(T\alpha_0)^j}{j!} \right\} \right] \text{ for } \tau \geq T_n$$

The exact form of $\pi(\tau|\mathbf{D})$ and $\pi'(\tau|\mathbf{D})$ being algebraically complicated we seek some simple asymptotic approximation of the marginal posterior of τ for an arbitrary prior $\pi(\alpha,\beta)$. This approximation would also be useful in proving desirable frequentist (or otherwise) properties of the marginal posterior of τ.

5. Asymptotic Analysis

We need to introduce some new notations before we get to the main results. We will first rewrite the likelihood function in 3.2 in terms of $\hat{\alpha}(\tau)$ and $\hat{\beta}(\tau)$ for $T_m \le \tau < T_{m+1}$, $\forall$ m = 0,...,n, where $\hat{\alpha}(\tau) = p_n/\overline{U}_n(\tau)$ and $\hat{\beta}(\tau) = p^n/\overline{V}_n(\tau)$, where p_n = m/n, p^n = (n-m)/n, $\overline{U}_n(\tau) = \frac{1}{n}\sum_{i=1}^{n}\left(T_i I_{[T_i \le \tau]} + \tau I_{[T_i > \tau]}\right)$ and $\overline{V}_n(\tau) = \frac{1}{n}\sum_{i=1}^{n}(T_i - \tau) I_{[T_i > \tau]}$. Also following the notation of §1 let T' be the lifetime variable with density function f(.), distribution function F(.), and survival function $\overline{F}$(.). Thus the likelihood function in 3.2 can be written as:

$$L(\alpha,\beta,\tau \mid \mathbf{D}) = \alpha^m \; e^{-m(\alpha/\hat{\alpha}(\tau))} \; \beta^{(n-m)} \; e^{-(n-m)\left(\beta/\hat{\beta}(\tau)\right)} \quad \ldots \quad (5.1)$$

For a fixed τ, we denote the sets $\{|\alpha-\hat{\alpha}(\tau)| \le c\}$ and $\{|\beta-\hat{\beta}(\tau)| \le c\}$ by A_c and B_c and the constants $\log(m)/\sqrt{m}$ and $\log(n-m)/\sqrt{n-m}$ by c_n and d_n respectively. Let,

$$J = \hat{\pi}\; \hat{\alpha}(\tau)^{(m+1)}\; \hat{\beta}(\tau)^{(n-m+1)}\; m^{-1/2}\; (n-m)^{-1/2}\; e^{-n}, \text{ where, } \hat{\pi} = \pi\,(\hat{\alpha}(\tau), \hat{\beta}(\tau)).$$

Let $(\alpha^*,\beta^*,\tau_0)$ be the true unknown value of (α,β,τ) and we tacitly assume that any mention of any expectations taken or of F(.) or $\overline{F}$(.) or any a.s. probability is to be evaluated at $(\alpha^*,\beta^*,\tau_0)$. Also let $\alpha(\tau)$ and $\beta(\tau)$ be as defined in 5.2 below.

$$\alpha(\tau) = \begin{cases} \alpha^* & \tau \le \tau_0 \\ \dfrac{\beta^*\left(1 - C_1 e^{-\beta^*\tau}\right)}{\left(C_2 - C_1 e^{-\beta^*\tau}\right)} & \tau > \tau_0 \end{cases} \quad \text{and} \quad \beta(\tau) = \begin{cases} \dfrac{\alpha^*}{1 + C_3 e^{\alpha^*\tau}} & \tau \le \tau_0 \\ \beta^* & \tau > \tau_0 \end{cases}$$

where $C_1 = e^{-(\alpha^*-\beta^*)\tau_0}$, $C_2 = (\beta^*/\alpha^*)(1 - e^{-\alpha^*\tau_0}) + e^{-\alpha^*\tau_0}$ and $C_3 = ((\alpha^*-\beta^*)/\beta^*)\, e^{-\alpha^*\tau_0}$ ... (5.2)

Regularity Conditions on $\pi(\alpha,\beta)$: Let us denote the set $\{|\alpha-\alpha^*| \le \varepsilon \times |\beta-\beta^*| \le \varepsilon\}$ by Π^*. We assume that there exists a Π^* bounded away from the origin such that the following holds for $\pi(\alpha,\beta)$.

RC1 : Π^* is in the support of the prior $\pi(\alpha,\beta)$ *i.e.* $\pi(\alpha, \beta) > 0\ \forall\ (\alpha, \beta) \in \Pi^*$.

RC2 : $\pi(\alpha, \beta)$ is continuously differentiable on Π^*.

RC3 : $\pi(\alpha, \beta)$ is bounded above by a constant B on its support.

Remark 1: Observe that any domain restriction imposed on $(\alpha^*, \beta^*, \tau_0)$ like $\alpha^* > \beta^*$ or $e^{-\alpha^*\tau_0} > q_0$ (known) can be taken care of by a suitable choice of the support of $\pi(\alpha,\beta)$ and thus Proposition 5.1 will be valid for any prior satisfying the above mentioned regularity conditions.

Assumptions : We need to make the following assumptions for finding an asymptotic analytical expression for $\pi(\tau|\mathbf{D})$.

A1 : m is such that both $m \to \infty$ and $(n-m) \to \infty$ with $m/n \to p$ for some $0<p<1$.

A2 : The convergent sequences $\hat{\alpha}(\tau)$ and $\hat{\beta}(\tau)$ (converging to $\alpha(\tau)$ and $\beta(\tau)$ respectively) satisfy the following. There exists a $\delta>0$ such that $\forall\ \tau \in (\tau_0 \pm \delta)$, $\{A_{c_n} \times B_{d_n}\} \subset \Pi^*$, for all m and (n-m) sufficiently large.

Since τ is fixed in the following treatment, ($|\tau-\tau_0|<\delta$), we suppress the dependence of $\hat{\alpha}(\tau)$ and $\hat{\beta}(\tau)$ on τ.

Lemma 1 : $I_1 = \int_{B_{d_n}} \int_{A_{c_n}} L(\alpha,\beta,\tau|\mathbf{D})\ \pi(\alpha,\beta)\ d\alpha\ d\beta = 2\pi\, J\, [\, 1 + o(1)\,]$

Proof : $I_1 =$

$$\int_{B_{d_n}} \int_{A_{c_n}} \exp[\ m\log\alpha\ -\ m\alpha/\hat{\alpha}\ +\ (n-m)\log\beta\ -\ (n-m)\beta/\hat{\beta}\ ]\ \pi(\alpha,\beta)\ d\alpha\ d\beta$$

Expanding the terms within the exponential and π around $(\hat{\alpha}, \hat{\beta})$ and noticing that the first derivatives evaluated at $(\hat{\alpha}, \hat{\beta})$ within the exponential vanish, we get the following

$$I_1 = \hat{\alpha}^m\, \hat{\beta}^{(n-m)}\, e^{-n} \int_{B_{d_n}} \int_{A_{c_n}} [\, 1 + R_\pi]$$

$$\exp\left[-\frac{1}{2}\left(\frac{\alpha-\hat{\alpha}}{\hat{\alpha}/\sqrt{m}}\right)^2 + \frac{1}{3}m\left(\frac{\alpha-\hat{\alpha}}{\alpha_r}\right)^3 - \frac{1}{2}\left(\frac{\beta-\hat{\beta}}{\hat{\beta}/\sqrt{n-m}}\right)^2 + \frac{1}{3}(n-m)\left(\frac{\beta-\hat{\beta}}{\beta_r}\right)^3 \right] d\alpha\ d\beta$$

Where $R_\pi = \left[\ (\alpha-\hat{\alpha})\pi_{\alpha_\pi} + (\beta-\hat{\beta})\pi_{\beta_\pi}\right]$, $\pi_{\alpha_\pi} = (1/\hat{\pi})\frac{\partial}{\partial\alpha}\pi(\alpha,\beta)\Big|_{(\alpha_\pi,\beta_\pi)}$, $\pi_{\beta_\pi} = (1/\hat{\pi})\frac{\partial}{\partial\beta}\pi(\alpha,\beta)\Big|_{(\alpha_\pi,\beta_\pi)}$ and (α_π, β_π) and $(\alpha_r, \beta_r) \in A_{c_n} \times B_{d_n}$. Now with the change of variable $(z_\alpha,z_\beta) = \left(\left(\frac{\alpha-\hat{\alpha}}{\hat{\alpha}/\sqrt{m}}\right)\left(\frac{\beta-\hat{\beta}}{\hat{\beta}/\sqrt{n-m}}\right)\right)$ we obtain the following :

$$I_1 = J \int_{-\log(n-m)}^{\log(n-m)} \int_{-\log(m)}^{\log(m)} \left[1 + m^{-1/2}\hat{\alpha}\pi_{\alpha_\pi} z_\alpha + (n-m)^{-1/2}\hat{\beta}\pi_{\beta_\pi} z_\beta \right] \exp -(1/2)\left[z_\alpha^2 + z_\beta^2\right]$$
$$\exp (1/3)\left[m^{-1/2}((\hat{\alpha}/\alpha_r)z_\alpha)^3 + (n-m)^{-1/2}((\hat{\beta}/\beta_r)z_\beta)^3\right] dz_\alpha \; dz_\beta$$

Now observe that by A2, $(\hat{\alpha},\hat{\beta})$, (α_r,β_r) and $(\alpha_\pi,\beta_\pi) \in \Pi^*$ for all m and (n-m) sufficiently large. Thus by RC1 and RC2, $\hat{\alpha}\pi_{\alpha_\pi}$ and $\hat{\beta}\pi_{\beta_\pi}$ are bounded. Also Π^* being bounded away from the origin, $(\hat{\alpha}/\alpha_r)^3$ and $(\hat{\beta}/\beta_r)^3$ are also bounded. Now by Taylor series expansion of the second exponential term, I_1 can be written as follows :

$$I_1 = J \int_{-\log(n-m)}^{\log(n-m)} \int_{-\log(m)}^{\log(m)} \exp -(1/2)\left[z_\alpha^2 + z_\beta^2\right] \left[1 + R_n(z_\alpha,z_\beta) \right] dz_\alpha \; dz_\beta$$

Where, $|R_n(z_\alpha,z_\beta)| \le n^{-1/2} C (|z_\alpha^3| + |z_\beta^3|)$ for some $0<C<\infty$, and for all m and (n-m) sufficiently large. Now since $\int_{-\infty}^{\infty} |z_u^3| \; e^{-(1/2)z_u^2} \; dz_u < \infty$, $u = \alpha, \beta$, we are justified in writing I_1 as $2\pi J [1 + o(1)]$ and thus proving the lemma. ●

Lemma 2 : $\int_{A_{c_n}^C} \exp m[\log\alpha - \alpha/\hat{\alpha}] \; d\alpha = \sqrt{2\pi} \; \hat{\alpha}^{(m+1)} m^{-1/2} e^{-m} o(1)$

Proof : Let $g(\alpha) = \log\alpha - \alpha/\hat{\alpha}$. Observe that $g(\alpha)$ has a global maximum at $\hat{\alpha}$. Thus for m sufficiently large, $g(\alpha) < g(\hat{\alpha}+c_n)$ for $\alpha>\hat{\alpha}+c_n$. Now by expanding $g(\hat{\alpha}+c_n)$ around $\hat{\alpha}$ by Taylor's series up to the second term, we obtain that $g(\hat{\alpha}+c_n) = g(\hat{\alpha}) - c_n^2/\alpha^{*2}$, for some α^* lying between $\hat{\alpha}$ and $\hat{\alpha}+c_n$. Now choose k such that $\int_{\alpha>0} e^{kg(\alpha)} \; d\alpha < c < \infty$. Thus $\int_{\alpha>\hat{\alpha}+c_n} e^{mg(\alpha)} \; d\alpha \le \exp [(m+1) \log\hat{\alpha} - m -(1/2)\log m] \exp[-\log m \; (\log m/\alpha^{*2} - 1/2)] \exp k[\log m/m\alpha^{*2} - \log\hat{\alpha} + 1] \; \hat{\alpha} \; c = \sqrt{2\pi} \; \hat{\alpha}^{(m+1)} \; m^{-1/2} \; e^{-m} \; o(1)$. The result follows in a similar manner for $\int_{\alpha<\hat{\alpha}-c_n} e^{mg(\alpha)} \; d\alpha$. ●

Proposition 5.1 : Under the above mentioned regularity conditions RC1, RC2 and RC3, and assumptions A1 and A2 and τ in the δ-neighborhood of τ_0 (where δ is as in A2),

$$\pi(\tau|\mathbf{D}) = 2\pi J [1 + o(1)] \qquad \ldots \qquad (5.3)$$

Proof : $$\pi(\tau|\mathbf{D}) = \int_0^\infty \int_0^\infty L(\alpha,\beta,\tau|D) \; \pi(\alpha,\beta) \; d\alpha \; d\beta$$

We are justified integrating over the whole positive quadrant by remark 1. Let I_1, I_2, I_3 and I_4 denote the integrals on $A_{c_n} \times B_{d_n}$, $A^C_{c_n} \times B_{d_n}$, $A_{c_n} \times B^C_{d_n}$ and $A^C_{c_n} \times B^C_{d_n}$ respectively, so that $\pi(\tau|\mathbf{D}) = \sum_{i=1}^{4} I_i$. By lemma 1, $I_1 = 2\pi J [\, 1 + o(1)\,]$.

$$I_2 = \int_{B_{d_n}} \int_{A^C_{c_n}} L(\alpha,\beta,\tau|\mathbf{D})\ \pi(\alpha,\beta)\ d\alpha\ d\beta$$

$$\leq B \int_{A^C_{c_n}} \exp m[\log\alpha - \alpha/\hat{\alpha}]\ d\alpha \int_{B_{d_n}} \exp(n-m)[\log\beta - \beta/\hat{\beta}]\ d\beta \qquad \text{(by RC3)}$$

Proceeding as in the proof of lemma 1 one can show that $\int_{B_{d_n}} \exp(n-m)[\log\beta - \beta/\hat{\beta}]\ d\beta = \sqrt{2\pi}\ \hat{\beta}^{(n-m+1)}\ (n-m)^{-1/2}\ e^{-(n-m)}\ [\, 1 + o(1)\,]$ and by lemma 2 $\int_{A^C_{c_n}} \exp m[\log\alpha - \alpha/\hat{\alpha}]\ d\alpha = \sqrt{2\pi}\ \hat{\alpha}^{(m+1)}\ e^{-m}\ m^{-1/2}\ o(1)$. Thus $I_2 = 2\pi J\ o(1)$. By a symmetric argument $I_3 = 2\pi J\ o(1)$. Repeating argument similar to lemma2 one can show that $I_4 = 2\pi J\ o(1)$. ●

By SLLN, assumption A1 is true with probability 1 with $p = F(\tau)$. Also $c_n, d_n \xrightarrow{a.s.} 0$. Now we will show that the assumption A2 will be true with probability tending to 1. which we formalize in terms of lemma 4.

Lemma 3 : $\hat{\alpha}(\tau) \xrightarrow{a.s.} \alpha(\tau)$ and $\hat{\beta}(\tau) \xrightarrow{a.s.} \beta(\tau)$.

Proof : Follows from straight forward application of the strong law.

Lemma 4 : There exists a $\delta>0$ such that $|\tau-\tau_0|<\delta \Rightarrow P\ (\ \{A_{c_n} \times B_{d_n}\} \subset \Pi^*,\ \forall\ n\geq N\ )$ tends to 1 as $N\to\infty$, for $c_n, d_n \xrightarrow{a.s.} 0$.

Proof : Observe from 5.2 that $\alpha(\tau)$ and $\beta(\tau)$ are continuous functions of τ. Thus there exists a $\delta>0$ such that $|\tau-\tau_0|<\delta \Rightarrow |\alpha(\tau)-\alpha^*|<\varepsilon/3$ and $|\beta(\tau)-\beta^*|<\varepsilon/3$. Also by lemma 3, $P\ (\ |\hat{\alpha}(\tau)-\alpha(\tau)| < \varepsilon/3$ and $|\hat{\beta}(\tau)-\beta(\tau)| < \varepsilon/3,\ \forall\ n\geq N\ )$ tends to 1, and since $c_n, d_n \xrightarrow{a.s.} 0$, $P\ (\ |c_n|<\varepsilon/3$ and $|d_n|<\varepsilon/3\ \forall\ n\geq N\ )$ tends to 1. Take any $(\alpha',\beta') \in \{|\alpha-\hat{\alpha}(\tau)|\leq c_n \times |\beta-\hat{\beta}(\tau)|\leq d_n\}$. Then $P\ (\ \{|\alpha'-\alpha^*|<\varepsilon \times |\beta'-\beta^*|<\varepsilon\},\ \forall\ n\geq N)$ tends to 1. ●

Proposition 5.2 : For τ in the δ-neighborhood of τ_0, $\frac{1}{n}\log \pi(\tau|\mathbf{D}) \xrightarrow{a.s.} A(\tau)-1$, where $A(\tau) = F(\tau)\log\alpha(\tau) + \overline{F}(\tau)\log\beta(\tau)$.

Proof: Follows directly from lemma 3, lemma 4 and 5.3 above.

Proposition 5.3 : $A(\tau)$ attains a global maximum at τ_0.

Proof : We will show that $A'(\tau) > 0$ for $\tau < \tau_0$ and $A'(\tau) < 0$ for $\tau > \tau_0$ and thus proving the result.

Case 1 : $\underline{\tau < \tau_0}$. Recall from 5.2 that for $\tau < \tau_0$ $\alpha(\tau) = \alpha^*$ and $\beta(\tau) = \dfrac{\alpha^*}{1 + C_3 e^{\alpha^*\tau}}$. Thus, $A'(\tau) = -[\, f(\tau) \log(\beta(\tau)/\alpha^*) - \bar{F}(\tau)\,(\beta'(\tau)/\beta(\tau))\,]$. Now $f(\tau) = \alpha^* e^{-\alpha^*\tau}$, $\bar{F}(\tau) = e^{-\alpha^*\tau}$, and it can be shown that $\dfrac{\beta'(\tau)}{\beta(\tau)} = -\dfrac{C_3 \alpha^* e^{\alpha^*\tau}}{1 + C_3 e^{\alpha^*\tau}}$. Thus, $A'(\tau) = -\alpha^* e^{-\alpha^*\tau}\,[\log(\beta(\tau)/\alpha^*) + \dfrac{C_3 e^{\alpha^*\tau}}{1 + C_3 e^{\alpha^*\tau}}]$. Now by substituting $Z = \log(\beta(\tau)/\alpha^*)$ one can rewrite $A'(\tau) = -\alpha^* e^{-\alpha^*\tau}\,[\, Z - e^Z + 1\,] > 0$.

Case 2 : $\underline{\tau > \tau_0}$. Recall from 5.2 that for $\tau > \tau_0$ $\alpha(\tau) = \dfrac{\beta^*\left(1 - C_1 e^{-\beta^*\tau}\right)}{\left(C_2 - C_1 e^{-\beta^*\tau}\right)}$ and $\beta(\tau) = \beta^*$. Thus, $A'(\tau) = [\, f(\tau) \log(\alpha(\tau)/\beta^*) + F(\tau)\,(\alpha'(\tau)/\alpha(\tau))\,]$. Now $f(\tau) = C_1 \beta^* e^{-\beta^*\tau}$, $F(\tau) = (1 - C_1 e^{-\beta^*\tau})$, and it can be shown that $\dfrac{\alpha'(\tau)}{\alpha(\tau)} = \dfrac{\left(C_1 \beta^* e^{-\beta^*\tau}\right)(C_2 - 1)}{\left(1 - C_1 e^{-\beta^*\tau}\right)\left(C_2 - C_1 e^{-\beta^*\tau}\right)}$. Thus, $A'(\tau) = C_1 \beta^* e^{-\beta'\tau}\,[\, \log(\alpha(\tau)/\beta^*) + \dfrac{C_2 - 1}{C_2 - C_1 e^{-\beta^*\tau}}\,]$. Now by substituting $Z = \log(\alpha(\tau)/\beta^*)$ one can rewrite $A'(\tau) = C_1 \beta^* e^{-\beta^*\tau}\,[\, Z - e^Z + 1\,] < 0$.

Thus for large n $\pi(\tau|\mathbf{D})$ attains a global maximum at τ_0, the true unknown value of τ. •

6. Simulation Results

This section compares the three different classical estimates of τ namely, $\hat{\tau}_{NRW}$, $\hat{\tau}_{BGJ1}$ and $\hat{\tau}_{BGJ2}$ with different Bayesian estimates through simulation study. For a given set of values of (α, β, τ) 100 samples of size 25 were generated and the mean and the MSE of the different estimates of τ across the 100 samples were obtained. In all the following tables the MSE's appear under the mean value within parentheses. All the simulations were done on an IBM 3090 machine. All the tables containing the simualtion results have been appended at the end of this section.

In Table 1A through 1D $\hat{\tau}_{MN}$ and $\hat{\tau}_{MD}$ respectively denote the posterior mean and the mode of $\pi(\tau|\mathbf{D})$ of §4 and τ'_{MN} and τ'_{MN} respectively denote the posterior mean and the mode of $\pi'(\tau|\mathbf{D})$ of §4. The prior ranges are $\beta_0 \le \beta < \alpha \le \alpha_0$ and $a \le \tau \le b$. The values of the prior parameters namely, $(\beta_0, \alpha_0, a, b)$ appear on the top of the tables.

We observe from Table 1A and 1B that the Bayesian estimates are not performing well as compared to $\hat{\tau}_{BGJ1}$ and $\hat{\tau}_{BGJ2}$. The effect of β_0 is also not very clear. One more point of interest is that α_0 hardly has any effect on the Bayesian estimates. The posterior mode in general is better behaved than the posterior mean. To improve upon the Bayesian estimates we now shrink the range of τ and the results are summarized in Table 1C and Table 1D.

Obviously the Bayesian estimates improved due to shrinking the prior range of τ. But recall from §2 that $\hat{\tau}_{BGJ1}$ and $\hat{\tau}_{BGJ2}$ use the additional information $F(\tau) \le p_0$, for some known p_0. Incorporating this information into the posterior yields $\alpha\tau \le \log(1/(1-p_0)) = c_0$(say). This restriction leads us to integrate α over $(\beta, c_0/\tau)$ for a fixed value of τ for finding $\pi(\tau|\mathbf{D})$ and it also restricts the upper bound of τ to c_0/β_0 for $\alpha > \beta > \beta_0$. A typical posterior arising with this restricted parameter space is denoted by $\tilde{\pi}(\tau|\mathbf{D})$ and is plotted in Figure 6. The Bayesian estimates as such being better behaved than $\hat{\tau}_{NRW}$, no more adjustments are necessary. (Recall from §2 that to evaluate $\hat{\tau}_{NRW}$ one needs to find the 0 of the Y_n process nearest to a $n^{1/2}$ consistent estimate of τ.) Let $\tilde{\tau}_{MN}$ and $\tilde{\tau}_{MD}$ respectively denote the mean and mode of $\tilde{\pi}(\tau|\mathbf{D})$ in the following tables. We will not consider $\pi'(\tau|\mathbf{D})$ any more.

As it can be seen from Tables 2A and 2B, the Bayesian estimates corresponding to the restricted posterior $\tilde{\pi}(\tau|\mathbf{D})$ significantly improved the estimates. From Table 1A-D the effect of β_0 is not very clear but as it can be observed from Tables 2A and 2B, increasing the value of β_0 significantly improves the estimates. The primary reason for this improvement is the reduction in the upper bound of τ. In general the Bayesian estimates tend to over-estimate τ compared to the classical estimates. Although biased estimates are not statistically desirable, considering the application of our theory in the problem of burn-in it is better to over-estimate τ than to under-estimate it.

Finally we make further improvement on the Bayesian estimates by observing the fact that for certain values of (α, β, τ) ((3.0, 2.0, 0.15) in particular) the posterior mode occurs at the largest order statistics with high probability. For large samples $P(T_n < b)$ is small and approaches 0 but consider the case of the finite sample of size 25 (sample size for the simulation study) and $b = 2.5$. Then for $(\alpha, \beta, \tau) = (3.0, 2.0, 0.15)$, $P(T_{25} < 2.5) = 0.8647$ (and indeed for about 85% of the generated samples T_{25} were less than b and among these samples in about 98% of the cases the posterior mode occurred at T_{25}!). We cannot explain this anomaly by asymptotic argument. For, as mentioned above $P(T_n < b)$ approaches 0 for large samples and thus the problem does not arise.

With the above observations we propose the following. For estimating τ from the posterior we take b, the upper bound of τ to be $(T_n+T_{n-1})/2$ (or any number less than T_n). Strictly speaking this is not justifiable from a Bayesian point of view for the prior range depending on the data, but since the likelihood has a flat tail for $\tau \geq T_n$ the only Bayesian way of handling this problem would be to consider a prior with a sharply decreasing tail, but in that case the tail behavior of the posterior would be completely prior dependent which is not at all desirable from a robustness consideration. Truncation seems desirable from this point of view. Another alternative would be to consider a less tail dependent estimate like the posterior median. Though this idea apparently seems reasonable, the median also would be inflated if T_n is small compared to b. In that case there will be a significant amount of probability content (for the posterior attains a maximum at T_n with high probability) in the region $[T_n, b)$ which in turn will pull the median up. Thus the only heuristic solution to the problem is to truncate the posterior before T_n and then work with the posterior and this solution works very well as it is shown in Tables 3A and 3B.

Observe that by truncation although $\tilde{\tau}_{MN}$ and $\tilde{\tau}_{MD}$improves significantly, the performance of $\hat{\tau}_{MN}$ and $\hat{\tau}_{MD}$ remains almost the same as before (but even for then significant improvement does occur at (3.0, 2.0, 0.15)). Thus we do not suggest truncation in general but in the presence of the additional knowledge of $F(\tau) \leq p_0 < 1$ significant improvement can be achieved by truncation and using the Bayesian estimates instead of the classical ones. As a matter of fact the truncation can be justified from a non-parametric point of view. If $F(\tau) \leq p_0 < 1$ for some known p_0 then using the empirical c.d.f. one may expect that any reasonable estimate of τ should be less than or equal to $\hat{\xi}_{p_0} < T_n$, where $\hat{\xi}_{p_0}$ is the sample estimate of ξ_{p_0}, the p_0-th population quantile. This would justify taking the upper bound of τ to be any number less than T_n. As the simulation study shows, incorporating this additional information by truncation only (as was done while computing $\hat{\tau}_{MN}$ and $\hat{\tau}_{MD}$ in Tables 3A and 3B) is not good enough. A full utilization of this additional information is accomplished only by a proper Bayesian treatment (as expected) which was done by considering the posterior $\tilde{\pi}(\tau|\mathbf{D})$. We summarize our findings through the simulation study below:

1. α_0 does not affect the posterior by a significant amount. So for computational simplicity one might take $\alpha_0 = \infty$.

2. Bayesian estimates can be improved by increasing β_0 and shrinking the prior / posterior range of τ, [a,b].

3. Posterior mode is a better estimate than the posterior mean. Posterior median is also an attractive choice as an estimate due to the flat tail of the posterior.

4. Bayesian estimates in general over-estimate τ than the classical ones, which is more desirable than under-estimation with the application to burn-in problem in mind.

5. In absence of any additional information the posterior $\pi(\tau|\mathbf{D})$ of §5.4 is our only option for inference. The estimates $\hat{\tau}_{MN}$ and $\hat{\tau}_{MD}$ are positively biased and could be alarmingly large for certain parameter values, in which case truncation might be useful.

6. In presence of the additional information of the form $F(\tau) \leq p_0 < 1$ for some known p_0, inference based on truncated $\tilde{\pi}(\tau|\mathbf{D})$ would be the best choice.

Table 1A

$\alpha_0 = 5.0/\infty$, $\beta_0 = 0.05$, a = 0.05, b = 3.0

α	β	τ	$\hat{\tau}_{BGJ1}$	$\hat{\tau}_{BGJ2}$	$\hat{\tau}_{NRW}$	τ'_{MN}	τ'_{MN}	$\hat{\tau}_{MN}$	$\hat{\tau}_{MD}$
3.0	2.0	0.15	0.068 (0.014)	0.048 (0.016)	0.352 (0.293)	1.768 (2.817)	0.905 (1.379)	1.768 (2.817)	0.905 (1.379)
3.0	1.0	0.10	0.141 (0.036)	0.064 (0.010)	0.781 (1.135)	1.281 (1.642)	0.217 (0.126)	1.281 (1.641)	0.217 (0.126)
2.0	1.0	0.20	0.114 (0.024)	0.066 (0.026)	0.604 (0.762)	1.520 (1.980)	0.640 (1.125)	1.520 (1.980)	0.640 (1.125)
2.0	1.0	0.10	0.109 (0.025)	0.056 (0.009)	0.392 (0.524)	1.434 (2.043)	0.763 (1.453)	1.434 (2.043)	0.763 (1.453)
2.0	0.5	0.20	0.186 (0.036)	0.079 (0.084)	0.310 (0.231)	0.589 (0.284)	0.240 (0.036)	0.589 (0.284)	0.240 (0.036)
2.0	0.5	0.10	0.194 (0.080)	0.066 (0.030)	0.480 (1.022)	0.790 (0.712)	0.240 (0.113)	0.790 (0.712)	0.240 (0.113)
1.0	0.5	0.40	0.204 (0.065)	0.202 (0.107)	1.030 (2.276)	1.077 (0.755)	0.662 (0.767)	1.083 (0.777)	0.662 (0.767)
1.0	0.5	0.10	0.099 (0.015)	0.155 (0.038)	1.693 (5.741)	1.057 (1.166)	0.149 (0.054)	1.057 (1.166)	0.107 (0.016)
0.5	0.25	1.50	0.434 (1.384)	0.720 (1.765)	3.167 (9.095)	1.337 (0.451)	0.630 (1.518)	1.337 (0.451)	0.630 (1.518)
0.5	0.25	0.10	0.179 (0.056)	0.253 (0.070)	3.223 (8.887)	0.818 (0.789)	0.064 (0.005)	0.818 (0.789)	0.064 (0.005)

Table 1B

$\alpha_0 = 5.0/\infty$, $\beta_0 = 0.20$, a = 0.05, b = 3.0

α	β	τ	$\hat{\tau}_{BGJ1}$	$\hat{\tau}_{BGJ2}$	$\hat{\tau}_{NRW}$	τ'_{MN}	τ'_{MD}	$\hat{\tau}_{MN}$	$\hat{\tau}_{MD}$
3.0	2.0	0.15	0.067 (0.014)	0.082 (0.014)	0.411 (0.228)	2.091 (3.916)	1.212 (1.638)	2.095 (3.930)	1.212 (1.638)
3.0	1.0	0.10	0.065 (0.008)	0.045 (0.007)	0.477 (0.781)	1.321 (1.975)	0.542 (0.805)	1.318 (1.975)	0.542 (0.805)
2.0	1.0	0.20	0.126 (0.023)	0.051 (0.029)	0.630 (0.705)	1.898 (3.492)	1.118 (1.943)	1.898 (3.491)	1.118 (1.943)
2.0	1.0	0.10	0.105 (0.017)	0.092 (0.011)	0.761 (0.971)	1.638 (2.951)	1.206 (2.541)	1.637 (2.951)	1.206 (2.541)
2.0	0.5	0.20	0.174 (0.021)	0.106 (0.053)	0.664 (1.881)	0.574 (0.244)	0.194 (0.011)	0.574 (0.244)	0.194 (0.011)
2.0	0.5	0.10	0.073 (0.014)	0.104 (0.022)	1.334 (3.156)	0.957 (0.996)	0.260 (0.344)	0.957 (0.996)	0.260 (0.344)
1.0	0.5	0.40	0.215 (0.057)	0.074 (0.128)	0.673 (1.047)	1.080 (0.652)	0.233 (0.057)	1.080 (0.652)	0.233 (0.057)
1.0	0.5	0.10	0.078 (0.009)	0.156 (0.016)	1.734 (5.382)	1.115 (1.387)	0.134 (0.106)	1.115 (1.387)	0.134 (0.106)
0.5	0.25	1.50	0.258 (1.597)	0.612 (2.036)	1.938 (4.189)	1.069 (0.511)	0.325 (1.835)	1.068 (0.515)	0.325 (1.835)
0.5	0.25	0.10	0.142 (0.027)	0.227 (0.050)	5.150 (9.999)	0.710 (0.670)	0.170 (0.234)	0.710 (0.670)	0.170 (0.234)

Table 1C

$\alpha_0 = 5.0/\infty$, $\beta_0 = 0.05$, a and b are given in parentheses under the value of τ

α	β	τ	$\hat{\tau}_{BGJ1}$	$\hat{\tau}_{BGJ2}$	$\hat{\tau}_{NRW}$	τ'_{MN}	τ'_{MD}	$\hat{\tau}_{MN}$	$\hat{\tau}_{MD}$
3.0	2.0	0.15 (0.10,1.0)	0.058 (0.019)	0.096 (0.014)	0.454 (0.250)	0.562 (0.188)	0.360 (0.142)	0.562 (0.188)	0.360 (0.142)
3.0	1.0	0.10 (0.05,1.0)	0.110 (0.028)	0.057 (0.014)	0.431 (0.442)	0.304 (0.066)	0.163 (0.045)	0.303 (0.066)	0.163 (0.045)
2.0	1.0	0.20 (0.10,1.0)	0.122 (0.063)	0.131 (0.031)	0.742 (0.865)	0.440 (0.079)	0.279 (0.076)	0.440 (0.079)	0.279 (0.076)
2.0	1.0	0.10 (0.05,1.0)	0.083 (0.016)	0.059 (0.008)	0.649 (0.948)	0.452 (0.151)	0.195 (0.069)	0.451 (0.151)	0.195 (0.069)
2.0	0.5	0.20 (0.10,1.0)	0.125 (0.021)	0.074 (0.048)	0.569 (1.211)	0.326 (0.029)	0.208 (0.007)	0.326 (0.029)	0.208 (0.007)
2.0	0.5	0.10 (0.05,1.0)	0.098 (0.027)	0.151 (0.058)	1.943 (7.369)	0.320 (0.075)	0.188 (0.056)	0.320 (0.075)	0.188 (0.056)
1.0	0.5	0.40 (0.25,1.0)	0.236 (0.060)	0.144 (0.127)	1.460 (4.675)	0.573 (0.037)	0.499 (0.055)	0.573 (0.037)	0.499 (0.055)
1.0	0.5	0.10 (0.05,1.0)	0.121 (0.024)	0.136 (0.019)	2.187 (7.422)	0.444 (0.130)	0.148 (0.031)	0.444 (0.130)	0.148 (0.031)
0.5	0.25	1.50 (0.50,2.0)	0.329 (1.493)	0.400 (1.729)	1.339 (1.362)	1.335 (0.056)	1.268 (0.285)	1.335 (0.056)	1.268 (0.285)
0.5	0.25	0.10 (0.05,1.0)	0.172 (0.042)	0.206 (0.059)	3.351 (9.982)	0.336 (0.075)	0.100 (0.017)	0.336 (0.075)	0.098 (0.017)

Table 1D

$\alpha_0 = 5.0/\infty$, $\beta_0 = 0.20$, a and b are given in parentheses under the value of τ

α	β	τ	$\hat{\tau}_{BGJ1}$	$\hat{\tau}_{BGJ2}$	$\hat{\tau}_{NRW}$	τ'_{MN}	τ'_{MD}	$\hat{\tau}_{MN}$	$\hat{\tau}_{MD}$
3.0	2.0	0.15	0.059	0.064	0.266	0.536	0.362	0.535	0.362
		(0.10,1.0)	(0.013)	(0.015)	(0.090)	(0.170)	(0.139)	(0.170)	(0.139)
3.0	1.0	0.10	0.129	0.026	0.477	0.332	0.088	0.332	0.088
		(0.05,1.0)	(0.034)	(0.008)	(0.587)	(0.074)	(0.003)	(0.074)	(0.003)
2.0	1.0	0.20	0.116	0.089	0.685	0.497	0.352	0.496	0.352
		(0.10,1.0)	(0.031)	(0.028)	(0.659)	(0.106)	(0.084)	(0.106)	(0.084)
2.0	1.0	0.10	0.121	0.063	1.060	0.384	0.183	0.384	0.183
		(0.05,1.0)	(0.028)	(0.009)	(2.068)	(0.100)	(0.040)	(0.100)	(0.040)
2.0	0.5	0.20	0.212	0.127	0.733	0.351	0.268	0.351	0.268
		(0.10,1.0)	(0.042)	(0.068)	(2.094)	(0.044)	(0.056)	(0.044)	(0.056)
2.0	0.5	0.10	0.163	0.073	1.029	0.313	0.137	0.313	0.137
		(0.05,1.0)	(0.046)	(0.012)	(3.259)	(0.068)	(0.030)	(0.068)	(0.030)
1.0	0.5	0.40	0.196	0.280	1.480	0.542	0.371	0.542	0.371
		(0.25,1.0)	(0.068)	(0.117)	(2.981)	(0.027)	(0.015)	(0.027)	(0.015)
1.0	0.5	0.10	0.091	0.095	1.800	0.375	0.127	0.359	0.113
		(0.05,1.0)	(0.013)	(0.008)	(6.223)	(0.095)	(0.015)	(0.088)	(0.011)
0.5	0.25	1.50	0.327	0.490	1.844	1.256	1.122	1.256	1.122
		(0.50,2.0)	(1.462)	(1.932)	(5.827)	(0.096)	(0.298)	(0.096)	(0.298)
0.5	0.25	0.10	0.134	0.186	5.152	0.306	0.107	0.295	0.105
		(0.05,1.0)	(0.016)	(0.058)	(8.812)	(0.060)	(0.034)	(0.057)	(0.034)

Table.2A

$\alpha_0 = \infty$, $\beta_0 = 0.05$, $a = 0.05$, $b = c_0/\beta_0$ are given in parentheses under the value of τ

α	β	τ	$\hat{\tau}_{BGJ1}$	$\hat{\tau}_{BGJ2}$	$\hat{\tau}_{NRW}$	$\tilde{\tau}_{MN}$	$\tilde{\tau}_{MD}$	$\hat{\tau}_{MN}$	$\hat{\tau}_{MD}$
3.0	2.0	0.15	0.085	0.088	0.324	1.727	1.319	1.853	1.326
		(2.50)	(0.019)	(0.015)	(0.145)	(3.023)	(1.882)	(3.022)	(1.827)
3.0	1.0	0.10	0.106	0.062	0.567	0.407	0.326	0.995	0.438
		(2.50)	(0.018)	(0.011)	(0.794)	(0.573)	(0.506)	(1.158)	(0.583)
2.0	1.0	0.20	0.128	0.114	0.612	0.475	0.350	0.492	0.475
		(2.50)	(0.027)	(0.033)	(0.753)	(0.545)	(0.431)	(0.540)	(0.545)
2.0	1.0	0.10	0.115	0.071	0.985	0.389	0.242	0.462	0.389
		(2.50)	(0.045)	(0.008)	(1.808)	(0.559)	(0.347)	(0.628)	(0.559)
2.0	0.5	0.20	0.247	0.063	0.966	0.221	0.158	0.696	0.283
		(2.50)	(0.064)	(0.042)	(2.686)	(0.004)	(0.005)	(0.419)	(0.178)
2.0	0.5	0.10	0.104	0.079	1.157	0.135	0.088	0.638	0.122
		(2.50)	(0.013)	(0.036)	(4.335)	(0.002)	(0.001)	(0.398)	(0.014)
1.0	0.5	0.40	0.205	0.225	1.096	0.314	0.197	0.416	0.341
		(2.50)	(0.069)	(0.143)	(2.428)	(0.014)	(0.056)	(0.301)	(0.014)
1.0	0.5	0.10	0.097	0.133	1.971	0.107	0.062	0.818	0.162
		(2.22)	(0.024)	(0.021)	(6.427)	(0.001)	(0.002)	(0.646)	(0.122)
0.5	0.25	1.50	0.348	0.548	1.889	0.991	0.703	0.382	0.991
		(2.50)	(1.468)	(1.853)	(5.478)	(0.340)	(0.861)	(1.604)	(0.340)
0.5	0.25	0.10	0.100	0.219	4.182	0.115	0.057	0.364	0.058
		(1.21)	(0.011)	(0.075)	(9.999)	(0.001)	(0.002)	(0.097)	(0.002)

Table 2B

$\alpha_0 = \infty$, $\beta_0 = 0.20$, a = 0.05, b = c_0/β_0 are given in parentheses under the value of τ

α	β	τ	$\hat{\tau}_{BGJ1}$	$\hat{\tau}_{BGJ2}$	$\hat{\tau}_{NRW}$	$\tilde{\tau}_{MN}$	$\tilde{\tau}_{MD}$	$\hat{\tau}_{MN}$	$\hat{\tau}_{MD}$
3.0	2.0	0.15	0.044	0.119	0.432	1.336	1.056	1.579	1.207
		(2.33)	(0.013)	(0.017)	(0.231)	(2.055)	(1.448)	(2.112)	(1.508)
3.0	1.0	0.10	0.091	0.079	0.527	0.135	0.076	0.527	0.178
		(1.57)	(0.023)	(0.016)	(0.792)	(0.004)	(0.001)	(0.262)	(0.094)
2.0	1.0	0.20	0.135	0.090	0.598	0.242	0.109	0.380	0.242
		(2.08)	(0.032)	(0.031)	(0.630)	(0.059)	(0.012)	(0.273)	(0.059)
2.0	1.0	0.10	0.071	0.082	0.694	0.126	0.070	0.446	0.196
		(1.06)	(0.010)	(0.009)	(0.888)	(0.003)	(0.001)	(0.145)	(0.067)
2.0	0.5	0.20	0.150	0.145	1.055	0.221	0.151	0.491	0.197
		(2.08)	(0.034)	(0.074)	(3.150)	(0.004)	(0.006)	(0.146)	(0.044)
2.0	0.5	0.10	0.081	0.073	1.227	0.119	0.078	0.104	0.119
		(2.08)	(0.012)	(0.014)	(3.182)	(0.001)	(0.001)	(0.008)	(0.001)
1.0	0.5	0.40	0.201	0.175	0.837	0.306	0.182	0.750	0.250
		(2.08)	(0.074)	(0.121)	(1.929)	(0.015)	(0.065)	(0.186)	(0.074)
1.0	0.5	0.10	0.137	0.105	2.088	0.117	0.064	0.225	0.108
		(2.08)	(0.049)	(0.010)	(6.805)	(0.001)	(0.002)	(0.020)	(0.011)
0.5	0.25	1.50	0.270	0.416	1.660	0.936	0.539	0.294	0.936
		(2.08)	(1.593)	(1.952)	(3.390)	(0.383)	(1.144)	(1.790)	(0.383)
0.5	0.25	0.10	0.106	0.214	3.211	0.103	0.058	0.145	0.064
		(2.08)	(0.018)	(0.069)	(9.999)	(0.000)	(0.002)	(0.003)	(0.003)

Table 3A

$\alpha_0 = \infty,\ \beta_0 = 0.05,\ a = 0.05,\ b = (T_n+T_{n-1})/2$

α	β	τ	$\hat{\tau}_{BGJ1}$	$\hat{\tau}_{BGJ2}$	$\hat{\tau}_{NRW}$	$\tilde{\tau}_{MN}$	$\tilde{\tau}_{MD}$	$\hat{\tau}_{MN}$	$\hat{\tau}_{MD}$
3.0	2.0	0.15	0.089 (0.020)	0.075 (0.015)	0.356 (0.180)	0.177 (0.010)	0.080 (0.006)	0.814 (0.505)	0.386 (0.206)
3.0	1.0	0.10	0.088 (0.017)	0.062 (0.010)	0.555 (0.697)	0.129 (0.003)	0.079 (0.001)	0.741 (0.585)	0.280 (0.250)
2.0	1.0	0.20	0.175 (0.041)	0.080 (0.030)	0.612 (0.663)	0.201 (0.003)	0.118 (0.010)	1.001 (0.769)	0.363 (0.213)
2.0	1.0	0.10	0.087 (0.022)	0.083 (0.009)	0.737 (1.056)	0.114 (0.003)	0.067 (0.002)	1.038 (0.990)	0.381 (0.427)
2.0	0.5	0.20	0.170 (0.029)	0.079 (0.050)	0.696 (1.399)	0.217 (0.004)	0.151 (0.005)	0.552 (0.258)	0.229 (0.071)
2.0	0.5	0.10	0.114 (0.025)	0.112 (0.030)	1.556 (5.659)	0.137 (0.003)	0.078 (0.001)	0.753 (0.590)	0.233 (0.220)
1.0	0.5	0.40	0.237 (0.080)	0.165 (0.129)	1.002 (1.944)	0.310 (0.015)	0.186 (0.061)	0.977 (0.496)	0.361 (0.248)
1.0	0.5	0.10	0.108 (0.024)	0.118 (0.015)	1.775 (5.657)	0.111 (0.001)	0.060 (0.002)	0.822 (0.649)	0.179 (0.147)
0.5	0.25	1.50	0.382 (1.425)	0.567 (1.935)	1.943 (4.892)	1.009 (0.300)	0.655 (0.974)	0.968 (0.554)	0.447 (1.465)
0.5	0.25	0.10	0.136 (0.030)	0.162 (0.041)	3.588 (9.999)	0.113 (0.001)	0.060 (0.002)	0.406 (0.127)	0.106 (0.026)

Table 3B

$\alpha_0 = \infty, \beta_0 = 0.20, a = 0.05, b = (T_n + T_{n-1})/2$

α	β	τ	$\hat{\tau}_{BGJ1}$	$\hat{\tau}_{BGJ2}$	$\hat{\tau}_{NRW}$	$\tilde{\tau}_{MN}$	$\tilde{\tau}_{MD}$	$\hat{\tau}_{MN}$	$\hat{\tau}_{MD}$
3.0	2.0	0.15	0.089 (0.017)	0.064 (0.016)	0.395 (0.252)	0.176 (0.006)	0.078 (0.006)	0.811 (0.498)	0.405 (0.221)
3.0	1.0	0.10	0.113 (0.026)	0.069 (0.013)	0.572 (0.760)	0.138 (0.003)	0.077 (0.001)	0.747 (0.530)	0.236 (0.141)
2.0	1.0	0.20	0.153 (0.046)	0.107 (0.030)	0.549 (0.659)	0.191 (0.003)	0.108 (0.011)	0.874 (0.562)	0.374 (0.194)
2.0	1.0	0.10	0.102 (0.023)	0.068 (0.008)	0.816 (1.187)	0.151 (0.125)	0.063 (0.002)	0.973 (0.904)	0.392 (0.368)
2.0	0.5	0.20	0.185 (0.035)	0.126 (0.068)	0.691 (1.748)	0.212 (0.004)	0.147 (0.005)	0.488 (0.164)	0.216 (0.053)
2.0	0.5	0.10	0.090 (0.013)	0.085 (0.021)	1.147 (3.848)	0.131 (0.003)	0.075 (0.001)	0.303 (0.066)	0.122 (0.022)
1.0	0.5	0.40	0.232 (0.070)	0.183 (0.137)	1.042 (2.536)	0.311 (0.014)	0.183 (0.059)	0.801 (0.267)	0.262 (0.086)
1.0	0.5	0.10	0.108 (0.031)	0.104 (0.010)	1.749 (5.639)	0.112 (0.001)	0.062 (0.002)	0.229 (0.021)	0.102 (0.013)
0.5	0.25	1.50	0.360 (1.427)	0.441 (1.935)	1.624 (4.219)	0.927 (0.377)	0.622 (0.985)	0.925 (0.648)	0.411 (1.513)
0.5	0.25	0.10	0.120 (0.021)	0.194 (0.054)	3.916 (9.999)	0.106 (0.000)	0.057 (0.002)	0.145 (0.003)	0.070 (0.003)

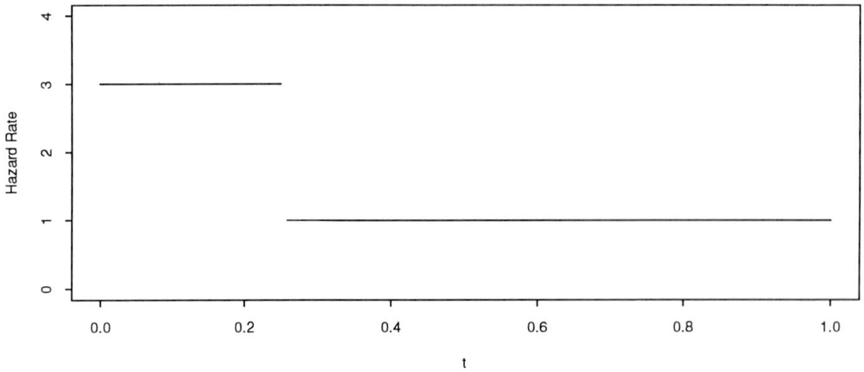

Figure 1 : A Typical Hazard Rate

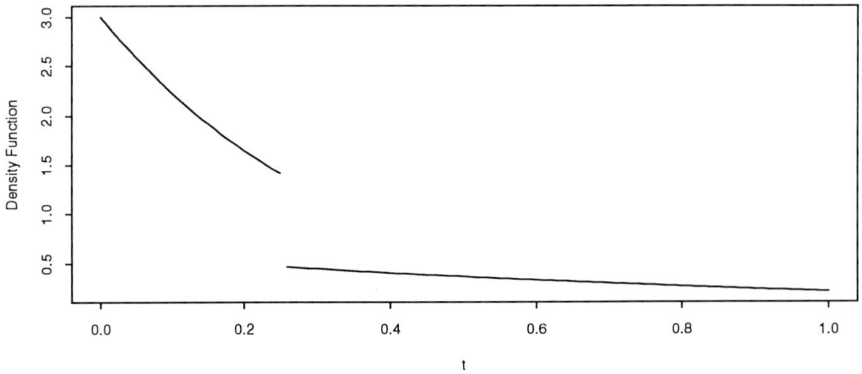

Figure 2 : A Typical Density Function

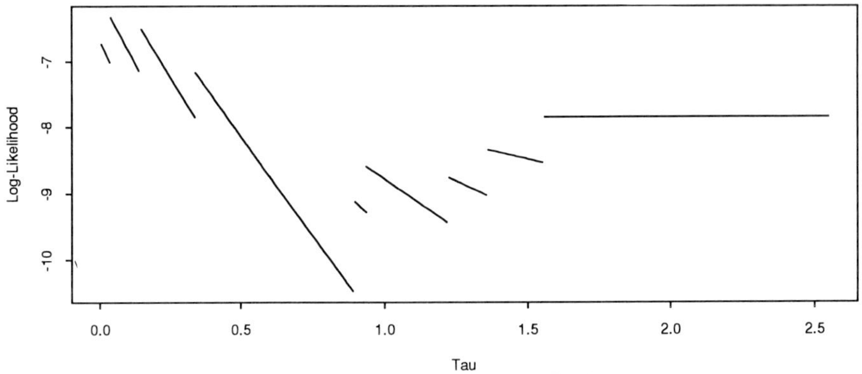

Figure 3 : A Typical Log-Likelihood Function

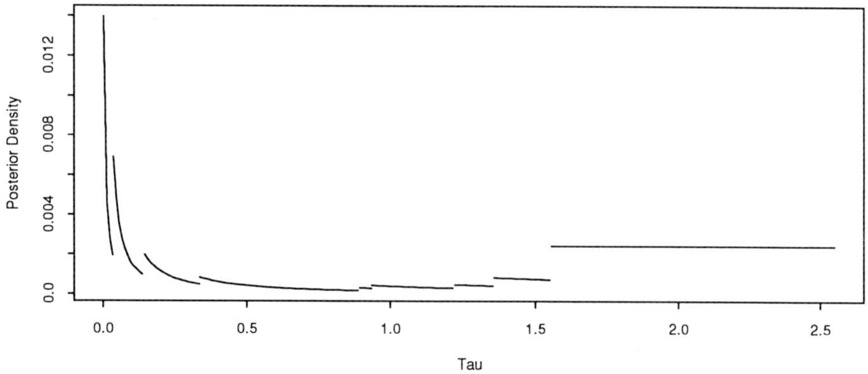

Figure 4 : A Typical Posterior Density of Tau with Unbounded Alpha

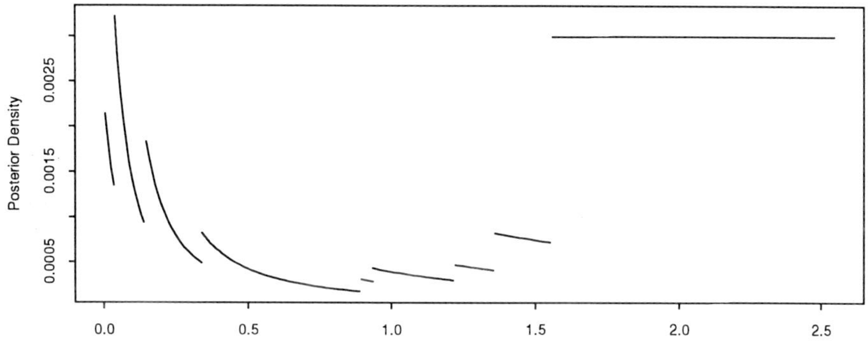

Figure 5 : A Typical Posterior Density of Tau with Bounded Parameter Space

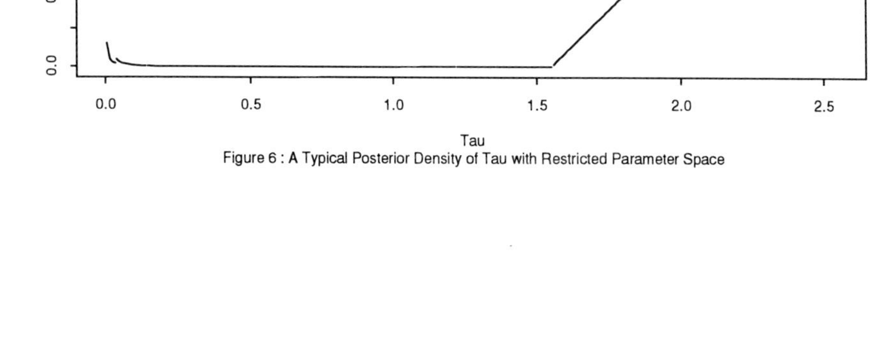

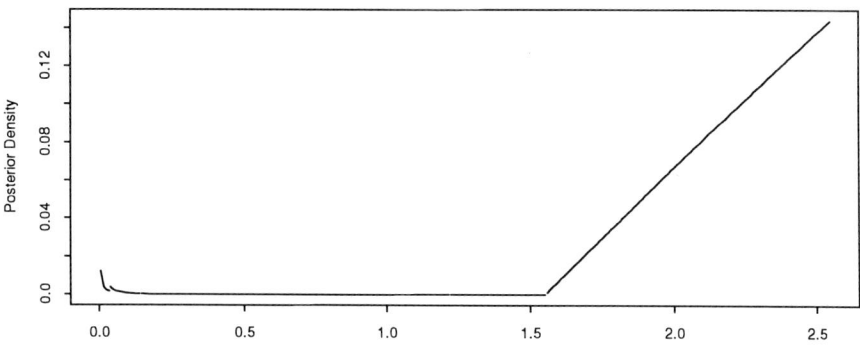

Figure 6 : A Typical Posterior Density of Tau with Restricted Parameter Space

References

Achcar J. A. & Bolfarine H. (1989) : "Constant hazard against a change-point alternative : A Bayesian approach with censored data." *Communication in Statistics A*. **18**. pp. 3801-3819.

Anderson J. A. & Senthilselvan A. (1982) : "A two-step regression model for the hazard function." *Applied Statistics*. **31**. pp. 44-51.

Basu A. P., Ghosh J. K. & Joshi S. N. (1988) : "On estimating change-point in a hazard rate." *Statistical Decision Theory and Related Topics IV*. Springer-Verlag. New-York. Gupta S. S. & Berger J. O. eds. **2**. pp. 239-252.

Bergman B. & Klefsjö B. (1984) : "Burn-in models and TTT-transforms." *Department of Mathematics , Luleå University, Technical Report no. 1984-5.*

Billingsley P. (1979) : *Probability and Measure*. New York : Wiley.

Chernoff H. and Rubin H. (1956) : "The Estimation of the Location of a Discontinuity in Density." *Proceedings of the Third Berkley Symposium on Mathematical Statistics and Probability* **1**. University of California Press. pp. 19-37.

Clarotti C. A. and Spizzichino F. (1990) : "Bayes burn-in Decision Procedure." *Probability Engineering and Information Science*. **4**. pp. 437-445.

Ghosh J. K. & Joshi S. N. (1989) : "On asymptotic distribution of an estimate of the change-point in a failure rate." *Indian Statistical Institute Technical Report no. SMU/2/1989*. (To appear in Communications in Statistics A)

Ibragimov I. A. and Has'minskii R. Z. (1981) : *Statistical Estimation Asymptotic Theory*. New York : Springer-Verlag.

Jensen F. & Petersen N. E. (1982) : *Burn-in : An Engineering Approach to the Design and Analysis of Burn-in Procedure*. New York : Wiley.

Kuo W. (1984) : "Reliability enhancement through optimal Burn-in." *IEEE Transactions on Reliability*. **R-33**. pp. 145-156.

Loader C. R. (1991) : "Inference for a Hazard Rate Change Point." *Biometrika*. **78**. pp. 749-757.

Matthews D. E. & Farewell V. T. (1982) : "On testing for a constant hazard against a change-point alternative." *Biometrics*. **38**. pp. 463-468.

Nguyen H. T., Rogers G. S. & Walker E. A. (1984) : "Estimation in change-point hazard rate models." *Biometrika*. **71**. pp. 299-304.

Serfling R. J. (1980) : *Approximation Theorems of Mathematical Statistics*. New-York : Wiley.

Spizzichino F. (1991) : "Sequential burn-in procedure." *Journal of Statistical Planning and Inference*. 29. pp. 187-197.

ADVANCES IN RELIABILITY
Edited by Asit P. Basu

Selecting the Best Exponential Population based on Type-I Censored Data: A Bayesian Approach

Shanti S. Gupta[a] and TaChen Liang[b]

[a]Statistics Department, 1399 Mathematical Sciences Building, Purdue University, West Lafayette IN 47907-1399, United States

[b]Mathematics Department, Wayne State University, Detroit MI 48202, United States

Abstract

We investigate the problem of selecting the exponential population having the largest mean life among several exponential populations. A Bayes rule based on type-I censored data is derived. A monotone property of the Bayes selection rule is discussed. An early selection rule is also proposed. Finally, an example is presented to illustrate the implementation of these two selection rules.

1. INTRODUCTION

Let $\pi_1, \ldots, \pi_k$ denote $k (k \geq 2)$ independent exponential populations with density functions $h(x|\theta_i) = \frac{1}{\theta_i} e^{-\frac{x}{\theta_i}}, x > 0$, where the values of the scale parameters $\theta_i, 1 \leq i \leq k$, are positive but unknown. Let $\theta_{[1]} \leq \ldots \leq \theta_{[k]}$ denote the ordered values of the parameters $\theta_1, \ldots, \theta_k$. It is assumed that the exact pairing between the ordered and the unordered parameters is unknown. The population associated with the largest value $\theta_{[k]}$ is considered as the best population. The problem of selecting the exponential population having the largest scale parameter $\theta_{[k]}$ has been studied by Sobel (1956) using sequential approach. Gupta (1963) studied some selection rules for gamma populations via subset selection approach. His selection rules can be applied for the exponential populations case. Recently, Huang and Huang (1980), and Berger and Kim (1985) also studied this selection problem using either subset selection approach or indifference zone approach with type-II censored data. The purpose of this paper is to derive Bayes rules to select the best exponential population based on type-I censored data.

Type-I censored data arise in many situations such as industrial life-testing, clinical trials and biological experiments. To motivate this study, consider a life-testing experiment, where m items from each of the k independent exponential populations are independently put on test at the outset and are not replaced on failure. Due to the time restriction, the experiment terminates at a prespecified time T. The failure time of an item is observable if it fails before time T. If an item still functions at the close of the experiment, its failure time is not observable. The item then is said to be censored at time T. This type of time censoring is known as type-I censoring. Type-I censoring scheme has received much attention in the statistical literature. See Bartholomew (1963), Yang and Sirvanci (1977), Spurrier and Wei (1980), and Mann, Schafer and Han (1982), among others.

Let $X_{ij}, 1 \leq j \leq m$, denote the failure times of the m items taken from population π_i based on a life-test experiment. According to the time censoring scheme, we only observe $\min(X_{ij}, T)$. Let $C_{ij} = 1$ if $X_{ij} < T$ and $C_{ij} = 0$ otherwise. Then, $N_i = \sum_{j=1}^{m} C_{ij}$ is the number of uncensored observations of the m items up to time T. Let $Y_{i1} \leq Y_{i2} \leq \ldots \leq Y_{iN_i}$ denote the ordered values of the N_i observable failure times and let $Y_i = \sum_{j=1}^{N_i} Y_{ij} + (m - N_i)T$. That is, Y_i is the total life time of the m items upto the time T. Then, $(Y_{i1}, \ldots, Y_{iN_i}, N_i)$ has a joint probability density function of the form:

$$f_i(y_{i1}, \ldots, y_{in}, n|\theta_i) = \frac{m!}{(m-n)!}\theta_i^{-n} exp\{-\theta_i^{-1}[\sum_{j=1}^{n} y_{ij} + (m-n)T]\}$$

$$= \frac{m!}{(m-n)!}\theta_i^{-n} exp\{-\theta_i^{-1} y_i\} \tag{1.1}$$

where $0 \leq n \leq m, 0 < y_{i1} \leq y_{i2} \leq \ldots \leq y_{in} < T, y_i = \sum_{j=1}^{n} y_{ij} + (m-n)T$ and $\sum_{j=1}^{n} \equiv 0$ if $n = 0$. Note that $(m-n)T \leq y_i \leq m\, T$. For convenience, we denote the expression at the right-hand-side of (1.1) by $f_i(y_i, n|\theta_i)$.

In this paper, we investigate the problem of selecting the best exponential population from among several exponential populations. A Bayes selection rule based on type-I censored data is derived in Section 2. A monotone property of the Bayes selection rule is discussed in Section 3. We also prove that the posterior density functions given the observed type-I censored data have the monotone likelihood ratio (MLR) property. Based on the MLR property, an early selection rule is proposed in Section 4. Finally, an example is presented to illustrate the implementation of the two selection rules.

2. A BAYES SELECTION RULE

Let $\underset{\sim}{Y} = (Y_1, \ldots, Y_k)$ and let $\underset{\sim}{N} = (N_1, \ldots, N_k)$ where $(Y_i, N_i), 1 \leq i \leq k$, are defined in Section 1. Let $\mathcal{N}$ be the sample space generated by $\underset{\sim}{N}$ and conditional on $\underset{\sim}{N} = \underset{\sim}{n} = (n_1, \ldots, n_k)$, let $\mathcal{Y}_n$ be the sample space generated by $\underset{\sim}{Y}$. Thus, for $\underset{\sim}{y} = (y_1, \ldots, y_k)\epsilon\mathcal{Y}_n, (m - n_i)T \leq y_i \leq m\, T, 1 \leq i \leq k$.

Let $\underset{\sim}{\theta} = (\theta_1, \ldots, \theta_k)$ and let $\Omega = \{\underset{\sim}{\theta}|\theta_i > 0, 1 \leq i \leq k\}$ be the parameter space. Let $\mathcal{A} = \{1, \ldots, k\}$ be the action space. Action i corresponds to the selection of population π_i as the best population. For a given $\underset{\sim}{\theta}\epsilon\Omega$, and an action i, the associated loss function $L^*(\underset{\sim}{\theta}, i)$ is defined by

$$L^*(\underset{\sim}{\theta}, i) = L(\theta_{[k]} - \theta_i) \tag{2.1}$$

where $L(x)$ is a nonnegative, nondecreasing function of x, $x \geq 0$, such that $L(0) = 0$.

Let $g(\underset{\sim}{\theta}) = \prod_{i=1}^{k} g_i(\theta_i)$ be the prior density function over the parameter space Ω. It is assumed that $\int L(\theta_{[k]})g(\underset{\sim}{\theta})d\underset{\sim}{\theta} < \infty$.

A selection rule $\underset{\sim}{\delta} = (\delta_1, \ldots, \delta_k)$ is defined to be a measurable mapping from the sample space $(\mathcal{N}, (\mathcal{Y}_{\underset{\sim}{n}})_{\underset{\sim}{n} \epsilon \mathcal{N}})$ to $[0,1]^k$ such that $0 \leq \delta_i(\underset{\sim}{n}, \underset{\sim}{y}) \leq 1$ and $\sum_{i=1}^{k} \delta_i(\underset{\sim}{n}, \underset{\sim}{y}) = 1$ for all $\underset{\sim}{y} \epsilon \mathcal{Y}_{\underset{\sim}{n}}, \underset{\sim}{n} \epsilon \mathcal{N}$. The value of $\delta_i(\underset{\sim}{n}, \underset{\sim}{y})$ is the probability of selecting population π_i as the best population given the observation $(\underset{\sim}{n}, \underset{\sim}{y})$.

Let $R(\underset{\sim}{\delta}, g)$ denote the Bayes risk associated with the selection rule $\underset{\sim}{\delta}$. Then,

$$R(\underset{\sim}{\delta}, g) = \int_{\Omega} \sum_{\underset{\sim}{n} \epsilon \mathcal{N}} \int_{\mathcal{Y}_{\underset{\sim}{n}}} \sum_{i=1}^{k} L(\theta_{[k]} - \theta_i)\delta_i(\underset{\sim}{n}, \underset{\sim}{y}) f(\underset{\sim}{y}, \underset{\sim}{n}|\underset{\sim}{\theta}) g(\underset{\sim}{\theta}) d\underset{\sim}{y} d\underset{\sim}{\theta} \tag{2.2}$$

where $f(\underset{\sim}{y}, \underset{\sim}{n}|\underset{\sim}{\theta}) = \prod_{i=1}^{k} f_i(y_i, n_i|\theta_i)$. Now, let

$$f_i(y_i, n_i) = \int_0^{\infty} f_i(y_i, n_i|\theta_i) g_i(\theta_i) d\theta_i, \; f(\underset{\sim}{y}, \underset{\sim}{n}) = \prod_{i=1}^{k} f_i(y_i, n_i),$$

$$g_i(\theta_i|y_i, n_i) = \frac{f_i(y_i, n_i|\theta_i) g_i(\theta_i)}{f_i(y_i, n_i)} \text{ and } g(\underset{\sim}{\theta}|\underset{\sim}{y}, \underset{\sim}{n}) = \prod_{i=1}^{k} g_i(\theta_i|y_i, n_i).$$

Using Fubini's theorem, it is easily seen that

$$R(\underset{\sim}{\delta}, g) = \sum_{\underset{\sim}{n} \epsilon \mathcal{N}} \int_{\mathcal{Y}_{\underset{\sim}{n}}} \sum_{i=1}^{k} \delta_i(\underset{\sim}{n}, \underset{\sim}{y}) \int_{\Omega} L(\theta_{[k]} - \theta_i) g(\underset{\sim}{\theta}|\underset{\sim}{y}, \underset{\sim}{n}) d\underset{\sim}{\theta} f(\underset{\sim}{y}, \underset{\sim}{n}) d\underset{\sim}{y}. \tag{2.3}$$

For each $(\underset{\sim}{n}, \underset{\sim}{y})$, let

$$\Delta_i(\underset{\sim}{n}, \underset{\sim}{y}) = \int_{\Omega} L(\theta_{[k]} - \theta_i) g(\underset{\sim}{\theta}|\underset{\sim}{y}, \underset{\sim}{n}) d\underset{\sim}{\theta}, \quad i = 1, \ldots, k, \tag{2.4}$$

and let

$$A(\underset{\sim}{n}, \underset{\sim}{y}) = \{i | \Delta_i(\underset{\sim}{n}, \underset{\sim}{y}) = \min_{1 \leq j \leq k} \Delta_j(\underset{\sim}{n}, \underset{\sim}{y})\}. \tag{2.5}$$

Then, a uniformly randomized Bayes rule is $\underset{\sim}{\delta}_G = (\delta_{G1}, \ldots, \delta_{Gk})$, where

$$\delta_{Gi}(\underset{\sim}{n}, \underset{\sim}{y}) = \begin{cases} |A(\underset{\sim}{n}, \underset{\sim}{y})|^{-1} & \text{if } i \in A(\underset{\sim}{n}, \underset{\sim}{y}), \\ 0 & \text{otherwise.} \end{cases} \tag{2.6}$$

3. A MONOTONICITY PROPERTY OF $\underset{\sim}{\delta}_G$

In this section, we claim that the Bayes selection rule $\underset{\sim}{\delta}_G$ has the following monotone property.

Theorem 3.1. For each $i = 1, \ldots, k$, $\delta_{Gi}(\underset{\sim}{n}, \underset{\sim}{y})$ is nondecreasing in y_i and also in $n_j, j \neq i$ when all other variables are kept fixed, and nonincreasing in n_i and also in $y_j, j \neq i$ when all other variables are kept fixed.

To prove this theorem, we need the following two lemmas.

Lemma 3.1. Let $0 \leq n_i^* \leq n_i \leq m$, $0 < y_i \leq y_i^* \leq m\ T$. Consider the likelihood ratio $r_i(\theta|y_i, n_i, y_i^*, n_i^*)$ defined by

$$r_i(\theta|y_i, n_i, y_i^*, n_i^*) = \begin{cases} \frac{g_i(\theta|y_i^*, n_i^*)}{g_i(\theta|y_i, n_i)} & \text{if } g_i(\theta|y_i, n_i) \neq 0 \\ 0 & \text{if both } g_i(\theta|y_i, n_i) = 0 \text{ and } g_i(\theta|y_i^*, n_i^*) = 0. \end{cases} \tag{3.1}$$

Then,

a) As $n_i = n_i^*$ and $y_i < y_i^*$, then $r_i(\theta|y_i, n_i, y_i^*, n_i^*)$ is nondecreasing in θ.

b) As $y_i = y_i^*$ and $n_i < n_i^*$, then $r_i(\theta|y_i, n_i, y_i^*, n_i^*)$ is nonincreasing in θ.

Proof: Note that as $g_i(\theta|y_i, n_i) \neq 0$, after simplification, we have

$$r_i(\theta|y_i, n_i, y_i^*, n_i^*) = g_i(\theta|y_i^*, n_i^*)/g_i(\theta|y_i, n_i)$$

$$= c(y_i, n_i, y_i^*, n_i^*)\theta^{n_i - n_i^*} \exp\{\theta^{-1}(y_i - y_i^*)\} \tag{3.2}$$

where

$$c(y_i, n_i, y_i^*, n_i^*) = \int_0^\infty \theta^{-n_i} \exp\{-\theta^{-1} y_i\} g_i(\theta) d\theta \Big/ \int_0^\infty \theta^{-n_i^*} \exp\{-\theta^{-1} y_i^*\} g_i(\theta) d\theta > 0.$$

Thus, the proof of this lemma is completed by following a straightforward argument.

Lemma 3.2. Let $\Delta_i(\underset{\sim}{n}, \underset{\sim}{y})$ be that defined in (2.4), for each $i = 1, \ldots, k$. Then, $\Delta_i(\underset{\sim}{n}, \underset{\sim}{y})$ is nonincreasing in y_i and also in n_j, $j \neq i$, when all the other variables are kept fixed, and nondecreasing in n_i and also in y_j, $j \neq i$, when all the other variables are kept fixed.

Proof: We prove that $\Delta_i(\underset{\sim}{n}, \underset{\sim}{y})$ is nonincreasing in y_i when all the other variables are kept fixed only. The others can be proved in a similar way.

Let $\underset{\sim}{\theta}^i = (\theta_1, \ldots, \theta_{i-1}, \theta_{i+1}, \ldots, \theta_k), \Omega^i = \{\underset{\sim}{\theta}^i | \theta_j > 0 \text{ for } j = 1, \ldots, k,\ j \neq i\}, \underset{\sim}{y} = (y_1, \ldots, y_k)$ and $\underset{\sim}{y}^* = (y_1^*, \ldots, y_k^*)$ where $y_j^* = y_j$ for $j \neq i$ and $y_i^* > y_i$. Then,

$$\Delta_i(\underset{\sim}{n}, \underset{\sim}{y}) = \int_{\Omega^i} \Big[\int_{\theta_i = 0}^\infty L(\theta_{[k]} - \theta_i) g_i(\theta_i | y_i, n_i) d\theta_i\Big] \prod_{\substack{j=1 \\ j \neq i}} g_j(\theta_j | y_j, n_j) d\underset{\sim}{\theta}^i.$$

Since for each fixed $\underset{\sim}{\theta}^i$, $L(\theta_{[k]} - \theta_i)$ is nonincreasing in θ_i and $g_i(\theta_i|y_i, n_i)$ has monotone likelihood ratio in y_i and θ_i (see Lemma 3.1.a), for $y_i^* > y_i$, we have

$$\int_{\theta_i=0}^{\infty} L(\theta_{[k]} - \theta_i) g_i(\theta_i|y_i, n_i) d\theta_i \geq \int_{\theta_i=0}^{\infty} L(\theta_{[k]} - \theta_i) g_i(\theta_i|y_i^*, n_i) d\theta_i,$$

and hence $\Delta_i(\underset{\sim}{n}, \underset{\sim}{y}) \geq \Delta_i(\underset{\sim}{n}, \underset{\sim}{y}^*)$.

Now, we see that Theorem 3.1 is a direct result of Lemma 3.2, (2.5) and (2.6).

4. AN EARLY SELECTION RULE

In this section, we consider the following linear loss function: $L^*(\underset{\sim}{\theta}, i) = \theta_{[k]} - \theta_i$, the difference between the parameters of the best and the selected populations. Thus, the set $A(\underset{\sim}{n}, \underset{\sim}{y})$ given in (2.5) turns out to be:

$$A(\underset{\sim}{n}, \underset{\sim}{y}) = \{i | E[\theta|y_i, n_i] = \max_{1 \leq j \leq k} E[\theta|y_j, n_j]\}, \tag{4.1}$$

where $E[\theta|y_i, n_i] = \int \theta g_i(\theta|y_i, n_i) d\theta$, the posterior mean of θ_i given $(Y_i, N_i) = (y_i, n_i)$. By Lemma 3.2, we have the following result.

Lemma 4.1. For each $i = 1, \ldots, k$, $E[\theta|y_i, n_i]$ is increasing in y_i and decreasing in n_i.

We will use this monotonicity property of $E[\theta|y_i, n_i]$ to derive a modified selection rule. This modified selection rule is designed to make a selection earlier than the termination time T of the life-testing experiment.

At time t, $0 < t < T$, let $N_i(t)$ denote the number of failures from population π_i up to time t. That is, $N_i(t) =$ number of $\{X_{ij} | 1 \leq j \leq m,\ X_{ij} < t\}$. Also, we let $Y_{i1} \leq \ldots \leq Y_{iN_i(t)}$ denote the $N_i(t)$ failure times up to the time t. At time t, exclude population π_i as a nonbest population if there exists some population π_h such that either

$$N_h(t) < m \text{ and } \int \theta g_h(\theta|y_h(t), m) d\theta \geq \int \theta g_i(\theta|y_i(t,T),\ N_i(t)) d\theta \tag{4.2.a}$$

or

$$N_h(t) = m \text{ and } \int \theta g_h(\theta|y_h(t), m) d\theta > \int \theta g_i(\theta|y_i(t,T), N_i(t)) d\theta \tag{4.2.b}$$

where

$$\begin{cases} y_h(t) = \sum\limits_{j=1}^{N_h(t)} y_{hj} + (m - N_h(t))t, \\ y_i(t,T) = \sum\limits_{j=1}^{N_i(t)} y_{ij} + (m - N_i(t))T. \end{cases} \tag{4.3}$$

We also let $S(t)$ denote the set of indices of the contending populations at time t. That is,

$$S(t) = \{i | N_h(t) < (=) m \text{ and } \int \theta g_i(\theta | y_i(t,T), N_i(t)) d\theta$$

$$> (\geq) \int \theta g_h(\theta | y_h(t), m) d\theta, h \neq i\}. \tag{4.4}$$

The life-testing experiment terminates as soon as there is a time t, $0 < t < T$, such that $|S(t)| = 1$ and in this situation, we select the population with the index in the set $S(t)$ as the best population. Otherwise, the experiment goes on until the time T. At the time T, let

$$S(T) = \{i | \int \theta g_i(\theta | y_i, N_i) d\theta = \max_{j \epsilon S(T^-)} \int \theta g_j(\theta | y_j, N_j) d\theta\}, \tag{4.5}$$

where $S(T^-)$ denotes the set of the indices of those populations having not been eliminated before the time T. Then, a uniformly randomized selection is made over the set $S(T)$.

From the above description, we see that this modified selection rule can make selection earlier than the termination time T. We denote this modified early selection rule by $\underset{\sim}{\delta}_G^*$ and let δ_{Gi}^* be the probability of selecting population π_i as the best population by applying the selection rule $\underset{\sim}{\delta}_G^*$. Note that the probability δ_{Gi}^*, $1 \leq i \leq k$, are functions of the data observed during the time interval $(0, T]$.

In the following, we will show

Theorem 4.1. Under the loss function $L^*(\underset{\sim}{\theta}, i) = \theta_{[k]} - \theta_i$, $\delta_{Gi}^* = \delta_{Gi}(\underset{\sim}{n}, \underset{\sim}{y})$ for all $1 \leq i \leq k$, $\underset{\sim}{y} \epsilon \mathcal{Y}_n$ and $\underset{\sim}{n} \epsilon \mathcal{N}$, where $\delta_{Gi}(\underset{\sim}{n}, \underset{\sim}{y})$ is defined by (4.1) and (2.6).

Note that $\delta_{Gi}(\underset{\sim}{n}, \underset{\sim}{y})$ is the probability of selecting population π_i as the best population based on the type-I censored data $(\underset{\sim}{n}, \underset{\sim}{y})$ obtained at the end of the time T.

Let $B = \{0 < t \leq T || S(t)| = 1\}$ and let

$$t_1 = \begin{cases} \inf B & \text{if } B \neq \phi, \\ T & \text{if } B = \phi, \end{cases} \tag{4.6}$$

where ϕ denotes an empty set. Note that if $B \neq \phi$, then $B = [t_1, T]$.

By a uniformly randomized selection over the set $S(T)$ when $t_1 = T$, Theorem 4.1 is equivalent to the following.

Theorem 4.2. $S(t_1) = A(\underset{\sim}{n}, \underset{\sim}{y})$ for all $(\underset{\sim}{n}, \underset{\sim}{y})$ where $A(\underset{\sim}{n}, \underset{\sim}{y})$ is defined in (4.1).

Proof: Case 1. As $t_1 < T$, then $|S(t_1)| = 1$. Without loss of generality, we let π_k be the population with index in the set $S(t_1)$. Since $A(\underset{\sim}{n}, \underset{\sim}{y})$ contains at least one element, it suffices to show that $i \notin A(\underset{\sim}{n}, \underset{\sim}{y})$ for all $i \neq k$. Since $i \notin S(t_1)$, it means that population π_i is eliminated at some time, say t_0, not later than t_1, by some population, say π_h. That is, at time t_0 either

$$N_h(t_0) < m \text{ and } \int \theta g_h(\theta | y_h(t_0), m) d\theta \geq \int \theta g_i(\theta | y_i(t_0, T), N_i(t_0)) d\theta \tag{4.7.a}$$

or

$$N_h(t_0) = m \text{ and } \int \theta g_h(\theta|y_h(t_0), m)d\theta > \int \theta g_i(\theta|y_i(t_0, T), N_i(t_0))d\theta. \tag{4.7.b}$$

Now, note that $N_i(t)$ is an nondecreasing function of $t\epsilon(0, T]$ and $N_i(t) \leq m$. Also, by (4.3), $y_h(t)$ is nondecreasing in t and $y_i(t, T)$ is nonincreasing in t. In fact, we have

$$N_h = N_h(T) \leq m, \quad N_i(t) \leq N_i(T) = N_i, \quad y_i(t_0, T) \geq y_i(T, T) \equiv y_i$$

$$y_h \equiv y_h(T) > (=) y_h(t_0) \text{ if } N_h(t_0) < (=) m.$$

Thus, when $N_h(t_0) = m$, then $N_h \equiv N_h(T) = m$. Then by Lemma 4.1, and (4.7.b),

$$\begin{aligned} \int \theta g_h(\theta|y_h, N_h)d\theta &= \int \theta g_h(\theta|y_h(t_0), m)d\theta \\ &> \int \theta g_i(\theta|y_i(t_0, T), N_i(t_0))d\theta \\ &\geq \int \theta g_i(\theta|y_i, N_i)d\theta. \end{aligned} \tag{4.8}$$

When, $N_h(t_0) < m$, then $y_h \equiv y_h(T) > y_h(t_0)$ and $N_h = N_h(T) \leq m$. Therefore, by Lemma 4.1 and (4.7.a),

$$\begin{aligned} \int \theta g_h(\theta|y_h, N_h)d\theta &> \int \theta g_h(\theta|y_h(t_0), m)d\theta \\ &\geq \int \theta g_i(\theta|y_i(t_0, T), N_i(t_0))d\theta \\ &\geq \int \theta g_i(\theta|y_i, N_i)d\theta. \end{aligned} \tag{4.9}$$

In either situations, we see that $i \notin A(\underset{\sim}{n}, \underset{\sim}{y})$.

Case 2. As $t_1 = T$, we need to prove that
(a) $i \notin S(T) \Rightarrow i \notin A(\underset{\sim}{n}, \underset{\sim}{y})$ and
(b) $i \epsilon S(T) \Rightarrow i \epsilon A(\underset{\sim}{n}, \underset{\sim}{y})$.

We prove part (a) first. Suppose $i \notin S(T)$. Then, π_i is eliminated at a time $t_0 \leq T$ by some other π_h.

If $t_0 < T$, this reduces to the situation discussed in Case 1.

If $t_0 = T$, then by (4.5), $\int \theta g_h(\theta|y_h, N_h)d\theta > \int \theta g_i(\theta|y_i, N_i)d\theta$. Therefore, by the definition of $A(\underset{\sim}{n}, \underset{\sim}{y})$, $i \notin A(\underset{\sim}{n}, \underset{\sim}{y})$.

Note that the statement in part (a) is equivalent to that

$$A(\underset{\sim}{n}, \underset{\sim}{y}) \subset S(T). \tag{4.10}$$

Now, part (b) is a direct consequence of (4.5) and (4.10). Therefore, we complete the proof of this theorem.

5. AN ILLUSTRATIVE EXAMPLE

We use the insulating fluid example (taken from Table 4.1, page 462 of Nelson (1982)) to illustrate the way to implement the selection rules $\underset{\sim}{\delta}_G$ and $\underset{\sim}{\delta}^*_G$. There are six groups of insulating fluid. The purpose is to identify which group of insulating fluid has the largest life-time when subjected to high voltage stress. Ten items from each group are put on a life-test experiment which is subjected to high voltage stress. The record of the times to breakdown in minutes is shown in Table 1. The result of Nelson (1982) indicates that the data in each group follows an exponential distribution.

Table 1
Times to Insulating Fluid Breakdown

Group	1	2	3	4	5	6
	1.89	1.30	1.99	1.17	8.11	2.12
	4.03	2.75	0.64	3.87	3.17	3.97
	1.54	0.00	2.15	2.80	5.55	1.56
	0.31	2.17	1.08	0.70	0.80	1.34
	0.66	0.66	2.57	3.82	0.20	1.49
	1.70	0.55	0.93	0.02	1.13	8.71
	2.17	0.18	4.75	0.50	6.63	2.10
	1.82	10.60*	0.82	3.72	1.08	7.21
	9.99*	1.63	2.06	0.06	2.44	3.83
	2.24	0.71	0.49	3.57	0.78	5.13

Suppose that time censoring scheme is adopted before the life-testing and the censoring time T is set to be 9 minutes. Therefore, in Table 1, the two failure times 9.99 and 10.60 marked with '*' should be censored data according to this censoring scheme. Then, we have

$$\begin{array}{llllll} y_1 = 25.36, & y_2 = 18.95, & y_3 = 17.48, & y_4 = 20.23, & y_5 = 29.89, & y_6 = 37.46, \\ n_1 = 9, & n_2 = 9, & n_3 = 10, & n_4 = 10, & n_5 = 10, & n_6 = 10. \end{array}$$

We also assume that the six scale random parameters $\theta_1, \ldots, \theta_6$ are iid with a common prior density function $g_i(\theta) = \theta^{-3}e^{-\frac{1}{\theta}}$. Therefore, $E(\theta_i|y_i, n_i) = \int_0^\infty \theta g_i(\theta|y_i, n_i)\, d\theta = (y_i + 1)/(n_i + 1)$, $1 \leq i \leq k$. Hence, $E(\theta_1|y_1, n_1) = 2.636$, $E(\theta_2|y_2, n_2) =$

1.995, $E(\theta_3|y_3,n_3) = 1.68$, $E(\theta_4|y_4,n_4) = 1.93$, $E(\theta_5|y_5,n_5) = 2.808, E(\theta_0|y_6,n_6) = 3.496$. According to the selection rule $\underset{\sim}{\delta}_G$, we select Group 6 as the best group.

However, if the modified selection rule $\underset{\sim}{\delta}^*_G$ is applied, for the same data set in Table 1, the selection can be made before the termination time T. According to the selection rule $\underset{\sim}{\delta}^*_G$, with the same prior distribution given above, at time t, $0 < t \leq T = 9$, remove π_i from further consideration if there exists some h such that either

a) $N_h(t) < 10$ and

$$(\sum_{j=1}^{N_h(t)} y_{hj} + (10 - N_i(t))t + 1)/11 \geq (\sum_{j=1}^{N_i(t)} y_{ij} + (10 - N_i(t))T + 1)(N_i(t) + 1)^{-1}, \text{ or}$$

b) $N_h(t) = 10$ and

$$(\sum_{j=1}^{10} y_{hj} + 1)/11 > (\sum_{j=1}^{N_i(t)} y_{ij} + (m - N_i(t))T + 1)(N_i(t) + 1)^{-1}.$$

Table 2 indicates the times (in minutes) at each of which some group is removed from the set of contending groups; and the life-testing experiment can be ended at time $t_1 = 6.63$. We then select Group 6 as the best group.

Table 2
Times to Reduce the Size of the Set of Contending Groups

Group	1	2	3	4	5	6
time	4.03	2.75	3.079	3.87	6.63	

Note that the modified experiment and the procedure lead to early selection and a saving of time $T - t_1 = 2.37$(minutes). Also, in Table 2, Group 3 is excluded as a non-best group at time $t = 3.079$ which is not a failure time for any item in Group 3. While for Group 1, 2, 4 and 5, the time at which the associated group is excluded as a non-best group is also a failure time of some item in that group.

6. ACKNOWLEDGEMENT

This research was partially supported by the Office of Naval Research Contract N00014-89-K-0170 and NSF Grants DMS–8606904, DMS–8923071 at Purdue University.

7. REFERENCES

Bartholomew, D. J. (1963). The sampling distribution of an estimate arising in life testing. *Technometrics* **5**, 361–374.

Berger, R. L. and Kim, J. S. (1985). Ranking and subset selection procedures for exponential populations with type-I and type-II censored data. *The Frontiers of Modern Statistical Inference Procedures* (Ed. E. J. Dudewicz), 425–455, American Sciences Press.

Gupta, S. S. (1963). On a selection and ranking procedure for gamma populations. *Ann. Instit. Statist. Math.* **14**, 199–216.

Huang, W. T. and Huang, K. C. (1980). Subset selections of exponential populations based on censored data. *Proceedings of Conference on Recent Development in Statistical Methods and Applications*, 237–254. Academia Sinica, Taipei, Taiwan.

Mann, N. R., Schafer, R. E. and Han, M. C. (1982). Confidence bounds for the exponential mean in time-truncated life tests. *Survival Analysis* (Eds. J. Crowley and R. A. Johnson), 152–165, Lecture Notes-Monograph Series, Vol. 2, Institute of Mathematical Statistics.

Nelson, W. (1982). *Applied Life Data Analysis*, Wiley, New York.

Sobel, M. (1956). Sequential procedures for selecting the best exponential population. *Proceedings of the Third Berkeley Symposium on Math. Statist. Probab. Vol. V*, 99–110, University of California Press, Berkeley.

Spurrier, J. D. and Wei, L. J. (1980). A test of the parameter of the exponential distribution in the type I censoring case. *J. Amer. Statist. Assoc.* **75**, 405–409.

Yang, G. and Sirvanci, M. (1977). Estimation of a time-truncated exponential parameter used in life testing. *J. Amer. Statist. Assoc.* **72**, 444–447.

ADVANCES IN RELIABILITY
Edited by Asit P. Basu

A BAYESIAN DECISION THEORY APPROACH TO SCREENING PROBLEMS

Richard A. Johnson[a] and A. Mouhab[b]

[a]Department of Statistics, University of Wisconsin, 1210 W. Dayton Street, Madison, Wisconsin 53706

[b]Department of Statistics, University of Wisconsin, 1210 W. Dayton Street, Madison, Wisconsin 53706

Abstract

We address the screening problem where an item is declared to be good if certain of its characteristics $\boldsymbol{y}$ lie within a specified ranges. Often these criterion variables cannot be measured without damaging the object. Consequently the screening procedure must be based on covariates. The characteristics of an experimental unit are modeled as a random vector $\boldsymbol{Z}$ which is partitioned into two sets $\boldsymbol{X}$ and $\boldsymbol{Y}$. The second set, $\boldsymbol{Y}$, is a set of criterion variables pertaining to the quality of the unit, and the first set of variables, $\boldsymbol{X}$, are covariates.

We investigate the screening problem using a Bayesian decision theory approach. The parameter values for the underlying distribution of $\boldsymbol{X}$ and $\boldsymbol{Y}$ are selected according to a prior distribution. Costs of misclassification are introduced and the Bayes rule is obtained. Our derivation of the Bayes rule justifies an existing procedure that was introduced in the framework of predictive distributions.

Simulations are given which compare, in the bivariate normal case, the predictive distribution based Bayes rule with the result of using the known parameter solution but with parameters replaced by their maximum likelihood estimates. The effects of varying correlation and of varying costs are considered. An example points out an important feature missed by Boys and Dunsmore (1986) in their proposed solution.

Returning to the general case, we give a careful derivation of the expansion of the predictive density in order to establish that the coefficients of powers of n^{-1} remain uniformly bounded in n when they are integrated over the entire sample space. Using this result and the first order expansion of the predictive density, we obtain an asymptotic bound for the Bayes risk.

1. INTRODUCTION

We consider the screening problem where an item is declared to be *acceptable* if certain of its characteristics $\boldsymbol{y}$ lie within specified ranges. When these criterion variables cannot be measured without damaging the object, the screening procedure should be based on covariates. We model the characteristics of an experimental unit as a random vector $\boldsymbol{Z}$ which is partitioned into two sets $\boldsymbol{X}$ and $\boldsymbol{Y}$. The second set, $\boldsymbol{Y}$, is a set of criterion variables pertaining to the quality of the unit, and the first set of variables, $\boldsymbol{X}$, are covariates. When $\boldsymbol{Y}$ is a single random variable, representing strength or life length, the criteria may be $\boldsymbol{Y} \geq b$. When $\boldsymbol{Y}$ has m components, the criteria may be based on the closeness to the target $\mathbf{0}$; for instance, on $\| \boldsymbol{Y} \|^2 = \sum_{i=1}^{m} Y_i^2 \leq b$.

In most of the literature, the screening problem has been solved without regard to a loss function. (Owen and Su (1977), Owen, Li and Chou (1981) and Wong, Meeker and Selwyn (1985)). Their goal is to determine a limit, x_*, on a single screening variable X so that the proportion of Y's above a specified limit L is raised to at least δ. All of the solutions presented for this formulation assume a joint normal distribution for X and Y.

We investigate the screening problem using a Bayesian decision approach which incorporates the (1) misclassification costs (2) conditional densities $f(\boldsymbol{x}, \boldsymbol{y}|\boldsymbol{\theta})$, given the parameters, and (3) a prior distribution for the parameters $\boldsymbol{\theta}$. In order to develop a screening procedure, a training set of data must be obtained which includes the values of both $\boldsymbol{X}$ and $\boldsymbol{Y}$.

For the case where the distribution of $(\boldsymbol{X}, \boldsymbol{Y})$ is bivariate normal and the parameters are known, Menzefricke (1984) used a decision theory approach for the problem where it is required that l future successes be obtained. When the parameters are unknown, he proposed, without any derivation, the use of the predictive model. Similarly, Boys and Dunsmore (1986), suggest a predictive distribution approach and even introduce costs. Given a data set(training sample), a specified criterion region for $\boldsymbol{Y}$, and the misclassification costs $c_{2|1}$ and $c_{1|2}$, they suggest, without derivation, solving for the optimal predictive region by minimizing the predictive loss. According to their development, such a rule would then be optimal among all rules based on the predictive distribution. Our decision theory approach established that the optimal predictive rule is indeed a Bayes rule and thus is optimal in a wider sense.

In Section 4, we give an example concerning the bivariate normal case.

Section 5 further investigates the form of the Bayes rules in the bivariate normal case using simulation. For the cases considered, an increase in $corr(\boldsymbol{X}, \boldsymbol{Y})$ increases the effectiveness of the Bayes rule with respect to the Bayes rule based on a flat prior and the rule based on known parameters but with maximum likelihood estimators substituted for the parameters. Another simulation looks at the effect of changing the ratio of misclassification costs.

Because of its similarity in structure to the Bayes factor, the denominator of the posterior distribution, it is well-known that the predictive distribution can be formally expanded in powers of n^{-1} as the training sample size n increases (cf. Dickey (1980)). In Section 6, we give a careful derivation of the expansion in order to establish that the coefficients of powers of n^{-1} remain uniformly bounded in n when they are integrated over the entire sample space. Using this result and the first order expansion of the predictive density, we obtain an asymptotic bound for the Bayes risk.

2. THE MAIN RESULT

A population has characteristics $\boldsymbol{Z}$ which consist of criterion variables $\boldsymbol{Y}$ and covariates $\boldsymbol{X}$. Let the population be described by the probability density function (pdf) $f(.|\boldsymbol{\theta})$, with respect to some σ-finite measure μ, where the parameter $\boldsymbol{\theta}$ may be vector valued. From a single population of items, we define two subpopulations π_1 and π_2 based upon whether or not the future criterion variables $\boldsymbol{y}_*$ belong to a specified Borel set B.

π_1 : $\boldsymbol{Z}_*$'s with $\boldsymbol{Y}_* \epsilon B$ (Acceptable items)

π_2 : $\boldsymbol{Z}_*$'s with $\boldsymbol{Y}_* \epsilon B^c$.

Screening becomes a problem when it is difficult or impossible to observe $\boldsymbol{Y}$ without damaging the item. We then want to use the covariates $\boldsymbol{X}$ to develop a screening procedure.

In order to develop an optimal screening procedure, we begin by collecting a random sample

$$\boldsymbol{Z}_i = \begin{bmatrix} \boldsymbol{X}_i \\ \boldsymbol{Y}_i \end{bmatrix} \quad \text{for } i = 1, 2, \ldots, n$$

that includes values of both the criterion variables $\boldsymbol{Y}_i$ and the covariates $\boldsymbol{X}_i$, for n items.

Let the misclassification costs be

$c_{1|2}$ = cost of misclassifying an item from π_2 as π_1

$c_{2|1}$ = cost of misclassifying an item from π_1 as π_2.

We assume that there is no loss for correct classification.

Let $\boldsymbol{Z}_*^t = (\boldsymbol{X}_*^t, \boldsymbol{Y}_*^t)$ be a future observation. We allow the decision rule to depend on both $\boldsymbol{X}_*$ and the training set $\boldsymbol{Z}_1, \ldots, \boldsymbol{Z}_n$.

$$\delta(\boldsymbol{X}_*) = \delta(\boldsymbol{X}_*; \boldsymbol{Z}_1, \cdots, \boldsymbol{Z}_n) = \begin{cases} a_1 & \text{if } \boldsymbol{X}_* \epsilon R_1 \\ a_2 & \text{if } \boldsymbol{X}_* \epsilon R_1^c \end{cases} \tag{2.1}$$

The region leading to an assignment to π_1 is also a function of the training set $R_1 = R_1(\boldsymbol{Z}_1, \ldots, \boldsymbol{Z}_n)$. In this notation, the loss function is

$$L(I_B(\boldsymbol{Y}_*), \delta(\boldsymbol{X}_*)) = L(1, \delta(\boldsymbol{X}_*)) I_B(\boldsymbol{Y}_*) + L(2, \delta(\boldsymbol{X}_*)) I_{B^c}(\boldsymbol{Y}_*) \tag{2.2}$$

where

$$L(1, \delta(\boldsymbol{X}_*)) = \begin{cases} c_{2|1} & \text{if } \delta(\boldsymbol{X}_*) = a_2 \\ 0 & \text{if } \delta(\boldsymbol{X}_*) = a_1 \end{cases}, \quad L(2, \delta(\boldsymbol{X}_*)) = \begin{cases} c_{1|2} & \text{if } \delta(\boldsymbol{X}_*) = a_1 \\ 0 & \text{if } \delta(\boldsymbol{X}_*) = a_2 \end{cases}$$

Here, $I_B(.)$ and $I_{B^c}(.)$ are indicator functions.

Because $\boldsymbol{Y}_*$ is not observed for the unit in question, we cannot obtain the explicit value for $L(I_B(\boldsymbol{Y}_*), \delta(\boldsymbol{x}_*))$. However, we can obtain a useful expression for the Bayes risk. Since Θ is random,

$$E(L) = E^{\Theta} E^{\boldsymbol{X}_*, \boldsymbol{Z}_1, \ldots, \boldsymbol{Z}_n | \Theta} L(I_B(\boldsymbol{Y}*), \delta(\boldsymbol{X}*)). \tag{2.3}$$

Under this structure, we obtain the Bayes screening rule.

Theorem 1.

Let $\boldsymbol{Z}_i$ $i = 1, \ldots, n$ and $\boldsymbol{Z}_*$ be independently and identically distributed each having pdf $f(.|\boldsymbol{\theta})$ with respect to μ. Let the future random vector $\boldsymbol{Z}_*$ be composed of $\boldsymbol{X}_*$ which is observable, and $\boldsymbol{Y}_*$ which is not. Let Θ have prior distribution $G(\boldsymbol{\theta})$ with pdf $g(\boldsymbol{\theta})$. Two subpopulations of items are defined as

$$\pi_1 : \boldsymbol{Z}_*\text{'s with } \boldsymbol{Y}_* \epsilon B$$

$$\pi_2 : \boldsymbol{Z}_*\text{'s with } \boldsymbol{Y}_* \epsilon B^c.$$

Let the costs of misclassifying an observation from π_i as from π_j be $c_{j|i}$. Then the screening region

$$R_1 = R_1(\boldsymbol{Z}_1, \ldots, \boldsymbol{Z}_n)$$

$$= \left\{ \boldsymbol{x}_* : P[\boldsymbol{Y}_* \epsilon B | \boldsymbol{x}_*, \boldsymbol{z}_1, \cdots, \boldsymbol{z}_n] \geq \frac{c_{1|2}}{c_{1|2} + c_{2|1}} \right\} \tag{2.4}$$

minimizes the Bayes risk (2.3).

If the predictive probability of equality in the equation (2.4) is zero, then the procedure is essentially unique.

Proof: From (2.3),

$$E(L) = c_{2|1} P[\boldsymbol{Y}_* \epsilon B] +$$

$$E^{\Theta} E^{\boldsymbol{Z}_1, \ldots, \boldsymbol{Z}_n | \Theta} \left[\ c_{1|2} P[\boldsymbol{X}_* \epsilon R_1, \boldsymbol{Y}_* \epsilon B^c | \Theta] - c_{2|1} P[\boldsymbol{X}_* \epsilon R_1, \boldsymbol{Y}_* \epsilon B | \Theta] \ \right]$$

$$= c_{2|1} P[\boldsymbol{Y}_* \epsilon B] + \int_{\Theta} \int_{\boldsymbol{z}_1} \cdots \int_{\boldsymbol{z}_n} \left[\ c_{1|2} \int_{R_1 \times B^C} f(\boldsymbol{z}_* | \boldsymbol{\theta}) d\mu(\boldsymbol{z}_*) - c_{2|1} \int_{R_1 \times B} f(\boldsymbol{z}_* | \boldsymbol{\theta}) d\mu(\boldsymbol{z}_*) \ \right]$$

$$(\prod_{i=1}^{n} f(\boldsymbol{z}_i | \boldsymbol{\theta})) g(\boldsymbol{\theta}) d\mu(z) \ldots d\mu(\boldsymbol{z}_n) d\boldsymbol{\theta}$$

Multiplying and dividing $(\prod_{i=1}^{n} f(\boldsymbol{z}_i|\boldsymbol{\theta}))g(\boldsymbol{\theta})$ by the marginal density $f(\boldsymbol{z}_1, \ldots, \boldsymbol{z}_n)$ $= \int_{\boldsymbol{\theta}} \prod_{i=1}^{n} f(\boldsymbol{z}_i|\boldsymbol{\theta}) g(\boldsymbol{\theta}) d\boldsymbol{\theta}$ and interchanging the order of integration we get an expression for $E(L)$ involving the posterior distribution, $g(\boldsymbol{\theta}|\boldsymbol{z}_1, \ldots, \boldsymbol{z}_n)$.

$$E(L) = c_{2|1} P[\boldsymbol{Y}_* \epsilon B]$$

$$+ \int_{\boldsymbol{z}_1} \cdots \int_{\boldsymbol{z}_n} \left[\ c_{1|2} \int_{R_1 \times B^c} \int_{\boldsymbol{\theta}} f(\boldsymbol{z}_* | \boldsymbol{\theta}) g(\boldsymbol{\theta} | \boldsymbol{z}_1, \ldots, \boldsymbol{z}_n) d\boldsymbol{\theta} d\mu(\boldsymbol{z}_*) \right.$$

$$-c_{2|1}\int_{R_1\times B}\int_{\theta} f(\boldsymbol{z}_*|\boldsymbol{\theta})g(\boldsymbol{\theta}|\boldsymbol{z}_1,\ldots,\boldsymbol{z}_n)d\boldsymbol{\theta}d\mu(\boldsymbol{z}_*)\ \Big]\ f(\boldsymbol{z}_1,\ldots,\boldsymbol{z}_n)\prod_{i=1}^{n}d\mu(\boldsymbol{z}_i)$$

$$= c_{2|1}P[\boldsymbol{Y}_*\epsilon B] + \int_{\boldsymbol{z}_1}\cdots\int_{\boldsymbol{z}_n}\Big[\ c_{1|2}\int_{R_1\times B^c} h(\boldsymbol{z}_*|\boldsymbol{z}_1,\ldots\boldsymbol{z}_n)d\ \mu(\boldsymbol{z}_*) \tag{2.5}$$

$$-c_{2|1}\int_{R_1\times B} h(\boldsymbol{z}_*|\boldsymbol{z}_1,\ldots,\boldsymbol{z}_n)d\mu(\boldsymbol{z}_*)\ \Big]\ f(\boldsymbol{z}_1,\ldots,\boldsymbol{z}_n)\prod_{i=1}^{n}d\mu(\boldsymbol{z}_i)$$

where

$$h(\boldsymbol{z}_*|\boldsymbol{z}_1,\ldots,\boldsymbol{z}_n) = \int_{\boldsymbol{\theta}} f(\boldsymbol{z}_*|\boldsymbol{\theta})g(\boldsymbol{\theta}|\boldsymbol{z}_1,\ldots,\boldsymbol{z}_n)d\boldsymbol{\theta}$$

is the predictive density of $\boldsymbol{Z}_*$ given the training set.

Next since

$$h(\boldsymbol{z}_*|\boldsymbol{z}_1,\ldots,\boldsymbol{z}_n) = h(\boldsymbol{x}_*|\boldsymbol{z}_1,\ldots,\boldsymbol{z}_n)h(\boldsymbol{y}_*|\boldsymbol{x}_*,\boldsymbol{z}_1,\ldots,\boldsymbol{z}_n)$$

the bracketed term in (2.5) can be expressed as

$$\int_{R_1} h(\boldsymbol{x}_*|\boldsymbol{z}_1,\ldots,\boldsymbol{z}_n)\Big[\ c_{1|2}\int_{B^c} h(\boldsymbol{y}_*|\boldsymbol{x}_*,\boldsymbol{z}_1,\ldots,\boldsymbol{z}_n)d\mu_2(\boldsymbol{y}_*)- \tag{2.6}$$

$$c_{2|1}\int_{B} h(\boldsymbol{y}_*|\boldsymbol{x}_*,\boldsymbol{z}_1,\ldots,\boldsymbol{z}_n)d\mu_2(\boldsymbol{y}_*)\ \Big]\ d\mu_1(\boldsymbol{x}_*).$$

Combining (2.5), and (2.6), we see that $E(L)$ is minimized by selecting R_1 such that

$$c_{1|2}\int_{B^c} h(\boldsymbol{y}_*|\boldsymbol{x}_*,\boldsymbol{z}_1,\ldots,\boldsymbol{z}_n)d\mu_2(\boldsymbol{y}_*) \le c_{2|1}\int_{B} h(\boldsymbol{y}_*|\boldsymbol{x}_*,\boldsymbol{z}_1,\ldots,\boldsymbol{z}_n)d\mu(\boldsymbol{y}_*) \tag{2.7}$$

or, equivalently

$$c_{1|2}P[\boldsymbol{Y}_*\epsilon B^c|\boldsymbol{x}_*,\boldsymbol{z}_1,\ldots,\boldsymbol{z}_n] \le c_{2|1}P[\boldsymbol{Y}_*\epsilon B|\boldsymbol{x}_*,\boldsymbol{z}_1,\ldots,\boldsymbol{z}_n]$$

Thus, we must choose R_1 so that the predictive probability that $\boldsymbol{Y}*$ belongs to B, given $(\boldsymbol{x}_*,\boldsymbol{z}_1,\ldots,\boldsymbol{z}_n)$, is large.

3. SCREENING ITEMS FROM MORE THAN ONE POPULATION

The underlying population which produces the observations $\boldsymbol{Z} = \begin{bmatrix} \boldsymbol{X} \\ \boldsymbol{Y} \end{bmatrix}$ may itself be a mixture of m populations $f_j(.|\boldsymbol{\theta}_j)$ with mixing proportions $q_j, j = 1, \ldots, m$. For instance, a warehouse could hold items manufactured at m different factories.

We assume that a training set $\boldsymbol{z}_1^j, \cdots, \boldsymbol{z}_{n_j}^j$ is available from the j-th subpopulation, for $j = 1, \ldots, m$. Further, for fixed j, the observations $\boldsymbol{z}_i^j$'s are independently distributed as $f_j(.|\boldsymbol{\theta}_j)$, given $\boldsymbol{\theta}_j$. Also, there is a different and independent prior distribution for each component population and the training sets are independent of each other, given $\boldsymbol{\theta}_1, \cdots, \boldsymbol{\theta}_m$.

According to this mixture model, a future observation

$\boldsymbol{v}_* = \boldsymbol{z}_*^j$ with probability q_j for $j = 1, \ldots, m$

where $\boldsymbol{z}_*^j$ is a future observation from the j-th population. Thus,

$$f_{\boldsymbol{v}_*}(\boldsymbol{v}_*|\boldsymbol{\theta}_1, \cdots, \boldsymbol{\theta}_m) = \sum_{j=1}^{m} q_j f_j(\boldsymbol{v}_*|\theta_j) \tag{3.1}$$

and an observation, $\boldsymbol{v}_*^t = (\boldsymbol{v}_{*_1}^t, \boldsymbol{v}_{*_2}^t)$, is called acceptable if $\boldsymbol{v}_{*_2} \epsilon B$.

We can now treat the problem as one of screening items from a single population and apply Theorem 1. Here the prior density is $\prod_{j=1}^{m} g_j(\boldsymbol{\theta}_j)$. Consequently, the optimal screening region R_1 takes the form:

$$R_1 = \left\{ \boldsymbol{v}_{*_1} : \sum_{j=1}^{m} \tilde{q}_j\,(\boldsymbol{v}_{*_1}) P_j(\boldsymbol{v}_{*_1}) \geq \frac{c_{1|2}}{c_{1|2} + c_{2|1}} \right\}. \tag{3.2}$$

where the updated proportions of the components are

$$\tilde{q}_j\,(\boldsymbol{v}_{*_1}) = \frac{q_j h_j(\boldsymbol{v}_{*_1}|\boldsymbol{z}_1^j, \ldots, \boldsymbol{z}_{n_j}^j)}{\sum_{k=1}^{m} q_k h_k(\boldsymbol{v}_{*_1}|\boldsymbol{z}_1^k, \ldots, \boldsymbol{z}_{n_k}^k)}$$

and

$$P_j(\boldsymbol{v}_{*_1}) = \int_B h_j(\boldsymbol{v}_{*_2}|\boldsymbol{v}_{*_1}, \boldsymbol{z}_1^j, \ldots, \boldsymbol{z}_{n_j}^j) d\mu_2(\boldsymbol{v}_{*_2}).$$

Note that, when screening items from a mixture of m populations, the Bayes screening rule is such that the weighted average of the predictive probabilities that $\boldsymbol{Y}_*^j$ belongs to B, given $\boldsymbol{X}_*^j = \boldsymbol{v}_{*_1}, \boldsymbol{z}_i^j, \ldots, \boldsymbol{z}_{n_j}^j$, is large. The weights are the updated proportions $\tilde{q}_j\,(\boldsymbol{v}_{*_1})$.

We state this result formally and include the details of proof.

Theorem 2.
Let $f_{\boldsymbol{V}_*}(\boldsymbol{v}_*|\boldsymbol{\theta}_1,\ldots,\boldsymbol{\theta}_m)$ be the mixture (3.1) of m populations with proportions q_j for $j=1,\ldots,m$, and let $\boldsymbol{Z}_i^j$, $i=1,\ldots,n_j$, be a training set and $\boldsymbol{Z}_*^j$ a future observation from the j-th population, distributed as $f_j(\boldsymbol{z}_*^j|\boldsymbol{\theta}_j)$ for $j=1,\ldots,m$. Let Θ_j have prior distribution $g_j(\boldsymbol{\theta}_j)$. The costs of misclassification and π_i, $i=1,2$ are defined as in the Theorem 1 above. Then, the screening region R_1 defined by equation (3.1) minimizes the Bayes Risk.

Proof: Applying Theorem 1 to the mixture model (3.1), we conclude that R_1 should consist of $\boldsymbol{v}_{*_1}$ such that

$$P[\boldsymbol{V}_{*_2}\epsilon B|\boldsymbol{v}_{*_1},\boldsymbol{z}_1^1,\ldots,\boldsymbol{z}_{n_1}^1,\ldots,\boldsymbol{z}_1^m,\ldots,\boldsymbol{z}_{n_m}^m]\geq\frac{c_{1|2}}{c_{1|2}+c_{2|1}}.$$

The predictive distribution $\boldsymbol{V}_*$, given the training sets, is

$$h(\boldsymbol{v}_*|\boldsymbol{z}_1^1,\ldots,\boldsymbol{z}_{n_1}^1,\ldots,\boldsymbol{z}_1^m,\ldots,\boldsymbol{z}_{n_m}^m)=\sum_{j=1}^{m}q_j\int_{\theta_j}f_j(\boldsymbol{v}_*|\boldsymbol{\theta}_j)g(\boldsymbol{\theta}_j|\boldsymbol{z}_1^j\cdots\boldsymbol{z}_{n_j}^j)d\boldsymbol{\theta}_j$$

$$=\sum_{j=1}^{m}q_jh_j(\boldsymbol{v}_*|z_1^j,\ldots,z_{n_j}^j)$$

and the conditional predictive distribution is

$$h(\boldsymbol{v}_{*_2}|\boldsymbol{v}_{*_1},z_1^1,\ldots,z_{n_1}^1,\ldots,z_1^m,\ldots,z_{n_m}^m)=\frac{h(\boldsymbol{v}_*|\boldsymbol{z}_1^1,\ldots,\boldsymbol{z}_{n_1}^1,\ldots,\boldsymbol{z}_1^m,\ldots,\boldsymbol{z}_{n_m}^m)}{h(\boldsymbol{v}_{*_1}|\boldsymbol{z}_1^1,\ldots,\boldsymbol{z}_{n_1}^1,\ldots,\boldsymbol{z}_1^m,\ldots,\boldsymbol{z}_{n_m}^m)}.$$

$$=\frac{\sum_{j=1}^{m}q_jh_j(\boldsymbol{v}_*|z_1^j,\ldots,z_{n_j}^j)}{\sum_{k=1}^{m}q_kh_k(\boldsymbol{v}_{*_1}|\boldsymbol{z}_1^k,\ldots,\boldsymbol{z}_{n_k}^k)}=\sum_{j=1}^{m}\frac{q_jh_j(\boldsymbol{v}_*|\boldsymbol{z}_1^j\ldots,\boldsymbol{z}_{n_j}^j)}{\sum_{k=1}^{m}q_kh_k(\boldsymbol{v}_{*_1}|\boldsymbol{z}_1^k,\ldots,\boldsymbol{z}_{n_k}^k)}$$

Multiplying and dividing the j-th term in the summation by

$$h_j(\boldsymbol{v}_{*_1}|\boldsymbol{z}_1^j,\ldots,\boldsymbol{z}_{n_j}^j)$$

we get

$$h(\boldsymbol{v}_{*_2}|\boldsymbol{v}_{*_1},\boldsymbol{z}_1^1,\ldots,\boldsymbol{z}_{n_1}^1,\ldots,\boldsymbol{z}_1^m,\ldots,\boldsymbol{z}_{n_m}^m)=\sum_{j=1}^{m}\tilde{q}_j\ (\boldsymbol{v}_{*_1})h_j(\boldsymbol{v}_{*_2}|\boldsymbol{v}_{*_1},\boldsymbol{z}_1^j,\ldots,\boldsymbol{z}_{n_j}^j).$$

Integrating over B, we obtain

$$P[\boldsymbol{V}_{*_2}\epsilon B|\boldsymbol{v}_{*_1}, z_1^1, \ldots, z_{n_1}^1, \ldots, z_1^m, \ldots, z_{n_m}^m] = \sum_{j=1}^{m} \tilde{q}\,(\boldsymbol{v}_{*_1})P_j(\boldsymbol{v}_{*_1})$$

where $P_j(\boldsymbol{v}_{*_1}) = \int_B h_j(\boldsymbol{v}_{*_2}|\boldsymbol{v}_{*_1}, z_1^j, \ldots, z_{n_j}^j)d\mu_2(\boldsymbol{v}_{*_2})$

so R_1 has the form asserted in (3.2).

4. EXAMPLE - SAMPLING FROM A BIVARIATE NORMAL

Consider the situation where the $\boldsymbol{Z}_i$'s have a bivariate normal population. Let $\boldsymbol{z}_*|\boldsymbol{\mu}, \boldsymbol{\Sigma}^{-1}$ be a future observation that is conditionally independent of the observations of the training set. The predictive density of $\boldsymbol{Z}_*$, given the training set, is the $p+m$-variate $\boldsymbol{t}$ with $v = n_0 + n + 1 - (p+m)$ degrees of freedom and matrix $\left(\frac{n+2}{n+1} \right) \frac{\boldsymbol{Q}}{v}$ (See Guttman (1970).

$$h(\boldsymbol{z}_*|\boldsymbol{z}_1, \ldots, \boldsymbol{z}_n) = k|\boldsymbol{Q}|^{\frac{-1}{2}} \left[1 + (\boldsymbol{z}_* - \boldsymbol{\mu}^*)^t \left(\frac{n+1}{n+2} \right) \boldsymbol{Q}^{-1}(\boldsymbol{z}_* - \boldsymbol{\mu}^*) \right]^{\frac{-(n_0+n+1)}{2}}, \tag{4.1}$$

where

$$\boldsymbol{Q} = \boldsymbol{V}_0 + \boldsymbol{V} + \frac{n}{n+1}(\overline{\boldsymbol{z}} - \boldsymbol{\mu}_0)(\overline{\boldsymbol{z}} - \boldsymbol{\mu}_0)^t \text{ and } \boldsymbol{\mu}^* = \frac{n}{n+1}\overline{\boldsymbol{z}} + \frac{1}{n+1}\boldsymbol{\mu}_0. \tag{4.2}$$

Thus, if we take $B = [b, \infty)$, as for instance when screening for strength, the optimal screening region R_1 is given by

$$R_1 = \left\{ x_*; P\left[Y_* \geq b|x_*, \boldsymbol{z}_1, \ldots, \boldsymbol{z}_n \right] \geq \delta_0 = \frac{c_{1|2}}{c_{1|2} + c_{2|1}} \right\} = \left\{ x_* : l(x_*) \leq 0 \right\}$$

where

$$l(x_*) = \gamma^{\frac{1}{2}}(b - \mu_2^* - \frac{q_{12}}{q_{11}}(x_* - \mu_1^*)) - t_{n_0+n,\delta_0}\left[1 + \frac{(n+1)(x_* - \mu_1^*)^2}{(n+2)q_{11}} \right]^{\frac{1}{2}} \tag{4.3}$$

t_{n_0+n,δ_0} is the $100(1-\delta_0)$-th percentile of the t-distribution with n_0+n degrees of freedom, $Q = (q_{ij})_{1\leq i,j\leq 2}$ is given by (4.2) and $\gamma = \dfrac{(n_0+n)(n+1)q_{11}}{(n+2)|\boldsymbol{Q}|}$.

Boys and Dunsmore (1986), using a strictly predictive approach and taking $r = \dfrac{q_{12}}{\sqrt{q_{11}q_{22}}} = .90$ and $n_0 + n = 20$, give an explicit expression for R_1 when $c_{2|1}/c_{1|2}$ ranges from 0.01 to 100.00. However, if for instance, $r = .60$ and $n_0 + n = 10$, the situation changes for the extreme cost ratios. In the first case R_1 is empty and in the second it is the whole real line.

To see this, we note that the first derivative of $l(x_*)$, vanishes only if

$$\left[\gamma\frac{q_{12}^2}{q_{11}}\left(\frac{n+1}{n+2}\right)-\left(\frac{n+1}{n+2}\right)^2 t^2_{n_0+n,\delta_0}\right](x_*-\mu_1^*)^2+\gamma q_{12}^2=0.$$

Consequently, if $\gamma\frac{q_{12}^2}{q_{11}}-\left(\frac{n+1}{n+2}\right)t^2_{n_0+n,\delta_0}\geq 0, l(\cdot)$ is strictly monotone in x_* and there is a unique finite solution x_{**} to $l(x_*)=0$. If $q_{12}>0$, the function $l(\cdot)$ is decreasing, so the optimal screening region R_1 has the form $R_1=[x_{**},\infty)$, where $x_{**}-\mu_1^*$ is equal to

$$\frac{\gamma b_*\frac{q_{12}}{q_{11}}\pm|t_{n_0+n,\delta_0}|\left[\gamma\left[\frac{q_{12}}{q_{11}}\right]^2+\left[\frac{n+1}{n+2}\right]\frac{\gamma b_*^2}{q_{11}}-\left[\frac{n+1}{n+2}\right]\frac{t^2_{n_0+n,\delta_0}}{q_{11}}\right]^{\frac{1}{2}}}{\gamma\left[\frac{q_{12}}{q_{11}}\right]^2-\left[\frac{n+1}{n+2}\right]\frac{t^2_{n_0+n,\delta_0}}{q_{11}}}$$

where $b_*=b-\mu_2^*$ and the appropriate choice of the sign. If $q_{12}<0$ the optimal screening region is $R_1=(-\infty,x_{**}]$.

However, if

$$\gamma\frac{q_{12}^2}{q_{11}}-\left(\frac{n+1}{n+2}\right)t^2_{n_0+n,\delta_0}<0 \tag{4.4}$$

the derivative of l vanishes for some x_{**}. It can be shown (see Mouhab (1991)) that there can be at most one root. When δ_0 is such that t_{n_0+n,δ_0} is positive the function $l(x_*)$ takes large negative values, because

$$l(x_*)\propto-\left[\gamma^{\frac{1}{2}}\frac{q_{12}}{\sqrt{q_{11}}}sgn(x_*)+t_{n_0+n,\delta_0}\left(\frac{n+1}{n+2}\right)^{\frac{1}{2}}\right]\frac{|x_*|}{\sqrt{q_{11}}}\rightarrow-\infty,\text{ as }|x_*|\rightarrow\infty.$$

When t_{n_0+n,δ_0} is negative, $l(x_*)\rightarrow\infty$ as $|x_*|\rightarrow\infty$. We conclude that, if $t_{n_0+n,\delta_0}>0$, the optimal screening region is the entire real line, that is $R_1=R$. If $t_{n_0+n,\delta_0}<0, R_1=\phi$, the empty set.

These somewhat surprising situations were missed by Boys and Dunsmore (1986). The condition (4.4) that leads to them is likely to hold when n and n_0 are small, so that γ is small, q_{12} is small, and the cost ratio is extreme so $t^2_{n_0+n}$ is large.

5. SIMULATION

In this section we describe Monte Carlo studies of the Predictive rule (see Theorem 1) and the screening rule (ML Estimated rule) based on a known distribution but with the parameters replaced by their maximum likelihood estimates. The screening rules are compared on the basis of (1) average cost and (2) expected total probability of misclassification.

The simulations were done on a VAX/11-751 computer using the CMLIB (1983) normal random number generator RNOR available as a Fortran subroutine. Independent data sets, consisting of three thousand repetitions, were generated for each case.

Since the screening rule depends on the misclassification costs (see (2.4)), in the first study, the prior distribution was held fixed and the misclassifications costs were varied. The parameters of the prior distribution were

$$\boldsymbol{\mu}_0 = \begin{bmatrix} 2.5 \\ 1.5 \end{bmatrix}, \boldsymbol{V}_0 = \begin{bmatrix} \nu_1^2 & \rho_0\nu_1\nu_2 \\ \rho_0\nu_1\nu_2 & \nu_2^2 \end{bmatrix} \text{ and } n_0 = 3.$$

First $\boldsymbol{\Sigma}^{-1}$ was selected from a Wishart distribution with matrix $\boldsymbol{V}_0$ and n_0 degrees of freedom, that is $\boldsymbol{\Sigma}^{-1} = \sum_{j=1}^{n_0+1}(\boldsymbol{V}_j - \overline{\boldsymbol{V}})(\boldsymbol{V}_j - \overline{\boldsymbol{V}})^t$ where $\boldsymbol{V}_j$'s are independent $N(\boldsymbol{0}, \boldsymbol{V}_0)$. We inverted $\boldsymbol{\Sigma}^{-1}$ to obtain $\boldsymbol{\Sigma}$ and then we selected $\boldsymbol{\mu}$ from a normal distribution with mean $\boldsymbol{\mu}_0$ and covariance matrix $\boldsymbol{\Sigma}/a_0$, where $a_0 = n_0 + 1$ for the simulation. Next, given $\boldsymbol{\mu}$ and $\boldsymbol{\Sigma}$, we generated a training sample of size n from a $N(\boldsymbol{\mu}, \boldsymbol{\Sigma})$ population where $n = 5, 15$ or 75.

The average costs of the Predictive rules, using the true prior ($TPPred.$) and using a flat prior ($FPPred.$), and the costs of the ML Estimated rule ($MLEst.$) were estimated from the simulations for various values of the misclassification costs, $c_{1|2}$ and $c_{2|1}$.

The results on estimated costs are summarized in Tables 1 and 2, with their standard errors ($st.error$) given in parentheses. We chose $\nu_1 = 1.75, \nu_2 = .50, b = .96 + \mu_{02}$ but numerous other choices are possible.

Table 1. Average cost comparison, $\rho_0 = -.75, \nu_1 = 1.75, \nu_2 = .50$.
Predictive rules ($TPPred.$ and $FPPred.$) vs ML Estimated rule ($MLEst.$).

	$c_{1\|2}$	1	1	1	1	1
	$c_{2\|1}$	1	3	7	9	19
$n = 5$	$MLEst - TPPred.$	.047	−.005	.258	.594	2.568
	$st.error$	(.009)	(.031)	(.126)	(.229)	(.536)
	$MLEst - FPPred.$	.000	−.015	.177	.409	1.655
	$st.error$	(.000)	(.024)	(.097)	(.175)	(.440)
$n = 15$	$MLEst - TPPred.$	.019	.002	.124	.260	1.002
	$st.error$	(.004)	(.022)	(.104)	(.184)	(.460)
	$MLEst - FPPred.$	.000	−.003	.175	.302	1.163
	$st.error$	(.000)	(.018)	(.089)	(.157)	(.399)
$n = 75$	$MLEst - TPPred.$	.004	.008	.093	.140	.496
	$st.error$	(.001)	(.017)	(.075)	(.129)	(.314)
	$MLEst - FPPred.$	.000	.002	.095	.149	.577
	$st.error$	(.000)	(.013)	(.065)	(.112)	(.289)

Table 2. Average cost comparison, $\rho_0 = -.05, \nu_1 = 1.75, \nu_2 = .50$.
Predictive rules ($TPPred.$ and $FPPred.$) vs ML Estimated rule ($MLEst.$).

	$c_{1\|2}$	1	1	1	1	1
	$c_{2\|1}$	1	3	7	9	19
$n=5$	$MLEst - TPPred.$	.067	$-.081$	.823	1.360	4.274
	$st.error$	(.010)	(.035)	(.122)	(.223)	(.499)
	$MLEst - FPPred.$	.000	$-.006$	.457	.922	3.285
	$st.error$	(.000)	(.027)	(.109)	(.197)	(.465)
$n=15$	$MLEst - TPPred.$	.033	.066	.642	1.031	3.313
	$st.error$	(.005)	(.029)	(.120)	(.215)	(.510)
	$MLEst - FPPred.$	.000	$-.008$	.445	.766	2.725
	$st.error$	(.000)	(.024)	(.108)	(.193)	(.474)
$n=75$	$MLEst - TPPred.$	.007	.070	.431	.712	2.098
	$st.error$	(.001)	(.026)	(.104)	(.186)	(.451)
	$MLEst - FPPred.$	.000	.005	.281	.503	1.728
	$st.error$	(.000)	(.021)	(.094)	(.169)	(.421)

From Tables 1 to 2, we see that for these particular cases.

1. The Predictives rules, using either a flat prior or the true prior, both significantly beat the ML Estimated rule.
2. As ρ_0, in $\boldsymbol{V}_0$ the constant matrix of the Wishart prior distribution, gets smaller in absolute value the improvement of the Predictive rules over the ML Estimated rule becomes larger.
3. As n, the training size increases, the differences in cost between the Predictive rules and the ML Estimated rule become smaller in absolute value.

In the second study, we held $\boldsymbol{\Sigma}$ fixed at

$$\boldsymbol{\Sigma} = \begin{bmatrix} \sigma_1^2 & \rho\sigma_1\sigma_2 \\ \rho\sigma_2\sigma_2 & \sigma_2^2 \end{bmatrix}$$

where $\sigma_1 = 1.75, \sigma_2 = .50$ and ρ could be $-.95, -.55, .55$ or $.95$. For each fixed ρ we let $\boldsymbol{\mu}$ have the prior distribution $N(\boldsymbol{\mu}_0, n_0^{-1}\boldsymbol{\Sigma})$, where $\boldsymbol{\mu}_0^t = (\mu_{01}, \mu_{02})$. We took $n_0 = 1$ for the simulation. From the model, the predictive density of $\boldsymbol{Z}_*|\ \underset{\sim}{\boldsymbol{Z}}$ is $N(\boldsymbol{\mu}^*, (n+2)(n+1)^{-1}\boldsymbol{\Sigma})$, with $\boldsymbol{\mu}^* = (n\overline{\boldsymbol{Z}} + \boldsymbol{\mu}_0)/(n+1)$. Consequently when $\rho > 0$,

$$ETPM = \boldsymbol{E}\left[\ \boldsymbol{I}_{[X_* - \mu_1^* > c_n(\overline{Y}, b, \boldsymbol{\Sigma})]}\boldsymbol{I}_{[Y_* < b]} + \boldsymbol{I}_{[X_* - \mu_1^* \leq c_n(\overline{Y}, b, \boldsymbol{\Sigma})]}\boldsymbol{I}_{[Y_* \geq b]}\ \right],$$

where

$$c_n(\overline{Y}, b, \Sigma) = \left(b - \mu_2^* - \left(\frac{n+2}{n+1} \right) z_{\delta_0} \sigma_2 \sqrt{1-\rho^2} \right) \frac{\sigma_1}{\rho \sigma_2}$$

and z_{δ_0} is the $100(1-\delta_0)$–th percentile of the standard normal distribution. Setting $U_* = X_* - \mu_1^*$, the $ETPM$ depends only on the bivariate normal $N(\begin{bmatrix} 0 \\ \mu_2^* \end{bmatrix}, \left(\frac{n+2}{n+1}\right)\Sigma)$ distribution which is free of μ_{01}. We took $\mu_{01} = 0$. Note that an item is labeled as acceptable if $Y_* > b$, where $b = .44 + \mu_{02}$. Another fact is that the $ETPM$ depends only on the difference $b - \mu_{02}$.

For each run we selected a $\boldsymbol{\mu}$ from the normal with mean $\boldsymbol{\mu}_0^t = (0, \mu_{02})$ and covariance matrix $\boldsymbol{\Sigma}$, where $\mu_{02} = -10, -9, \ldots, +10$. Then, given $\boldsymbol{\mu}$, we generated a training sample of size n and calculated the Expected Total Probability of Misclassification of the Predictive rule and the ML Estimated rule, denoted by $ETPM$ and $ET\hat{P}M$ respectively. The differences $ET\hat{P}M - ETPM$ in the estimated total probability of misclassification are reported in Tables 3 and 4 along with the standard errors which appear in parentheses,

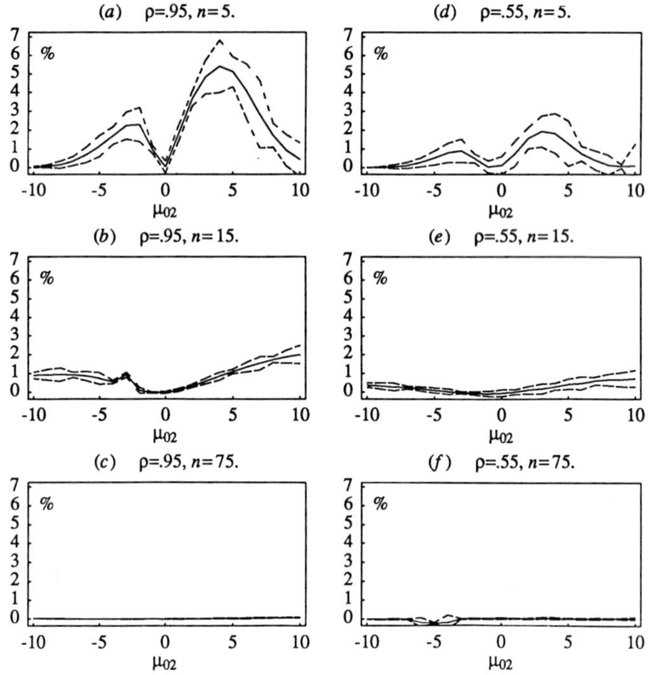

Figure 1. $(ET\hat{P}M - ETPM) \times 10^2$, $\rho > 0$

Table 3. $\rho = .95, \sigma_1 = 1.75, \sigma_2 = .50$
Predictive rule ($ETPM$) vs ML Estimated rule ($ET\hat{P}M$).
Table gives differences ($ET\hat{P}M - ETPM$) $\times 10^2$.

μ_{02}	−10	−9	−8	−7	−6	−5	−4
	.018	.063	.147	.338	.675	1.146	1.703
$n = 5$							
	(.018)	(.042)	(.098)	(.122)	(.203)	(.289)	(.231)
	.888	.918	.928	.912	.846	.719	.529
$n = 15$							
	(.088)	(.142)	(.179)	(.075)	(.126)	(.155)	(.044)
	.014	.007	.000	−.004	−.008	−.015	−.013
$n = 75$							
	(.002)	(.003)	(.003)	(.001)	(.002)	(.003)	(.003)

μ_{02}	−3	−2	−1	0	1	2	3
	2.253	2.298	.967	.021	1.837	3.661	4.837
$n = 5$							
	(.359)	(.450)	(.104)	(.154)	(.218)	(.197)	(.459)
	.298	.084	−.042	−.032	.104	.325	.586
$n = 15$							
	(.069)	(.075)	(.014)	(.028)	(.047)	(.047)	(.075)
	−.014	−.014	−.013	−.010	−.007	−.002	.003
$n = 75$							
	(.004)	(.006)	(.004)	(.007)	(.009)	(.006)	(.003)

μ_{02}	4	5	6	7	8	9	10
	5.394	5.098	4.081	2.839	1.727	.903	.414
$n = 5$							
	(.708)	(.401)	(.722)	(.894)	(.315)	(.425)	(.453)
	.840	1.091	1.321	1.533	1.726	1.880	1.995
$n = 15$							
	(.106)	(.062)	(.125)	(.177)	(.082)	(.175)	(.243)
	.009	.016	.024	.032	.041	.050	.060
$n = 75$							
	(.011)	(.007)	(.011)	(.014)	(.009)	(.013)	(.017)

Table 4. $\rho = .55, \sigma_1 = 1.75, \sigma_2 = .50$
Predictive rule $(ETPM)$ vs ML Estimated rule $(ET\hat{P}M)$.
Table gives differences $(ET\hat{P}M - ETPM) \times 10^2$.

μ_{02}	-10	-9	-8	-7	-6	-5	-4
$n = 5$	.004	.007	.044	.110	.275	.509	.790
	(.008)	(.014)	(.056)	(.086)	(.117)	(.189)	(.256)
$n = 15$	.364	.323	.270	.241	.161	.078	.019
	(.065)	(.089)	(.105)	(.043)	(.060)	(.068)	(.074)
$n = 75$	$-.020$	$-.020$	$-.021$	$-.023$	$-.200$	$-.232$	$-.200$
	(.007)	(.011)	(.013)	(.009)	(.124)	(.015)	(.186)
μ_{02}	-3	-2	-1	0	1	2	3
$n = 5$	0.898	0.482	.027	.114	0.743	1.572	1.939
	(.315)	(.117)	(.164)	(.237)	(.338)	(.282)	(.416)
$n = 15$	$-.035$	$-.072$	$-.087$	$-.053$	$-.007$	.095	.154
	(.030)	(.048)	(.067)	(.088)	(.066)	(.103)	(.134)
$n = 75$	$-.021$	$-.020$	$-.020$	$-.017$	$-.015$	$-.016$	.014
	(.010)	(.015)	(.019)	(.022)	(.012)	(.018)	(.022)
μ_{02}	4	5	6	7	8	9	10
$n = 5$	1.827	1.272	0.738	0.392	0.128	.057	.098
	(.528)	(.603)	(.196)	(.254)	(.268)	(.054)	(.587)
$n = 15$	.269	0.406	0.448	0.576	0.644	0.656	0.710
	(.090)	(.136)	(.170)	(.102)	(.152)	(.192)	(.225)
$n = 75$	$-.007$	$-.004$	$-.006$	$-.003$	.004	.004	.006
	(.026)	(.014)	(.020)	(.025)	(.015)	(.022)	(.027)

To better understand Tables 3 to 4, we graph the results as Figure 1. The solid line is the scaled estimate $(ET\hat{P}M - ETPM) \times 10^2$ and the dashed lines are the control limits $(ET\hat{P}M - ETPM) \times 10^2 \pm 2$ standard errors.

From Tables 3 and 4, and Figure 1, for the particular cases considered, we see that

1. For $n = 5$ and 15, the estimated $ETPM$ is always significantly smaller than the estimated $ET\hat{P}M$ for some values of μ_{02}. Typically, these μ_{02} values are on both sides of zero.
2. As the correlation of (X, Y) increases in absolute value the difference $ET\hat{P}M - ETPM$ becomes larger.
3. As the training sample size, n, increases the estimates of $ETPM$ and $ET\hat{P}M$ approach one another. For $n = 75$ even allowing for standard errors reveals very little difference.

Similar conclusions were reached for the cases of negative correlation.

6. AN ASYMPTOTIC EXPANSION OF THE PREDICTIVE DENSITY

Since the Bayes screening rule and Bayes risk involve the predictive density, we first derive an asymptotic expansion for the predictive density and then use it to get an expansion of the risk. Formal expansions have been proposed (see for instance Dickey (1980)), but in order to expand the Bayes risk in this and other contexts, we need to establish some integrability conditions for the terms in the expansion.

Recall the notation where $g(\boldsymbol{\theta})$ is the prior density, $f(\boldsymbol{z}_i|\boldsymbol{\theta})$ is the conditional density of $\boldsymbol{Z}_i$ given $\boldsymbol{\theta}$, and the predictive density of any future random vector $\boldsymbol{V} \epsilon R^k$, given a training sample $\underset{\sim}{\boldsymbol{Z}} = (\boldsymbol{z}_1, \ldots, \boldsymbol{z}_n)$, is defined as

$$h(\boldsymbol{v}|\ \underset{\sim}{\boldsymbol{Z}}) = \frac{\int f(\boldsymbol{v}|\boldsymbol{\theta}) g(\boldsymbol{\theta}) \prod_{i=1}^{n} f(\boldsymbol{z}_i|\boldsymbol{\theta}) d\boldsymbol{\theta}}{\int g(\boldsymbol{\theta}) \prod_{i=1}^{n} f(\boldsymbol{z}_i|\boldsymbol{\theta}) d\boldsymbol{\theta}} \tag{6.1}$$

To simplify the presentation, we restrict our attention to the case where $f(\boldsymbol{z}_i|\boldsymbol{\theta})$ belongs to a multivariate exponential family with a possibly multivariate parameter. That is.

$$f(\boldsymbol{z}_i|\boldsymbol{\theta}) = \exp[\boldsymbol{A}(\boldsymbol{\theta})\boldsymbol{z}_i - C(\boldsymbol{\theta})] \tag{6.2}$$

is the pdf with respect to some measure μ. Similar asymptotic results hold for other regular cases that are not exponential families. We assume that

Assumptions B

(B.1) The dominating measure μ is not supported on a flat.
(B.2) The interior of the parameter space $\Theta(\mu)$, denoted by $\Theta^0(\mu)$, is not empty.
(B.3) The parametric function $\boldsymbol{A}(\boldsymbol{\theta})$ is homeomorphism and has continuous partial derivatives of order $s + 3$, in a neighborhood of $\boldsymbol{\theta}_0$, for $s \geq 2$.
(B.4) The prior, $g(\boldsymbol{\theta})$, has continuous partial derivatives of order $s + 1$, and is positive at $\boldsymbol{\theta}_0$.

It is well known (cf Johnson (1976)) that both the numerator and the denominator in (6.1) can be expanded. Let $[x]$ be the integer part of x.

Theorem 3.

Let $f(\boldsymbol{v}|\boldsymbol{\theta})$ belong to an exponential family which satisifies *Assumptions B*. Then, with maximum likelihood estimate $\hat{\theta}_n$,

$$h(\boldsymbol{v}|\tilde{\boldsymbol{Z}}) = f(\boldsymbol{v}|\hat{\boldsymbol{\theta}}_n) + \sum_{i=1}^{[\frac{S}{2}]} n^{-i}\gamma_i(\boldsymbol{v},\tilde{\boldsymbol{Z}}) + n^{-(s+1)/2}K(\boldsymbol{v},\tilde{\boldsymbol{Z}}) \tag{6.3}$$

for $\omega = (\boldsymbol{z}_1, \boldsymbol{z}_2, \cdots)$ belonging to an almost sure set of outcomes. Further, $|K(\boldsymbol{v},\tilde{\boldsymbol{Z}})| \leq \boldsymbol{M}_\omega(\boldsymbol{v})$ for sufficiently large n and the coefficients satisfy,

$$\text{Sup}_{n\geq N_\omega} \int |\gamma_i(\boldsymbol{v},\tilde{\boldsymbol{Z}})| d\mu(\boldsymbol{v}) \leq c < \infty \qquad i = 1, \ldots, [s/2] \tag{6.4}$$

$$\text{Sup}_{n\geq N_\omega} \int |K(\boldsymbol{v},\tilde{\boldsymbol{Z}})| d\mu(\boldsymbol{v}) \leq c < \infty$$

for some N_ω.

Remark. In the single parameter case, the term $\gamma_1(\boldsymbol{v},\tilde{\boldsymbol{Z}})$ becomes

$$\frac{1}{2}\left[\frac{\partial^2}{\partial\theta^2} f(\boldsymbol{v}|\hat{\theta}_n) + 2\frac{\partial}{\partial\theta} f(\boldsymbol{v}|\hat{\theta}_n)\frac{\partial}{\partial\theta}\log(g(\hat{\theta}_n))\right]\left(-\frac{\partial^2}{\partial\theta^2}\log(f(\overline{Z}|\hat{\theta}_n))\right)^{\frac{-1}{2}}$$

$$+\frac{1}{2}\frac{\partial}{\partial\theta} f(\boldsymbol{v}|\hat{\theta}_n)\frac{\partial^3}{\partial\theta^3}\log(f(\overline{Z}|\hat{\theta}_n)\left(-\frac{\partial^2}{\partial\theta^2}\log(f(\overline{Z}|\hat{\theta}_n))\right)^{\frac{-3}{2}}$$

Proof: By changing variables, we can rewrite the predictive density as

$$h(\boldsymbol{v}|\tilde{\boldsymbol{Z}}) = \frac{\int f(\boldsymbol{v}|\boldsymbol{t})g(\boldsymbol{t})R_n(\boldsymbol{t})d\boldsymbol{t}}{\int g(\boldsymbol{t})R_n(\boldsymbol{t})d\boldsymbol{t}}$$

where

$$\boldsymbol{t} = \sqrt{n}(\boldsymbol{\theta} - \hat{\boldsymbol{\theta}}_n), R_n(\boldsymbol{t}) = \frac{\prod_{i=1}^{n} f(\boldsymbol{z}_i|\boldsymbol{t})}{\prod_{i=1}^{n} f(\boldsymbol{z}_i|\hat{\boldsymbol{\theta}}_n)}$$

and $\hat{\boldsymbol{\theta}}_n$ is the maximum likelihood estimator of $\boldsymbol{\theta}$. Using Corollary 3.1.3 in Johnson and Ladalla (1978), we have the following expansion for the denominator

$$\int g(\boldsymbol{t})R_n(\boldsymbol{t})d\boldsymbol{t} = \sum_{i=0}^{[\frac{S}{2}]} n^{-i}\beta_i(\tilde{\boldsymbol{Z}}) + n^{-\frac{s+1}{2}}k(\tilde{\boldsymbol{Z}}) \tag{6.5}$$

where $|k(\tilde{\boldsymbol{Z}})| \leq \boldsymbol{M}_\omega$ for sufficiently large n, almost at all ω. A similar expansion holds for the numerator because it has the same form as the denominator except that $f(\boldsymbol{v}|\boldsymbol{\theta})g(\boldsymbol{\theta})$ is now treated as a new prior. This is possible since

$\int f(\boldsymbol{v}|\boldsymbol{\theta})g(\boldsymbol{\theta})d\boldsymbol{\theta} = f(\boldsymbol{v}) < \infty$

for almost all $\boldsymbol{v}$. Replacing $g(\boldsymbol{t})$ in (6.5) by $f(\boldsymbol{v}|\boldsymbol{t})g(\boldsymbol{t})$, for fixed $\boldsymbol{v}$, we obtain

$$\int f(\boldsymbol{v}|\boldsymbol{t})g(\boldsymbol{t})R_n(\boldsymbol{t})d\boldsymbol{t} = \sum_{i=0}^{[\frac{s}{2}]} n^{-i}\beta_i(\boldsymbol{v}, \tilde{\boldsymbol{Z}}) + n^{-\frac{s+1}{2}} k(\boldsymbol{v}, \tilde{\boldsymbol{Z}}) \tag{6.6}$$

The predictive density, h is the ratio of the left hand sides of (6.6) and (6.5), so its asymptotic expansion is just the ratio of the right hand sides.
Consequently the $\gamma_i(\boldsymbol{v}, \tilde{\boldsymbol{Z}})$ in (6.3) are the solutions of a system of $[\frac{s}{2}]+1$ equations. The first term is $\gamma_0(\boldsymbol{v}, \tilde{\boldsymbol{Z}}) = \beta_0(\boldsymbol{v}, \tilde{\boldsymbol{Z}})/\beta_0(\tilde{\boldsymbol{Z}}) = f(\boldsymbol{v}|\hat{\boldsymbol{\theta}}_n)$, as asserted, and the second term is $\gamma_1(\boldsymbol{v}, \tilde{\boldsymbol{Z}}) = \dfrac{\beta_1(\boldsymbol{v}, \tilde{\boldsymbol{Z}}) - f(\boldsymbol{v}|\hat{\boldsymbol{\theta}}_n)\beta_1(\tilde{\boldsymbol{Z}})}{\beta_0(\tilde{\boldsymbol{Z}})}$.

Next, we proceed to establish the first bound in (6.4). We know that the numerator of the predictive density can be expanded as

$$\int f(\boldsymbol{v}|\boldsymbol{t})g(\boldsymbol{t})R_n(\boldsymbol{t})d\boldsymbol{t} = \int \exp[-\tfrac{1}{2}\boldsymbol{t}'Q(\hat{\boldsymbol{\theta}}_n)\boldsymbol{t}]P_s^T(\boldsymbol{t}, \omega)d\boldsymbol{t} + n^{[\frac{s+1}{2}]}k(\boldsymbol{v}, \tilde{\boldsymbol{Z}})$$

where the matrix $Q(\boldsymbol{\theta})$ is defined as $\left(-\dfrac{\partial^2 \log(f(\overline{\boldsymbol{Z}}|\boldsymbol{\theta}))}{\partial\theta_i \partial\theta_j} \right)_{1 \le i,j \le h}$ and

$$P_s^T(\boldsymbol{t}, \boldsymbol{v}, \omega) = \sum_{j=0}^{s} n^{-j/2} \sum_{l_1+\cdots+l_h+m_1+\cdots+m_H=j} C_{l_1 \cdots m_H}(\boldsymbol{v}, \omega) \frac{\prod_{i=1}^{h} u_i^{l_i} w_l^{m_i}}{\prod_{i=1}^{h} l_i! \prod_{l=1}^{H} m_l!}$$

where w_l is of the form $w_{i_1 \cdots i_h} = n\prod_{j=1}^{h}(\theta_j - \hat{\theta}_j)^{i_j} = n\prod_{j=1}^{h} u_j^{i_j}$,

$$C_{l_1 \cdots m_H}(\boldsymbol{v}, \omega) = \frac{\partial^j P_s(\boldsymbol{u}, \boldsymbol{w}, \boldsymbol{v}, \omega)}{\partial^{l_1}u_1 \cdots \partial^{l_h}u_h \partial^{m_1}w_1 \cdots \partial^{m_H}w_H}\Big|_{\|\boldsymbol{u}\|=\|\boldsymbol{w}\|=0},$$

$$P_s(\boldsymbol{u}, \boldsymbol{w}, \boldsymbol{v}, \tilde{\boldsymbol{Z}}) = fg_s^T(\hat{\boldsymbol{\theta}}_n + \boldsymbol{u})R_{n,s}^T(\boldsymbol{\theta})\exp[\tfrac{1}{2}n\boldsymbol{u}'Q(\hat{\boldsymbol{\theta}}_n)\boldsymbol{u}], \tag{6.7}$$

$$fg_s^T(\hat{\boldsymbol{\theta}}_n + \boldsymbol{u}) = \sum_{j=0}^{s} \sum_{i_1+\cdots+i_h=j} \frac{\partial^j f(\boldsymbol{v}|\hat{\boldsymbol{\theta}}_n)g(\hat{\boldsymbol{\theta}}_n)}{\partial\boldsymbol{\theta}_1^{i_1} \cdots \partial\boldsymbol{\theta}_h^{i_h}} \prod_{l=1}^{h} \frac{(\boldsymbol{\theta}_l - \hat{\boldsymbol{\theta}}_l)^{i_l}}{i_l!},$$

and $\log(R_{n,s}^T(\boldsymbol{\theta}))$ is the $(s+3)$-rd order truncated Taylor expansion of $\log(R_n(\boldsymbol{\theta}))$ without the remainder term, evaluated at $\hat{\boldsymbol{\theta}}_n$ (See pages 11-16 Johnson and Ladalla (1976)).

Observe that the $\beta_j(\boldsymbol{v}, \tilde{\boldsymbol{Z}})$'s in (6.6) depend on $\boldsymbol{v}$ only through the $C_{l_1 \cdots m_H}(\boldsymbol{v}, \omega)$'s, which in turn depend on $\boldsymbol{v}$ only through the first $(s+1)$ partial derivatives of $f(\boldsymbol{v}|\boldsymbol{\theta})g(\boldsymbol{\theta})$ evaluated at $\hat{\boldsymbol{\theta}}_n$. By the exponential family structure (6.2), the derivatives of $\exp(C(\boldsymbol{\theta}))$ can be obtained by differentiating under the integral sign. Evaluating these at $\hat{\boldsymbol{\theta}}_n$, we get

$$\text{Sup}_{n\geq N_\omega}\left|\frac{\partial^j \exp[C(\hat{\boldsymbol{\theta}}_n)]}{\partial\theta_1^{i_1}\cdots\partial\theta_h^{i_h}}\right| = \tag{6.8}$$

$$\text{Sup}_{n\geq N_\omega}\left|\int \frac{\partial^j \exp[\boldsymbol{A}(\boldsymbol{\theta})\boldsymbol{v}]}{\partial\theta_1^{i_1}\cdots\partial\theta_h^{i_h}}\Big|_{\theta=\hat{\theta}_n} d\mu(\boldsymbol{v})\right| \leq c < \infty$$

Further, notice that, for $j = 1, \ldots, s+1$, the positive part of the integrand satisfies

$$\int \left[\frac{\partial^j \exp\big(\boldsymbol{A}(\boldsymbol{\theta})\boldsymbol{v}\big)}{\partial\theta_1^{i_1}\cdots\partial\theta_h^{i_h}}\right]^+ d\mu(\boldsymbol{v}) = \int \frac{\partial^j \exp\big(\boldsymbol{A}(\boldsymbol{\theta})\boldsymbol{v}\big)}{\partial\theta_1^{i_1}\cdots\partial\theta_h^{i_h}} d\mu_j^+(\boldsymbol{v}) \tag{6.9}$$

and the integral is continuous in $\boldsymbol{\theta}$, where μ_j^+ is the restriction of the measure μ to the set $\left[\boldsymbol{v}\epsilon R^k; \left(\frac{\partial^j \exp[\boldsymbol{A}(\boldsymbol{\theta})\boldsymbol{v}]}{\partial\theta_1^{i_1}\cdots\partial\theta_h^{i_h}}\right) \geq 0\right]$. Since $\hat{\boldsymbol{\theta}}_n$ remains within a compact set containing $\boldsymbol{\theta}_0$, with probability one,

$$\text{Sup}_{n\geq N_\omega}\int \left[\frac{\partial^j \exp[\boldsymbol{A}(\boldsymbol{\theta})\boldsymbol{v}]}{\partial\theta_1^{i_1}\cdots\partial\theta_h^{i_h}}\Big|_{\theta=\hat{\theta}_n}\right]^+ d\mu(\boldsymbol{v}) \leq c < \infty \tag{6.10}$$

almost surely and sufficiently large N_ω. A similar result holds for the negative part, so

$$\text{Sup}_{n\geq N_\omega}\left|\frac{\partial^j \exp[\boldsymbol{A}(\boldsymbol{\theta})\boldsymbol{v}]}{\partial\theta_1^{i_1}\cdots\partial\theta_j^{i_h}}\Big|_{\theta=\hat{\theta}_n}\right| d\mu(\boldsymbol{v}) \leq c < \infty \tag{6.11}$$

Since each $C_{l_1\cdots m_H}(\boldsymbol{v},\omega)$ in (6.7) is a linear combination of the products of the terms in (6.8) and (6.11),

$$\text{Sup}_{n\geq N_\omega}\int |C_{l_1\cdots m_H}(\boldsymbol{v},\omega)|d\mu(\boldsymbol{v}) \leq c < \infty \text{ almost surely} \tag{6.12}$$

and so

$$\text{Sup}_{n\geq N_\omega}\int |\boldsymbol{\beta}_j(\boldsymbol{v},\tilde{\boldsymbol{Z}})|d\mu(\boldsymbol{v}) \leq c < \infty \text{ almost surely.}$$

Finally, since the $\gamma_j(\boldsymbol{v},\tilde{\boldsymbol{Z}})$'s are linear functions of the $\boldsymbol{\beta}_j(\boldsymbol{v},\tilde{\boldsymbol{Z}})$, where the coefficients depend on $\beta_j(\tilde{\boldsymbol{Z}})$ in (6.4) but not on $\boldsymbol{v}$, the first result in (6.4) follows.

We also need to establish the second bound in (6.4). The proof relies on a slight extension of Lemma 3.1.1 and Lemma 3.1.2 in Johnson and Ladalla (1976), where we again replace the original prior by $g(\boldsymbol{t})f(\boldsymbol{v}|\boldsymbol{t})$. We first show that there exists a function $\boldsymbol{M}_{\omega,1}^s(\boldsymbol{v})$ and a positive integer $N_{\omega,1}^s$ such that

$$\int_{\|\boldsymbol{t}\|\leq n^{1/6}} |f(\boldsymbol{v}|\boldsymbol{t})g(\boldsymbol{t})R_n(\boldsymbol{t}) - \exp[-\tfrac{1}{2}\boldsymbol{t}'Q(\hat{\boldsymbol{\theta}}_n)\boldsymbol{t}]P_s^T(\boldsymbol{t},\boldsymbol{v},\omega)|dt \tag{6.13}$$

$$\leq M^s_{\omega,1}(\boldsymbol{v})n^{-\frac{s+1}{2}}$$

and

$\mathrm{Sup}_{n\leq N^s_{\omega,1}} \int \boldsymbol{M}^s_{\omega,1}(\boldsymbol{v})d\mu(\boldsymbol{v}) \leq c < \infty$ almost surely.

To begin,

$$|\log R_n(\boldsymbol{t}) - \log R^T_{n,s}(\boldsymbol{t})| \leq cn^{-\frac{s+1}{2}} \left(\sum_{l=1}^{h} |t_l|^{s+3} \right) \tag{6.14}$$

where c is a generic constant. Next, using the truncated series in (6.7)

$$|f(\boldsymbol{v}|\boldsymbol{t})g(\boldsymbol{t})R_n(\boldsymbol{t}) - fg^T_s(\boldsymbol{t})R^T_{n,s}(\boldsymbol{t})| \leq R_n(\boldsymbol{t})|fg^T_s(\boldsymbol{t}) \| \frac{R^T_{n,s}(\boldsymbol{t})}{R_n(\boldsymbol{t})} - 1|$$

$$+R_n(\boldsymbol{t})|f(\boldsymbol{v}|\boldsymbol{t})g(\boldsymbol{t}) - fg^T_s(\boldsymbol{t})|$$

where

$$|f(\boldsymbol{v}|\boldsymbol{t})g(\boldsymbol{t}) - fg^T_s(\boldsymbol{t})| = | \sum_{i_1+\cdots+i_h=s+1} \frac{\partial^{s+1}(f(\boldsymbol{v}|\boldsymbol{\theta}^*_n)g(\boldsymbol{\theta}^*_n))}{\partial\theta_1^{i_1}\cdots\partial\theta_h^{i_h}} \prod_{l=1}^{h} \frac{u_l^{i_l}}{i_l!}|$$

$$\leq n^{-\frac{s+1}{2}} \sum_{i_1+\cdots+i_h=s+1} |\frac{\partial^{s+1}(f(\boldsymbol{v}|\boldsymbol{\theta}^*_n)g(\boldsymbol{\theta}^*_n))}{\partial\theta_1^{i_1}\cdots\partial\theta_h^{i_h}}| \prod_{l=1}^{h} |t_l|^{i_l} \tag{6.15}$$

and $\| \boldsymbol{\theta} - \boldsymbol{\theta}^*_n \| \leq \| \boldsymbol{\theta} - \hat{\boldsymbol{\theta}}_n \|$.

From our equation (6.14) and (6.15) and Lemma 2.3.2 of the reference, we deduce that

$$|f(\boldsymbol{v}|\boldsymbol{t})g(\boldsymbol{t})R_n(\boldsymbol{t}) - fg^T_s(\boldsymbol{t})R^T_{n,s}(\boldsymbol{t})| \leq cn^{-\frac{s+1}{2}} \exp(-\tfrac{1}{2}\boldsymbol{t}'(Q(\boldsymbol{\theta}_0) - \eta \boldsymbol{I})\boldsymbol{t}] \times$$

$$\left(c|fg^T_s(\boldsymbol{t})| \sum_{l=1}^{h} |t_l|^{s+3} + c \sum_{i_1+\cdots+i_h=s+1} |\frac{\partial^{s+1}(f(\boldsymbol{v}|\boldsymbol{\theta}^*_n)g(\boldsymbol{\theta}^*_n))}{\partial\theta_1^{i_1}\cdots\partial\theta_h^{i_h}}| \prod_{l=1}^{h} |t_l|^{i_l} \right)$$

for sufficiently large n, where η is a positive real number smaller than the smallest eigenvalue of $Q(\boldsymbol{\theta}_0)$.

Integrating over the set $\| \boldsymbol{t} \| \leq n^{1/6}$, we get

$$\int_{\|\boldsymbol{t}\|\leq n^{1/6}} |f(\boldsymbol{v}|\boldsymbol{t})g(\boldsymbol{t})R_n(\boldsymbol{t}) - fg^T_s(\boldsymbol{t})R^T_{n,s}(\boldsymbol{t})|d\boldsymbol{t} \leq$$

$$cn^{-\frac{s+1}{2}}\sum_{j=0}^{s}\sum_{i_1+\cdots+i_h=j}\left|\frac{\partial^j(f(\boldsymbol{v}|\hat{\boldsymbol{\theta}}_n)g(\hat{\boldsymbol{\theta}}_n))}{\partial\theta_1^{i_1}\cdots\partial\theta_h^{i_h}}\right|+ \tag{6.16}$$

$$cn^{-\frac{s+1}{2}}\sum_{i_1+\cdots+i_h=s+1}\left|\frac{\partial^{s+1}(f(\boldsymbol{v}|\boldsymbol{\theta}_n^*)g(\boldsymbol{\theta}_n^*))}{\partial\theta_1^{i_1}\cdots\partial\theta_h^{i_h}}\right| = M_1(\boldsymbol{v},\omega)n^{-\frac{s+1}{2}}$$

for sufficiently large n.
Rewriting $P_s(\boldsymbol{u},\boldsymbol{w},\boldsymbol{v},\omega)$ in (6.7) by expressing $\boldsymbol{u}$ and $\boldsymbol{w}$ in terms of $\boldsymbol{t}$ we get $P_s(\boldsymbol{t},\boldsymbol{v},\omega)$, where

$$|P_s(\boldsymbol{t},\boldsymbol{v},\omega)-P_s^T(\boldsymbol{t},\boldsymbol{v},\omega)| \le cn^{-\frac{s+1}{2}}\sum_{i=1}^{h}|c_i(\boldsymbol{v},\omega)||t_i|^{s+1}+ \tag{6.17}$$

$$cn^{-\frac{s+1}{2}}\sum_{i_1+\cdots+i_h=3}|c_{i_1\cdots i_h}(\boldsymbol{v},\omega)|\left(\prod_{l=1}^{h}|t_l|^{i_l}\right)^{s+1}$$

where the $c_i(\boldsymbol{v},\omega)$'s and $c_{i_1\cdots i_h}(\boldsymbol{v},\omega)$ are of the form (6.7) and satisfy

$\mathrm{Sup}_{n\ge N_\omega}\int|c_{i_1\cdots i_j}(\boldsymbol{v},\omega)|d\mu(\boldsymbol{v}) \le c < \infty$ almost surely.

Now, by the definition of P_s in (6.7),

$$\int_{\|\boldsymbol{t}\|\le n^{1/6}}|fg_s^T(\boldsymbol{t})R_{n,s}^T(\boldsymbol{t})-\exp[-\tfrac{1}{2}\boldsymbol{t}'Q(\hat{\boldsymbol{\theta}}_n)\boldsymbol{t}]|P_s^T(\boldsymbol{t},\boldsymbol{v},\omega)|d\boldsymbol{t} =$$

$$\int_{\|\boldsymbol{t}\|\le n^{1/6}}\exp[-\tfrac{1}{2}\boldsymbol{t}'Q(\hat{\boldsymbol{\theta}}_n)\boldsymbol{t}]|P_s(\boldsymbol{t},\boldsymbol{v},\omega)-P_s^T(\boldsymbol{t},\boldsymbol{v},\omega)|d\boldsymbol{t}$$

From (6.7) and the usual arguments the last integral is dominated by

$$cn^{-\frac{s+1}{2}}\int_{\|\boldsymbol{t}\|\le n^{1/6}}\exp[-\tfrac{1}{2}\boldsymbol{t}'(Q(\boldsymbol{\theta}_0)-\eta\boldsymbol{I})\boldsymbol{t}]\Bigg(\sum_{i=1}^{h}|c_i(\boldsymbol{v},\omega)||t_i|^{s+1}+$$

$$c\sum_{i_1+\cdots+i_h=3}|c_{i_j\cdots i_h}(\boldsymbol{v},\omega)|\left(\prod_{l=1}^{h}|t_l|^{i_l}\right)^{s+1}\Bigg)d\boldsymbol{t} \le cn^{-\frac{s+1}{2}}M_2(\boldsymbol{v},\omega) \tag{6.18}$$

where $M_2(\boldsymbol{v},\omega)=\sum_{i=1}^{h}|c_i(\boldsymbol{v},\omega)|+\sum_{i_1+\cdots+i_h=3}|c_{i_1\ldots i_h}(\boldsymbol{v},\omega)|$
Finally by the triangle inequality and the bounds (6.16) and (6.18)

$$\int_{\|\boldsymbol{t}\|\leq n^{1/6}} |f(\boldsymbol{v}|\boldsymbol{t})g(\boldsymbol{t})R_n(\boldsymbol{t}) - \exp[-\tfrac{1}{2}\boldsymbol{t}'Q(\hat{\boldsymbol{\theta}}_n)\boldsymbol{t}]P_s^T(\boldsymbol{t},\boldsymbol{v},\omega)|d\boldsymbol{t}$$

$$\leq \int_{\|\boldsymbol{t}\|\leq n^{1/6}} |f(\boldsymbol{v}|\boldsymbol{t})g(\boldsymbol{t})R_n(\boldsymbol{t}) - fg_s^T(\boldsymbol{t})R_{n,s}^T(\boldsymbol{t})|d\boldsymbol{t}+$$

$$\int_{\|\boldsymbol{t}\|\leq n^{1/6}} |fg_s^T(\boldsymbol{t})R_{n,s}^T(\boldsymbol{t}) - \exp[-\tfrac{1}{2}\boldsymbol{t}'Q(\hat{\boldsymbol{\theta}}_n)\boldsymbol{t}]P_s^T(\boldsymbol{t},\boldsymbol{v},\omega)|d\boldsymbol{t}$$

$$\leq n^{-\frac{s+1}{2}}(M_1(\boldsymbol{v},\omega) + M_2(\boldsymbol{v},\omega))$$

Thus, we take $M_{\omega,1}^s(\boldsymbol{v}) = M_1(\boldsymbol{v},\omega) + M_2(\boldsymbol{v},\omega)$ and (6.13) follows.
Next we consider the integral over $\| \boldsymbol{t} \| > n^{1/6}$. As in Johnson and Ladalla (1976), we bound the integral $\boldsymbol{I}$ by $\boldsymbol{I}_1 + \boldsymbol{I}_2 + \boldsymbol{I}_3$, where

$$\boldsymbol{I}_1 = \int_{n^{1/6} < \|\boldsymbol{t}\| \leq \sqrt{n}\delta_2} f(\boldsymbol{v}|\boldsymbol{t})g(\boldsymbol{t})R_n(\boldsymbol{t})d\boldsymbol{t}$$

$$\boldsymbol{I}_2 = \int_{\|\boldsymbol{t}\| > \sqrt{n}\delta_2} f(\boldsymbol{v}|\boldsymbol{t})g(\boldsymbol{t})R_n(\boldsymbol{t})dt$$

$$\boldsymbol{I}_3 = \int_{\|\boldsymbol{t}\| > n^{1/6}} \exp[-\tfrac{1}{2}\boldsymbol{t}'Q(\hat{\boldsymbol{\theta}}_n)\boldsymbol{t}]|P_s^T(\boldsymbol{t},\boldsymbol{v},\omega)|d\boldsymbol{t}$$

We consider each integral separately. Using Lemma 2.3.2 of the reference, we get

$$\boldsymbol{I}_1 = \int_{n^{1/6} < \|\boldsymbol{t}\| \leq \sqrt{n}\delta_2} f(\boldsymbol{v}|\boldsymbol{t})g(\boldsymbol{t})R_n(\boldsymbol{t})d\boldsymbol{t} \leq$$

$$\int_{\|\boldsymbol{t}\| > n^{1/6}} f(\boldsymbol{v}|\boldsymbol{t})g(\boldsymbol{t})\exp[-\tfrac{1}{2}\boldsymbol{t}'(Q(\boldsymbol{\theta}_0) - \eta\boldsymbol{I})\boldsymbol{t}]d\boldsymbol{t}$$

$$\leq \int_{\|\boldsymbol{t}\| > n^{1/6}} f(\boldsymbol{v}|\boldsymbol{t})g(\boldsymbol{t})\exp[-\tfrac{1}{2}d_1\boldsymbol{t}'\boldsymbol{t}]dt$$

where $\eta < d =$ smallest eigenvalue of the positive definite matrix $Q(\boldsymbol{\theta}_0)$, and d_1 is the smallest eigenvalue of $Q(\boldsymbol{\theta}_0) - \eta\boldsymbol{I}$.
On the set $\{\boldsymbol{t} :\| \boldsymbol{t} \|> n^{1/6}\} = \{\boldsymbol{t} : \boldsymbol{t}'\boldsymbol{t} > n^{1/3}\}$, we have $\exp[-\frac{1}{2}d_1\boldsymbol{t}'\boldsymbol{t}] < \exp[-\frac{1}{2}d_1 n^{1/3}]$, so

$$\boldsymbol{I}_1 \leq f(\boldsymbol{v})\exp[-\tfrac{1}{2}d_1 n^{1/3}]$$

For $\boldsymbol{I}_2$, we proceed as in the proof of Lemma 3.1.2 in Johnson and Ladalla (1976),

$$\boldsymbol{I}_2 = \int_{\|\boldsymbol{t}\| > \sqrt{n}\delta_2} f(\boldsymbol{v}|\boldsymbol{t})g(\boldsymbol{t})R_n(\boldsymbol{t})dt < e^{-n\epsilon}\int_{\|\boldsymbol{t}\| > \sqrt{n}\delta_2} f(\boldsymbol{v}|\boldsymbol{t})g(\boldsymbol{t})d\boldsymbol{t} \leq f(\boldsymbol{v})e^{-n\epsilon}$$

Next,

$I_3 = \int_{\|t\|>n^{1/6}} \exp[\frac{1}{2}\boldsymbol{t}'Q(\hat{\boldsymbol{\theta}}_n)\boldsymbol{t}]|P_s^T(\boldsymbol{t},\boldsymbol{v},\omega)|d\boldsymbol{t}$

where $P_s^T(\boldsymbol{t},\boldsymbol{v},\omega)$ is a polynomial in the t_i's whose constants $C_{l_1\cdots m_H}(\boldsymbol{v},\omega)$, given in (6.7), satisfy (6.12). Except that the constants $C_{l_1\cdots m_H}(\boldsymbol{v},\omega)$ are functions of $\boldsymbol{v}$, we proceed as in the reference to get

$$I_3 \le \sum_{j=0}^{s} \sum_{l_1+\cdots+m_H=j} |C_{l_1\cdots m_H}(\boldsymbol{v},\omega)|cn^{-\frac{s+1}{2}} = n^{-\frac{s+1}{2}} M_1^s(\boldsymbol{v},\omega)$$

By (6.12) $M_1^s(\boldsymbol{v},\omega)$ has its integrals uniformly bounded. Thus, we take the constant $M_{\omega,2}^s(\boldsymbol{v}) = M_1^s(\boldsymbol{v},\omega) + 2f(\boldsymbol{v})$ to obtain

$$I = \int_{\|\boldsymbol{t}\|>n^{1/6}} |f(\boldsymbol{v}|\boldsymbol{t})g(\boldsymbol{t})R_n(\boldsymbol{t}) - \exp[-\frac{1}{2}\boldsymbol{t}'Q(\hat{\boldsymbol{\theta}}_n)\boldsymbol{t}]P_s^T(\boldsymbol{t},\boldsymbol{v},\omega)|d\boldsymbol{t} \tag{6.19}$$

$$\le M_{\omega,2}^s(\boldsymbol{v})n^{-\frac{s+1}{2}} \qquad \text{for } n > N_{\omega,2}^s$$

There $N_{\omega,2}^s$ is a positive integer and

$\mathrm{Sup}_{n\ge N_{\omega,2}}^s \int M_{\omega,2}^s(\boldsymbol{v})d\mu(\boldsymbol{v}) \le c < \infty$ almost surely.

As a consequence of (6.13) and (6.19), we conclude that

$$\int |f(\boldsymbol{v}|\boldsymbol{t})g(\boldsymbol{t})R_n(\boldsymbol{t}) - \exp[-\frac{1}{2}\boldsymbol{t}'Q(\hat{\boldsymbol{\theta}}_n)\boldsymbol{t}]P_s^T(\boldsymbol{t},\boldsymbol{v},\omega)|d\boldsymbol{t}$$

$$\le (M_{\omega,1}^s(\boldsymbol{v}) + M_{\omega,2}^s(\boldsymbol{v}))n^{-\frac{s+1}{2}}$$

so, from (6.6), we get

$$|k(\boldsymbol{v},\tilde{\boldsymbol{Z}})| \le M_{\omega,1}^s(\boldsymbol{v}) + M_{\omega,2}^s(\boldsymbol{v})$$

for some sufficiently large N_ω, on almost sure set. Thus $k(\boldsymbol{v},\tilde{\boldsymbol{Z}})$ has its integrals uniformly bounded. Moreover, $K(\boldsymbol{v},\tilde{\boldsymbol{Z}})$ in (6.3) is some linear combination of $\beta_j(\boldsymbol{v},\tilde{\boldsymbol{Z}})$'s and $k(\boldsymbol{v},\tilde{\boldsymbol{Z}})$, where the coefficients depend on $\beta_j(\tilde{\boldsymbol{Z}})$'s and $k(\tilde{\boldsymbol{Z}})$. These latter quantities do not depend on $\boldsymbol{v}$ and, as functions of n, converge to finite constants. Consequently $|K(\boldsymbol{v},\tilde{\boldsymbol{Z}})|$ has its integrals uniformly bounded.

7. ASYMPTOTIC BOUND FOR THE BAYES RISK OF THE PREDICTIVE RULE

Under the assumed cost structure, the Bayesian risk can be expressed as (see Section 2)

$$r_p = c_{2|1}P\left[\ \boldsymbol{V}\epsilon B\ \right] - E\int_{[g_p>0]} h_1(\boldsymbol{v}_1|\ \widetilde{\boldsymbol{Z}})g_p(\boldsymbol{v}_1)d\mu_1(\boldsymbol{v}_1) \tag{7.1}$$

where,

$$g_p(\boldsymbol{v}_1) = (c_{1|2}+c_{2|1})\int_B h(\boldsymbol{v}_2|\boldsymbol{v}_1,\widetilde{\boldsymbol{Z}})d\mu_2(\boldsymbol{v}_2) - c_{1|2}, \tag{7.2}$$

$h_1(.|\ \widetilde{\boldsymbol{Z}})$ is the marginal predictive density of $\boldsymbol{V}_1$, and $h(\boldsymbol{v}_2|\boldsymbol{v}_1,\widetilde{\boldsymbol{Z}})$ is the conditional predictive density of $\boldsymbol{V}_2$ given $\boldsymbol{v}_1$.

We expand g_p as

$$g_p(\boldsymbol{v}_1) + c_{1|2} = \tag{7.3}$$

$$(c_{1|2}+c_{2|1})\int_B \Big(\ f(\boldsymbol{v}_2|\boldsymbol{v}_1,\hat{\boldsymbol{\theta}}_n) + n^{-1}\gamma(\boldsymbol{v}_2|\boldsymbol{v}_1,\widetilde{\boldsymbol{Z}}) + n^{-2}\overline{\boldsymbol{K}}(\boldsymbol{v}_2|\boldsymbol{v}_1)\ \Big)\, d\mu_2(\boldsymbol{v}_2)$$

where

$$\gamma(\boldsymbol{v}_2|\boldsymbol{v}_1) = \frac{\gamma(\boldsymbol{v}) - f(\boldsymbol{v}_2|\boldsymbol{v}_1,\hat{\boldsymbol{\theta}}_n)\gamma_1(\boldsymbol{v}_1)}{f_1(\boldsymbol{v}_1|\hat{\boldsymbol{\theta}}_n)},$$

$\gamma_1(.)$ and $\gamma(.)$ are the first correction terms of the expansions of $h_1(.|\ \widetilde{\boldsymbol{Z}})$ and $h(.|\ \widetilde{\boldsymbol{Z}})$, respectively. Equivalently

$$g_p(\boldsymbol{v}_1) = \hat{g}(\boldsymbol{v}_1) + n^{-1}\ \widetilde{\gamma}_1\ (\boldsymbol{v}_1) + n^{-2}\overline{k}_1(\boldsymbol{v}_1)$$

where

$$\hat{g}(\boldsymbol{v}_1) = (c_{1|2}+c_{2|1})\int_B f(\boldsymbol{v}_2|\boldsymbol{v}_1,\hat{\theta}_n)d\mu_2(\boldsymbol{v}_2) - c_{1|2},$$

$$\widetilde{\gamma}_1\ (\boldsymbol{v}_1) = \int_B (c_{1|2}+c_{2|1})\gamma(\boldsymbol{v}_2|\boldsymbol{v}_1)d\mu_2(\boldsymbol{v}_2) \text{ and}$$

$$\overline{k}_1(\boldsymbol{v}_1) = \int_B (c_{1|2}+c_{2|1})\overline{K}(\boldsymbol{v}_2|\boldsymbol{v}_1,\widetilde{\boldsymbol{Z}})d\mu_2(\boldsymbol{v}_2).$$

Following the usual steps (see Mouhab (1991)), we get

Theorem 4.

Under assumptions of Theorem 3,

$$0 \leq \hat{r} - r_p \leq n^{-1}\overline{\gamma} + O(n^{-2}) \tag{7.4}$$

where

$$\hat{r} = c_{2|1}P\Big[\ \boldsymbol{V}\epsilon B\ \Big] - E\int_{[\hat{g}>0]} h_1(\boldsymbol{v}_1|\ \widetilde{\boldsymbol{Z}})g_p(\boldsymbol{v}_1)d\mu_1(\boldsymbol{v}_1)$$

is the risk of the ML Estimated screening rule, and

$$\overline{\gamma} = E\int|\ \widetilde{\gamma}_1\ (\boldsymbol{v}_1)|h_1(\boldsymbol{v}_1|\ \widetilde{\boldsymbol{Z}})d\mu_1(\boldsymbol{v}_1)$$

is finite.

Proof: We break the integral into positive and negative parts.

$$r_p = c_{2|1}P\Big[\ \boldsymbol{V}\epsilon B\ \Big] - E\int_{[g_p>0]} h_1(\boldsymbol{v}_1|\ \widetilde{\boldsymbol{Z}})g_p(\boldsymbol{v}_1)d\mu_1(\boldsymbol{v}_1)$$

$$= \hat{r} - E\int_{[g_p>0]/[\hat{g}>0]} h_1(\boldsymbol{v}_1|\ \widetilde{\boldsymbol{Z}})g_p(\boldsymbol{v}_1)d\mu_1(\boldsymbol{v}_1) -$$

$$E\int_{[\hat{g}>0]/[g_p>0]} h_1(\boldsymbol{v}_1|\ \widetilde{\boldsymbol{Z}})\Big(\ -g_p(\boldsymbol{v}_1)\ \Big)\ d\mu_1(\boldsymbol{v}_1).$$

Consequently,

$$0 \leq \hat{r} - r_p \leq E\int h_1(\boldsymbol{v}_1)\Big(\ g_p(\boldsymbol{v}_1) - \hat{g}(\boldsymbol{v}_1)\ \Big)^+\ d\mu_1(\boldsymbol{v}_1)$$

$$+E\int h_1(\boldsymbol{v}_1)\Big(\ g_p(\boldsymbol{v}_1) - \hat{g}(\boldsymbol{v}_1)\ \Big)^-\ d\mu_1(\boldsymbol{v}_1)$$

$$= E\int h_1(\boldsymbol{v}_1|\ \widetilde{\boldsymbol{Z}})|n^{-1}\ \widetilde{\gamma}_1\ (\boldsymbol{v}_1) + n^{-1}\overline{k}_1(\boldsymbol{v}_1)|d\mu_1(\boldsymbol{v}_1).$$

So the result follows.

References

Boys, R.J. and Dunsmore, I.R. (1986). "Screening in a Normal Model", *Journal of the Royal Statistical Society, B* 48, 60-69.

CMLIB, Kahaner D. (1983). Scientific Computing Division, NBS Marsaglia, George, Computer Science Department, Washington State University.

Dickey, J.M. (1980). "Approximate Broadbalk Wheatfield Example". *Bayesian Analysis in Economics and Statistics*, North Holland Publishing Company, 333-354.

Guttman, I. (1970). "Statistical Tolerance Regions: Classical and Bayesian". Darien, Conn., Hafner Pub. Co.

Johnson, R.A. (1967). "Asymptotic Expansions Associated with the n-th Power of a Density". *Ann. Math. Statist.*, 38, 1266-1272.

Johnson, R.A. (1970). "Asymptotic Expansions Associated with Posterior Distribution", *Ann. Math. Statist.,* 41, 851-864.

Johnson, R.A. and Ladalla, J. (1976). "The Large Sample Behavior of Posterior Distributions When Sampling From Multivariate Exponential Family Models, and Allied Results". Part I, Technical Report, University of Wisconsin-Madison.

Johnson, R.A. and Ladalla, J. (1976). "The Large Sample Behaviour of Posterior Distributions When Sampling From Multivariate Exponential Family Models, and Allied Results". Pat II, Technical Report, University of Wisconsin-Madison.

Johnson, R.A. and Ladalla, J. (1978). "The Large Sample Behavior of Posterior Distributions When Sampling From Multivariate Exponential Family Models, and Allied Results". *Sankhya: The Indian Journal of Statistics*, 41, 196-215.

Ladalla, J. (1976). "The Large Sample Behavior of Posterior Distributions When Sampling From Multivariate Exponential Family Models, and Allied Results". Ph.D. Thesis, University of Wisconsin-Madison.

Lehmann, E.L. (195). "Testing Statistical Hypothesis". John Wiley, New York.

Li, L. and Owen, D.B. (1970). "Two-Sided Screening Procedures in the Bivariate". *Technometrics*, 21.

Mee, R.W. (1990). "An Improved Procedure for Screening Based on a Corollated, Normally Distributed Variable". *Technometrics*, 32, 83-92.

Menzefricke, U. (1984). "A Decision-Theoretic Approach to Some Screening Problems". *Annals of the Institute of Statistical Mathematics*, Part A, 36, 485-497.

Mouhab, A. (1991). "A Bayesian Decision Theory Approach to Screening and Classification, With Large Samples Results". Ph.D. Thesis, University of Wisconsin-Madison.

Owen, D.B. and Su, Y.H. (1977). "Screening Based on Normal Variables". *Technometrics*, 19, 65-68.

Owen, D.B. (1980). "A Table of Normal Integrals". *Communication in Statist. Part B: Simulation and Computation*, 9, 389-419.

Owen, D.B., Li, L. and Chou, Y. (1981). " Prediction Intervals for Screening Using a Measured Correlated Variate". *Technometrics*, 23, 165-170.

Wong, A., Meeker, J.B. and Selwyn, M.R. (1985). "Screening on Correlated Variables: A Bayesian Approach". *Technometrics*, 27, 423-431.

ADVANCES IN RELIABILITY
Edited by Asit P. Basu

Macroscale activation energy in integrated circuit reliability

J. Kitchin

Semiconductor Operations Group, Digital Equipment Corporation, 77 Reed Road, Hudson MA 01749, USA

Abstract

Consider a element or device in an integrated circuit having a thermally activated failure mechanism yielding a time to failure with nominal value $\tau = \Psi(\mathbf{\Gamma}, A) \exp(E_a/kT)$ where E_a is the actual "microscale" activation energy for the mechanism. This paper compares the effect on the "macroscale" activation energy determined by Arrhenius plots of failure data from units, such as integrated circuits, containing N such elements in series failing independently for two classes of statistical models—one where E_a varies randomly across failure elements versus one where the pre-exponential parameter A varies randomly across the failure elements. It is shown that for commonly used functional forms for $\Psi(\mathbf{\Gamma}, A)$ the random microscale activation energy model will yield a significantly reduced value for the macroscale activation energy of the units when N is large. The electromigration failure model of Lloyd and Kitchin [*J. App. Phys.*, 69 (1991)] for length scaling of fine-line interconnect reliability is shown to be an example of this effect. It is speculated that the wide range of reported values for activation energy for some failure mechanisms may be partly explained by this N-element effect on macroscale activation energies.

1. INTRODUCTION

Consider the following model for the (nominal) time to failure τ of a component having a thermally-activated failure mechanism:

$$\tau = \Psi(\mathbf{\Gamma}, A) \exp(E_a/kT) \tag{1}$$

where E_a is the activation energy for the mechanism, k is Boltzmann's constant, T is absolute temperature, $\mathbf{\Gamma}$ is a vector of model parameters, which may depend on T and include additional accelerating stresses (such as current density) and failure criteria (such as concentration level of a failure species), and A is a parameter independent of T.

Several time to failure models employed in the prediction of integrated circuit reliability are special cases of Equation (1). The simplest

$$\tau = A \exp(E_a/kT), \tag{2}$$

is known as the Arrhenius model. Klinger, Nakada, and Menendez (1990) point to more general models where $\Psi(\mathbf{\Gamma}, A) = AT^{-n}$ with n depending on the detailed model of the reaction dynamics driving the failure mechanism. Nelson (1990) refers to the case of $n = 1$ as the Eyring model. The literature on the electromigration failure mechanism for integrated circuit interconnect has dealt extensively with models using $\Psi(\mathbf{\Gamma}, A) = Aj^{-n}$, where j is current density. The model $n = 2$ argued by Black (1969) is commonly used. These electromigration models are equivalent to Equation (2) for the purpose of studying thermal activation, but Shatzkes and Lloyd (1986) argue for $\Psi(\mathbf{\Gamma}, A) = AT^2j^{-2}$ obtained from solving a diffusion equation for the failure species. Peck (1986) proposes $\Psi(\mathbf{\Gamma}, A) = AH^{-n}$, where H is relative humidity, to (empirically) model an ensemble of data on the effect of humidity on interconnect reliability in non-hermetic integrated circuits. Equation (2) is often found to be adequate in temperature-accelerated lifetesting because of the relatively weak dependence of $\Psi(\mathbf{\Gamma}, A)$ on T compared to that of $\exp(E_a/kT)$.

At least two features are common to all attempts to apply the model of Equation (1): (a) "determination" of E_a is necessary for accurate prediction of integrated circuit reliability at use temperatures from time to failure data from integrated circuits or test vehicles (which we will call "units") typically taken at accelerated temperatures, and (b) time to failure τ will be observed to vary randomly from unit to unit, so a statistical model must be superimposed.

In this paper we examine the effect on the determination of E_a of considering each unit to be composed of N independently failing "elements" (failure elements) and each failure element having an activation energy drawn at random such that

$$\Pr\{E_a \leq e\} = W_{\mathbf{\Theta}}(e), \tag{3}$$

where

$$\frac{\partial \mathbf{\Theta}}{\partial T} \equiv \mathbf{0}, \tag{4}$$

that is, the parameters $\mathbf{\Theta}$ of the distribution family W do not depend on T. Only the earliest element failure time in a unit is observable. We show how the activation energy "determined" from the commonly used Arrhenius plot technique depends on $\mathbf{\Theta}$ and N, and suggest this dependence may partly explain the wide ranges in activation energy determined for some integrated circuit failure mechanisms.

2. MICROSCALE AND MACROSCALE

In studying the reliability of objects whose failure behavior is governed by physical and chemical interactions it is useful to distinguish a "macroscale" of events and model parameters from a "microscale" of such, a distinction made in statistical physics. (See, for example, Reif (1965), Chapter 1.)

The microscale refers to the scale of atomic and molecular interaction, where statistical thermodynamics and kinetics are the basis for models. A microscale activation energy would be a physically meaningful parameter in such models, and refer to some difference in free energy between initial states and resultant states of atoms and molecules.

The macroscale refers to the scale of events observable with modern but common instrumentation, such a failures detectable by microscopy or a change in an electrical parameter such as resistance.

Often, macroscale models of integrated circuit failure are refined by detailing a microscale model of the physics of failure, say by use of a model such as Equation (1) with E_a determined by separate means that can quantify a microscale activation energy.

We refer to the single-value activation energy estimated through repeated observation of τ analyzed with respect to Equation (1) as a macroscale activation energy, to distinguish it from both the microscale activation energy concept and the parameters of any distribution of microscale activation energy. Such an activation energy determined in this way is necessarily the result of macroscale failure criteria and observation, rather than by direct examination of the microscale mechanism that leads to the failure. Other authors have used the term "apparent" activation energy (e.g., Blanks (1980), Klinger et al. (1990)).

In the Arrhenius plot technique for determining E_a, one plots the log median of a sample of unit lifetimes for each of two or more values of T versus $1/kT$ and applies the inversion of Equation (2)

$$E_a = \frac{\partial \ln(\tau)}{\partial(1/kT)} \tag{5}$$

See, for example, page 84 of Towner (1985). Frequently in reliability analysis (and unlike the Towner example) E_a is introduced *only* at the macroscale as a fitting parameter in a model such as Equation (2), without reference to a specific microscale process.

3. ACTIVATION ENERGY AS A RANDOM VARIABLE

Some researchers in reliability physics have considered microscale activation energy as a randomly varying quanitity. Dutta and Horn (1981) empirically derive a bell-shaped distribution of energies (whose peak is used a fixed activation energy) for noise in thin Ag films from studies of the change in noise magnitude with temperature. Joyce et al. (1985) assume that device-to-device random variations in the composition x of the $Al_xGa_{1-x}As$ confining layer in semiconductor lasers lead to $W_{\Theta}(e) = \Phi((e-\mu)/\sigma)$, where Φ is the standard normal cumulative density function, and point out that a lognormal distribution for laser lifetime τ modeled by Equation (2) results. Lloyd and Kitchin (1991) take $W_{\Theta}(e) = \Phi((e-\mu)/\sigma)$ for activation energy for grain boundary diffusion as a starting point in a statistical model for electromigration-induced failure in Al interconnects. Physically reasonable values of μ and σ are postulated for a Monte Carlo study of the sampling variation of macroscale activation energy for electromigration.

4. AN ALTERNATE STATISTICAL MODEL

Many of the statistical failure time models described in Nelson (1990) and commonly used in modeling integrated circuit reliability are a less general form of Equation (1)

$$\tau = A\psi(\mathbf{\Gamma}) \exp(E_a/kT) \tag{6}$$

in which the scaling parameter A has the distribution

$$\Pr\{A \leq a\} = W_{\Theta}(a), \tag{7}$$

and E_a is fixed. In the next section we compare this common statistical model to that in which E_a is random to study the difference the two statistical models have in their effect on the determination of a macroscale activation energy.

5. MACROSCALE ACTIVATION ENERGY IN THE TWO STATISTICAL MODELS

Consider an accelerated test to failure of a series system of N independent elements with failure distribution $\rho F(t)$ where ρ is the probability that the failure mechanism modeled by Equation (1) is present and $F(t) = \Pr\{\tau \leq t\}$. Then distribution of system time to failure is

$$G(t) = 1 - [1 - \rho F(t)]^N. \tag{8}$$

The above model with $\rho = 1$ is sometimes referred to as a "weakest link" model in the non-statistics literature.

The goal of this section is to quantify, for each of the two statistical models introduced, the effect of increasing N (that is, increasing the scale of the test vehicle or the integrated ciruit) on the macroscale activation energy as determined via Equation (5).

For generality let t_q be such that $G(t_q) = q$ and assume that the number of identical test units is large enough to make the sampling variability of estimator for the quantile t_q insignificant reletive to the effects we are trying to uncover. For convenience let

$$q_{N,\rho} \stackrel{\text{def}}{=} (1 - (1-q)^{(1/N)})/\rho, \tag{9}$$

$$\Delta(\mathbf{\Gamma}, A, T) \stackrel{\text{def}}{=} \frac{T}{\Psi(\mathbf{\Gamma}, A)} \frac{\partial \Psi(\mathbf{\Gamma}, A)}{\partial T}. \tag{10}$$

Under the E_a random model of Equations (3) and (4) the expression for the macroscale activation energy is

$$\frac{\partial \ln(t_q)}{\partial(1/kT)} = -kT\Delta(T, \mathbf{\Gamma}, A) + W_{\Theta}^{-1}[q_{N,\rho}] \tag{11}$$

so that the activation energy determined decreases as N increases. Note in particular that for large N and ρ the shape of the left tail of the distribution W_{Θ} has the dominant effect on the activation energy determined. The example in the next section gives some specific values to illustrate this effect.

Compare Equation (11) to

$$\frac{\partial \ln(t_q)}{\partial(1/kT)} = -kT\Delta\left(T, \mathbf{\Gamma}, W_{\Theta}^{-1}[q_{N,\rho}]\right) + E_a \tag{12}$$

which results from the statistical model of Equations (7) and (4), where A is random and E_a is constant.

Note that $kT \approx 0.04$ electron-volts (eV), the usual unit of activation energy, at $T = 473K$, in the range of common accelerated test conditions. We can focus on the quantity $\partial\Delta(T, \mathbf{\Gamma}, A)/\partial A$ as a gross measure of the sensitivty of the macroscale activation energy determination to N. Note that for the failure time model of Equation (6), (which holds for all examples given in the Introduction) it follows that $\partial\Delta(T, \mathbf{\Gamma}, A)/\partial A \equiv 0$, and thus (unlike the E_a random model) N has no effect on the macroscale activation energy. If, more generally, for T in the range of accelerated testing, $\Delta(T, \mathbf{\Gamma}, A)$ is slowly varying in A over the range of A's random variability relative to 0.04 eV, N will have little effect on the macroscale activation energy. Should both A and E_a be (independently) random, the insensitivity argument still applies, and offers a way to separate the (now) two sources of variability in τ.

Thus for large N the estimated macroscale activation energy can differ considerably from the typical microscale activation energy. This difference may explain some of the range in reported values of activation energies, as the test vehicle used by each experimenter would likely have a different value of N.

6. EXAMPLE

Lloyd and Kitchin (1991) model thin flim interconnects (lines) for integrated circuits as a series system of N independently failing segments (N to be determined initially by grain size). Each segment (element) is modeled with an extension of the Shatzkes and Lloyd (1986) diffusion model for the failure species accumulating in grain boundaries due to flux divergences arising from randomly varying diffusivities in the grain boundaries of a segment. Each grain boundary is modeled as having a microscale activation energy drawn from a normal distribution with parameters μ and σ.

Lloyd and Kitchin find for "fine" lines, that is, lines where line width equals or exceeds median grain size, that Equation (8) with the $\Psi(\mathbf{\Gamma}, A) = AT^2 j^{-2}$ version of Equation (1) as an element failure model and $W_{\Theta}(e) = \Phi((e-\mu)/\sigma)$ is (theoretically) a good approximation when ρ equals to the fraction of segments with exactly 1 grain boundary present. For their model

$$\frac{\partial \ln(t_{.50})}{\partial(1/kT)} \approx -2kT + \mu + \sigma\Phi^{-1}\left[1 - 0.5^{(1/N\rho)}\right] \tag{13}$$

A typical electromigration test vehicle might have $N\rho \approx 500$, and thus a macroscale activation energy of $\approx -2kT + \mu - 3\sigma$. They propose $\mu = 0.75$ eV and $\sigma = 0.053$ eV as physically reasonable values for the microscale activation energy distribution in Al lines. Thus Equation (13) predicts a macroscale activation energy of $-0.08 + 0.75 - 0.16 = 0.51$ eV, similar to values found in others electromigration studies (such as Towner (1985).

For $W_{\Theta}(e) = \Phi((e-\mu)/\sigma)$ (and in any location-scale family of distributions for E_a) a large number of (acting) failure elements means that the macroscale activation energy is a value deep in the left tail of the distribution. Hence the tail behavior more than the overall shape of distribution is important in accurate characterzation of the temperature scaling model, unlike the models where A is modeled as random. (The E_a random model also leads to a temperature dependence in the standard deviation of τ.)

Kitchin and Lloyd (1991) propose the relationship of Equation (13) as a way to falsify their model. The difference in macroscale activation energy from $N = 10$ to $N = 1000$ is about 0.1 eV, thought to be detectable in a well-designed and controlled study. They also speculate that it may be the source for some of the descrepancies in published determinations of the (seemingly) microscale activation energy for electromigration.

7. SUMMARY

We propose a general family $\tau = \Psi(\mathbf{\Gamma}, A) \exp{(E_a/kT)}$ of failure time models and draw a distinction in such models between a microscale activation energy E_a based in the physics and chemistry of a stressed failure element and a macroscale activation energy determined by Arrhenius plots of failure data of stressed integrated circuits containing N such elements in series failing independently. We compare the effect of N on the determination of the macroscale activation energy in two classes of statistical models—one where E_a varies randomly across failure elements versus one where A varies randomly across the failure elements. Under commonly used functional forms for $\Psi(\mathbf{\Gamma}, A)$, the former model class leads to a significantly reduced value for the macroscale activation energy of an integrated circuit when N is large, compared to the insentivity to N of the latter model class. The electromigration failure model of Lloyd and Kitchin (1991) for length scaling of fine-line interconnect reliability is an example of this N-element effect, an effect that may partly explain the wide ranges of reported values for activation energy of some failure mechanisms.

8. ACKNOWLEDGEMENTS

The author is greatful to J. R. Lloyd of Digital Equipment Corporation and to M. J. Luvelle and D. J. Klinger of AT&T Bell Laboratories for helpful discussions and comments.

References

[1] Black, J. R. (1969), "Electromigration—A Brief Survey and Some Recent Results," *IEEE Trans. Electron Dev.*, ED-16, 338-347.

[2] Blanks, H. S. (1980), "The Temperature Dependence of Component Failure Rate," *Microelectron. Reliab.*, 20, 297-307.

[3] Dutta, P. and Horn, P. A. (1981), "Low-frequency Fluctuations in Solids: 1/f Noise," *Rev. Mod. Physics*, 53, 497-516.

[4] Joyce, W. B., Liou, K-Y., Nash, F. R., Bossard, P. R., and Hartman, R. L. (1985), "Methodology of Accelerated Aging" *AT&T Technical Journal*, 64, 717-764.

[5] Kitchin, J. and Lloyd, J. R. (1991), "Investigations on a New Failure Model for Electromigration," *Materials Research Society Symposium Proceedings*, 225, 27-33.

[6] Klinger, D. J., Nakada, Y., and Menendez, M. A. (1990), *AT&T Reliability Manual*, New York: Van Nostrand Reinhold.

[7] Lloyd, J. R., and Kitchin, J. (1991), "The Electromigration Failure Distribution: The Fine-line Case," *Journal of Applied Physics*, 69, 2117-2127.

[8] Nelson, W. (1990), *Accelerated Testing*, New York: Wiley.

[9] Peck, D. S. (1986), "Comprehensive Model for Humidity Testing Correlation," *Proc. International Reliability Physics Symp.*, 24, 44-50.

[10] Reif, F. (1965) *Fundamentals of Statistical and Thermal Physics*, New York: McGraw-Hill.

[11] Shatzkes, M. and Lloyd, J. R. (1986), "A Model for Conductor Failure Considering Diffusion Concurrently with Electromigration Resulting in a Current Exponent of 2," *Journal of Applied Physics*, 59, 3890-3892.

[12] Towner, J. M. (1985), "Electromigration-induced Short Circuit Failure," *Proc. International Reliability Physics Symp.*, 23, 81-86.

ADVANCES IN RELIABILITY
Edited by Asit P. Basu

ON MAXIMUM LIKELIHOOD ESTIMATION BASED ON RANKED SET SAMPLES, WITH APPLICATIONS TO RELIABILITY

P.H. Kvam[a] and F. J. Samaniego[b]

[a]P.H. Kvam is Statistician, Los Alamos National Laboratory, Los Alamos, New Mexico 87545.

[b]F.J. Samaniego is Professor, Division of Statistics, University of California, Davis, CA, 95616. His work is supported in part by grant MDA904-90-H-4031 from the National Security Agency.

Abstract

A ranked set sample from an underlying probability distribution F consists of a set of independently observed order statistics from F. Ranked set samples occur in life testing when one may observe the number of failed components in a system in conjunction with system failure time. The problem of estimating the common distribution F of component lifetimes based on such data is addressed via the method of maximum likelihood. In both parametric and nonparametric settings, MLE's are compared to unbiased estimators proposed in the literature. In both settings, the MLE's are found to be superior.

1. INTRODUCTION

The problem of making inferences about the lifetime distributions of the components of a complex system when only system lifetime data is available is one of the most difficult open problems in Reliability Theory. Some versions of this problem are clearly unsolvable because of the lack of identifiability of the system lifetime distribution as a function of the lifetime distributions of components. When auxiliary information regarding components is available (for example, when one may observe the failure times of all components which fail at or before the time the system fails), lifetime distributions of all components and of the system as a whole can be estimated consistently (see Doss, Freitag and

Proschan (1989)]; in the absence of such information, there has been little progress on the problem.

This paper is dedicated to a special case of this general problem which can actually be treated quite completely. We will show that likelihood methods can be sucessfully applied to the problem of estimating a component lifetime distribution when system lifetime data is available from one or several systems based on components with lifetimes that are independent and identically distributed according to a common underlying distribution F. When the available data consists of the iid lifetimes of systems of the same design, the problem of estimating the common distribution F of the component lifetimes is simply a matter of inverting the functional relationship between F and the system lifetime. The problem of combining lifetime data from systems of varying design is more complicated -- this is the problem to which our attention will be directed.

Our treatment will require some, but actually very little, auxiliary information on component lifetimes; we will assume that when a system fails, one can ascertain precisely <u>how many</u> components have failed. A simple autopsy (see Meilijson (1981) for a discussion) of a failed system will typically reveal which components have failed, even though their actual failure times will generally be unknown. The auxiliary information provided by such an autopsy, together with our assumption of iid component lifetimes, makes it possible to identify the system lifetime as a specific order statistic drawn from the (mostly) unobserved random sample of component lifetimes (see Samaniego (1985) for further details). Actually, for r-out-of-k systems, it will be known with certainty, without need for an autopsy, that the system failure time is in fact the $(k-r+1)^{st}$ smallest order statistic among the k component lifetimes. Let us assume that a collection of n systems, the ith system having k_i components, are independently placed on test, and that the system lifetime, together with the number of failed components at system failure, is recorded for each. We will be interested in the following inference problem: If $X_{s:t}$ represents the sth smallest order statistic in a random sample of size t from the distribution F, then, given the independent observations $\{X_{r_i:k_i}, i=1,\ldots,n\}$, how might we best estimate F? Before presenting parametric and nonparametric treatments of this problem, we pause briefly to review the general framework within which the present problem is nested.

The process of "ranked set sampling" was introduced by McIntyre (1952) for application in an agricultural experiment. The technique was developed in order to exploit the fact that measuring pasture yields from small research plots was time consuming and costly, while it was easy for observers examining these yields to rank them visually according to size as long as their number was small. In this initial intended application, yields were to be groupd into random subsamples of size k. The smallest yield was chosen and measured from the first group, the second smallest from the next group, and so on. The result was a collection of independently observed order statistics. When each of the k order statistics is replicated in the sample the same number of times (say m), the ranked set sample is said to be balanced. Such a sample contains mk observations, and may be displayed in a rectangular array as follows:

$$X_{1:k:1},\ X_{1:k:2},\ldots,X_{1:k:m}$$
$$X_{2:k:1},\ X_{2:k:2},\ldots,X_{2:k:m}$$
$$\ldots\ldots$$
$$X_{i:k:1},\ X_{i:k:2},\ldots,X_{i:k:m}$$
$$\ldots\ldots$$
$$X_{k:k:1},\ X_{k:k:2},\ldots,X_{k:k:m} \tag{1}$$

Each row of (1) is assumed to be independent of every other row, and, assuming a common underlying distribution F, the random variables $\{X_{i:k:j},\ j=1,\ldots,m\}$ are iid with distribution function given by

$$F_{i:k}(x) = \sum_{r=i}^{k} \binom{k}{r} F(x)^r (1-F(x))^{k-r} . \tag{2}$$

Early work on ranked set sampling concentrated on estimation of the population mean. Takahasi and Wakimoto (1968) proved McIntyre's assertion that the estimation of a population mean will be improved by using the resulting sample average ($\bar{X}_{rss}$) of the mk measurements in a balanced ranked set sample instead of the sample average ($\bar{X}_{srs}$) of a simple random sample of the same size. They proved that $\bar{X}_{rss}$ is the only unbiased linear estimator of the mean of an arbitrary distribution F that can be constructed from a balanced ranked set sample; they also proved, using the calculus of variations on the parameter F, that the variances of the two estimators $\left(\sigma^2_{rss}, \sigma^2_{srs}\right)$ satisfy the inequalities

$$\frac{2}{k+1}\sigma^2_{srs} \le \sigma^2_{rss} \le \sigma^2_{srs} . \qquad (3)$$

These bounds were shown to be sharp. The improvement of $\bar{X}_{rss}$ over $\bar{X}_{srs}$ can be traced to the fact that the order statistics comprising a ranked set sample are independent and contain more information than a simple random sample of the same size (since, in the latter case, the ordered values are correlated).

More recently, the problem of estimating a distribution function nonparametrically based on a ranked set sample has been examined. Stokes and Sager (1988) showed that the empirical distribution function F_n based on the n = mk measurements from a balanced ranked set sample has several desirable properties as an estimator of the underlying distribution function F. In particular, they showed that F_n is an unbiased estimator of F, and is uniformly more efficient in estimating F than the empirical distribution function based on a simple random sample of the same size.

In reliability applications, there are two good reasons to consider alternative approaches to the estimation of μ and F based on a ranked set sample. First, the estimators $\hat{\mu}$ and F_n above are well-defined only for balanced ranked set samples; in reliability studies, it would often be the case that lifetime data is available on a collection of systems for which certain order statistics cannot be observed. For example, designed redundancy might make it impossible to observe the smallest order statistic, since the system never fails upon the first component failure. Further, the order of the failure time of the component which causes a system to fail is typically random, making it quite unlikely that a ranked set sample obtained in practice would be balanced. The need for estimation procedures applicable to unbalanced ranked set samples has motivated us to examine the performance of likelihood methods in this problem. Collateral motivation is provided by the desire to exploit any additional information that the experimentor might have regarding the life- testing experiment of interest. For example, would one wish to use $\overline{X}_{rss}$ as an estimator of the mean of an exponential distribution in a balanced ranked set sample? We will show that the answer to this question is <u>no</u>.

In section 2, we treat a parametric version of the problem of estimating mean life based on a ranked set sample. In section 3, the nonparametric version of this problem is considered, with the goal of constructing a reasonable estimator of the underlying distribution F which applies equally to the balanced and unbalanced cases. Because of space limitations, we will emphasize here the likelihood approach to the analysis of balanced ranked set samples; we refer the reader to Kvam (1990) and Kvam and Samaniego (1991) for details on the extension of our results to the unbalanced case. It should, however, be clear from our derivations that the extension to unbalanced data causes no essential difficulties.

2. THE EXPONENTIAL DISTRIBUTION

Because of its prevalance in life testing and other reliability related experiments, the family of exponential distributions seems worthy of individual analysis. This section is devoted to the study of maximum likelihood estimation of the mean of an exponential distribution given a balanced ranked set sample. For simplicity, we will express the likelihood function in terms of the distribution's failure rate rather than in terms of its mean. Accordingly, we define

$$\mathcal{L}(\lambda) = \prod_{i=1}^{k} \prod_{j=1}^{m} \lambda\, e^{-\lambda(k-i+1)x_{i:k:j}} \left(1-e^{-\lambda x_{i:k:j}}\right)^{i-1} \tag{4}$$

Let $L(\lambda)$ denote $\log \mathcal{L}(\lambda)$, and denote its first and second derivative with respect to λ as L' and L'':

$$L'(\lambda) = \frac{mk}{\lambda} - \sum_{i=1}^{k} \sum_{j=1}^{m} \left[(k-i+1)x_{i:k:j} + \frac{(i-1)x_{i:k:j}e^{-\lambda x_{i:k:j}}}{1-e^{-\lambda x_{i:k:j}}}\right],$$

$$L''(\lambda) = -\frac{mk}{\lambda^2} - \sum_{i=1}^{k} \sum_{j=1}^{m} \frac{(i-1)x^2_{i:k:j}e^{-\lambda x_{i:k:j}}}{\left(1-e^{-\lambda x_{i:k:j}}\right)^2}.$$

Since $L''(\lambda) < 0$ for all values of $\lambda > 0$, the unique solution of $L'(\lambda) = 0$ is the MLE. To calculate the MLE, several alternative numerical methods may be applied without computational difficulty; we suggest Newton's algorithm, summarized by the equation:

$$\lambda_{new} = \lambda_{old} - L'(\lambda_{old})/L''(\lambda_{old}).$$

In our experience, the iterations produced by this method converge quite rapidly.

We now turn to a comparison of the natural competing estimators of the mean of an exponential distribution: the "traditional" estimator $\bar{X}_{rss}$ and the MLE discussed above $\left(\text{i.e., } \hat{\mu} = \frac{1}{\hat{\lambda}}\right)$. Both estimators will be asymptotically normal; we will denote the asymptotic variance of $\bar{X}_{rss}$ and $\hat{\mu}$ as V_{rss} and V_{mle}, respectively. The following lemma and theorem suggest a natural ordering of these competing estimators.

<u>Lemma 1</u>. Let $I_r(\lambda)$ denote the Fisher information corresponding to $f_{r:k}(x;\lambda)$, the probability density function of the rth smallest order statistic from a random sample of k exponential random variables with mean $\mu = \frac{1}{\lambda}$. For fixed $k \geq 3$,

$$I_r(\lambda) = \begin{cases} \lambda^{-2} & r=1 \\ \lambda^{-2}\left(1+k(k-1)\int_0^1 (\log t)^2 \frac{t^{k-1}}{1-t}\,dt\right) & r=2 \\ \lambda^{-2} + \frac{k(k-r+1)}{r-2} E\left(X^2_{r-2:k-1}\right) & 3\leq r\leq k. \end{cases} \tag{5}$$

If k=2, the Fisher information is defined by the terms corresponding to r=1 and r=2.

<u>Proof</u>: $I_r(\lambda)$ can be written $-E\left\{\frac{\partial^2}{\partial\lambda^2} \log f_{r:k}(x;\lambda)\right\}$ where $f_{r:k}(x;\lambda) = r\binom{k}{r} \lambda e^{-\lambda x(k-r+1)}(1-e^{-\lambda x})^{r-1}$, $x > 0$, $1\leq r\leq k$. If r=1, we simply obtain $I_1(\lambda) = -E\left\{\frac{\partial^2}{\partial\lambda^2}(\log\lambda - \lambda Xk)\right\} = \lambda^{-2}$. If r=2, the transformation $t=e^{-\lambda x}$ in the resulting integral simplifies the numerical calculations:

$$I_2(\lambda) = \lambda^{-2} + k(k-1)\int_0^\infty x^2\lambda e^{-\lambda kx}(1-e^{-\lambda x})^{-1}\,dx$$

$$= \lambda^{-2}\left(1 + k(k-1)\int_0^1 (\log t)^2 \frac{t^{k-1}}{1-t}\,dt\right).$$

If $3 \le r \le k$, $\frac{\partial}{\partial\lambda} \log f_{r:k}(x;\lambda) = \lambda^{-1} - x(k-r+1) + \frac{(r-1)xe^{-\lambda x}}{1-e^{-\lambda x}}$ and

$\frac{\partial^2}{\partial\lambda^2} \log f_{r:k}(x;\lambda) = -\lambda^{-2} - (r-1)\, x^2 e^{-\lambda x}\left(1-e^{-\lambda x}\right)^{-2}$ so that

$$I_r(\lambda) = \lambda^{-2} + (r-1)\, E\left\{ \frac{X_{r:k}^2 e^{-\lambda X_{r:k}}}{\left(1-e^{-\lambda X_{r:k}}\right)^2} \right\},$$

from which we can extract the final representation of $I_r(\lambda)$ in (5). □

Theorem 1. Suppose a balanced ranked set sample is drawn from an exponential distribution with mean μ. Then, as $n \to \infty$:

$$\sqrt{mk}\left(\bar{X}_{rss} - \mu\right) \xrightarrow{d} N\left(0, V_{rss}\right),$$

and

$$\sqrt{mk}(\hat{\mu}-\mu) \xrightarrow{d} N\left(0, V_{mle}\right), \tag{6}$$

where the asymptotic variances in (6) are given by

$$V_{rss} = \frac{\mu^2}{k} \sum_{r=1}^{k} \frac{1}{r}, \tag{7}$$

and $V_{mle} = \mu^2 / \left(1+(k-1)b_k + \sum_{r=3}^{k} \frac{k-r+1}{r-2} \left(c_{r-2:k-1}^2 + d_{r-2:k-1}\right)\right)$,

with the constants b, c and d given by

$$b_k = \int_0^1 (\log t)^2 \frac{t^{k-1}}{1-t}\, dt,$$

$$c_{i:k} = \sum_{r=1}^{i} \frac{1}{k-r+1}$$

and

$$d_{i:k} = \sum_{r=1}^{i} \frac{1}{(k-r+1)^2}$$

Proof: Since $\sigma_{i:k}^2 = \mu^2 \sum_{r=1}^{i} \frac{1}{(k-r+1)^2}$, it follows that

$$\begin{aligned}
\sigma_{rss}^2 &= \mathrm{Var}\left(\frac{1}{mk} \sum_{i=1}^{k} \sum_{j=1}^{m} X_{i:k:j}\right) \\
&= \frac{\mu^2}{mk^2} \sum_{i=1}^{k} \sum_{r=1}^{i} \left(\frac{1}{k-r+1}\right)^2 \\
&= \frac{\mu^2}{mk^2} \sum_{r=1}^{k} \sum_{i=1}^{r} \frac{1}{r^2} \\
&= \frac{\mu^2}{mk^2} \sum_{r=1}^{k} \frac{1}{r}\,.
\end{aligned} \tag{8}$$

Since $\bar{X}_{rss}$ is a linear combination of independent random variables, its asymptotic distribution follows by standard arguments. Define $I_r(\lambda)$ as in (5) and let $I(\lambda) = \frac{1}{k}\sum_{r=1}^{k} I_r(\lambda)$. If $\hat{\lambda}$ is the unique solution of the likelihood equation $L'(\lambda) = 0$, then under typical regularity conditions (satisfied by the exponential distribution)

$$\sqrt{mk}\ (\hat{\lambda}-\lambda) \xrightarrow{d} N(0, I^{-1}(\lambda)) \tag{9}$$

(see, for example, Lehmann (1983)). In terms of λ, this asymptotic variance may be written

$$V_{mle} = \frac{1}{\lambda^2} + \frac{(k-1)}{\lambda^2} b_k + \sum_{r=3}^{k} \frac{k-r+1}{r-2} E(X_{r-2:k-1}^2). \tag{10}$$

Note that $EX_{r-2:k-1}^2 = \lambda^2(c_{r-2:k-1}^2 + d_{r-2:k-1})$, where c and d are defined in (8) above. The expression in (7) may be obtained from (10) by using the delta method on the one-to-one mapping $g(\lambda) = \frac{1}{\lambda}$, $\lambda>0$. □

Because calculation of the asymptotic variance of the MLE is computationally intensive, variances for $1 \le k \le 100$ were obtained numerically and compared to asymptotic variances of $\bar{X}_{rss}$ each value of k. Table 1 lists V_{rss}, V_{mle}, and their ratio for selected integers between 2 and 100; this range should cover most applications involving ranked set samples. The integral $\int_0^1 (\log t)^2 \frac{t^{k-1}}{1-t} dt$ was replaced by a tight lower bound using numerical integration, so the relative efficiencies shown are slightly conservative.

As the Table 1 below clearly shows, the MLE $\hat{\mu}$ systematically outperforms $\bar{X}_{rss}$ as an estimator of the mean μ based on a balanced ranked set sample from an exponential distribution. This fact, together with the applicability of $\hat{\mu}$ to unbalanced ranked set samples, suggests that the MLE is strongly preferable to $\bar{X}_{rss}$ in the exponential case.

Table 1

Asymptotic variances of $\bar{X}_{rss}$ and the MLE $\tilde{\mu}$ for a balanced ranked set sample of exponential random variables with mean = 1. The column on the right reports the efficiency of $\hat{\mu}$ relative to $\bar{X}_{rss}$.

k	V_{rss}	V_{mle}	$RE(\hat{\mu}\|\bar{X}_{rss})$
2	0.7500	0.7125	1.0526
3	0.6111	0.5532	1.1047
4	0.5208	0.4521	1.1520
5	0.4567	0.3823	1.1947
6	0.4083	0.3311	1.2332
7	0.3704	0.2920	1.2684
8	0.3397	0.2612	1.3006
9	0.3143	0.2363	1.3304
10	0.2929	0.2157	1.3581
11	0.2745	0.1984	1.3839
12	0.2586	0.1837	1.4080
13	0.2446	0.1710	1.4308
14	0.2323	0.1599	1.4523
15	0.2212	0.1502	1.4727
16	0.2113	0.1416	1.4920
17	0.2023	0.1340	1.5105
18	0.1942	0.1271	1.5281
19	0.1867	0.1209	1.5449
20	0.1799	0.1152	1.5610
25	0.1526	0.0935	1.6330
30	0.1332	0.0786	1.6937
50	0.0900	0.0481	1.8711
100	0.0519	0.0244	2.1270

3. NONPARAMETRIC ESTIMATION

The method of maximum likelihood can also be used to estimate the underlying distribution function. We will give a brief overview of nonparametric maximum likelihood estimation (NPMLE) as it applies to a general (possibly unbalanced) ranked set sample. For a full treatment of the NPMLE problem in this setting, see Kvam and Samaniego (1991).

Let us write the ordered values within the ranked set sample as $X_{(1)} < X_{(2)} <\ldots< X_{(n)}$, so that each observed order statistic has now been ranked relative to the others. If $X_{(j)}$ denotes the jth smallest observation in the ranked set sample, we can also identify it as the $r_{(j)}^{th}$ smallest item from an independent subsample of size $k_{(j)}$ from which only the observation $X_{(j)}$ was made. If $k_{(j)} = k$, $j=1,\ldots,n$, and if $r_{(j)}$ are observed in equal numbers among the integers $1,\ldots,k$, then the ranked set sample is balanced; however, we do not rely on this assumption in our discussion of maximum likelihood estimation of F.

If the underlying distribution F were absolutely continuous with density function f, the marginal density of $X_{(i)}$ could be expressed as

$$f_{r_{(i)}:k_{(i)}}(x) = r_{(i)}\binom{k_{(i)}}{r_{(i)}} F^{r(i)-1}(x)\, \bar{F}^{s(i)}(x) f(x)$$

where $s_{(i)} = k_{(i)} - r_{(i)}$. Given the observed data, we seek to maximize the likelihood of F over the class of all probability distributions. Since there is no dominating measure with respect to which all probability distributions are absolutely continuous, the familiar version of the likelihood function as a product of densities is inapplicable. Instead, we will use a generalized maximum likelihood approach of Kiefer and Wolfowitz (1956), and maximize the function

$$\mathcal{L}(F) = \prod_{i=1}^{n} F\left(X_{(i)}\right)^{r(i)-1} \bar{F}\left(X_{(i)}\right)^{s(i)} dF\left(X_{(i)}\right), \tag{11}$$

where $dF(x) = F(x) - F(x-)$, and refer to the function F^* that maximizes $\mathcal{L}$ as the nonparametric maximum likelihood estimator of F. It can be shown that the NPMLE assigns mass only to the set $(X_{(1)},\ldots,X_{(n)})$ unless $r_{(n)} < k_{(n)}$, in which case F^* also assigns mass to the interval $\{x: x > X_{(n)}\}$. Like the empirical distribution function, the NPMLE is a step function with jumps only at the observed data points; however, these two estimators tend to distribute their mass differently.

Since the likelihood function can be reduced to a function of n parameters corresponding to the n distinct jumps of the NPMLE, it is convenient to define $\phi_j = F^*\left(X_{(j)}\right)$, $j=1,\ldots,n$ to rewrite the likelihood function in terms of $\underset{\sim}{\phi} = (\phi_1,\ldots,\phi_n)$. Using this notation, we seek to find $\underset{\sim}{\phi}^*$ in the parameter space $\phi = \{\underset{\sim}{\phi}: 0 \le \phi_1 \le \ldots \le \phi_n \le 1\}$ that maximizes the likelihood function

$$\mathcal{L}(\underset{\sim}{\phi}) = \prod_{i=1}^{n} \phi^{r(i)-1}\left(1 - \phi_i\right)^{s(i)}\left(\phi_i - \phi_{i-1}\right) \tag{12}$$

where we define $\phi_0 = 0$. Since $\mathcal{L}$ is continuous, positive and bounded in Φ, and because $\mathcal{L} = 0$ on the boundary of the parameter space Φ, the existence of an NPMLE is guaranteed.

The NPMLE may be found by solving the likelihood equations iteratively. The n-variable problem reduces to an (n-1)-variable problem if $r_{(n)} = k_{(n)}$, since in that case $\phi_n = 1$. We will sketch the solution of the log-likelihood equations assuming that $r_{(n)} < k_{(n)}$; the special case in which $r_{(n)} = k_{(n)}$ can be solved in the same manner. The NPMLE must satisfy the following equations:

$$\frac{\partial}{\partial\phi_1}\log\mathcal{L}(\underset{\sim}{\phi}) = \frac{r_{(1)}}{\phi_1} - \frac{s_{(1)}}{1-\phi_1} - \frac{1}{\phi_2-\phi_1} = 0,$$

$$\frac{\partial}{\partial\phi_j}\log\mathcal{L}(\underset{\sim}{\phi}) = \frac{r_{(j)}-1}{\phi_j} - \frac{s_{(j)}}{1-\phi_j} + \frac{1}{\phi_j-\phi_{j-1}} - \frac{1}{\phi_{j+1}-\phi_j} = 0, \quad j=2,\ldots,n-1,$$

$$\frac{\partial}{\partial\phi_n}\log\mathcal{L}(\underset{\sim}{\phi}) = \frac{r_{(n)}-1}{\phi_n} - \frac{s_{(n)}}{1-\phi_n} + \frac{1}{\phi_n-\phi_{n-1}} = 0. \tag{13}$$

By examining the Hessian matrix directly, it can be shown that the vector $\underset{\sim}{\phi}^* = \left(\phi_1^*,\ldots,\phi_n^*\right)$ that satisfies these log-likelihood equations is the unique nonparametric maximum likelihood estimator. The proof of this claim, along with proofs of subsequent results stated in this section can be found in Kvam and Samaniego (1991). Although the NPMLE exists uniquely, the solution of the equations in (13) is difficult to obtain analytically. One can obtain an iterative solution, however, using a numerical method generated by the log-likelihood equations and an initial estimate of ϕ_1.

Specifically, we use the first equation in (13) and an initial estimate $\hat{\phi}_1$ to solve for ϕ_2, then we use the second equation to solve for ϕ_3, and so on. After $\phi_2,\phi_3,\ldots,\phi_n$ are generated from our initial $\hat{\phi}_1$ and the first (n-1) log-likelihood equations, we are left to solve the last equation

$$H\left(\phi_n,\phi_{n-1}\right) = \frac{\partial}{\partial\phi_n}\log\mathcal{L}(\underset{\sim}{\phi}) = 0.$$

It can be shown that $H\left(\hat{\phi}_n,\hat{\phi}_{n-1}\right) > 0$ implies that $\hat{\phi}_1 < \phi_1^*$, so that one should increase the estimate of ϕ_1 before repeating the procedure. Similarly, if $H < 0$, then $\hat{\phi}_1 > \phi_1^*$ and $\hat{\phi}_1$ should be decreased. The estimate

$\hat{\phi}_1$ should be updated in the following manner. With a given set of bounds for ϕ_1^* (initially we can select LB = 0 and UB = $r_{(1)}/\hat{k}_{(1)}$), choose $\hat{\phi}_1$ as the midpoint (i.e., $\hat{\phi}_1 = \frac{1}{2}$ LB + $\frac{1}{2}$ UB), and generate $\hat{\underset{\sim}{\phi}}$ using $\hat{\phi}_1$ in the log-likelihood equations. If $H(\hat{\phi}_n, \hat{\phi}_{n-1}) > 0$, $\hat{\phi}_1$ underestimates ϕ_1^*, so we reassign LB = $\hat{\phi}_1$. If $H < 0$ or $\hat{\underset{\sim}{\phi}} \notin \Phi$, $\hat{\phi}_1$ overestimates ϕ_1^*, so we reassign LB = $\hat{\phi}_1$. In either case, the size of the interval is halved. It can be shown that there exists an interval around ϕ_1^* for which any $\hat{\phi}_1$ in that interval leads to a solution in the parameter space by using this iterative method. Furthermore, this method produces a sequence $\hat{\underset{\sim}{\phi}}^{(1)}$, $\hat{\underset{\sim}{\phi}}^{(2)}, \ldots, \hat{\underset{\sim}{\phi}}^{(n)}$ for which $\hat{\underset{\sim}{\phi}}^{(n)} \in \Phi$ for all sufficiently large n and such that $\hat{\underset{\sim}{\phi}}^{(n)}$ converges to $\hat{\underset{\sim}{\phi}}^*$.

Because the NPMLE has not been expressed in closed form, investigation of its properties as an estimator is not straightforward. However, a nonparametric version of the EM algorithm used to solve for the NPMLE iteratively leads to a consistency argument (see Kvam and Samaniego (1991)). It can be shown that the estimator generated by the EM algorithm converges to $\underset{\sim}{\phi}^*$, and if the ranked set sample is balanced and the EDF is used as a seed, the MLE is a consistent estimator of F. The consistency argument can be extended to cover the unbalanced case as well.

As is evident from the likelihood equations in (13), the complexity of the computation of the NPMLE of F is no greater for the unbalanced case than it is for a balanced ranked set sample. The fact that the estimator exists and can be shown to be consistent in both balanced and unbalanced cases represents an important advantage of the NPMLE over the empirical cdf of the ranked set sample. In the balanced case, it is natural to ask: which estimator is better? We present in the figures that follow the outcome of several simulations showing that, even in the balanced case, the NPMLE tends to outperform the empirical cdf. The distributions used were as follows: Weibull: shape parameter = 2, scale parameter = 1/2; Lognormal: μ of log X = -1; σ^2 of log X = .16.

4. ACKNOWLEDGEMENT

Dr. Kvam's work was supported in part by grant DAAL03-89K-0010 from the Office of Army Research during a visit to the Department of Statistics, Iowa State University, in the summer of 1991.

Professor Samaniego's work was supported in part by grant MDA904-90-H-4031 from the National Security Agency.

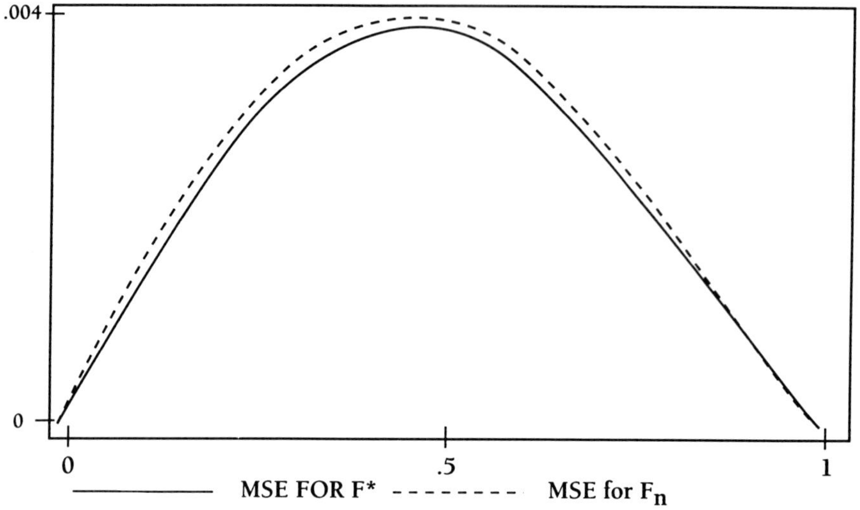

Figure 1. Average MSE of F^* and F_n; F = W(2,1/2); 10,000 Replications.

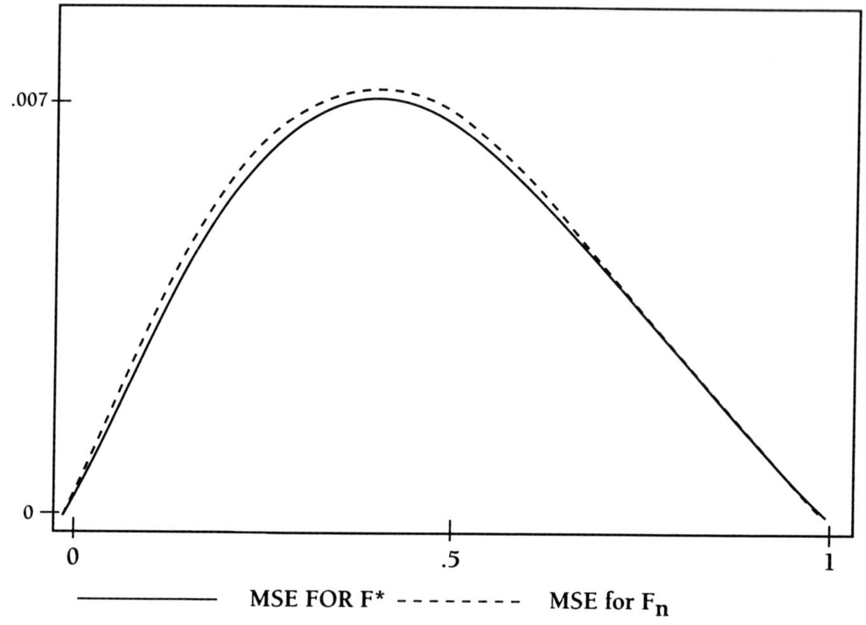

Figure 2. Average MSE of F^* and F_n; F = LN(-1,.16); 10,000 Replications.

5. REFERENCES

Doss, H., Freitag, S. and Proschan, F. (1989). "Estimating Jointly System and Component Reliability Using a Mutual Censorship Approach." *Annals of Statistics* 17, 764-782.

Kiefer, J. and Wolfowitz, J. (1956). "Consistency of the Maximum Likelihood Estimator in the Presence of Infinitely Many Incidental Parameters." *Annals of Mathematical Statistics*, 27, 887-906.

Kvam, Paul H. (1990). "Estimation Based on Ranked Set Sampling." Doctoral Dissertation, Division of Statistics, University of California, Davis.

Kvam, P.H. and Samaniego, F.J. (1991). "Nonparametric Maximum Likelihood Estimation Based on Ranked Set Sampling." Technical Report No. 221, Division of Statistics, University of California, Davis.

Lehmann, E.L. (1983). Theory of Point Estimation, New York:Wiley and Sons.

McIntrye, G.A. (1952). "A Method of Unbiased Selective Sampling Using Ranked Sets." *Australian Journal of Agricultural Research*, 3, 385-390.

Meilijson, I. (1981). "Estimation of the Lifetime Distribution of the Parts From the Autopsy Statistics of the Machine." *Journal of Applied Probability*, 18, 829-838.

Samaniego, F.J. (1985). "On Closure of the IFR Class Under Formation of Coherent Systems." *IEEE Transactions on Reliability*, R-34, 69-72.

Stokes, S.L. and Sager, T.W. (1988). "Characterization of a Ranked Set Sample With Applications to Estimating Distribution Functions." *Journal of the American Statistical Association*, 83, 374-381.

Takahasi, K. and Wakimoto, K. (1968). "On Unbiased Estimates of the Population Mean Based on Samples Stratified by Means of Ordering." *Annals of the Institute of Statistical Mathematics* 20, 1-31.

ADVANCES IN RELIABILITY
Edited by Asit P. Basu

Mixtures of Lifetime Distributions

Mei-Ling Ting Lee

Statistics Department, Harvard University, Cambridage, MA 02138

Presented at International Research Conference on Reliability

Abstract:

In this paper several forms of mixtures of lifetime distributions are reviewed. Reliability applications and dependence measures of the frailty model are discussed.

Key words: Decreasing failure rate, Exponential distribution, Frailty model, Generalized gamma distribution, Laplace transform, Inverse Gaussian distribution, Local correlation function, Overdispersion, Poisson mixtures, System reliability, Positive dependence, Proportional hazards models, Weibull distribution.

1. INTRODUCTION

Mixture distributions have been discussed by many authors. Feller(1943) considered a general class of *"contagious" distributions* which includes the compound Poisson distribution discussed by Greenwood and Yule (1920), the Polya-Eggenberger distribution, as well as Neyman Type A distributions (Neyman, 1939). Robbins (1948) considered a general measure theoretic form of the fundamental theorem for mixtures of distributions. Teicher (1959) investigated mixtures of additively closed families. The problem of identifiability was discussed by Teicher (1961, 1963, 1967), Tallis (1969), Redner (1981), and Redner and Walker(1984).

This survey article focuses on mixtures of lifetime distributions and their applications. Mixtures of lifetime distributions occur frequently in the theory and applications of statistical analysis. In an electronic device, components made with the same materials may be susceptible to the same but unknown level of environmental effects. In survival analysis, when the population under study is heterogeneous, one can consider the population as a mixture of individuals with different risks and the heterogenity can be modeled by a frailty factor. Random effect models, overdispersion models, and some competing risks models also involve mixture distributions.

Gumbel (1960) considered mixed exponential distributions. Harris (1968) considered mixtures of exponentials in a queue service discipline. Hutchison (1979) discussed bivariate mixtures of exponentials distributions. Based on mixtures of exponentials, Lindley and Singpurwalla (1986) derived a multivariate distribution with univariate Pareto marginals. Lindley and Singpurwalla's model was generalized by Nayak (1987) to multivariate Lomax distributions. Roy and Mukherjee (1988) derived general formulas for mixtures of exponential distributions.

Kao (1959) and Rider (1961) investigated finite mixtures of Weibull distributions. Takahasi (1965) derived gamma mixtures of independent Weibull distributions and called it the multivariate Burr distribution. Jewell (1982) considered nonparametric mixtures of exponential and Weibull(fixed shape) distributions. Crowder (1985, 1989) considered a distribution model for repeated failure time measurements and derived a multivariate distirbution with Weibull marginals. Hougaard (1986) investigated a similar class of multivariate mixture distributions. Lee and Gross (1989, 1991) considered mixtures of generalized gamma distributions and discussed moment inequalities and dependence properties.

Mixtures of discrete lifetime distributions have been discussed in the literature in the context of overdispersion models. A lot of research has been done on applications of negative binomial distributions, i.e. gamma mixtures of Poisson distributions, and beta-binomial distributions. See Paul and Plackett (1978), Cox (1983) and Cox and Snell (1989) for a review. For the bivariate case, Marshall and Olkin (1985) introduced a class of distributions generated by the bivariate Bernoulli distributions.

Sichel (1971,1975) introduced a distribution on the nonnegative integers which is obtained by mixing the Poisson distribution with the inverse Gaussian distribution. A multivariate extension of the Sichel distribution has been investigated by Stein, Zucchini and Juritz (1987) and Barndorff-Nielsen, Blaesild and Sesadri (1991). Mixtures by the inverse Gaussian distribution have been discussed by Sankaran (1968), Bhattacharya and Kumar (1986), Ord and Whitmore (1986), Willmot (1987), Whitmore (1988), and Whitmore and Lee (1991).

A partial list of forms of mixture models that have appeared recently in the literature is given in section 2. A bibliography for some distributional properties and limit results of mixture models is given in section 3. A bibliography for statistical inference is given in section 4. Reliability applications and dependence properties of the frailty model is reviewed in section 5.

2. MIXTURE MODELS

Listed below are some forms of mixture models that have appeared recently. This list is by no means complete and it should be pointed out that many distributions can be consided in several different forms.

2.1 A general form

Most mixture distributions (mixture of survival functions), have the following general form

$$F(x) = \int_{-\infty}^{\infty} F(x;\alpha)dG(\alpha), \tag{1}$$

where $F(x;\alpha)$ is a cumulative distribution (survival distribution), and $G(\alpha)$ is a cumulative distribution function.

Suppose that an item has lifetime distribution $F(x;\alpha)$. Parameter α represents the unknown environmental effect which varies randomly across items according to distribution $G(\alpha)$. The population lifetime of this type of items then has the mixture distribution given in (1). Overdispersion models such as negative binomial and beta-binomial distributions also have this form.

2.2 Finite mixtures

If the mixing distribution G in (1) has a finite support, then the resulting mixture distribution F can be written as

$$F(x) = \sum_{i=1}^{k} p_i F_i(x). \tag{2}$$

Assume that products manufactured on k assembly lines have lifetime distributions $F_i(x)$, $1 \leq i \leq k$. Then the outgoing lot consists of a random mixture and the resulting lifetime distribution for products from this lot is of form (2), where p_i is the percentage of products manufactured on the ith assembly line.

2.3 Mixture of convolutions:

Suppose that $X_1, \ldots, X_n$ are i.i.d. with distribution F, and suppose that $\Theta \geq 0$ is a random nonnegative integer having distribution G. Denote the θ-th convolution of F with itself by $F^{(\theta)}$. The random variable $U = X_1 + \ldots + X_\Theta$ has the distribution function given by

$$P(U \leq x) = \int F^{(\theta)}(x) dG(\theta). \tag{3}$$

Distributions of the form (3) are often called compound distributions.

For example, in a cumulative damage shock model (see, e.g., Esary, Marshall and Proschan, 1973), a device is subject to shocks occurring randomly according to a Poisson process with rate λ. Assume that (1) the ith shock causes a random damage X_i and (2) $X_1, X_2, \ldots,$ are i.i.d with distribution $F(x)$. Then the probability that the device survives in time interval $[0, t]$ is a mixture of convolutions given by

$$\sum_{k=0}^{\infty} e^{-\lambda t} \frac{(\lambda t)^k}{k!} F^{(k)}(x). \tag{4}$$

2.4 Mixtures of product families:

Suppose that $X_1, \ldots, X_n$ are i.i.d. with distribution F, and suppose that $\Theta \geq 1$ is a random positive integer having distribution G. Then the random variable V, defined by $V = \min(X_1, \ldots, X_\Theta)$, has a survival function of the form

$$P(V \geq x) = \int \{1 - F(x)\}^{\theta} dG(\theta). \tag{5}$$

Similarly, the random variable W, defined by $W=\max(X_1, \ldots, X_\Theta)$, has a distribution function of the form

$$P(W \leq x) = \int \{F(x)\}^{\theta} dG(\theta). \tag{6}$$

2.5 Frailty Model, Mixed Proportional Hazard Model

Assume that, conditional on a frailty factor Z which describes the individual's relative risk, the population lifetime random variable X has a conditional survival function of the form

$$P(X > x|Z = z) = \{S_0(x)\}^z, \tag{7}$$

where $S_0(t)$ is the baseline survival function.

If the frailty factor Z is a nonnegative random variable and has a distribution function $G(z)$, then, unconditionally, X has the survival function

$$P(X > x) = \textstyle\int_0^\infty \{S_0(x)\}^z dG(z). \tag{8}$$

In a frailty model, the dependence between individuals in a group is modelled by a group specific quantity called *frailty*. Hougaard (1986a) proposed a family of frailty distributions which includes the positive stable distributions, the gamma distributions, and the inverse Gaussian distributions.

Survivor functions of the form (7), which is the Lehmann family generated from baseline survivor function $S_0(x)$, is often used in the proportional hazards (PH) regression model introduced by Cox (1972). If one of the covariates in the PH model is not observed, then the resulting survivor function is a mixture of the form (8).

2.6 Multivariate mixture distributions

Various multivariate mixture distributions have been discussed in the form of conditional independence. Assume that, conditional on some random variable $Z = z$ say, random variables $X_1, \ldots, X_n$ are independently distributed. Then, unconditionally, the joint distribution of $X_1, \ldots, X_n$ is a multivariate mixture distribution. If, in addition, $X_1, \ldots, X_n$ are conditionally i.i.d., given $Z = z$, then the unconditional joint distribution of $X_1, \ldots, X_n$ is exchangeable.

2.6.1 Random additive effects

Let $X_i = \alpha_i Y_i + Z$, $i = 1, \ldots, n$, where Y_i, $i = 1, \ldots, n$, are independent random variables, Z is a baseline random variable which is independent of the Y's, and $\alpha_1, \ldots, \alpha_n$ are parameters. This kind of mixture distributions often arises in statistical inference as a random effects model.

2.6.2 Random multiplicative effects

Let $X_i = Y_i Z$, $i = 1, \ldots, n$, where $Y_1, \ldots, Y_n$ and Z are independent. Then the vector $(X_1, \ldots, X_n)$ has a multivariate mixture distributions. In particular, if Z has a standard Normal distribution and the Y_i are i.i.d. with positive stable distributions, the resulting joint distribution of $(X_1, \ldots, X_n)$ is multivariate symmetric stable. This kind of random vectors arise naturally in stochastic processes directed by randomized time. See Lee and Whitmore (1992).

2.6.3 Mixtures of convolutions and product families

Extending models considered in (3), (5) and (6) by allowing components X_i having distributions F_i, for $i = 1, 2, \ldots$, Marshall and Olkin (1991) discussed general properties for multivariate mixtures of convolutions and product families. They also gave many examples such as the bivariate Poisson, bivariate negative binomial, bivariate normal, bivariate logistic, and bivariate Pareto distributions.

2.6.4 Multivariate distributions generated by frailty models

Assume that given $Z = z$, X_i, $i = 1, \ldots, n$, are conditionally i.i.d. with proportional hazard distributions. That is,

$$P(X_i > x_i | Z = z) = \{S_0(x_i)\}^z, \quad i = 1, \ldots, n$$

Then, unconditionally, the joint distribution is given by

$$P(X_1 > x_1, \ldots, X_n > x_n) = \int \ldots \int \{S_0(x_1) \cdots S_0(x_n)\}^z dG(z). \tag{9}$$

Multivariate mixtures of PH models also has the form (9). Moreover, (9) is a special case of mixtures of survival product families.

Gumbel (1960) proposed two bivariate exponential distributions which is a special case of (9). Clayton (1978) considered a frailty model for association in bivariate lifetables. His model was further discussed by Oakes (1982). Vaupel, Manton and Stallard (1979) and Manton, Stallard and Vaupel (1986) considered an infinite mixture model in which an unobserved nonnegative random variable represents all individual differences in endowment for longevity. Flinn and Heckman (1982, 1983) considered mixed PH model in analyzing labor force and unemployment. Clayton and Cuzick (1985) considered multivariate generalizationn of the proportional hazards model by includidng a random effect representing heterogeneity of frailty. Notice that the frailty model is a special case of the mixture of product families. The frailty model for heterogeneous populations was also discussed by Hougaard (1984, 1986b), Crowder (1989) and Oakes (1989), among others.

Mixtures of exponential distributions and mixtures of Weibull distributions discussed by Lindley and Singpurwalla (1986), Nayak (1987), Roy and Mukherjee (1988) which were reviewed in section 1 can be considered as special cases of the frailty model. Genest and Mackay (1987) discussed the relationship between the class of Archimedean distributions and bivariate distributions generated by the frailty model. Oakes (1989) showed that the observable bivariate distribution determines the unobserved frailty distribution up to a scale parameter. He also showed that any bivariate frailty model leads to an archimedean survivor function, but the converse statement is false. In a more general setting, Marshall and Olkin (1988) discussed the dependence property of a multivariate mixture model and derived new families of bivariate distributions with marginals as parameters. In the context of competeing risks, Heckman and Honore (1989) considered the consequences for identifiability of a mixed PH model. Lindeboom and van den Berg (1991) discussed the importance of the mixing distribution. van den Berg and Steerneman (1991) investigated properties of correlation cefficients in multivariate mixed PH models.

3. GENERAL PROPERTIES OF MIXTURE DISTRIBUTIONS

Keilson and Steutel (1974) considered scale parameter mixtures and exponent parameter mixtures and discussed measures and moments inequalities. Shaked (1977) considered exchangeable distributions and introduced the concept of positive dependence by mixture. Tong (1977) and Shaked and Tong (1985) discussed ordering theorems for conditionally independent and identically distributed random variables. Shaked (1980) discussed crossing properties for exponential families. The geometry of mixture likelihoods was discussed by Lindsay (1983). Lynch (1988) discussed how the dispersiveness of the mixing distribution carries over to the mixed model in terms of generalized convex functions. Boland, Proschan and Tong (1989) discussed crossing properties of mixture distributions. They made stochastic and variability comparisons for binomial-beta, mixed Weibull, and mixed gamma distributions. Invariance results under dependence by mixing, such as limit theory and Berry-Essen bounds, were discussed by Jensen (1991).

4. STATISTICAL INFERENCES

Mendenhall and Hader (1958) and Tallis and Light (1968) considered parameter estimation for mixtures of exponential failure times. For a good review of the statistical analysis of finite mixture distributions see Everitt (1985), Titterington, Smith and Makov (1985), and McLachlan and Basford (1988).

Blum and Susarla (1977) considered estimation of a mixing distribution function. Laird (1978) considered nonparametric maximum likelihood estimation of a mixing distribution. Jewell (1982) investigated maximum likelihood estimate of the mixing distribution of arbitrary nonparametric mixtures of exponential and Weibull(fixed shape) distributions. Lesperance and Kalbfleisch (1992) discussed a fast algorithm for calculating the nonparametric MLE of a mixing distribution.

Heckman, Robb and Walker (1990) introduced nonparametric methods for testing the hypothesis that duration data can be represented by a mixture of exponential distributions. Crowder (1985,1989) and Kimber and Crowder (1991) considered mixture models in the context of repeated measurements applications. Gross and Lam (1981) and Cantor and Knapp (1985) discussed hypothesis testing on paired observations modeled by conditionally independent exponentials. Lee and Klein (1988,1989) considered statistical methods for combining laboratory and field data based on a random environmental stress model. Inference procedures for mixtures of Weibull distributions and mixtures of gamma distributions were discussed by Kao (1959), Falls (1970), and Dickinson (1974), among others.

Although mixture models are generally rather difficult to handle statistically by frequentist procedures, mixture models fits naturally into Bayes framework. Robbins (1964, 1980) discussed decision problems and inference procedures using the empirical Bayes approach. Currit and Singpurwalla (1988) considered Bayesian inferences for the reliability function in the exponential-mixture model. Whitmore and Lee (1991) discussed an inverse Gaussian mixture of exponential distributions and considered inference procedures using empirical Bayes method.

In the discrete case, Paul and Plackett (1978) examined the effect of Poisson mixtures on statistical inference. Cox (1983) showed that, in some situations, maximum likelihood estimation of a mixture model retains high efficiency in the presence of modest amounts of overdispersion. Lawless (1987) considered inference procedures for negative-binomial and mixed Poisson regression models. Follmann and Lambert (1989) discussed generalized logistic regression by nonparametric mixing.

Simar (1976) investigated maximum likelihood estimation of a compound Poisson process. Doyle, Hansen and McNolty (1980) investigated the mixed exponential failure process.

5. RELIABILITY APPLICATIONS OF THE FRAILTY MODEL

Lee (1991) investigated some following general properties and applications for the frailty model defined in (8) and (9).

5.1 A Laplace transform representation

Assume that, conditional on a common but unknown environmental effect $Z = z$, nonnegative lifetime random variables $X_1, \ldots, X_n$ are independently distributed with individual survivor functions of the form

$$S(x_i|Z = z) = P(X_i > x_i|Z = z) = \exp[-zH_i(x_i)],\ i = 1, \ldots, n \tag{10}$$

where $H_i(x_i) = \int_0^{x_i} h_i(u)du$, and $h_i(u)$ is a nonnegative function such that $\lim_{x\to\infty} H_i(x) = \infty$.

Also assume that the mixing random variable Z is nonnegative and has a cumulative distribution function $G(z)$ and a corresponding Laplace transform

$\mathcal{L}_G(\xi) = \int_0^\infty e^{-\xi z} dG(z) = E[\exp(-\xi Z)]$.

Then, unconditionally, the joint survival distribution function of $X_1, \ldots, X_n$ is given by

$$\overline{F}(x_1, \ldots, x_n) = \mathcal{L}_G(\textstyle\sum_{i=1}^n H_i(x_i)), \tag{11}$$

and the unconditional joint p.d.f. of $X_1, \ldots, X_n$, if exists, is given by

$$f(x_1, \ldots, x_n) = (-1)^n \left(\textstyle\prod_{i=1}^n h_i(x_i)\right) \mathcal{L}_G^{(n)}(\textstyle\sum_{i=1}^n H_i(x_i)), \tag{12}$$

where $\mathcal{L}_G^{(n)}(\xi) = \frac{d^n}{d\xi^n}\mathcal{L}_G(\xi) = \int_0^\infty (-z)^n e^{-\xi z} dG(z)$.

Under this general representation, marginal hazard rates and crude hazard rates can be easily obtained. Also, unconditionally, the joint distribution of $X_1, \ldots, X_n$ is dependent by total positivity of order infinity in each pair of arguments, with the remaining arguments fixed.

5.2 MOMENTS AND REGRESSIONS

Using the Laplace transform representation discussed above, Lee (1991) generalized results by Nayak (1987) and Roy and Mukherjee (1988) and derived the following moment formulas which are useful in regression analysis.

$$E(X_n|X_1 = x_1, \ldots, X_{n-1} = x_{n-1})$$

$$= \int_0^\infty \frac{\mathcal{L}_G^{(n-1)}(\sum_{i=1}^n H_i(x_i))}{\mathcal{L}_G^{(n-1)}(\sum_{i=1}^{n-1} H_i(x_i))} dx_n \tag{13}$$

$$E(X_n^2|X_1 = x_1, \ldots, X_{n-1} = x_{n-1})$$

$$= \int_0^\infty 2x_n \frac{\mathcal{L}_G^{(n-1)}(\sum_{i=1}^n H_i(x_i))}{\mathcal{L}_G^{(n-1)}(\sum_{i=1}^{n-1} H_i(x_i))} dx_n \tag{14}$$

5.3 Multivariate decreasing failure rates

Also, Lee (1991) generalized Roy and Mukherjee (1988)'s results and show that if each h_i is nonincreasing for $i = 1, \ldots, n$, then the vector $(X_1, \ldots, X_n)$ has a multivariate decreasing hazard rate according to the definition of Brindley and Thompson (1972).

5.4 Imperfect repair

In this section we show that models of the form (11) have been discussed before by Berg and Cleroux (1982), Brown and Proschan (1983), Whitaker and Samaniego (1989) in the context of imperfect repair.

Let $S_0(t)$ denote the baseline survival function of a new system. Assume that at failure the system is repaired to a condition as good as new with probability p and is minimally repaired to a condition as good as just prior to failure with probability $1-p$. Let Y_1 be the system age at the 1st perfect repair, then

$$P(Y_1 > t) = \{S_0(t)\}^p. \tag{15}$$

Since the probability parameter p of perfect repair varies from repairman to repairman, the value of p is random. Assume that the repair probability P has a distribution $G(p)$, then

$$P(Y_1 > t) = \int \{S_0(t)\}^p dG(p). \tag{16}$$

It is clear that equation (15) and equation (11) belong to the same model.

5.5. Influence of the environment on the system

For the special case where the joint components lifetimes are a mixture of n independent exponential distributions, Lefevre and Malice (1989) discussed how the mixing distribution can affect the number of components functioning and the system reliability for a k-out-of-n coherent system. Their results were extended to the general mixture model by Lee (1991). To present these results, first we will review some definitions of partial orderings discussed in Stoyan (1983).

Definition If X and Y have distributions F and G, respectively, then $X \overset{st}{\geq} Y$ if and only if $\overline{F}(a) \geq \overline{G}(a)$ for all a. Note that $X \overset{st}{\geq} Y$ iff $E[f(x)] \geq E[f(Y)]$ for all increasing f.

Definition $X \overset{c}{\geq} Y$ if and only if $E[f(X)] \geq E[f(Y)]$ for all increasing convex functions f.

Definition $X \overset{cv}{\geq} Y$ if and only if $E[f(X)] \geq E[f(Y)]$ for all increasing concave functions f.

Note that it's clear from the above definitions that $X \overset{cv}{\geq} Y$ iff $-X \overset{c}{\leq} -Y$.

Assume that, given a unknown environmental effect Z, component lifetimes $X_1, \ldots,$ X_n of a k-out-of-n system are conditionally independent. Let $N(t, Z)$ denote the number of components working at time t, $t \geq 0$, under the environmental effect Z. Lee (1991) shows that if Z_1 and Z_2 denote distinct unknown environmental effects, then (a) If $Z_1 \overset{st}{\geq} Z_2$, then $N(t, Z_1) \overset{st}{\leq} N(t, Z_2)$, for all $t \geq 0$. and (b) If $Z_1 \overset{cv}{\geq} Z_2$, then $N(t, Z_1) \overset{c}{\leq} N(t, Z_2)$, for all $t \geq 0$. Furthermore, if $X_1, \ldots, X_n$ are conditionally i.i.d., and $R_k(t, Z)$ denotes the reliability of a k-out-of-n system, then $Z_1 \overset{st}{\geq} Z_2$ implies that $R_k(t, Z_1) \leq R_k(t, Z_2)$.

5.6 Dependence

Dependence concepts related to the frailty model have been discussed by several authors. Marshall and Olkin (1988) show that if the multivariate survival function is of the form

$$S(t_1, \ldots, t_n) = \int \prod_{i=1}^{n} S_i^{\theta}(t_i) dG(\theta),$$

then S is TP_∞ in each pair of arguments. Oakes (1989) considered the function

$$\frac{S(t_1,t_2)\frac{\partial^2}{\partial t_1 \partial t_2} S(t_1,t_2)}{\frac{\partial}{\partial t_1} S(t_1,t_2) \frac{\partial}{\partial t_2} S(t_1,t_2)} \tag{17}$$

as a local measure of dependence. In this section we consider an alternative measure. The local correlation function, introduced by Bjerve and Doksum (1991), can be used as a local measure of dependence. The local correlation function between X_2 and X_1, near $X_1 = x_1$, is defined by

$$\rho(x_1) = \frac{\sigma_1 \beta(x_1)}{[(\sigma_1 \beta(x_1))^2 + \sigma^2(x_1)]^{\frac{1}{2}}}, \tag{18}$$

where $\beta(x_1) = \frac{d}{dx_1} E(X_2 | X_1 = x_1)$, $\sigma_1^2 = \mathrm{Var}(X_1)$, and $\sigma^2(x_1) = \mathrm{Var}(X_2 | X_1 = x_1)$. The following example demonstrates how the local correlation function describes the local dependence.

Example (Gamma Mixture of Exponentials)

Assume that the conditional survival function of X_i given Z is of the form $S(x_i | Z = z) = e^{-zx_i}$, for $i = 1, 2$, and that Z has a gamma (p, λ) distribution. Then $E(X_2 | X_1 = x_1) = (\lambda + x_1)/p$, and $\mathrm{Var}(X_2 | X_1 = x_1) = (p+1)(\lambda + x_1)^2 / [p^2(p-1)]$, and the local correlation function is given by

$$\rho(x_1) = \frac{\lambda^2 (p-1)(p-3)}{\{[\lambda^2(p-1)(p-3)]^2 + (p+1)(\lambda + x_1)^2/(p-1)^2\}^{\frac{1}{2}}}, \quad \text{if } p > 3.$$

Note that the derivative $\frac{d}{dx_1}\rho(x_1) < 0$ for all x_1 which shows that the local dependence of X_2 on X_1 decreases as the value of X_1 increases.

ACKNOWLEDGEMENT: I thank the referee for helpful comments.

REFERENCES

1. Barlow, R.E. and Proschan, F. (1981). "Statistical Theory of Reliability and Life Testing," Holt, Reinhart and Winstorn, Inc.

2. Barndorff-Nielsen, O.E., Blaesild, P. and Seshadri, V. (1991). Multivariate distributions with generalized inverse Gaussian marginals, and associated Poisson mixtures, preprint.

3. Bhattacharya, S.K. and Kumar, S. (1986). E-IG model in life testing, *Calcutta Statistical Association Bulletin*, **35**, 85-90.

4. Bjerve, S. and Doksum, K. (1990) Correlation curves: measures of association as functions of covariate values, preprint.

5. Blum, J.R. and Susarla, V. (1977). Estimation of a mixing distribution, *Ann. Prob.*, **5**, 200-209.

6. Boland, P.J., Proschan, F. and Tong, Y. L. (1989). Crossing properties of mixture distributions. *Probability in the Engineering and Informational Sciences*, **3**, 355-366.

7. Brindley, E.C. and Thompson, Jr, W.A. (1972), Dependence and aging aspects of multivariate survival, *JASA*, **67**, 822-830.

8. Brown, M. and Proschan, F. (1983). Imperfect repair, *J. Appl. Prob.*, **20**, 851-859.

9. Cantor, A.B. and Knapp, R.G. (1985). A test of the equality of survival distributions, based on paired observations from conditionally independent exponential distributions. *IEEE transactions on Reliability*, Vol. R-34, No.4, 342-346.

10. Clayton, D. (1978). A model for association in bivariate life tables and its applications in epidemiological studies of familial tendendy in chronic disease incidence. *Biometrika*, **65**, 141-151.

11. Clayton, D. and Cuzick, J. (1985). Multivariate generalizations of the proportional hazards model, *J. R. Statist. Soc.*, **A148**, 82-117.

12. Cox, D.R. (1972). Regression models and life tables (with discussion), *J. R. Stat. Soc.*, **B, 26**, 187-202.

13. Cox, D.R. (1983). Some remarks on overdispersion, *Biometrika*, **70**, 269-274.

14. Cox, D.R. and Snell, E.J. (1989). *Analysis of binary data*, 2nd edition, Chapman and Hall, New York.

15. Crowder, M. (1985). A distribution model for repeated failure time measurements, *JRSS*, **B47**, 447-452.

16. Crowder, M. (1989). A multivariate distribution with Weibull connections, *JRSS,* **B51**, 93-107.

17. Currit, A. and Singpurwalla, N.D. (1988). On the reliability function of a system of components sharing a common environment, *J. Appl. Prob.*, **26**, 763-771.

18. Dickinson, J.P.(1974). On the resolution of a mixture of observations from two gamma distributions by the method of maximum likelihood. *Metrika*, **21**, 133-141.

19. Doyle, J. Hansen, E. and McNolty, F. (1980). Properties of the mixed exponential failure process. *Technometrics*, **22**, 555-565.

20. Esary, J.D., A.W. Marshall, and Proschan, F. (1973). Shock models and wear processes, *Ann. Prob.*, **1**, 627-649.

21. Everitt, B.S. (1985). Mixture distributions, In *Enchclopedia of Statistical Sciences*, **Vol.5**, S. Kotz and N.L. Johnson (Eds.). New York: Wiley, 559-569.

22. Falls, L. W. (1970). Estimation of parameters in compound Weibull distributions. *Technometrics*, **12**, 399-407.

23. Feller, W. (1943). On a general class of contagious distributions, *Ann. Math. Stat*, **14**, 389-399.

24. Flinn, C.J. and Heckman, J.J. (1982), Models for the analysis of labor force dynamics, in *Advances in Econometrics*, **V.1**, Basmann, R. and Rhodes, G. eds., (JAI Press, Greenwich).

25. Flinn, C.J. and Heckman, J.J. (1983), Are unemployment and out of the labor force behaviorally distinct labor force states? *Journal of Labor Economics*, **1**, 28-42.

26. Follmann, D. A., and Lambert, D. (1989). Generalizing logistic regression by non-parametric mixing, *JASA*, **84**, 295-300.

27. Genest, C. and Mackay, R.J. (1986). The joy of copulas: bivariate distributions with given marginals. *The American Statistician*, **40**, 280-283.

28. Greenwood, M. and Yule, G.U. (1920). An inquiry into the nature of frequency distribution representative of multiple happenings with particular reference to the occurrence of multiple attacks of disease or of repeated accidents, *J. Roy. Stat. Soc.*, **83**, 255-279.

29. Gross, A.J. and Lam, C.F. (1981). "Paired observations from a survival distribution." *Biometrics* **37**, 505-511.

30. Gumbel, E.J. (1960). Bivariate exponential distributions, *JASA*, **55**, 698-707.

31. Harris, C.M. (1968). The Pareto distribution as a queue service discipoine. *Operat. Res.*, **16**, 307-313.

32. Harris, C. M. and Singpurwalla, N. (1968). Life distributions derived from stochastic hazard functions, *IEEE Trans. on Reliability*, **R-17**, 70-79.

33. Heckman, J.J., Robb, R., and Walker, J.R. (1990). Testing the mixture of exponentials hypothesis and estimating the mixing distribution by the method of moments, *JASA*, **85**, 582-589.

34. Heckman, J.J. and Honore, B.E. (1989). The identifiability of the competing risks model, *Biometrika*, **76**, 325-330.

35. Hougaard, P. (1984). Life table methods for heterogeneous populations: distributions describing the heterogenity. *Biometrika*, **71**, 75-83.

36. Hougaard, P. (1986a). Survival models for heterogeneous populations derived from stable distributions, *Biometrika*, **73(2)**, 387-396.

37. Hougaard, P. (1986b). A class of multivariate failure time distributions, *Biometrika*, **73(3)**, 671-678.

38. Hougaard, P. (1990). Fitting a multivariate failure time distribution. *IEEE Reliability*, **38**, 444-448.

39. Hutchison, T.P. (1979). Four applications of a bivariate distribution. *Biometrical J.*, 553-563.

40. Jensen, D.R. (1991). Invariance under dependence by mixing. in *Topics in Statistical Dependence*, IMS Lecture Notes-Monograph Seriew, **16**, eds, Block, Sampson, and Savits.

41. Jewell, N.P. (1982). Mixtures of exponential distributions. *Ann. Stat.*, **10**, 479-484.

42. Kao, J.H.K. (1959). A graphical estimation of mixed Weibull parameters in life testing of electron tubes. *Technometrics*, **7**, 639-643.

43. Keilson, J. and Steutel, F. W. (1974). Mixtures of distributions, moment ineuqlities and measures of exponentiality and normality, *Ann. Prob.*, **2**, 112-130.

44. Kimber, A.C. and Crowder, M.J. (1990), A repeated measurements model with applications in psychology, *Brit. J. Math. Stat. Pschol.*, preprint.

45. Laird, N. (1978). Nonparametric maximum likelihood estimation of a mixing distribution, *JASA*, **73**, 805-811.

46. Lawless, J.F. (1987). Negative binomial and mixed Poisson regression, *The Canadian Journal of Statistics*, **15**, 209-225.

47. Lee, M.-L.T. (1991). Reliability applications of the mixed proportional hazards model, (submitted paper).

48. Lee, M.-L.T. and Gross, A.J. (1989). Properties of conditionally independent generalized gamma distributions. *Probability in the Engineering and Informational Sciences*, **3**, 289-297.

49. Lee, M.-L.T. and Gross, A.J. (1991). Lifetime distributions under random environments, *Journal of Statistical Planning and Inference*, **29**, 137-143.

50. Lee, M.-L.T. and Whitmore, G.A. (1992). Stochastic processes directed by randomized time, *Journal of Applied Probability*, to appear.

51. Lee, S. and Klein, J. (1988). Bivariate models with a random environmental factor. *IAPQR*, **13**, 1-18, (1988).

52. Lee, S. and Klein, J. (1989). Statistical methods for combining laboratory and field data based on a random environmental stress model, in *Recent Developments in Statistics and Their Applications*, eds., J. Klein and J. Lee, Freedom Adademy pub. Co., Seoul.

53. Lefevre, C. and Malice, M.-P. (1989). On a system of components with joint lifetimes distributed as a mixture of independent exponential laws, *J. Appl. Prob.*, **26**, 202-208.

54. Lesperance, M. L. and Kalbfleisch, J. D. (1992). An algorithm for computing the nonparametric MLE of a mixing distribution. *JASA*, **87**, 120-126.

55. Lindeboom, M. and van den Berg, G.J. (1991). Heterogeneity in bivariat duration models, the importance of the mixing distribution, Research Memorandum, Leiden University.

56. Lindley, D.V. and Singpurwalla, N.D. (1986). Multivariate distributions for the life lengths of components of a system sharing a common environment. *J. Appl. Prob.* **23**, 418-431.

57. Lindsay, B.G. (1983). The geometry of mixture likelihoods: a general theory. *Annals of Statistics*, **11**, 86-94.

58. Lynch, J. (1988). Mixtures, generalized convexity and Balayages, *Scand J. Statist*, **15**, 203-210.

59. Manton, K.G., Stallard, E. and Vaupel, J.W. (1981). Methods for comparing the mortality experience of heterogeneous populations. *Demography*, **18**, 389-410.

60. Manton, K.G., Stallard, E. and Vaupel, J.W. (1986). Alternative models for the heterogeneity of mortality risks among the ages, *JASA*, **81**, 635-644.

61. Marshall, A.W. and Olkin, I. (1985). A family of bivariate distributions generated by the bivariate Bernoulli distribution. *JASA*, **80**, 332-338.

62. Marshall, A.W. and Oklin, I. (1988). Families of multivariate distributions. *JASA*, **83**, 834-841.

63. Marshall, A.W. and Olkin, I.(1991). Distributions generated by mixtures. in *Topics in Statistical Dependence*, IMS Lecture Notes-Monograph Series, **16**, eds., Block, Sampson, and Savits.

64. Mendenhall, W. and Hader, R.J. (1958). Estimation of parameters of mixed exponential distributed failure times from censored life test data. *Biometrika*, **45**, 504-520.

65. McLachlan, G.J. and Basford, K.E. (1985). *Mixture Models*, Marcel Dekker, Inc., New York.

66. Nayak, T.K. (1987). Multivariate Lomax distribution: properties and usefulness in reliability theory. *J. Appl. Prob.*, **24**, 170-177.

67. Neyman, J. (1939). On a new class of contagious distributions, applicable in entomology and bacteriology, *Annals of Math. Stat.*, **10**, 35-57.

68. Oakes, D. (1982). A model for association in bivariate data. *J. R. Statist. Soc.*, **B 44**, 414-422.

69. Oakes, D. (1989). Bivariate survival models induced by frailties,. *JASA*, **84**, 487-489.

70. Ord, J.K. and Whitmore, G.A. (1986). The Poisson-inverse Gaussian distribution as a model of species abundance, *Communications in Statistics*, **15**, 853-871.

71. Paul, S.R. and Plackett, R.L. (1978), Inference sensitivity for Poisson mixtures, *Biometrika*, **65**, 591-602.

72. Redner, R.A. (1981). Note on the consistency of the maximum likelihood estimate for nonidentifiable distributions. *Ann. Statist*, **9**, 225-228.

73. Redner, R.A. and Walker, H.F. (1984). Mixture densities, maximum likelihood and the EM Algorithm, *SIAM Review*, **26**, 195-239.

74. Rider, P.R. (1961). The method of moments applied to a mixture of two exponential distributions. *Ann. Math. Stat.*, **32**, 142-147.

75. Robbins, H. (1948). Mixture of distributions, *Ann. Math. Stat*, **19**, 360-369.

76. Robbins, H. (1980). Estimation and prediction for mixtures of the exponential distribution, *Proc. Nat. Adad. Sciences*, **77**, 2382-2388.

77. Roy, D. and Mukherjee, S.P. (1988). Generalized mixtures of exponential distributions, *J. Appl. Prob.*, **25**, 510-518.

78. Sankaran, M. (1968). Mixtures by the inverse Gaussian distribution. *Sankhya* (B), **30**, 455-458.

79. Shaked, M. (1977). A concept of positive dependence for exchangeable random variables, *Ann. of Statistics*, **5**, No.3, 505-515.

80. Shaked, M. (1980). On mixtures from exponential families. *J. R. Stat. Soc.*, **B 42**, 192-198.

81. Shaked, M. and Tong, Y.L. (1985). Some partial orderings of exchangeable random variables by positive dependence. *Journal of Multivariate Analysis*, **17**, 333-349.

82. Sichel, H.S. (1971). On a family of discrete distributions particularly suited to represent long-tailed frequency data, in *Proceedings of the Third Sumposium on Mathematical Statistics*, Ed. N.F. Laubscher, 51-57, Pretoria, S.A.: C.S.I.R.

83. Sichel, H.S. (1975). On a distribution law for word frequencies. *JASA* **70** (351), 542-547.

84. Simar, L. (1976). Maximum likelihood estimation of a compound Poisson process, *Ann. Stat.*, **4**, 1200-1209.

85. Stein, G.Z., Zucchini, W. and Juritz, J.M. (1987). Parameter estimation for the Sichel distribution and its multivariate extension, *JASA*, **82**, 938-944.

86. Stoyan, D. (1983). *Comparison methods for queues & other stochastic models*, Wiley, N.Y.

87. Takahasi, K. (1965). Note on the multivariate Burr's distribution, *Ann. Inst. Statist. Math.*, **17**, 257-260.

88. Tallis, G.M. (1969). The identifiability of mixtures of distributions. *J. Appl. Prob.*, **6**, 389-398.

89. Tallis, G.M. and Light, R. (1968). The use of fractional moments for estimating the parameters of a mixed exponential distribution. *Technometrics*, **10**, 161-175.

90. Teicher, H. (1960). On the mixture of distributions. *Ann. Math. Stat.*, **31**, 55-73.

91. Teicher, H. (1961). Identifiability of mixtures. *Ann. Math. Stat.*, **32**, 244-248.

92. Teicher, H. (1963). Identifiability of finite mixtures. *Ann. Math. Stat.*, **34**, 1265-1269.

93. Titterington, D.M., Smith, A.F.M. and Makov, U.E. (1988). *Statistical Analysis of Finite Mixture Distributions*, Wiley, New York.

94. Tong, Y.L. (1977). An ordering theorem for conditionally independent and identically distributed random variables, **5**, 274-277.

95. van den Berg, G.J. and Steerneman, T. (1991). The correlation of durations in multivariate hazard rate models. Discussion Paper No.960, The Center for mathematical Studies in Economics and Management Science, Northwestern University.

96. Whitaker, L.R. and Samaniego, F.J. (1989). Estimating the reliability of systems subject to imperfect repair. *JASA*, **84**, 301-309.

97. Whitmore, G.A. (1988). Inverse Gaussian mixtures of exponential distributions, Unpublished, McGill University, Montreal.

98. Whitmore, G.A., and Lee, M.-L.T. (1991). A multivariate survival distribution generated by an inverse Gaussian mixture of exponentials. *Technometrics*, **33**, 39-50.

99. Willmot, G.E. (1987). The Poisson-inverse Gaussian distribution as an alternatives to the negative binomial, *Scand. Actuarial J.*, 113-127.

ADVANCES IN RELIABILITY
Edited by Asit P. Basu

Variate Generation for Monte Carlo Analysis of Reliability and Lifetime Models

Lawrence M. Leemis[a] and Li-Hsing Shih[b]

[a] Department of Mathematics, College of William and Mary, Williamsburg, VA 23187 USA

[b] Department of Mining and Petroleum Engineering, National Cheng Kung University, Tainan, Taiwan R.O.C.

Abstract

Monte Carlo methods have increasing importance as reliability and lifetime models become less mathematically tractable. This paper considers basic and more advanced methods for generating random lifetimes. The basic methods are the inverse-chf technique, competing risks and thinning. These methods parallel the inverse-cdf, composition and acceptance/rejection methods, which are density-based. The accelerated life and proportional hazards lifetime models are used to account for the effects of covariates on a random lifetime. Variate generation algorithms for Monte Carlo simulation in both the renewal and nonhomogeneous Poisson process cases are a simple extension of the inverse-chf technique. Finally, a nonparametric method for generating event times for a nonhomogeneous Poisson process using a piecewise-linear cumulative intensity function is presented.

1. INTRODUCTION

In reliability models, a continuous nonnegative random variable typically represents the lifetime of a component or system. There are several functions which completely specify the distribution of the lifetime. Four of these functions are useful in describing variate generation algorithms: the survivor function, the probability density function, the hazard function and the cumulative hazard function. Other functions, which will not be used here, are the characteristic function, the Mellin transform, the mean residual life function, the moment generating function, the density quantile function and the total time on test transform. The density quantile function is discussed in Parzen (1979), the mean residual life function is discussed in Swartz (1973), and the total time on test transform is discussed in Barlow (1979).

This paper considers techniques that are useful for generating random variates for Monte Carlo simulation analysis. Recent textbooks (e.g., Devroye (1986) and Dagpunar (1988)) are devoted entirely to the topic. It is assumed that all distributions are continuous, although all of these algorithms can be modified to accommodate discrete distributions. Notation used in this paper is shown below.

T	a continuous nonnegative random variable (lifetime)
$\sim$	means "is distributed as"
$S(t)$	the survivor function for T
$f(t)$	the probability density function for T
$h(t)$	the hazard function for T
$H(t)$	the cumulative hazard function for T
U	a random number
$\mathbf{z}$	a $q \times 1$ vector of covariates
$\boldsymbol{\beta}$	a $q \times 1$ vector of regression coefficients
$\psi(\mathbf{z})$	a link function
$\lambda(t)$	the intensity function
$\Lambda(t)$	the cumulative intensity function.

The survivor function, also known as the reliability function and complementary cdf, is defined by

$$S(t) = P[T \geq t] \qquad\qquad t \geq 0,$$

which is a nonincreasing function of t satisfying $S(0) = 1$ and $\lim_{t \to \infty} S(t) = 0$. The reliability is important in the study of *systems* of components since it is the appropriate argument in the structure function to determine system reliability (Barlow and Proschan, 1981). $S(t)$ is the fraction of the population which will survive to time t, as well as the probability that a single item will survive to time t. A result that is associated with the survivor function that is useful for variate generation is that $S(T)$ is uniformly distributed between zero and one by the probability integral transformation. Thus, $S^{-1}(1-U)$ generates a lifetime variate for Monte Carlo simulation, where $U \sim U(0, 1)$. This is analogous to the inverse-cdf technique, where $F^{-1}(U)$ is used to generate a random variate, and $F(x)$ is the cumulative distribution function (cdf).

When the survivor function is differentiable,

$$f(t) = -S'(t) \qquad\qquad t \geq 0$$

is the associated probability density function. Finite mixture models for m populations of items may be modeled using the probability density function

$$f(t) = \sum_{i=1}^{m} p_i \, f_i(t) \qquad\qquad t \geq 0,$$

where $f_i(t)$ is the probability density function for the ith population and p_i is the probability of selecting an item from the ith population, $i = 1, 2, ..., m$. Mixture models are used in composition, a density- based variate generation technique.

The hazard function, also known as the rate function and force of mortality, can be defined by

$$h(t) = \frac{f(t)}{S(t)} \qquad\qquad t \geq 0.$$

The hazard function is popular in reliability work because it has the intuitive interpretation as the amount of *risk* associated with an item that has survived to time t. The hazard function is a special form of the complete intensity function at time t for a point process (Cox and Oakes, 1984). In other words, the hazard function is mathematically equivalent to the intensity function for a nonhomogeneous Poisson process, and the failure time corresponds to the first event time in the process. Competing risks models are easily formulated in terms of $h(t)$, as will be seen in the next section.

The cumulative hazard function can be defined by

$$H(t) = \int_0^t h(\tau)d\tau \qquad t \geq 0.$$

As indicated by Griffith (1982) and others, if T is a random lifetime with cumulative intensity function H, then $H(T)$ is an exponential random variable with a mean of one. Thus, $H^{-1}(-\log(1-U))$ generates a variate for Monte Carlo simulation when $U \sim U(0, 1)$ and log is the natural logarithm.

The remainder of this paper is organized as follows. Section 2 gives three algorithms that can be used to generate random lifetimes when a hazard function or cumulative hazard function is specified. Section 3 shows how to generate random lifetimes via inversion when the accelerated life or proportional hazards models are used to model the lifetimes. Finally, Section 4 considers an algorithm for generating failure times under a nonhomogeneous Poisson process with a piecewise-linear intensity function.

2. RANDOM LIFETIME GENERATION

Three hazard-based techniques are presented in this section: inverse-chf, competing risks and thinning. These three techniques parallel the associated density-based techniques described in Schmeiser (1990), Deak (1990) and other simulation textbooks: inverse-cdf, composition and acceptance/rejection. In all of the techniques described below, it is assumed that there is a source of independent random numbers (e.g., a stream of $U(0,1)$ random variates). Examples of the use of these techniques are given in Leemis and Schmeiser (1985). All of the techniques reviewed here concern *univariate* random variables. Generating pairs of random variables is considered in the recent article by Grimlund (1992).

The inverse cumulative hazard function technique, or *inverse-chf* technique, is based on the observation alluded to earlier that $H(T)$ is exponentially distributed with a mean of one. So

$$T \leftarrow H^{-1}(-\log(1-U))$$

generates a single random lifetime T. This algorithm is easiest to implement if H can be inverted in closed-form. Some properties of the inverse-chf technique are that it is synchronized (i.e., one random number produces one lifetime), monotone (i.e., larger random numbers produce larger lifetimes), equivalent to the inverse-cdf technique and accommodates truncated distributions. It can also be modified to generate order statistics, which are useful for generating the lifetime of a k-out-of-n system.

The second basic technique is *competing risks,* which can be applied when the hazard function can be written as the sum of hazard functions, each corresponding to a "cause" of failure

$$h(t) = \sum_{j=1}^{k} h_j(t)$$

where $h_1(t)$, $h_2(t)$, ..., $h_k(t)$ are the hazard functions associated with the k causes of failure acting in a population. The minimum of the lifetimes from each of these risks corresponds to the system lifetime. Competing risks is most commonly used to analyze series systems, but it can also be applied in actuarial models. The algorithm is

generate T_j from $h_j(t)$, $j = 1, 2, ..., k$
$T \leftarrow \min\{T_1, T_2, ..., T_k\}$

The third basic technique is *thinning,* which was originally used by Lewis and Shedler (1979a) for generating the event times in a nonhomogeneous Poisson process. It can be adapted for producing a single lifetime by ignoring all but the first event time generated. In order to use this technique, a majorizing hazard function $h^*(t)$ must be found that satisfies $h^*(t) \geq h(t)$ for all $t \geq 0$. The algorithm is

$T \leftarrow 0$
repeat
  generate Y from $h^*(t)$ given $Y > T$
  $T \leftarrow T + Y$
  generate $S \sim U(0, h^*(T))$
until
$S \leq h(T)$

Generating Y may be done by inversion or any other method.

A fourth technique, special properties, is neither density-based nor hazard-based, so it will not be considered separately here. An example of the use of a special property is generating an Erlang random variable as the sum of independent exponential random variables.

3. ACCELERATED LIFE & PROPORTIONAL HAZARDS MODELS

The effect of covariates on survival often complicates the analysis of a set of lifetime data. In a medical setting, these covariates are usually patient characteristics, such as age, gender or blood pressure. In reliability, covariates such as the turning speed of a machine tool or the stress applied to a component affect the lifetime of an item. Two models that are often used to incorporate the effect of these covariates on lifetimes are the *accelerated life* and *proportional hazards* models. This section describes algorithms for the generation of lifetimes for Monte Carlo simulation that are described by one of these models.

The $q \times 1$ vector $\mathbf{z}$ contains covariates associated with a particular item or individual. The covariates are linked to the lifetime by the function $\psi(\mathbf{z})$, which satisfies $\psi(\mathbf{O}) = 1$ and $\psi(\mathbf{z}) \geq 0$ for all $\mathbf{z}$. A popular choice is $\psi(\mathbf{z}) = e^{\boldsymbol{\beta}'\mathbf{z}}$, where $\boldsymbol{\beta}$ is a $q \times 1$ vector of regression coefficients.

The cumulative hazard function for T in the *accelerated life* model is (Cox and Oakes, 1984)

$$H(t) = H_0(t\, \psi(\mathbf{z}))$$

where H_0 is a baseline cumulative hazard function. Note that when $\mathbf{z} = \mathbf{0}$, $H_0 \equiv H$. In this model, the covariates accelerate ($\psi(\mathbf{z}) > 1$) or decelerate ($\psi(\mathbf{z}) < 1$) the rate at which the item moves through time. The *proportional hazards* model

$$H(t) = \psi(\mathbf{z})\, H_0(t)$$

increases ($\psi(\mathbf{z}) > 1$) or decreases ($\psi(\mathbf{z}) < 1$) the failure rate of the item by the factor $\psi(\mathbf{z})$ for all values of t.

All of the algorithms are based on the fact that $H(T)$ is exponentially distributed with a mean of 1. Therefore, equating the cumulative hazard function to $-\log(1-U)$ and solving for t yields the appropriate generation technique. In the accelerated life model, since time is being expanded or contracted by a factor $\psi(\mathbf{z})$, variates are generated by

$$T \leftarrow \frac{H_0^{-1}(-\log(1-U))}{\psi(\mathbf{z})}.$$

In the proportional hazards model, equating $-\log(1-U)$ to $H(t)$ yields the variate generation formula

$$T \leftarrow H_0^{-1}\left(\frac{-\log(1-U)}{\psi(\mathbf{z})}\right).$$

In addition to generating individual lifetimes, these variate generation techniques may also be applied to point processes. A renewal process, for example, with time between events having a cumulative hazard function $H(t)$, can be simulated by using the appropriate generation formula for the two cases shown above. These variate generation formulas must be modified, however, to generate variates from a nonhomogeneous Poisson process (NHPP). Properties of an NHPP are described in Cinlar (1975).

In an NHPP, the hazard function, $h(t)$, is replaced by the intensity function, which governs the rate at which events occur. Note that the two functions are "equivalent" in the sense that they will generate the same initial event time. In an NHPP, subsequent events are generated according to the intensity function. To determine the appropriate method for generating variates from an NHPP, assume that the last event in a point process has occurred at time a. The cumulative hazard function for the time of the next event conditioned on survival to time a is

$$H_{T \mid T>a}(t) = H(t) - H(a) \qquad t \geq a.$$

In the accelerated life model, where $H(t) = H_0(t\, \psi(\mathbf{z}))$, the time of the next event is generated by

$$T \leftarrow \frac{H_0^{-1}(H_0(a\, \psi(\mathbf{z})) - \log(1-U))}{\psi(\mathbf{z})}.$$

Equating the conditional cumulative hazard function to $-\log(1-U)$, the time of the next event in the proportional hazards case is generated by

$$T \leftarrow H_0^{-1}(H_0(a) - \frac{\log(1-U)}{\psi(\mathbf{z})}).$$

An example of the application of these algorithms to a particular parametric distribution is given in Leemis (1987). Extensions to the case where the covariates are time dependent are given in Leemis, Shih and Reynertson (1990) and Shih and Leemis (1993). Table 1 summarizes the variate generation algorithms for the accelerated life and proportional hazards models, where the last event is assumed to have occurred at time a. Note that $1 - U$ has been replaced with U in this table to save a subtraction, although the sense of the monotonicity is reversed.

Table 1
Lifetime generation in regression models

	Renewal	NHPP
Accelerated life	$T \leftarrow a + \frac{H_0^{-1}(-\log(U))}{\psi(\mathbf{z})}$	$T \leftarrow \frac{H_0^{-1}(H_0(a\psi(\mathbf{z})) - \log(U))}{\psi(\mathbf{z})}$
Proportional hazards	$T \leftarrow a + H_0^{-1}(\frac{-\log(U)}{\psi(\mathbf{z})})$	$T \leftarrow H_0^{-1}(H_0(a) - \frac{\log(U)}{\psi(\mathbf{z})})$

The renewal and NHPP algorithms are equivalent when $a = 0$ (since a renewal process is equivalent to an NHPP restarted at zero after each event), the accelerated life and proportional hazards models are equivalent when $\psi(\mathbf{z}) = 1$, and all four cases shown in the above table are equivalent when $H_0(t) = \lambda t$ (the exponential case) because of the memoryless property.

4. GENERATING A NONHOMOGENEOUS POISSON PROCESS

This section contains the description of a nonparametric technique for estimating the cumulative intensity function of an NHPP from one or more realizations and the associated algorithm for generating random variates. Unlike many existing techniques, this method does not require the modeler to specify any parameters or weighting functions. An NHPP is often suggested as an appropriate model for modeling the failure times of repairable systems whose rate of occurrence of failures varies over time (Ascher and Feingold, 1984). The nonparametric technique presented here estimates the cumulative intensity function of an NHPP on the time interval $(0, S]$ from one or more realizations.

Many authors suggest the use of NHPP's for modeling systems with inputs whose rates vary over time. Fishman and Kao (1977), Kaminsky and Rumph (1977), Klein and Roberts (1984), Lee, Wilson and Crawford (1991), Lewis and Shedler (1979a and 1979b) and Shanthikumar (1986) all consider the generation of event times for a nonhomogeneous Poisson process. Similar expositions are given in many of the simulation textbooks given in the discrete-event simulation overview in Schmeiser (1990). Inverse-transformation algorithms for some common stochastic processes are considered in Schmeiser and Song (1989).

An NHPP is a generalization of a homogeneous Poisson process where events occur randomly over time at the rate of λ events per unit time. The rate at which events occur in an NHPP varies over time as determined by the *intensity function,* $\lambda(t)$. The *cumulative intensity function* is defined by

$$\Lambda(t) = \int_0^t \lambda(\tau) d\tau \qquad t \geq 0$$

and is interpreted as the expected number of events by time t.

The intensity function, $\lambda(t)$, for an NHPP is assumed to be positive for all $t \in (0, S]$ and is continuous for almost every $t \in (0, S]$. The cumulative intensity function is to be estimated from k realizations of the NHPP on $(0, S]$, where S is a known constant. Let n_i $(i = 1, 2, ..., k)$ be the number of observations in the ith realization, $n = \sum_{i=1}^{k} n_i$, and let $t_{(1)}, t_{(2)}, ..., t_{(n)}$ be the order statistics of the superposition of the k realizations, $t_{(0)} = 0$ and $t_{(n+1)} = S$. Setting $\hat{\Lambda}(S) = \frac{n}{k}$ yields a process where the expected number of events by time S is the average number of events in k realizations, since $\Lambda(S)$ is the expected number of events by time S. The piecewise-linear estimator of the cumulative intensity function between the time values in the superposition is

$$\hat{\Lambda}(t) = \frac{in}{(n+1)k} + [\frac{n(t - t_{(i)})}{(n+1)k(t_{(i+1)} - t_{(i)})}] \qquad t_{(i)} < t \leq t_{(i+1)};\ i = 0, 1, 2, ..., n$$

given in Leemis (1991). This estimator passes through the points $(t_{(i)}, \frac{in}{(n+1)k})$, for $i = 1, 2, ..., n+1$.

The assumption that there will not be any ties, (i.e., $t_{(i)} < t_{(i+1)}$ for $i = 0, 1, ..., n$) may not always be satisfied in practice due to rounding. The estimate for $\Lambda(t)$ given above should be modified so that there is a discontinuity at the value where tied values occur. For example, if $t_{(m)} = t_{(m+1)}$ for some m, then $\hat{\Lambda}(t_{(m)}) = \hat{\Lambda}(t_{(m+1)}) = \frac{mn}{(n+1)k}$, and $\lim_{t \downarrow t_{(m+1)}} \hat{\Lambda}(t) = \frac{(m+1)n}{(n+1)k}$. In other words, there is a jump in the estimate of the cumulative intensity function of $\frac{n}{(n+1)k}$ where the tie occurs. Multiple tied values are handled analogously.

The rationale for using a linear function between the data values is that inversion can be used for generating realizations without having tied events. If the usual step-function estimate of $\Lambda(t)$ is used, only the $t_{(i)}$ values could be generated.

Since the number of events that occur in the NHPP of interest by time t has the Poisson distribution with mean $\Lambda(t)$, a strong consistency result is obtained, i.e.,

$$\lim_{k \to \infty} \hat{\Lambda}(t) = \Lambda(t) \quad \text{with probability one.}$$

Also, an asymptotically exact $100(1-\alpha)\%$ confidence interval for $\Lambda(t)$ is

$$\hat{\Lambda}(t) - z_{\alpha/2}\sqrt{\frac{\hat{\Lambda}(t)}{k}} < \Lambda(t) < \hat{\Lambda}(t) + z_{\alpha/2}\sqrt{\frac{\hat{\Lambda}(t)}{k}},$$

where $z_{\alpha/2}$ is the $1 - \alpha/2$ fractile of the standard normal distribution.

The cumulative intensity function for an NHPP is often estimated in order to generate variates for Monte Carlo simulation. Using a time transformation (Cinlar, 1975, page 96), the event times from a unit Poisson process, E_1, E_2, ..., can be transformed to the event times of an NHPP via $T_i = \Lambda^{-1}(E_i)$. For the NHPP estimate considered here, the events at times $T_1, T_2, \ldots$ can be generated for Monte Carlo simulation by the algorithm below, given n, k, S and the superpositioned values.

$i \leftarrow 1$
generate $U_i \sim U(0,1)$
$E_i \leftarrow -\log(1 - U_i)$
while $E_i < \frac{n}{k}$ **do**
    begin
        $m \leftarrow \left\lfloor \frac{(n+1)\,k\,E_i}{n} \right\rfloor$
        $T_i \leftarrow t_{(m)} + [t_{(m+1)} - t_{(m)}]\left(\frac{(n+1)k\,E_i}{n} - m\right)$
        $i \leftarrow i + 1$
        generate $U_i \sim U(0,1)$
        $E_i \leftarrow E_{i-1} - \log(1 - U_i)$
    end

Thus, it is a straightforward procedure to obtain a realization of $i - 1$ events on $(0, S]$ from the superpositioned process and $U(0,1)$ values $U_1, U_2, \ldots, U_i$. Inversion has been used to generate this NHPP, so certain variance reduction techniques, such as antithetic variates or common random numbers, may be applied to the simulation output. Replacing $1 - U_i$ with U_i in generating the exponential variates will save CPU time although the direction of the monotonicity is reversed. Tied values in the superposition do not pose any problem to this algorithm although there may be tied values in the realization. As n increases, the amount of memory required increases, but the amount of CPU time required to generate a realization depends only on the ratio n/k, the average number of events per realization. Thus collecting more realizations (resulting in narrower confidence intervals) increases the amount of memory required, but does not impact the CPU time for generating a realization.

ACKNOWLEDGEMENT

The authors thank Jim Tyson and Mike Oltmanns for their help in proofreading this manuscript.

REFERENCES

1 Ascher, H. and Feingold, H., 1984. *Repairable Systems Reliability*. Marcel Dekker.
2 Barlow, R.E., 1979. Geometry of the Total Time on Test Transform. *Naval Research Logistics Quarterly*. 26, 393-402.
3 Barlow, R.E. and Proschan, F., 1981. *Statistical Theory of Reliability and Life Testing*. To Begin With.
4 Cinlar, E., 1975. *Introduction to Stochastic Processes*. Prentice-Hall.
5 Cox, D.R. and Oakes, D., 1984. *Analysis of Survival Data*. Chapman and Hall.
6 Dagpunar, J., 1988. *Principles of Random Variate Generation*. Oxford Science Publications.
7 Deak, I., 1990. Random Number Generators and Simulation. Akademisai Kiado, Budapest.
8 Devroye, L., 1986. *Non-Uniform Random Variate Generation*. Springer-Verlag.
9 Fishman, G.S. and Kao, E.P.C., 1977. A Procedure for Generating Time-Dependent Arrivals for Queueing Simulations. *Naval Research Logistics Quarterly*. 24, 4, 661-666.
10 Griffith, W., 1982. Representation of Distributions Having Monotone or Bathtub-Shaped Hazard Functions. *IEEE Transactions on Reliability*. R-31, 95-96.
11 Grimlund, R.A., 1992. Generating Statistically Dependent Pairs of Random Variables: A Marginal Distribution Approach. *European Journal of Operations Research*. 57, 1, 39-53.
12 Kaminsky, F.C. and Rumph, D.L., 1977. Simulating Nonstationary Poisson Processes: A Comparison of Alternatives Including the Correct Approach. *Simulation*. 29, 1, 17-20.
13 Klein, R.W. and Roberts, S.D., 1984. A Time-varying Poisson Arrival Process Generator. *Simulation*. 43, 4, 193-195.
14 Lee, S., Wilson, J.R., and Crawford, M.M., 1991. Modeling and Simulation of a Nonhomogeneous Poisson Process with Cyclic Features. *Communications in Statistics--Simulation and Computation*. 20, 2&3, 777-809.
15 Leemis, L.M. and Schmeiser, B.W., 1985. Random Variate Generation for Monte Carlo Experiments. *IEEE Transactions on Reliability*. R-34, 1, 81-85.
16 Leemis, L.M., 1987. Variate Generation for the Accelerated Life and Proportional Hazards Models. *Operations Research*. 35, 6, 892-894.
17 Leemis, L.M., Shih, L.H. and Reynertson, K., 1990. Variate Generation for the Accelerated Life and Proportional Hazards Models with Time Dependent Covariates. *Statistics and Probability Letters*. 10, 1, 335-339.
18 Leemis, L.M., 1991. Nonparametric Estimation of the Intensity Function for a Nonhomogeneous Poisson Process. *Management Science*. 37, 7, 886-900.

19 Lewis, P.A.W. and Shedler, G.S., 1979a. Simulation of Nonhomogeneous Poisson Processes by Thinning. *Naval Research Logistics Quarterly*. 26, 3, 403-413.

20 Lewis, P.A.W. and Shedler, G.S., 1979b. Simulation of Nonhomogeneous Poisson Processes with Degree-Two Exponential Polynomial Rate Function. *Operations Research*. 27, 5, 573-580.

21 Parzen, E., 1979. Nonparametric Statistical Data Modeling. *Journal of the American Statistical Association*. 74, 365, 105-131.

22 Schmeiser, B. and Song, T., 1989. Inverse-transformation Algorithms for Some Common Stochastic Processes. *Proceedings of the 1989 Winter Simulation Conference,* MacNair, E.A., Musselman, K.J. and Heidelberger, P. (eds.), 490-496.

23 Schmeiser, B., 1990. Simulation Experiments. From *Stochastic Models* (D.P. Heyman and M.J. Sobel, editors). North-Holland.

24 Shanthikumar, J., 1986. Uniformization and Hybrid Simulation/Analytic Models of Renewal Processes. *Operations Research*. 34, 4, 573-580.

25 Shih, L. and Leemis, L., 1993. Variate Generation for a Nonhomogeneous Poisson Process with Time Dependent Covariates. *Journal of Statistical Computation and Simulation*. Forthcoming.

26 Swartz, G.B., 1973. The Mean Residual Life Function. *IEEE Transactions on Reliability*. R-22, 108-109.

ADVANCES IN RELIABILITY
Edited by Asit P. Basu
1993 Elsevier Science Publishers B.V.

Experiment design to explore multi-step, multi-stress failure modes

M. J. LuValle

AT&T Bell Laboratories, Room 2K-511, Crawfords Corner Road, Holmdel, New Jersey, 07733-1988

Abstract

Engineering procedures using accelerated stress are becoming increasingly important in engineering practice. Two such procedures are Environmental Stress Screening (ESS) and Stress-Life (STRIFE) testing. Both procedures are instituted at the system level, and rely strongly on cyclic and variable stress using many stresses. These tests are both meant as "elephant tests" (Nelson, 1990), in which any failure modes induced by the test should be fixed by changing the design or manufacturing processes. Unfortunately, some fixes can still be very expensive so judgement about impact at operating conditions is necessary.

If the rate of a process leading to failure is governed by a single rate limiting step, then simple acceleration factors can be used for understanding how time under even complicated acceleration like that used in ESS and STRIFE converts to time in the field. This is not true when several steps in a process vie for domination in the range of stresses being investigated. Such competition is especially likely when whole systems rather than simple components are being accelerated.

In this paper some theory for experiment design for multi-step, multi-stress failure modes is begun. The theory is based on studying the behavior of a large class of physical processes that specify how multiple steps and multiple stresses interact in producing failure. Restrictions to the physical model that allows designs analogous to lower resolution designs in classical design theory are discussed. Finally the applicability of the current theory is discussed along with necessary and interesting problems for further research.

1. INTRODUCTION

Accelerated stress has a long history of use in manufacturing, and in the telephone industry in particular (Jones, 1927). The essential idea behind accelerated stress is to compress time as seen by the physical system under study so that the engineer can get a preview of the reliability of that system. In the past, conventional wisdom has been to keep the patterns of stress acceleration relatively simple (e.g. constant stress or simple patterns of increasing stress) so that simple acceleration models would allow comparison of the survival time under accelerated stress to survival time under operating conditions (Nelson, 1990). Unfortunately, even with simple stress patterns, the acceleration models can become very complicated (LuValle, Welsher, and Mitchell, 1986). Interestingly, introducing a little more complexity into the stress patterns can allow better identification of these complex models (LuValle, 1990)

For different reasons, procedures using more complicated stress patterns are gaining popularity with engineers. Two new engineering procedures that are playing an increasingly important role are:

1. Environmental Stress Screening (ESS): Used to identify manufacturing defects in early production at the system level (Kirklewski, 1990, Hough and Messer, 1990, Vellmure and Wooley, 1990). This procedure relies strongly on monitored thermal and power cycling. Often ESS is discussed as a screen to be used on 100% of production, but this is not an economically attractive alternative for high volume products.

2. Stress-Life (STRIFE) tests: Used to identify design defects leading to potential reliability problems in the prototype stage (Haibel, 1990). This procedure relies on cycling procedures in which the amplitude and/or cycling rate increases over time.

These two procedures are unusual in that they are instituted at the system level, and they make strong use of variable stress. The practical use of either of these procedures requires judgement both about how they should be designed to be most revealing, and about whether any given failure mode identified by them should be eliminated by product or manufacturing process redesign.

A vital piece of information in making these judgements is the probability that a failure mode identified under accelerated conditions will occur in the lifetime of the product under normal operating conditions. In some cases the cost of eliminating the failure mode is so low that it is unnecessary to evaluate this probability. However often it is necessary to make some hard economic decisions, and some estimate of this probability is required.

The traditional method of making this judgement is to assume that failures caused by a given failure mode follow some standard failure time distribution (e.g. exponential, lognormal, Wiebull), and that the relationship between time to failure under accelerated conditions and time to fail under operating conditions can be expressed in a single number called an acceleration factor. The statistical model corresponding to the notion of an acceleration factor is known either as the accelerated life model (Cox and Oakes, 1984) or the accelerated failure time model (Kalbfliesch and Prentice, 1980) depending upon whose book you are reading. Both of these models reduce to the relationship.

$$F_s(t) = F_{s_0}(\theta_s t) \tag{1}$$

Here θ_s is the acceleration factor taking time under stress s to time under operating stress (stress s_0), and $F_s()$ and $F_{s_0}()$ are the cumulative failure distributions at stress s and stress s_0 respectively. Interestingly, under slightly stronger specifications, this model is the simplest physically reasonable model for acceleration. In particular change the model so that instead of simply scaling time in the failure distribution, the acceleration factor is used to scale failure time for each unit in the population (as in equation 2). Then this model corresponds to the situation in which the process leading to failure is controlled by a single rate limiting step over the whole stress regime of interest (Klinger, 1990, LuValle, Welsher, and Mitchell, 1986, LuValle, Welsher, and Svoboda, 1987).

$$t_i(s_0) = \theta_s t_i(s) \tag{2}$$

In the above equation $t_i(s)$ is the failure time for unit i under stress s.

An example in which a physical process leading to failure has a single rate limiting step is the following. Suppose failure is caused by the accumulation of the final product of a sequence of chemical reactions, and there is one step of that sequence serving as a bottleneck for the whole process over the range of stresses of interest. That bottleneck step is the rate limiting step.

When the process does have a single rate limiting step, prediction of life under variable stress can be done by using the cumulative exposure model (Nelson, 1990) in which time at a given stress is simply weighted by the appropriate acceleration factor. Thus the equivalent age of the device under operating conditions can be calculated simply as:

$$\text{age at time } T = \int_0^T \theta_{s(t)}\, dt$$

In many cases (LuValle, Welsher, and Mitchell, 1986, LuValle, Welsher, and Svoboda, 1987) the physical processes leading to failure do not have a single rate limiting step. In these cases simple acceleration factors are inappropriate. When multiple stresses are operating it is often the case that more than one rate limiting step is operating, and hence the accelerated life model doesn't hold. Perhaps more important in interpreting the results of ESS and STRIFE, it is not possible to use simple cumulative exposure models to determine if a failure mode is going to have significant effect at operating conditions.

A reasonable engineering strategy in this situation is to use ESS and STRIFE to identify failure modes, eliminate those failure modes that can be eliminated easily, don't consider those failure modes that simply don't happen under operating conditions (e.g. the frame melted), and for the remaining 1-2% of the failure modes left over do supplementary accelerated life tests. To simplify experimentation these life test should be run using special failure mode specific test vehicles (Kaplan and Mitchell, 1986). This presumes that there is sufficient knowledge of the material systems that ESS regimes can be designed so as not to cause adverse aging (recall that ESS is a SCREEN on early production, so the stresses should not appreciably age the devices that pass it). The purpose of this paper is to begin laying a scientific framework for designing accelerated life tests both for the purpose of investigating possible ESS regimes to make sure that they do not cause premature aging, and to investigate new failure modes that fall in the 1 to 2% category above.

In a previous paper (LuValle, 1990) it was shown that step stress tests are necessary in distinguishing between multi-step models, and that single step step stress tests provide as much information as multi step step stress tests for a large class of dynamic models.

In this paper those results are extended to tackle the combinatorial problem of how to design step stress experiments when multiple stress factors are being used to accelerate failures. In particular a correspondence will be shown between optimum step stress design and a minimum spanning tree between points in stress space representing constant stress experiments. This result, and the theoretical background leading up to it will be presented in the next section. The third section contains two important extensions of the result, one to multiple failure modes and/or paths, and the other to failure processes governed by a particular kind of nonlinear dynamics. The fourth section discusses what the current state of the theory is with respect to application, and poses important problems for further research.

The particular problem addressed in this paper is, given a collection of constant stress experiments in stress space, and no model of the dependence of the physical parameters of the failure causing process on the particular parametrization of stress variables, what is the maximum number of step stress experiments required in order to predict the result of any step stress experiment between those constant stress levels.

The theory developed here provides a foundation upon which experiment design theory involving models of the dependence between stress and the physical parameters of the failure causing process can be built.

2. BACKGROUND AND THEORETICAL RESULTS

In classical experiment design, the size of the experiment necessary to estimate an effect is a rough function of the number and type of "interactions" present in the effect of the factors on the response. A roughly analogous result can be stated for step stress experiments, where here the size of the step stress experiment is defined as the number of constant stress levels that have to be stepped between. In this section we will develop a theoretical framework that we can use to develop a basic theorem for step stress experiments. In the next section we discuss how we can define rough "levels of interaction" corresponding to the complexity of physical models.

In this section we will confine ourselves to the following linear kinetic model for a failure causing process. Assume that we have available a sample from a population of "devices" D_i for $i = 1,\ldots,n$. Each device has associated with it an unobservable vector denoting its internal state at time t, $A_{i,t}$. Assume that except for multiplication by a scalar constant each $A_{i,0}$ is identical, and that for a given constant stress s $A_{i,t}$ evolves according to the equation:

$$\frac{dA_{i,t}}{dt} = K(s)A_{i,t} \qquad (3)$$

Here K(s) is a square matrix function of stress, s, with real distinct nonpositive eigenvalues. We assume that K(s) is the same across all devices. Justification for the restrictions on K(s) can be provided by chemistry in many cases.

In the concrete world of devices, each component of the vector $A_{i,t}$ represents the concentration of a physical species such as a chemical, or dislocations in a crystal. For example, in a simple model of corrosion caused printed wiring board failure, the first component of $A_{i,t}$ may be the concentration of Cl_2 (Chlorine molecules), and the second component could be the concentration of $CuCl_2$. The chemical process leading to failure is a competition between outgassing of the chlorine and combination of chlorine and copper to form a conductive filament. The matrix representing this process would have the form:

$$\begin{bmatrix} -(k_1+k_2) & 0 \\ k_1 & 0 \end{bmatrix}$$

Here k_1 is the rate at which Cl_2 combines with copper to form $CuCl_2$ and k_2 is the rate of outgassing. Both constants depend on stress. Similar mathematical models can be built of laser degradation and failure and degradation processes in other devices.

We assume that the observable for any given device takes the form

$$Y_{i,t} = I[V'A_{i,t} > C_i] \qquad (4)$$

where $I[x,t]$ is an indicator function that is 0 until the first time that x is true and then 1 after and V is a vector with a positive scalars in the places corresponding to the failure causing states in A, and 0's elsewhere. Thus $Y_{i,t}$ denotes when a failure occurs.

In our example V' is the vector $(0,1)$ and C_i is the amount of $CuCl_2$ necessary to bridge the path between two conductors in device i.

The time to failure distribution at any given stress condition is determined by equations (3) and (4), and the distribution of the terms $C_i/\|A_{i,t}\|$ However the transformation taking time from one stress condition to another depends only on equation (3). Solving for this transformation we find that it takes the form of the implicit function defined by:

$$\sum_{i=1}^{m} b_i(s_2)\exp(\lambda_i(s_2)t_2) = \sum_{i=1}^{m} b_i(s_1)\exp(\lambda_i(s_1)t_1) \tag{5}$$

modulo the regularity conditions necessary to make each side of equation (5) monotone increasing. The $\lambda_i(s)$ are the eigenvalues of $K(s)$. Again these regularity conditions are often met by real physical systems (LuValle, 1990).

A more revealing representation that can be used to motivate the usefulness of step stress procedures is given in equation 5' below.

$$V'\exp(K(s_2)t_2)A_0 = V'\exp(K(s_1)t_1)A_0 \tag{5'}$$

Here $A_0 = A_{i,0}/\|A_{i,0}\|$ If we step from stress s_1 to stress s_2 at time t_1', and t_2' is the time corresponding to t_1' from equation 5', then the time transformation after the step between the experiment that was held constantly at stress s_2 and the experiment stepped to s_2 is:

$$V'\exp(K(s_2)(\Delta t_{2,2}))\exp(K(s_2)t_2')A_0 \tag{6}$$
$$= V'\exp(K(s_2)(\Delta t_{1,2}))\exp(K(s_1)t_1')A_0$$

Where $\Delta t_{i,j}$ is the time elapsed since the step. What is particularly interesting here is that while

$$V'\exp(K(s_2)t_2')A_0 = V'\exp(K(s_1)t_1')A_0 \tag{7}$$

it is not true that

$$\exp(K(s_2)t_2')A_0 = \exp(K(s_1)t_1')A_0 \tag{8}$$

Thus the version of equation (5) that is appropriate after such a step is:

$$\sum_{i=1}^{m} b_i(s_2,s_2,t_2')\exp(\lambda_i(s_2)\Delta t_{2,2}) = \sum_{i=1}^{m} b_i(s_1,s_2,t_1')\exp(\lambda_i(s_2)\Delta t_{1,2}) \tag{9}$$

Here we define $b_i(s_j,s_k,t_j')$ as the coefficient of $\exp(\lambda_i(s_k)\Delta t_{j,k})$ after the step from s_j to s_k at t_j'.

In particular we see that the time transformation between a population of devices that have been sitting at stress s_2 and one recently stepped to s_2 is not the identity function, because the weights of the exponential functions will be different. This should be contrasted to the case when there is a single rate limiting step in the physical process, and the time transformation is an identity function.

The main theoretical results in LuValle 1990 were that under the additional condition that none of the b_i are 0,

i. step stress tests are necessary to identify the difference between different models (proven using a simple example).
ii. one step step stress tests provide as much information as multistep step stress tests.

A third result was stated allowing the approximation of continuously changing stress by many steps, but the proof given in that paper is incorrect due to a confusion in notation. We conjecture and assume that this third result is true, and that a sufficiently fast ramp can be well approximated by a sudden step. However we will make no attempt to prove those results here. The mathematics that follows does not depend on this assumption, but whether that mathematics is connected to real step stress experimentation does.

The key to the proof of the second result is also the key to the proof of the result that we will state shortly here. It is shown in the appendix of this paper (and nearly in LuValle, 1990) that there is a matrix $\beta(s_1,s_2)$ satisfying the following.

Define $b(s_1,t_1')$ as the vector of $b_i(s_1)\exp(\lambda_i(s_1)t_1')$, $b(s_1,s_2,t_1')$ as the vector of $b_i(s_1,s_2,t_1')$, then:

$$b(s_1,s_2,t_1') = \beta(s_1,s_2)b(s_1,t_1') \tag{10}$$

Further it is shown that:

$$\beta(s_1,s_3) = \beta(s_2,s_3)\beta(s_1,s_2) \tag{11}$$

For any stresses s_1, s_2, s_3.

Finally it is shown elsewhere (LuValle, 1990) that $\beta(s_1,s_2)$ can in principle be estimated by 1 constant stress experiment at stress s_2 and m step stress experiments from s_1 to s_2 over the range of t_1' and t_2' where $\beta(s_1,s_2)$ can be defined. This is easily seen by an application of the implicit function theorem to equations (5) and (9) under the regularity conditions given in (LuValle, 1990).

The main result of the section now follows immediately. To be able to predict the result of a step stress experiment between point s_1 and point s_2, all that is necessary is that you know the values of the $\beta(s_i,s_j)$ for any path connecting the two. Thus we simply need to do constant stress experiments at each point, and know any minimal spanning tree created of edges made up of the $\beta(s_i,s_j)$. Stated more formally:

Theorem: Under the regularity conditions stated here and in (LuValle, 1990) if there are M different points in stress space at which constant stress experiments are being run, then the maximum number of step stress experiments required in order to predict the result of any step stress experiment between the M points is (M-1)*m with an additional M constant stress experiments.

Proof: M-1 is the number of edges on the minimal spanning tree of a graph with M nodes. Since the minimal spanning tree provides a path between any two points, the result follows from equations (9), (10), and (11).

In the next section we go on to show that there are general classes of models where it is possible to get away with fewer experiments.

3. EXTENSIONS TO SEPARABLE PROCESS PATHS, AND SIMPLE NONLINEAR MODELS

Imagine a staircase of models corresponding to the number of step stress designs necessary between a given collection of constant stress experimental points to predict the result of any step stress experiment within that collection. The general linear kinetic model given in the previous section falls somewhere in the middle. In the worst case of a general nonlinear kinetic model, even m, the number of experiments between two constant stresses necessary to predict the results of further experiments between those two stresses may not be finite. In this section we exhibit a class of models requiring fewer step stress experiments than the general linear kinetic model, and discuss how a particular model of interaction allows some nonlinear kinetic models to be dealt with in this theory.

In this section we will apply one restriction to the set of M constant stress points that we are interested in. Define a "controlled" array of points in stress space to be an array such that if we pick any point in the array, and any dimension in stress space, there is a collection of other points in the array that vary only in that dimension. The simplest way to construct such a controlled array is to take sets of values in each stress dimension, and take the set including the cross product of those points as the array. This last design is of course a full factorial experiment. There are controlled arrays that are sparser than full factorials, although a 2^k full factorial is the minimal controlled array in k dimensions. The reduction in number of step stress experiments that we now show for a particular class of physical models should hold for any controlled array. For simplicity of presentation we will discuss the result only for a full factorial design.

There are actually two classes of models that correspond to fewer step stress experiments being required (apart from the trivial case in which a single rate limiting step dominates the failure causing process over the whole stress regime and step stress is unnecessary). The simplest class of models is those in which there are two independent failure modes and the failure modes are effected by orthogonal subspaces in stress space. In the simplest possible case there are two dimensions in stress (e.g. temperature and humidity), two failure modes, and each failure mode is only effected by one stress dimension. If there are n1 values of stress 1 (n1 temperatures) and n2 values of stress 2 (n2 relative humidities) then there are n1*n2 total constant stress points. But instead of m*(n1*n2-1) step stress experiments, there are only m*((n1-1)+(n2-1)) step stress experiments required. Note that for simplicity we assume that m the dimension of the vector describing the internal state of the system is the same for each failure mode. To see why this reduction is possible, note that the effect of step stress on the uncensored survival distribution of failure mode 1 can be completely predicted by m*(n1-1) stepstress experiments, by the results of the previous section. Similarly failure mode 2 can be completely predicted from m*(n2-1) step stress experiments. The combined survival distribution can be calculated simply by taking the product of the two individual survival functions. Since under random censoring, the Kaplan-Meier (Kaplan and Meier, 1958) estimate of a failure distribution is unbiased and consistent, the transformations for each failure mode can be calculated separately and the result follows.

The extension to the case in which the stress subspaces corresponding to each failure mode have a non-trivial intersection is not difficult.

Independent failure modes are not the only way that such a reduction in the number of step stress experiments can happen. Another possibility is that the linear kinetic process giving rise to failure can be separated into physically independent paths. Strictly speaking this happens if there is an ordering of the vector A_t such that the β matrices can be represented as a partitioned matrix with the non zero submatrices lying only on the diagonal, for any step stress experiment. One simple way to guarantee this partitioning can be described in terms of another representation for chemical processes based on directed graphs.

The simplest way to construct a directed graph representing a chemical process is to lay out the states, A_i (here i represents the different components of the vector) as the nodes of a graph, and use the "rate constant" matrix (denoted by K in the previous section) to draw the directed edges. Each non zero off diagonal K_{ij} represents a directed edge on the graph between state i and state j. A sufficient condition for the process to be separable is:

i. All edges connected to the failure causing state, say state f, point only into f, so there are no arrows with their root at state f.

ii. If state f is removed along with the edges pointing into it, then the graph decomposes into a set of orthogonal (in the graph theoretic sense) graphs. The number of (physically) independent process is the number of orthogonal graphs.

While it is relatively easy to define a difference between independent failure modes, and independent process paths, it is very difficult to come up with an example of a failure mode which can be purely identified as a single failure mode with independent process paths that would not be considered, on a closer look, as two dependent failure modes. In particular, corrosion failures in printed wiring boards can be caused by an excess of chlorine, or bromine in processing. Either will result in the formation of conductive filaments that are indistinguishable without elemental analysis. However the rate limiting portion of the failure causing process may proceed at different rates for the two chemicals. Thus when both chlorine and bromine are present, filaments containing both can be formed but the contributions of each are different. This might be identified either as a failure mode with two independent process paths, or as two dependent failure modes. The language is unclear, but at least the mathematics follows.

The rules of design for physically independent process paths will be the same as for independent failure modes. Thus we group both into a general category of separable failure processes, and can state the following corollary to the theorem of the previous section.

Corollary 1: Suppose that there are k separable failure processes that are of concern, and the stress subspaces effecting each process are mutually orthogonal. Assume m_i is the dimension of the state vector (A_t in the previous section) corresponding to process i, the constant stress experiments are arranged in a full factorial structure, and the projection of the full factorial experiment onto the relevant subspace for failure process i has M_i points. Then the total number of step stress experiments required is:

$$\sum_{i=1}^{k} m_i (M_i - 1)$$

The proof is trivial following the discussion. The important point here is that the number of step stress designs necessary in the separable failure process case grows arithmetically with the number of stress dimensions being considered, rather than geometrically, thus the notion of separable processes corresponds to considering a reduced set of interactions in classical experiment design.

To define the class of nonlinear failure processes that we wish to discuss we assume now that one of the stresses can be applied in a cyclic fashion. Further assume that the failure causing process will proceed without the cyclic stress ever being applied, but each application of the cyclic stress (regardless of amplitude, or the shape of the cycle) increases the absolute values of the nonzero elements of the K matrix. For example mechanical cycling might increase the rate of diffusion through a particular material by changing the physical structure of the grain boundaries.

If we ignore the cyclic stress, the theory of section two allows us to estimate the number of step stress experiments required. However the addition of the cyclic stress adds an additional complication. Typically the step stress procedures described before were used to change the rate constant (K) matrix in the middle of the process. Now each application of the cyclic stress results in an irreversible change in K. In particular, under this model of interaction it is necessary to consider each application of a cycle as stepping the material to a different stress. With this transformation of level of stress to number of cycles, the theory of section 2 will follow.

While the theory as it stands is too cumbersome for immediate application, there is some useful ideas that can be derived in terms of testing systems to see what "level of interaction" is present between the effects of the stresses on given failure modes, and whether the underlying physical process can be well approximated by a linear physical process such as was described in section 2. In particular, a test for linearity in the process could be based on equation (11). If a direct estimate of $\beta(s_1,s_3)$ departs sufficiently from one based on $\beta(s_1,s_2)$ and $\beta(s_2,s_3)$ for any three points in stress space, then either the dynamic system is not linear or the estimate of m, the dimension of the state vector of the kinetic process, is too low. A similar test based on the β's and the kind of separability assumed, could be made for separable vs non-separable physical processes.

4. FURTHER RESEARCH AND APPLICATION

The use of ESS and STRIFE testing, and other developments along that line will undoubtedly improve the reliability and quality of manufactured products in the future. However without some theory to explain what varying stress is doing to the products, they will remain entirely empirical procedures with a high frequency of misuse and disappointment in the results. A good practical theory of experiment design and data analysis for step stress experiments can aid the engineer, both in the initial design of such procedures, and in interpreting their results.

As mentioned in the previous section, while there is some potential applicability of the theory in terms of testing departures from assumed theoretical models, for the most part the theory demands too many experiments to be practical at this point. Probably the most important use of step stress experiments given the current state of the theory is in testing for departure from the accelerated life model. Equations (7), (8), and (9) and the discussion around them provide theoretical motivation for this, while a recent paper studies the power of graphical data analysis procedures plus step stress for detecting departures from the accelerated life model (LuValle, 1991).

The major theoretical difficulty standing in the way of practical use of step stress in estimating failure processes, is the incorporation of parametric dependencies of the elements of the matrix K on the stress into the theory. In a study in which step stress experiment designs were used in conjunction with parametric models of stress dependence it was possible to identify nonlinear structure in the failure causing process (LuValle and Hines, 1991) much more easily than is implied by the theory developed here. It appears that for most models, the models are identifiable even when step stress is done between only two stresses, but a rigorous proof of this is difficult.

Assuming that the statistical theory can be built, a major problem in applied physics that the applied statistician should be aware of is the proper definition of the stress dimensions. For example there is often a controversy as to whether relative humidity or partial vapor pressure is the appropriate thermodynamic variable for studying the effects of humidity.

Different answers will result in different experiments, with attention paid to different details of control.

Finally, this whole paper brings up the connection between graph theory and dynamic processes. The connection between topology in general and dynamic processes is very deep however I am not aware of specific connections to graph theory, other than the arrow representation of chemical processes, and the theory brought up here. Further exploration of this interface seems worthwhile.

MATHEMATICAL APPENDIX

Define $D_\lambda(s)$ to be the diagonal matrix of eigenvalues of $K(s)$, and $P(s)$ the corresponding matrix with each column an eigenvector. Using this notation, the solution to equation (3) has the form

$$A_t = P(s)\exp(D_\lambda(s)\times t)P^{-1}(s)A_0 \tag{A1}$$

and under step stress from stress s_1 to s_2 at time $t_1{}'$ the solution has the form

$$\begin{aligned} A_t &= (P(s_2)\exp(D_\lambda(s_2) \times (t-t_1{}'))P^{-1}(s_2)) \times \\ &\quad P(s_1))\exp(D_\lambda(s_1) \times t_1{}')P^{-1}(s_1)A_0 \\ &= P(s_2)\exp(D_\lambda(s_2) \times (t-t_1{}'))P^{-1}(s_2)A_{t_1{}'} \end{aligned} \tag{A2}$$

implying

$$V'A_t = \sum_{i=1}^{m} b_i(s_1,s_2,t_1{}')\exp(\lambda_i(s_2)(t-t_1{}')) \tag{A3}$$

It is useful at this point to look at the fine structure of the $b_i(s_1)$ and $b_i(s_1,s_2,t_1{}')$. To do this we will use the following notational conventions.

Let $M_i^c(s)$ denote the ith column of matrix $M(s)$, and let $M_j^r(s)$ denote its jth row. Then from equation (A1) we see that $b_j(s)$ has the form

$$b_j(s) = V'P_j^c(s)(P^{-1}(s))_j^r A_0 \tag{A4}$$

Similarly from equation (A2) we see that

$$b_i(s_1,s_2,t_1{}') = V'P_i^c(s_2)[(P^{-1}(s_2)P(s_1))_i^r]\exp(D_\lambda(s_1) \times t_1{}')P^{-1}(s_1)A_0 \tag{A5}$$

$$= V'P_i^c(s_2)[(P^{-1}(s_2)P(s_1))_i^r]\begin{bmatrix} (P^{-1}(s_1))_2^r A_0\exp(\lambda_2(s_1)\times t_1{}') \\ \cdot \\ \cdot \\ \cdot \\ (P^{-1}(s_1))_m^r A_0\exp(\lambda_m(s_1)\times t_1{}') \end{bmatrix}$$

By combining equations (A4) and (A5) we see that

$$b_i(s_1,s_2,t_1{}') = \sum_{j=1}^{m} \beta_{ij}(s_1,s_2)b_j(s_1)\exp(\lambda_j(s_1)\times t_1{}') \qquad \text{A6)}$$

where

$$\beta_{ij}(s_1,s_2) = \frac{(V'P_i^c(s_2))(P^{-1}(s_2)P(s_1))_{ij}}{(V'P_j^c(s_1))} \tag{A7}$$

Define $b(s_1,t_1')$ as the vector of $b_i(s_1)\exp(\lambda_i(s_1)t_1')$, $b(s_1,s_2,t_1')$ as the vector of $b_i(s_1,s_2,t_1')$, and $\beta(s_1,s_2)$ as the matrix of $\beta_{ij}(s_1,s_2)$. Then we may rewrite equation A6 as:

$$b(s_1,s_2,t_1') = \beta(s_1,s_2)b(s_1,t_1') \tag{A8}$$

Further, defining $D(s)$ as the diagonal matrix with elements $V'P_i^c(s)$ then

$$\beta(s_1,s_2) = D(s_2)(P^{-1}(s_2)P(s_1))(D(s_1))^{-1} \tag{A9}$$

Thus it is easy to see:

$$\beta(s_1,s_3) = \beta(s_2,s_3)\beta(s_1,s_2) \tag{A10}$$

Which provides us with the mathematical background necessary to prove the main result of section 2.

BIBLIOGRAPHY

Cox, D. R. and Oakes, D. *Analysis of Survival Data,* Chapman and Hall, New York, 1984.

Haibel, C. (Hewlett Packard) viewgraphs from the sematech workshop on accelerated testing, Albuquerque, New Mexico, 1990.

Hough, D. L., and Messer C.R., "Reliability stress testing of a commercial product", Proceedings of the Institute of Environmental Sciences, pp. 770-778, 1990.

Jones, P. C., "Crowding years of service into weeks: Accelerated Life Tests at Bell Laboratories determine the durability of telephone apparatus", Wisconsin Telephone News, 22, (1927).

Kalbfleisch, J.D., and Prentice, R.L., *The Statistical Analysis of Failure Time Data* Wiley, New York 1980.

Kaplan, E. L., and Meier, P., "Nonparametric estimation from incomplete observations", Journal of the American Statistical Association, 53, pp 457-481, (1958).

Kaplan, B. A., and Mitchell, J. P., "Use of Technology Test Vehicles for Accelerated Life Testing of Printed Circuit Products", Proceedings of the Sixth Annual International Electronics Packaging Conference, pp. 745-756, 1986.

Kirklewski, D. D. "Is the word integrity in your ESS vocabulary", Proceedings of the Institute of Environmental Sciences, pp. 767-769, 1990.

Klinger, D. J., "Failure time and rate constant of degradation: An argument for the inverse relationship", submitted to IEEE transactions on Reliability (1990).

LuValle, M. J., "A note on experiment design for accelerated life tests", Microelectronics and Reliability, v. 30, No. 3, pp591-603, (1990).

LuValle, M. J., "Experiment design and graphical analysis for checking acceleration models", To appear in Microelectronics and Reliability.

LuValle, M. J., and Hines L. L., "Using step stress to explore the kinetics of failure", To appear in Quality and Reliability Engineering International.

LuValle, M. J., Welsher, T. L., and Mitchell, J. P., "A new approach to the extrapolation of accelerated life test data", Proccedings of the Fifth International Conference on Reliability and Maintainability, Biarritz, France, 1986, pp 620-635.

LuValle, M. J., Welsher, T. L., and Svoboda, K. (1988), "Acceleration Transforms and Statistical Kinetic Models", Journal of Statistical Physics, Vol. 52, pp. 311-320, 1988.

Nelson, W. *Accelerated Testing: Statistical Models, Test Plans, and Data Analysis* Wiley, New York, 1990.

Vellmure, W., and Wooley, D. T., "ESS effectiveness- Improved screening equals program cost reductions and long term reliability", Proceedings of the Institute of Environmental Sciences, pp. 789-793, 1990.

ADVANCES IN RELIABILITY
Edited by Asit P. Basu

Determination of stopping criteria during product development

T. A. Mazzuchi[a] and R. Soyer[b]

[a]Department of Operations Research, The George Washington University, Washington, DC 20052, USA

[b]Department of Management Science, The George Washington University, Washington, DC 20052, USA

Abstract

A Bayesian approach is presented for analyzing the reliability of a product during its development phase. Prior information for both product reliability and the reliability growth process is used to motivate the prior joint distribution for product reliability over the specified range of tests. This subjective information as well as attribute test information at each stage of testing are combined to update the knowledge about the product reliability not only at the current stage of testing, but also at subsequent stages of testing and at the end of testing as well. After development of the model, the important question of determining when to stop the development process is addressed. Use of different stopping criteria is discussed and a simplified approach for analyzing the stopping problem is offered.

1. INTRODUCTION AND OVERVIEW

During product development, testing is performed in stages and at the end of each test stage, design changes\modifications are made to the product in hopes of improving its performance. This testing – modification process, often termed "reliability growth", has been the focus of much attention in the reliability literature. A common way of classifying reliability growth models is by the type of data testing yields, i.e. attribute or variable. While variable data provides more information, there are circumstances where attribute data is the only information available (due to measurement limitations) or the only information which is relevant (due to the product's intended use). Fard and Dietrich (1983) give a brief review of some of the more well known attribute reliability growth models. A distinction of these models is the classification of failure type into one [for example Lloyd and Lipov (1962)] or two [for example Barlow and Scheuer (1966)] categories. The latter authors classified failure types as *inherent* – those whose elimination is not possible; and *assignable cause* – those whose elimination is possible through product modification.

In a modern environment characterized by high cost, high reliability products, program managers are often asked to assess the product development without the benefit of a large amount of test results. As pointed out in Kaplan, Cunha, Dykes, and Shaver (1990), the above situation necessitates incorporation of other relevant information into the assessment process. A formal framework for incorporating test data and other relevant information into an analysis is provided via the Bayesian paradigm. Smith (1972) [later corrected by Fard and Dietrich (1987)] presented a Bayesian analysis of the attribute reliability growth problem for a single failure type.

A joint uniform distribution, defined over an ordered region, is the assumed prior distribution for the (nondecreasing) reliabilities after each test stage. Exact expressions were obtained for the posterior distribution of the system reliability after each testing stage. A Bayesian version of the Barlow – Scheuer model, was presented in Weinrich and Gross (1978) who use a Dirichlet distribution as the prior joint distribution for the initial probability of inherent and assignable cause failures. Assuming that, at the end of each stage of testing, all or none of the effects of assignable cause failure could have been removed, bounds are developed for posterior quantities relating to system reliability.

The main drawback of the aforementioned Bayesian approaches is their focus on system reliability only *after* each test stage. Addressing other relevant issues such as the required testing effort, the projected system reliability upon conclusion of the development program, and the termination of the development program, requires the ability to forecast future system reliability. In this paper, we develop a framework to address these issues and show how this framework provides a unification of past Bayesian work in reliability growth modeling.

2. PROPOSED RELIABILITY GROWTH MODEL

We assume a testing – modification scenario is conducted so that identical replications of the product are tested until a failure is observed. Furthermore, it is assumed that all failures are of the assignable cause type. Upon the discovery of a failure, a modification is made to the product to remove the cause and therefore increases product reliability. The test – modification scenario is repeated for some specified number of times and then after the last modification the product is released.

Let m denote the total number of testing – modification stages, R_i, $i = 1, \ldots, m$, denote the product reliability for the ith stage of testing (that is, prior to the ith product modification), and R_{m+1} denote the final (field) reliability. Because engineering modifications are made to the item upon the discovery of a failure, it is reasonable to assume that

$$0 \le R_1 \cdots \le R_{m+1} \le 1. \tag{1}$$

A natural and mathematically tractable prior distribution for $\underset{\sim}{R} = (R_1, \ldots, R_{m+1})$ is the ordered Dirichlet distribution given as

$$\Pi(\underset{\sim}{R} \mid D^{(0)}) = \frac{\Gamma(\beta)}{\prod_{j=1}^{m+2} \Gamma(\beta\alpha_j)} \prod_{j=1}^{m+2} (R_j - R_{j-1})^{\beta\alpha_j - 1}, \tag{2}$$

where $R_0 \equiv 0$, $R_{m+2} \equiv 1$, and $D^{(0)}$ represents the prior information captured by the prior parameters β, $\alpha_i > 0$, $\sum_{i=1}^{m+2} \alpha_i = 1$. The distribution is defined over the simplex $\{\underset{\sim}{R} \mid 0 \le R_1 \cdots \le R_{m+1} \le 1\}$ and thus embodies the restrictions in (1) and imposes no additional restrictions in the analysis.

A major advantage of the ordered Dirichlet distribution is that all relevant marginal distributions are Beta distributions in the interval [0,1]. For example, if $\alpha_i^* = \sum_{j=1}^{i} \alpha_j$, then it can be shown that

$$[\, R_i \mid D^{(0)}] \sim \text{Beta}\big(\beta\alpha_i^*, \beta(1-\alpha_i^*)\big), \tag{3}$$

$$[\, R_i - R_j \mid D^{(0)}\,] \sim \text{Beta}\big(\beta(\alpha_j^* - \alpha_i^*), \beta(1-\alpha_j^* + \alpha_i^*)\big), \quad \text{for } i<j, \tag{4}$$

$$[\, \frac{R_i}{R_j} \mid D^{(0)}\,] \sim \text{Beta}\big(\beta\alpha_i^*, \beta(\alpha_j^* - \alpha_i^*)\big), \quad \text{for } i<j. \tag{5}$$

Note that (3) describes the stage reliabilities whereas (4) and (5) convey the additive reliability improvement and the ratio of stage reliabilities, respectively. This offers a wide variety of possibilities for different methods of encoding and feedback and makes it more convenient to elicit and incorporate prior judgment into the analysis.

It has been common practice to elicit measures such as the mean, mode, and\or variance in order to specify the prior distribution parameters. That is, the elicited quantities would be compared with the analytical expressions in order to specify the prior parameters. These expressions can be obtained in closed form for all of the above distributions. For example, from (3) we obtain

$$E[R_i \mid D^{(0)}] = \alpha_i^* \tag{6}$$

$$VAR[R_i \mid D^{(0)}] = \frac{\alpha_i^*(1-\alpha_i^*)}{\beta+1} \tag{7}$$

$$MODE[R_i \mid D^{(0)}] = \frac{\beta\,\alpha_i^* - 1}{\beta - 2} \qquad (\text{exists if } \beta\alpha_i^* \geq 1,\ \beta > 2). \tag{8}$$

As pointed out in Mosleh and Apostolakis (1982), however, research in experimental psychology has indicated that, for skewed distributions, assessors tend to bias their estimates of the mean towards the median and it is therefore advisable to avoid assessments of distribution moments. Furthermore, for the problem at hand, because interest centers around the perception of the <u>overall</u> reliability growth process, it is possibly more reasonable to consider the elicited reliabilities collectively by equating them to the joint modal expressions obtained from (2) as

$$\left(\frac{\beta\,\alpha_1^* - 1}{\beta - (m+3)}, \frac{\beta\,\alpha_2^* - 2}{\beta - (m+3)}, \ldots, \frac{\beta\,\alpha_{m+1}^* - (m+1)}{\beta - (m+3)} \right) \tag{9}$$

which exists when $\beta\,\alpha_i^* \geq i$ and $\beta > m+3$.

Either the joint or marginal modal values can be used to specify the values of the α_i. To determine the parameter β, we note that β reflects the level of uncertainty about the reliability growth process as a whole as well as controlling the spread of the individual distributions given in (3), (4), and (5). Thus, it can be obtained in an interactive fashion by noting its effect on these distributions.

We note that the above framework can be used to unify the previous Bayesian approaches to reliability growth modeling. For example the model proposed by Smith (1979) can be seen as a special case of (2) with $\alpha_i = 1/(m+2)$, $i = 1, \dots,$ m+2 , and $\beta = m+2$. It follows from (6), (7), and (8) that for the Smith model,

$$E[R_i \mid D^{(0)}] = \frac{i}{m+2},$$

$$MODE[R_i \mid D^{(0)}] = \frac{i-1}{m}.$$

and

$$VAR[R_i \mid D^{(0)}] = \frac{i(m+2-i)}{(m+2)^2(m+3)},$$

Note that the Smith model implies an assumption of linear reliability growth and that all the stage reliability distributions are a function of only the number of testing stages. Thus the model does not allow for the incorporation of any prior judgment into the analysis.

The framework can be extended to encompass the Barlow-Scheuer model by extending the lines of reasoning presented in Weinrich and Gross (1978) who specify a Dirichlet prior for the inherent failure probability, q, and the stage 1 assignable cause failure probability, p_1. A Dirichlet-type prior can be specified on q and assignable cause failure probabilities for all the stages , p_i, $i = 1, \dots, m+1$, by preserving the desired ordering $p_1 \geq \dots \geq p_{m+1}$ (in current notation $R_i = 1 - q - p_i$). The prior distribution has the form

$$\Pi(q, \underset{\sim}{p} \mid D^{(0)}) = \frac{\Gamma(\beta)}{\prod_{j=0}^{m+2} \Gamma(\beta\alpha_j)} (1 - q - p_1)^{\beta\alpha_0 - 1} \prod_{j=1}^{m+1} (p_j - p_{j+1})^{\beta\alpha_j - 1} (q)^{\beta\alpha_{m+2} - 1} \tag{10}$$

where $p_{m+2} \equiv 0$, β, $\alpha_i > 0$, and $\sum_{i=0}^{m+2} \alpha_i = 1$.

A proper Bayesian analysis of the Barlow-Scheuer with full distributional results for p_i and $R_i = 1 - q - p_i$ at any stage is presented in Mazzuchi and Soyer (1992).

3. PREDICTIVE ANALYSIS

After the specification of the prior parameters, reliability predictions for any (future) test stage i are given by the prior marginal distribution in (3). These distributions can also be used to make probability statements about N_i, the number of products tested in each stage i, via

$$Pr\{N_i = n_i \mid D^{(0)}\} = \int_0^1 Pr\{N_i = n_i \mid R_i\} \, \Pi(R_i \mid D^{(0)}) \, dR_i \tag{11}$$

where $\Pi(R_i \mid D^{(0)})$ is the prior marginal distribution given by (3) and the observation model is given by

$$\Pr\{N_i = n_i \mid R_i\} = (1 - R_i)R_i^{n_i - 1} \qquad n_i = 1, 2, \ldots \quad . \tag{12}$$

Thus the predictive distribution for the number of test products is given by

$$\Pr\{N_i = n_i \mid D^{(0)}\} = \frac{\beta(1-\alpha_i^*) \prod_{j=0}^{n_i-2} (\beta\alpha_i^* + j)}{\prod_{j=0}^{n_i-1} (\beta + j)} \qquad n_i = 1, 2, \ldots , \tag{13}$$

where $\prod_{j=0}^{-1} \{\bullet\} \equiv 1$. Furthermore it can be shown that

$$E[N_i \mid D^{(0)}] = E\Big[E[N_i \mid D^{(0)}, R_i]\Big] = \frac{\beta - 1}{\beta(1-\alpha_i^*) - 1} \tag{14}$$

for $\beta > 1/(1-\alpha_i^*)$. The above distributions may be used to assess testing effort both locally (within each stage) and globally (over the entire test\development program).

Furthermore, if the goal of the test\development program is to eventually release a finished batch of products of size B, an important issue is the assessment of the number of field failures. This can be achieved using the predictive distribution for X_{i+1}, the number of field failures when the product is released after stage i, given by

$$\Pr\{X_{i+1} = x \mid D^{(0)}\} = \int_0^1 \Pr\{X_{i+1} = x \mid R_{i+1}\}\, \Pi(R_{i+1} \mid D^{(0)})\, dR_{i+1} \tag{15}$$

where

$$\Pr\{X_{i+1} = x \mid R_{i+1}\} = \binom{B}{x}(1 - R_{i+1})^x (R_{i+1})^{B-x} \qquad x = 1, 2, \ldots, B. \tag{16}$$

Thus the predictive distribution for the number of field failures when the product is released after stage i is given by the Polyá distribution

$$\Pr\{X_{i+1} = x \mid D^{(0)}\} = \binom{B}{x} \frac{\prod_{j=0}^{B-x-1} \left(\beta\alpha_{i+1}^* + j\right) \prod_{j=0}^{x-1} \left(\beta(1-\alpha_{i+1}^*) + j\right)}{\prod_{j=0}^{B-1} (\beta + j)} \tag{17}$$

for x = 1, 2, ... B and with

$$E[X_{i+1} \mid D^{(0)}] = E\Big[E[X_{i+1} \mid D^{(0)}, R_{i+1}]\Big] = B(1 - \alpha^*_{i+1}). \tag{18}$$

4. POSTERIOR RESULTS

We define $D^{(i)} \equiv \{D^{(0)}, n_1, \ldots, n_i\}$ and $D_k^{(i)} \equiv D^{(i-1)} \cup \{$k successes during i^{th} testing stage$\}$. Then after i (complete) stages of testing the likelihood function of $\underset{\sim}{R}$ is given by

$$\mathcal{L}(D^{(i)}; \underset{\sim}{R}) = \prod_{j=1}^{i} (1 - R_j) R_j^{n_j - 1} \tag{19}$$

and thus via a standard application of Bayes theorem, the posterior distribution of $\underset{\sim}{R}$, $\Pi(\underset{\sim}{R} \mid D^{(i)})$, is obtained proportional to

$$\left\{\prod_{j=1}^{i} (1 - R_j) R_j^{n_j - 1} (R_j - R_{j-1})^{\beta\alpha_j - 1}\right\}\left\{\prod_{j=i+1}^{m+2} (R_j - R_{j-1})^{\beta\alpha_j - 1}\right\}. \tag{20}$$

Expanding the $(1 - R_j)$ terms in (20) yields

$$\sum_{\ell_1=0}^{1} \cdots \sum_{\ell_i=0}^{1} (-1)^{\sum_{j=1}^{i} \ell_j} \prod_{j=1}^{i} R_j^{\ell_j + n_j - 1} (R_j - R_{j-1})^{\beta\alpha_j - 1} \times \left\{\prod_{j=i+1}^{m+2} (R_j - R_{j-1})^{\beta\alpha_j - 1}\right\}. \tag{21}$$

After observing $D^{(i)}$, we may wish to make inference about the current product reliability, $R_i \mid D^{(i)}$, to determine if modification is necessary or cost effective. After modifying the product, however, it is $R_{i+1} \mid D^{(i)}$ which reflects the current product reliability and is used to determine the necessity\impact of additional testing. Thus, after the ith testing stage, we are only interested in the quantities $R_i, \ldots R_{m+1}$, the current and future product reliabilities. We therefore integrate out $(R_1, \ldots R_{i-1})$ over the region $0 \leq R_1 \leq \cdots \leq R_{i-1} \leq R_i$. After some manipulations, we obtain $\Pi(R_i, \ldots R_{m+1} \mid D^{(i)})$ as equal to

$$\sum_{\ell_1=0}^{1}\cdots\sum_{\ell_i=0}^{1} \mathcal{W}(\underset{\sim}{\ell}) \frac{\Gamma(S_i+\beta)}{\Gamma(S_i+\beta\alpha_i^*)\prod_{j=i+1}^{m+2}\Gamma(\beta\alpha_j)} (R_i)^{S_i+\beta\alpha_i^*-1}\prod_{j=i+1}^{m+2}(R_j-R_{j-1})^{\beta\alpha_j-1} \tag{22}$$

where $S_j = \sum_{z=1}^{j} \ell_z + n_z - 1$, $\underset{\sim}{\ell} = (\ell_1, \ldots \ell_i)$ and

$$\mathcal{W}(\underset{\sim}{\ell}) = \frac{(-1)^{\sum_{j=1}^{i}\ell_j}\left\{\prod_{j=1}^{i-1}\frac{\Gamma(S_j+\beta\alpha_j^*)}{\Gamma(S_j+\beta\alpha_{j+1}^*)}\right\}\left\{\frac{\Gamma(S_i+\beta\alpha_i^*)}{\Gamma(S_i+\beta)}\right\}}{\sum_{\ell_1=0}^{1}\cdots\sum_{\ell_i=0}^{1}(-1)^{\sum_{j=1}^{i}\ell_j}\left\{\prod_{j=1}^{i-1}\frac{\Gamma(S_j+\beta\alpha_j^*)}{\Gamma(S_j+\beta\alpha_{j+1}^*)}\right\}\left\{\frac{\Gamma(S_i+\beta\alpha_i^*)}{\Gamma(S_i+\beta)}\right\}}. \tag{23}$$

Note that (22) is a mixture of ordered Dirichlet distributions of the form as in (2) and thus all prior distributional characteristics are preserved in the posterior as mixtures. Thus, given the available test information, $D^{(i)}$, we can assess stage reliabilities, future testing effort, and number of field failures upon release, by simply taking weighted averages of the closed form expressions in Section 2 evaluated for revised parameter values.

We also note that during a test stage, each single observation results in a revision of our uncertainty concerning the product reliability. For example, if during the ith test stage, we observe k successful tests of the product, then the current posterior distribution is given by (22) (with weights defined in (23)) with S_i replaced by $S_{i-1}+k$ and the summation terms over ℓ_1 to ℓ_{i-1}. The occurrence of a failure, however, not only causes us to update our uncertainty about the current product reliability, but also to undertake a product modification. After the modification, the assessment of current product reliability involves not only the test results but also the perceived effect of the reliability growth process (see Figure 1).

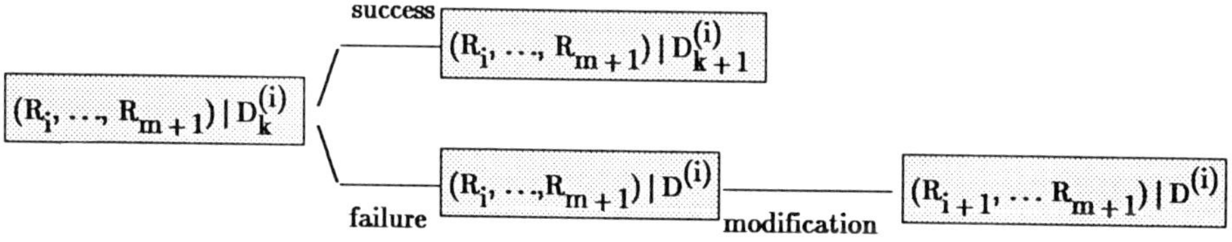

Figure 1. Effect of a Single Test Result.

5. EXPLORATION OF STOPPING CRITERIA

In order to determine when to stop the development program, we must know the criteria by which the final product will be evaluated. There are several possible criteria including the final reliability, the deviation from some specified target reliability, the number of field failures occurring in a released batch of a specified size, etc. In general, we may say that we should stop the development program when the expected utility of stopping given all the available information is greater than or equal to the expected utility of continuing the development program. The utility function is often defined in terms of the testing and modification costs (including labor), the cost of delaying the release of the product, and the criteria by which the final product is evaluated. In general, obtaining an explicit theoretical solution to this problem is difficult unless there is a limit on the number of items which can be tested or the number of testing stages. In what follows we assume that the goal of the development program is to eventually produce a fixed batch of items of size B and that the utility is expressed via the trade-off between testing cost and the cost of field failure. Specifically, we let C_T denote the cost of testing a single product and let C_F denote the cost of a single field failure. Furthermore, we consider the case where there is a limit, m, on the number of testing stages and after each stage (and the subsequent modification), a decision is made to either stop testing or continue testing. If testing is stopped, then the batch of B products is released.

For the case of a two stage testing program (m=2), the decision tree is given in Figure 2 where, following the usual convention, square nodes denote decision points and circular nodes represent uncertain quantities. Note that the decision node, $\mathfrak{D}_1$, denotes the decision to test or stop the testing program and release the batch, B, without testing. If the decision is to stop, then X_1 failures will be observed in the field. As X_1 is a random quantity, the occurrence of field failures is marked by random node $\mathfrak{R}_1$. If at $\mathfrak{D}_1$, the decision is to test (perform the first stage of testing), then a random number of items, N_1, will be tested until a failure is found. The random occurrence of test items is denoted by random node $\mathfrak{R}_2$. After observing the failure and performing a modification, the next decision is whether to continue to the next stage of testing or not (denoted by decision node $\mathfrak{D}_2$). Again, if the decision is to stop the testing program and release, then X_2 field failures will be observed (denoted by $\mathfrak{R}_3$). If the decision is to continue testing, then a random number of items, N_2, will be tested until a failure is found (denoted by random node $\mathfrak{R}_4$). After observing the failure and performing a modification, the batch is released and a random number of field failures, X_3 is observed (denoted by random node $\mathcal{R}_5$).

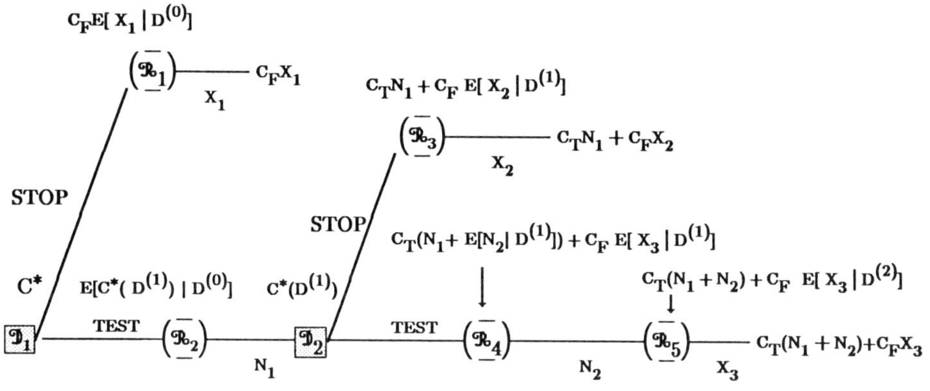

Figure 2. Decision Tree for the Example Scenario.

Note that the payoffs associated with the terminal nodes are a linear function of the total number tested and the total number of field failures. Solving the sequential decision problem merely involves the usual "folding back the tree" by using the expected values of the appropriate forecast distributions. For example, at $\mathcal{R}_5$ we obtain

$$E_{X_3}\left[C_T(N_1 + N_2) + C_F X_3 \,|\, D^{(2)}\right] = C_T(N_1 + N_2) + C_F \, E[X_3 \,|\, D^{(2)}] \tag{24}$$

where $E[X_3 \,|\, D^{(2)}]$ is the analog of (18) after observing $D^{(2)}$. At $\mathcal{R}_4$ we take the expectation of (24) with respect to N_2 given $D^{(1)}$ and obtain

$$\begin{aligned} &E_{N_2}\left[C_T(N_1 + N_2) + C_F E_{X_3}\left[X_3 \,|\, D^{(2)}\right] \,|\, D^{(1)}\right] \\ &\qquad = C_T(N_1 + E[N_2 \,|\, D^{(1)}]) + C_F \, E[X_3 \,|\, D^{(1)}] \end{aligned} \tag{25}$$

where $E[N_2 \,|\, D^{(1)}]$ and $E[X_3 \,|\, D^{(1)}]$ are the analogs of (14) and (18) after observing $D^{(1)}$. In a like fashion to (24) we obtain at $\mathcal{R}_3$

$$E_{X_2}\left[C_T N_1 + C_F X_2 \,|\, D^{(1)}\right] = C_T N_1 + C_F \, E[X_2 \,|\, D^{(1)}]. \tag{26}$$

At $\mathfrak{D}_1$, the decision is made based on maximization of expected utility (or equivalently minimization of expected cost) and the optimal cost is given as

$$C^*(D^{(1)}) = \text{MIN}\left\{ \left\{C_T N_1 + C_F\, E[X_2 \mid D^{(1)}]\right\},\right.$$
$$\left. \left\{C_T(N_1 + E[N_2 \mid D^{(1)}]) + C_F\, E[\, X_3 \mid D^{(1)}]\right\} \right\}. \tag{27}$$

Next, we taking the expectation of (27) with respect to N_1 given $D^{(0)}$ at $\mathfrak{R}_2$ thus yielding

$$E_{N_1}[C^*(D^{(1)}) \mid D^{(0)}] = C_T E[N_1 \mid D^{(0)}] + \sum_{n_1 \in S_1} C_F\, E[X_2 \mid N_1 = n_1, D^{(0)}] \Pr\{N_1 = n_1 \mid D^{(0)}\}$$
$$+ \sum_{n_1 \in \bar{S}_1} \left(C_T E[N_2 \mid N_1 = n_1, D^{(0)}] + C_F\, E[\, X_3 \mid N_1 = n_1, D^{(0)}]\right) \Pr\{N_1 = n_1 \mid D^{(0)}\} \tag{28}$$

where S_1 is the set of all values of n_1 which will result in the decision to stop after one test and $\bar{S}_1$ is its complement. Finally, the initial decision at $\mathfrak{D}_1$ is again based on the minimal cost. The minimal expected overall cost is defined as

$$C^* = \text{MIN}\left\{\left\{C_F E[\, X_1 \mid D^{(0)}]\right\}, \left\{E_{N_1}[C^*(D^{(1)}) \mid D^{(0)}]\right\}\right\} \tag{29}$$

where $C_F E[\, X_1 \mid D^{(0)}]$ is obtained from $\mathfrak{R}_1$ and thus the initial decision to test (or stop) is based on comparison of the expected cost of the two alternatives. Note that though (28) is a complicated expression, it is easily evaluated using a computer. Also note from (28) that

$$C^*(D^{(1)}) \le C_T\, E[N_1 \mid D^{(0)}]$$
$$+ \text{MIN}\left\{\left\{C_F\, E[X_2 \mid D^{(0)}]\right\}, \left\{C_F\, E[X_3 \mid D^{(0)}] + C_T\, E[N_2 \mid D^{(0)}]\right\}\right\} \tag{30}$$

which provides the following bounding strategy. If

$$C_F E[\, X_1 \mid D^{(0)}] \ge C_T\, E[N_1 \mid D^{(0)}]$$
$$+ \text{MIN}\left\{\left\{C_F\, E[X_2 \mid D^{(0)}]\right\}, \left\{C_F\, E[X_3 \mid D^{(0)}] + C_T\, E[N_2 \mid D^{(0)}]\right\}\right\} \tag{31}$$

then the optimal decision is to continue testing. If the reverse is true then we must perform the comparison in (29). This provides a computational advantage as the terms in (31) are more easily computed than (28).

In general, for any value of m, the decision to test or stop at the initial decision node, requires a comparison of the expected cost of no testing, with the expected minimum cost of testing one more stage, two more stages, and so on up to testing m more stages. Furthermore, after i stages of testing, the decision to continue testing or stop at decision node $\mathfrak{D}_{i+1}$ is based on comparison of the expected cost of stopping given $D^{(i)}$ with the minimum expected cost of testing 1, 2, ..., $m-i$ more stages. Following the reasoning for the m=2 example then the bounding strategy involves a comparison of the expected additional cost of stopping given $D^{(i)}$,

$$C_F E[X_{i+1} \mid D^{(i)}] \tag{32}$$

with the expected additional cost of testing 1, 2, ..., $m-i$ more stages given $D^{(i)}$. The expected additional cost of testing δ more stages is given as

$$C_F E[X_{i+1+\delta} \mid D^{(i)}] + \sum_{j=1}^{\delta} C_T E[N_{i+j} \mid D^{(i)}] \qquad \delta = 1, \ldots, m-i. \tag{33}$$

Thus at, $\mathfrak{D}_{i+1}$, if

$$C_F E[X_{i+1} \mid D^{(i)}] \geq \underset{\delta \in \{1, \ldots, m-i\}}{\text{MIN}} \left\{ C_F E[X_{i+1+\delta} \mid D^{(i)}] + \sum_{j=1}^{\delta} C_T E[N_{i+j} \mid D^{(i)}] \right\} \tag{34}$$

then the optimal decision at $\mathfrak{D}_{i+1}$ is to continue testing, otherwise a more complicated expression must be evaluated.

6. REFERENCES

1. R. E. Barlow, E. M. Scheuer, "Reliability growth during a development test program", *Technometrics*, Vol. 8, 1966, pp. 53-60.

2. N. S. Fard, D. L. Dietrich, "A Bayes reliability growth model for a development test program", *IEEE Trans. on Reliability*, Vol. R-36, 1987, pp. 568 – 572.

3. N. S. Fard, D. L. Dietrich, "Comparison of attribute reliability growth models", *Proceedings of the Annual Reliability and Maintainability Symposium*, 1983, pp. 24 – 29.

4. S. Kaplan, G. D. N. Cunha, A. A. Dykes, D. Shaver, "A Bayesian methodology for assessing reliability during product development", in *Proceedings of the Annual Reliability and Maintainability Symposium,* 1990, pp. 205 – 209.

5. D. K. Lloyd, M. Lipov, *Reliability: Management, Methods, and Mathematics*, Prentice Hall, Englewood Cliffs, N. J., 1977, 2nd edition.

6. T. A. Mazzuchi, R. Soyer, "A Bayesian analysis of product reliability during the development phase", accepted for publication , *IEEE Trans. on Reliability,* 1992.

7. A. Mosleh, G. Apostolakis, "Some properties of distributions useful in the study of rare events", *IEEE Trans. on Reliability*, Vol. R-26, 1982, pp. 87 – 94.

8. A. F .M. Smith, "A Bayesian note on reliability growth during a development testing program", *IEEE Trans. on Reliability*, Vol. R-26, 1977, pp. 346 – 347.

9. C. M. Weinrich, A. J. Gross, "The Barlow-Scheuer reliability growth model from a Bayesian viewpoint", *Technometrics,* Vol. 20, 1978, pp. 249-254.

ADVANCES IN RELIABILITY
Edited by Asit P. Basu

MULTIVARIATE STOCHASTIC DOMINANCE WITH SOME IMPLICATIONS

S.P.Mukherjee [a] and A.Chatterjee [b]

[a] Department of Statistics, University of Calcutta, Calcutta-700 019, India

[b] Department of Statistics, University of Burdwan, Burdwan-713 104, India

Abstract

Stochastic dominance between two multivariate probability distributions as applicable in reliability theory has been considered in terms of their survival functions. Its relationship with several multivariate ageing properties has been explored. The interrelations between hazard gradient dominance, multivariate mean remaining life dominance and stochastic dominance (as considered here) between two life (failure time) vectors have been investigated. A unification of several versions of bivariate ageing properties has also been attempted.

1. INTRODUCTION

Stochastic ordering between two univariate probability distributions, as introduced by Lehmann (1955), is a basic tool of probability and statistics. For two random variables (r.v.'s) X and Y, one says 'X dominates Y stochastically (written $X \underset{d}{>} Y$)' iff

$$P[X > t] \geq P[Y > t] \; ; \; \forall \; t \, \varepsilon (\infty, \infty)$$

Marshall and Olkin (1979) stated few equivalent relations to characterize stochastic ordering between two r.v.'s X and Y. In reliability and life testing, such ordering plays an important role. Apart from characterizing several ageing properties (c.f. Stoyan (1983)) through this tool, various hypothesis testing problems in reliability theory use this tool quite extensively and studies in this regard can be found in Barlow et. al. (1972), Shorack and Wellner (1987), Robertson et. al. (1988) to mention a few.

In the present paper the multivariate (m.v.) version of such dominance as applicable to the study of life distributions will be considered. It is expected that modelling of m.v. distributions through such dominance and the study of various m.v. ageing properties through such dominance will necessarily facilitate to some extent the study of various hypotheses testing problems related to m.v. ageing in reliability theory. Attempt in this regard will be made elsewhere.

Let $\underline{X} = (X_1,\ldots, X_p)'$ and $\underline{Y} = (Y_1,\ldots, Y_p)'$, integer $p \geq 1$, be two non-negative random p - vectors (r.p.v.) defined over $\mathbb{R}_p^+$, the non-negative orthant of $\mathbb{R}_p$. Let the distribution function (d.f.) and survival function (s.f.) of $\underline{X}$ be $F(\underline{x}) = P[\underline{X} \leq \underline{x}]$ and $\bar{F}(\underline{x}) = P[\underline{X} > \underline{x}]$; $\forall\ \underline{x} = (x_1,\ldots,x_p)' \in \mathbb{R}_p^+$, respectively, where all the vector inequalities stated here and in subsequent discussions are in lexicographic sense. Let the d.f. and s.f. of $\underline{Y}$ be similarly defined as $G(\underline{x})$ and $\bar{G}(\underline{x})$, respectively.

Definition 1.1 : In the above set-up, we say F (or $\underline{X}$) dominates G (or $\underline{Y}$) stochastically (written $F \underset{K}{\geq} G$ or $\underline{X} \underset{K}{\geq} \underline{Y}$) iff

$\bar{F}(\underline{x}) \geq \bar{G}(\underline{x})$; $\forall\ \underline{x} \in \mathbb{R}_p^+$, with inequality strict for at least one $\underline{x}$.

Such dominance was also considered by Cambanis et.al.(1976) and Bergman (1978). Marshall and Olkin (1979) proposed several non-equivalent versions of stochastic dominance (including the one stated in Definition 1.1 and not necessarily with respect to r.p.v. defined over $\mathbb{R}_p^+$) indicating the complete chain of inter-relations among various versions and exploring some of the consequences.

Before elaborating what is contained in the present article, let us consider the residual (generic) life vector $\underline{X}_\gamma$ of $\underline{X}$ with s.f.

$$\bar{F}_\gamma(\underline{x}) = P[\underline{X}_\gamma > \underline{x}] \overset{\Delta}{=} P[\underline{X} > \underline{x} + \underline{\gamma} \mid \underline{X} > \underline{\gamma}]$$

$$= \bar{F}(\underline{x} + \underline{\gamma}) / \bar{F}(\underline{\gamma}) \ ; \ \forall\ \underline{x},\underline{\gamma} \in \mathbb{R}_p^+,$$ where $\underline{\gamma} = (\gamma_1,\ldots,\gamma_p)'$ and Δ-stands for 'by definition'. Let the corresponding d.f. of $\underline{X}_\gamma$ be F_γ.

In section 2 of the present article the dominance relation as given in Definition 1.1 is explored. Several m.v. ageing properties as introduced by Buchanan and Singpurwalla (1977) have been obtained by considering the m.v. stochastic dominance as in Definition 1.1 between the original life and residual life (eventually both of them are r.p.v.'s). In section 3, hazard gradient dominance and multivariate mean remaining life dominance between two m.v. life distributions have been introduced and their relationships with stochastic dominance have been investigated. Finally, in section 4, in a bivariate (b.v.) set up, the relationship between TP_2 functions (c.f. Karlin (1968)), distributions with vector b.v. increasing/decreasing hazard rate (VBIHR/VBDHR) (c.f. Johnson and Kotz (1975)) and distributions with b.v. increasing/decreasing failure rate (BIFR/BDFR) (c.f. Buchanan and Singpurwalla (1977)) has been established.

Note : Only transitivity and reflexivity properties hold for the relation in Definition 1.1. Its antisymmetry property can not be properly defined, which is usual for any m.v. order relations. As such, this type of stochastic dominance can be termed as pre ordering rather than partial ordering.

2. RELATIONSHIP WITH AGEING PROPERTIES

In the present section a neat documentation of not so surprising relationships between m.v. stochastic dominance and various m.v. ageing properties (as introduced by Buchanan and Singpurwalla (1977)) will be attempted through the following theorems.

Theorem 2.1 :

i) F is MIFR - VS provided $F_{\underline{x}} \underset{K}{\geq} F_{\underline{y}}$; $\forall\ \underline{x}, \underline{y} \in \mathbb{R}_p^+$ with $\underline{x} < \underline{y}$ and vice-versa.

ii) F is MNBU - VS provided $F \underset{K}{\geq} F_{\underline{y}}$; $\forall\ \underline{y} \in \mathbb{R}_p^+$ and vice-versa.

Proof : i) Observe that $F_{\underline{x}} \underset{K}{\geq} F_{\underline{y}}$; $\forall\ \underline{x}, \underline{y} \in \mathbb{R}_p^+$ with $\underline{x} < \underline{y}$

$$<=> \quad \frac{\bar{F}(\underline{t} + \underline{x})}{\bar{F}(\underline{x})} \geq \frac{\bar{F}(\underline{t} + \underline{y})}{\bar{F}(\underline{y})} \quad ; \ \forall\ \underline{t}, \underline{x}, \underline{y} \in \mathbb{R}_p^+ \text{ with } \underline{x} < \underline{y}$$

$<=>$ $\bar{F}(\underline{x} + \underline{t}) / \bar{F}(\underline{x})$ is non-increasing in $\underline{x}$; $\forall\ \underline{x}, \underline{t} \in \mathbb{R}_p^+$

$<=>$ F is MIFR - VS.

ii) Observe that $F \underset{K}{>} F_{\underline{y}}$; $\forall\ \underline{y} \in \mathbb{R}_p^+$

$$<=> \quad \bar{F}(\underline{x}) \geq \bar{F}(\underline{x} + \underline{y}) / \bar{F}(\underline{y}) \ ; \ \underline{x}, \underline{y} \in \mathbb{R}_p^+$$

$$<=> \quad \bar{F}(\underline{x} + \underline{y}) \leq \bar{F}(\underline{x})\ \bar{F}(\underline{y}) \ ; \ \forall\ \underline{x}, \underline{y} \in \mathbb{R}_p^+$$

$<=>$ F is MNBU - VS Q.E.D.

For two independent r.p.v.'s $\underline{X}$ and $\underline{Y}$, let $\underline{X}_Y$ be the generic r.p.v. (life beyond a random survival point) with s.f. given by

$$\bar{F}_Y(\underline{x}) \overset{\Delta}{=} P[\underline{X} > \underline{x} + \underline{Y} \mid \underline{X} > \underline{Y}]$$

$$= \int_{\mathbb{R}_p^+} \bar{F}(\underline{x} + \underline{y})\, dG(\underline{y}) \Big/ \int_{\mathbb{R}_p^+} \bar{F}(\underline{y})\, dG(\underline{y})$$

Let the corresponding d.f. of $\underline{X}_Y$ be given by F_Y.

Theorem 2.2 : $F \underset{K}{>} F_{\gamma}$ provided F is MIFR - VS

Proof : Define the generic r.p.v. $\underline{X}_{\underline{Y}+\underline{\gamma}}$ with s.f.

$$\bar{F}_{Y+\gamma}(\underline{x}) \overset{\Delta}{=} P[\underline{X} > \underline{x} + \underline{Y} + \underline{\gamma} \mid \underline{X} > \underline{Y} + \underline{\gamma}]$$

$$= P[\underline{X} > \underline{x} + \underline{Y} + \underline{\gamma}] / P[\underline{X} > \underline{Y} + \underline{\gamma}]$$

We will show that if F is MIFR - VS, then $F_{\gamma} \underset{K}{>} F_{Y+\gamma}$ (2.1)

Consider

$$\bar{F}_{\gamma}(\underline{x}) - \bar{F}_{Y+\gamma}(\underline{x}) = \bar{F}(\underline{x}+\underline{\gamma})/\bar{F}(\underline{\gamma}) - \bar{F}(\underline{x}+\underline{Y}+\underline{\gamma}) / \bar{F}(\underline{Y}+\underline{\gamma}) ; \quad \forall\ \underline{x},\underline{\gamma} \in \mathbb{R}_p^+ \tag{2.2}$$

Now (2.2) has the same sign as

$$\int_{\mathbb{R}_p^+} (\bar{F}(\underline{x} + \underline{\gamma}). \bar{F}(\underline{y} + \underline{\gamma}) - \bar{F}(\underline{\gamma}).\bar{F}(\underline{x} + \underline{y} + \underline{\gamma}))\, dG(\underline{y})) \tag{2.3}$$

assuming $\bar{F}(\underline{\gamma}). \bar{F}(\underline{Y} + \underline{\gamma}) = \bar{F}(\underline{\gamma}). \int_{\mathbb{R}_p^+} \bar{F}(\underline{y} + \underline{\gamma})\, dG(\underline{y}) > 0.$

Since F is MIFR - VS, we have

$$\bar{F}(\underline{x} + \underline{\gamma})/ \bar{F}(\underline{\gamma}) - \bar{F}(\underline{x} + \underline{y} + \underline{\gamma}) / \bar{F}(\underline{y} + \underline{\gamma}) \geq 0; \quad \forall\ \underline{y} \in \mathbb{R}_p^+$$

$$<=> \bar{F}(\underline{x} + \underline{\gamma})\, \bar{F}(\underline{y} + \underline{\gamma}) - \bar{F}(\underline{\gamma})\, \bar{F}(\underline{x} + \underline{y} + \underline{\gamma}) \geq 0; \quad \forall\ \underline{y} \in \mathbb{R}_p^+$$

Thus the integrand of (2.3) is non-negative and hence

$\bar{F}_{\gamma}(\underline{x}) \geq \bar{F}_{Y+\gamma}(\underline{x}); \ \forall\ \underline{x} \in \mathbb{R}_p^+$ which means (2.1) is true.

In view of (2.1) by putting $\underline{\gamma} = \underline{0} = (0 \ldots 0)'$, we have

$\bar{F}(\underline{x}) \geq \bar{F}_{Y}(\underline{x}); \ \forall\ \underline{x} \in \mathbb{R}_p^+$ i.e., $F \underset{K}{>} F_{Y}$, Q.E.D.

Notes :

1) The converse of theorem 2.2 is not necessarily true, although that of theorem 2.1 is evidently valid.

2) By reversing the dominance relations, replacing '$\geq(\leq)$' by '$\leq(\geq)$' and 'non-increasing' by'non-decreasing' in the respective cases corresponding results for various dual versions of theorems 2.1 and 2.2 could be proved easily.

3) By substituting $\underline{x}^* = (x,\ldots,x)'$, $\underline{y}^* = (y,\ldots,y)'$, $\underline{t}^* = (t,..,t)'$and $\underline{\gamma}^* = (\gamma, \ldots, \gamma)'$ in place of $\underline{x}$, $\underline{y}$, $\underline{t}$ and $\underline{\gamma}$ in the respective cases, the results concerning 'S','W', 'VW' versions as referred in Buchanan and Singpurwalla (1977) could be easily obtained.

3. HAZARD GRADIENT DOMINANCE, MULTIVARIATE MEAN REMAINING LIFE DOMINANCE, AND THEIR RELATIONSHIP WITH M.V. STOCHASTIC DOMINANCE

In this section we will use the notations introduced by Mukherjee and Chatterjee (1991) in defining hazard gradient (HG) dominance (a generalization of failure rate ordering as stated in Ross (1983)) and m.v. mean remaining life (MMRL) dominance (a generalization of mean remaining life ordering as stated in Gupta and Kirmani (1987)) between two life vectors $\underline{X}$ and $\underline{Y}$ having s.f. $\bar{F}$ and $\bar{G}$ and d.f. F and G respectively.

Definition 3.1 : Let $\underline{X}$ be the r.p.v. defined on $\mathbb{R}_p^+$. Define the Borel measurable function on $\mathbb{R}_p^+$:

$$\underline{\lambda}_F(\underline{x}) = (\lambda_{1F}(\underline{x}) \ldots \lambda_{pF}(\underline{x}))' \quad \text{where}$$

$$\lambda_{iF}(\underline{x}) \stackrel{\Delta}{=} f_i\,(x_i \mid \underline{X}_{(i)} > \underline{x}_{(i)}) \,/\, \bar{F}_i\,(x_i \mid \underline{X}_{(i)} > \underline{x}_{(i)})$$

$$= -\left(\frac{\delta}{\delta x_i}\bar{F}\,(\underline{x})\right) / \bar{F}\,(\underline{x}); \quad \forall\, \underline{x} \in \mathbb{R}_p^+ \text{ with } \bar{F}(\underline{x}) > 0,\ i = 1(1)p.$$

Here $f_i(x_i \mid \underline{X}_{(i)} > \underline{x}_{(i)})$ is the conditional density of $X_i \mid \underline{X}_{(i)} > \underline{x}_{(i)}$ and $\bar{F}_i(x_i \mid \underline{X}_{(i)} > \underline{x}_{(i)})$ is the conditional s.f. of $X_i \mid \underline{X}_{(i)} > \underline{x}_{(i)}$ with $\underline{X}_{(i)} = (X_1 \ldots X_{i-1}, X_{i+1}, \ldots, X_p)'$ - the ith deleted r.$\overline{p-1}$.v. obtained from $\underline{X}$, $i = 1(1)p$.

Johnson and Kotz (1975) and Marshall (1975) called $\underline{\lambda}_F(\underline{x})$ as the HG vector of F. It is to be noted that on the set $\{\, \underline{x} : \bar{F}(\underline{x}) = 0 \,\}$ the function $\underline{\lambda}_F(\underline{x})$ can be defined in an arbitrary fashion.

Definition 3.2 : Let $\underline{X}$ be the r.p.v. defined on $\mathbb{R}_p^+$. Define the Borel measurable function on $\mathbb{R}_p^+$:

$$\underline{r}_F\,(\underline{x}) = (\, r_{1F}\,(\underline{x}) \ldots\ldots r_{pF}\,(\underline{x}))', \quad \text{where}$$

$$r_{iF}\,(\underline{x}) \stackrel{\Delta}{=} E\,(X_i - x_i \mid \underline{X} > \underline{x}\,)$$

$$= \int_{x_i}^{\infty} \bar{F}\,(t_i, \underline{x}_{(i)})\, dt_i \,/\, \bar{F}(\underline{x}); \quad \forall\, \underline{x} \in \mathbb{R}_p^+ \text{ with } \bar{F}(\underline{x}) > 0,\ i = 1(1)p.$$

Here $(t_i, \underline{x}_{(i)}) = (x_1, \ldots x_{i-1}, t_i, x_{i+1}, \ldots\ldots x_p)'$, $i = 1(1)p$.

Zahedi (1985) called $\underline{r}_F(\underline{x})$ the MMRL function of F. Here also it is to be noted that on the set $\{\, \underline{x} : \bar{F}(\underline{x}) = 0 \,\}$ the function $\underline{r}_F(\underline{x})$ can be defined in an arbitrary fashion.

Also define the corresponding HG vector and MMRL function for G(or $\underline{Y}$) as $\underline{\lambda}_G$ and $\underline{r}_G$ respectively in the similar way.

Definition 3.3 : In the above set up, F has HG-dominance over G (written $F \underset{HG}{>} G$ or $\underline{X} \underset{HG}{>} \underline{Y}$) provided

$$\underline{\lambda}_F(\underline{x}) \leq \underline{\lambda}_G(\underline{x}), \ \forall \ \underline{x} \in \mathbb{R}_p^+$$

Definition 3.4 : In the above set up, F has MMRL - dominance over G (written $F \underset{MMRL}{>} G$ or $\underline{X} \underset{MMRL}{>} \underline{Y}$) provided

$$\underline{r}_F(\underline{x}) \geq \underline{r}_G(\underline{x}); \ \forall \ \underline{x} \in \mathbb{R}_p^+$$

We are now in a position to establish several results connecting HG-dominance, MMRL -dominance and m.v. stochastic dominance.

Theorem 3.1 : The following statements are equivalent on $\mathbb{R}_p^+$.

a) $F \underset{HG}{>} G$

b) $\bar{F}(x) / \bar{G}(x)$ is non-decreasing in every component of $\underline{x}$

c) $F_\gamma \underset{K}{>} G_\gamma \ ; \ \forall \underline{\gamma} \in \mathbb{R}_p^+$.

Proof : Observe that $(\delta/\delta x_i) \ \log (\bar{F}(\underline{x}) / \bar{G}(\underline{x}))$

$= (\bar{G}(\underline{x}) / \bar{F}(\underline{x}).\bar{G}^2(\underline{x})) \ [\bar{G}(\underline{x}).(\delta/\delta x_i) \bar{F}(\underline{x}) - \bar{F}(\underline{x}).(\delta/\delta x_i) \bar{G}(\underline{x})]$

$= -\lambda_{iF}(\underline{x}) + \lambda_{iG}(\underline{x}); \ \forall \ \underline{x} \in \mathbb{R}_p^+$.

Thus (b) $\Leftrightarrow (\delta/\delta x_i) \ \log (\bar{F}(\underline{x}) / \bar{G}(\underline{x})) \geq 0; \ \forall \ i = 1(1)p$

$\Leftrightarrow \lambda_{iF}(\underline{x}) \leq \lambda_{iG}(\underline{x}); \forall \ \underline{x} \in \mathbb{R}_p^+, \ i = 1(1)p$

$\Leftrightarrow$ (a)

Again (c) $\Leftrightarrow \bar{F}_\gamma(\underline{x}) \geq \bar{G}_\gamma(\underline{x}); \ \forall \ \underline{x}, \underline{\gamma} \in \mathbb{R}_p^+$

$\Leftrightarrow \bar{F}(\underline{x} + \underline{\gamma}) / \bar{G}(\underline{x} + \underline{\gamma}) \geq \bar{F}(\underline{\gamma}) / \bar{G}(\underline{\gamma}); \forall \ \underline{x}, \underline{\gamma} \in \mathbb{R}_p^+$

$\Leftrightarrow$ (b) Q.E.D.

Theorem 3.2 : (i) $F \underset{HG}{>} G \Rightarrow F \underset{K}{>} G$

(ii) $F \underset{HG}{>} G \Rightarrow F \underset{MMRL}{>} G$

Proof : (i) We have $\underline{\lambda}_F(\underline{x}) \leq \underline{\lambda}_G(\underline{x})$; $\forall\ \underline{x} \in \mathbb{R}_p^+$

Integrating both sides over the piecewise smooth closed path in $(\underline{0},\underline{t})$ (cfs.Block (1977), Galambos and Kotz (1978), Mukherjee and Chatterjee (1988)) we have

$$_{\underline{o}}\int^{\underline{t}} \underline{\lambda}_F(\underline{x})\, d\underline{x} \leq {}_{\underline{o}}\int^{\underline{t}} \underline{\lambda}_G(\underline{x})\, d\underline{x};\ \forall\ \underline{t} \in \mathbb{R}_p^+$$

$$<=> \ -\log \bar{F}(\underline{t}) \leq -\log \bar{G}(\underline{t});\ \forall\ \underline{t} \in \mathbb{R}_p^+ <=> F \underset{K}{>} G$$

(ii) By virtue of theorem 3.1 we have

$$F \underset{HG}{>} G <=> \bar{F}(\underline{x} + \underline{t}) / \bar{F}(\underline{x}) \geq \bar{G}(\underline{x} + \underline{t}) / \bar{G}(\underline{x});\ \forall\ \underline{x}, \underline{t} \in \mathbb{R}_p^+ \tag{3.1}$$

Taking $\underline{t} = (t_i, 0. \underline{t}_{(i)})$, in particular, (3.1) reduces to

$$\bar{F}(x_i + t_i, \underline{x}_{(i)}) / \bar{F}(\underline{x}) \geq \bar{G}(x_i + t_i, \underline{x}_{(i)}) / \bar{G}(\underline{x})$$

$$=> {}_o\int^{\infty} \bar{F}(x_i + t_i, \underline{x}_{(i)})\, dt_i / \bar{F}(\underline{x}) \geq {}_o\int^{\infty} \bar{G}(x_i + t_i, \underline{x}_{(i)})\, dt_i / \bar{G}(\underline{x})$$

$$<=> \int_{x_i}^{\infty} \bar{F}(y_i, x_{(i)})\, dy_i / \bar{F}(x) \geq \int_{x_i}^{\infty} \bar{G}(y_i, \underline{x}_{(i)})\, dy_i / \bar{G}(\underline{x})$$

$<=> r_{iF}(\underline{x}) \geq r_{iG}(\underline{x});\ \forall\ \underline{x} \in \mathbb{R}_p^+,\ i = 1(1)p$, which means $F \underset{MMRL}{>} G$. Q.E.D.

Note : The implications like '$F \underset{K}{>} G => F \underset{MMRL}{>} G$' or '$F \underset{MMRL}{>} G => F \underset{K}{>} G$' are not valid, since in the univariate set-up the corresponding implications are <u>not</u> true. For counter examples in univariate set-up we refer to Gupta and Kirmani (1987).

4.BIFR/BDFR DISTRIBUTIONS IN RELATION TO VBIHR/VBDHR DISTRIBUTIONS VIA TP_2 PROPERTY

In the univariate set-up an one-to-one correspondence exists between PF_2 functions and IFR distributions (Barlow and Proschan (1975),p.76). Several multivariate versions of IFR distribution have so far been proposed, but the ones due to Johnson and Kotz (1975) [Vector Multivariate increasing/decreasing hazard rate (VMIHR/VMDHR)] and Buchanan and Singpurwalla (1977) [Multivariate increasing/decreasing failure rate (MIFR/MDFR)] received special attention. Also the bivariate version of PF_2 function known as TP_2 function is well known (Karlin (1968)). Naturally a question arises as to how far the IFR properties in bivariate set-up are related with TP_2 property.

Aparently it is believed that there may not exist any connection between VMIHR/VMDHR distributions and MIFR/MDFR distributions. We will now relate them through some additional condition (via TP_2 property) at least in the bivariate set-up. We consider only the increasing (non-decreasing to be more specific) case. Also, as we are concerned with a single distribution F, we will use λ_i in place of λ_{iF} to designate different components of hazard gradient of F.

Theorem 4.1 : A s.f. $\bar{F}(x_1,x_2)$ [> 0] of the random life vector (X_1,X_2) is BIFR-VS provided it is VBIHR and $1/\bar{F}$ is TP_2 in (x_1,x_2).

Proof : F is BIFR-VS <=> $\bar{F}(x_1+t_1, x_2+t_2)/\bar{F}(x_1,x_2)$ is non-increasing in $x_1,x_2 \geq 0$; $\forall$ $t_1,t_2 \geq 0$.

Thus for fixed $x_1 \geq 0$; we have (since $\bar{F}(x_1,x_2) > 0$)

$\bar{F}(x_1,x_2).(\delta/\delta x_2)\bar{F}(x_1+t_1,x_2+t_2) - \bar{F}(x_1+t_1,x_2+t_2).(\delta/\delta x_2)\bar{F}(x_1,x_2) \leq 0$, yielding

$$\lambda_2(x_1+t_1,x_2+t_2) \geq \lambda_2(x_1,x_2);\ \forall\ x_2,t_1,t_2 \geq 0 \qquad (4.1.a)$$

Similarly for fixed $x_2 \geq 0$, we have

$$\lambda_1(x_1+t_1,x_2+t_2) \geq \lambda_1(x_1,x_2);\ \forall\ x_1,t_1,t_2 \geq 0 \qquad (4.1.b)$$

Again $1/\bar{F}(x_1,x_2)$ is TP_2 in (x_1,x_2) <=>

For $t_1,t_2 \geq 0$, we have
$$\begin{vmatrix} 1/\bar{F}(x_1,x_2) & 1/\bar{F}(x_1+t_1,x_2) \\ 1/\bar{F}(x_1,x_2+t_2) & 1/\bar{F}(x_1+t_1,x_2+t_2) \end{vmatrix} \geq 0$$

<=> $\bar{F}(x_1+t_1,x_2+t_2) / \bar{F}(x_1,x_2+t_2) \leq \bar{F}(x_1+t_1,x_2) / \bar{F}(x_1,x_2)$

<=> $\bar{F}(x_1+t_1,x_2)/\bar{F}(x_1,x_2)$ is non-increasing in x_2 for fixed $x_1 \geq 0$ and $t_1 \geq 0$.

<=> $\bar{F}(x_1,x_2).(\delta/\delta x_2)\bar{F}(x_1+t_1,x_2) - \bar{F}(x_1+t_1,x_2).(\delta/\delta x_2)\bar{F}(x_1,x_2) \geq 0$

with $t_1 \geq 0$, for fixed $x_1 \geq 0$ (since $\bar{F}(x_1,x_2) > 0$).

<=> $\lambda_2(x_1,x_2) \leq \lambda_2(x_1+t_1,x_2)$; $\forall$ $x_2, t_1 \geq 0$, for fixed $x_1 \geq 0$ (4.2.a)

Similarly, we can have from $1/\bar{F}$ TP_2 in (x_1,x_2), the following

$$\lambda_1(x_1,x_2) \leq \lambda_1(x_1,x_2+t_2);\ \forall\ x_1, t_2 \geq 0, \text{ for fixed } x_2 \geq 0 \qquad (4.2.b)$$

Also F is VBIHR

<=> $\lambda_2(x_1,x_2+t_2) \geq \lambda_2(x_1,x_2)$ for fixed $x_1 \geq 0$; $\forall$ $t_2, x_2 \geq 0$

<=> $\lambda_2(x_1+t_1,x_2+t_2) \geq \lambda_2(x_1+t_1,x_2)$ for fixed $x_1 \geq 0$; $\forall$ $x_2, t_1,t_2 \geq 0$ (4.3.a)

Similarly F is VBIHR

$\Leftrightarrow \quad \lambda_1(x_1+t_1, x_2+t_2) \geq \lambda_1(x_1, x_2+t_2)$ for fixed $x_2 \geq 0$; $\forall\ x_1, t_1, t_2 \geq 0$ (4.3.b)

Thus it is evident that

(4.2.a) and (4.3.a) => (4.1.a), for fixed $x_1 \geq 0$; $\forall\ x_2, t_1, t_2 \geq 0$ and

(4.2.b) and (4.3.b) => (4.1.b), for fixed $x_2 \geq 0$; $\forall\ x_1, t_1, t_2 \geq 0$ (4.4)

Since F is a BIFR-VS, $1/\bar{F}$ is TP_2 and F is VBIHR are assured by paired conditions (4.1.a) & (4.1.b), (4.2.a) & (4.2.b) and (4.3.a) & (4.3.b) respectively the desired result follows from the observation (4.4). Q.E.D.

Notes :

1) The dual version of the above theorem can be proved similarly and may be stated as :

A s.f. $\bar{F}(>0)$ of random life vector (X_1,X_2) is BDFR-VS provided it is VBDHR and $\bar{F}$ is TP_2 in (x_1,x_2).

2) The 'S' version of Buchanan and Singpurwalla (1977) follows by taking $t_1=t_2=t$. It should be noted, however, that 'W' and 'VW' versions do not follow similarly, since in that case we require $x_1=x_2=x$ and thus by fixing $(x_2)x_1$, $(x_1)x_2$ is automatically fixed and as such λ_i's, i=1,2 can not be properly defined.

3) Among the BIFR-VS, VBIHR and TP_2 properties (or their respective duals) imposed on F (or $\bar{F}$) only the last two jointly imply the first one. No other two can be claimed to imply the third one.

4) The case for more than two variables is not easy since the multivariate version of TP_2 property may not come to our rescue.

5. ACKNOWLEDGEMENT

The authors are thankful to the referee for useful comments, which are invaluable in preparing the current revised version.

6. REFERENCES

1 Barlow,R.E. and Proschan,F.(1975): Statistical Theory of Reliability and Life Testing: Probability Models. Holt, Rinehart and Winston. NY.

2 Barlow,R.E., Bartholomew,D.J., Bremner,J.M. and Brunk,H.D. (1972): Statistical Inference Under Order Restrictions. Wiley. NY.

3 Bergmann,R. (1978): Some classes of semi-ordering relations for random vectors and their use for comparing covariances. Math.Nachr. 82. 103-114.

4 Block,H.W. (1977): Multivariate reliability classes. In Applications of Statistics. Ed.P.R.Krishnaiah. North Holland. NY. 79-88.

5 Buchanan,H.B. and Singpurwalla,N.D. (1977): Some stochastic characterization of multivariate survival. In The Theory and Application of Reliability. Vol.1. Ed.C.P.Tsokos and I.N.Shimi. Academic Press. NY. 329-348.

6 Cambanis,S., Simons,G. and Stout,W. (1976): Inequalities of Ek(X,Y) when the marginals are fixed. Zeit.Wahrscheinlichkeitsth. 36. 285-294.

7 Galambos,J. and Kotz,S. (1978): Characterizations of Probability Distributions. Lecture Notes in Mathematics. Ed. A.Doldand and B.Eckmann. Springer-Verlag. Berlin, Hidelberg, NY.

8 Gupta,R.C. and Kirmani,S.N.U.A. (1987): On order relations between reliability measures. Communications in Statistics: Stochastic Models. 3. 149-156.

9 Johnson,N.L. and Kotz,S. (1975): A vector multivariate hazard rate. Journal of Multivariate Analysis. 5. 53-66, Erratum, ibid 498.

10 Karlin,S. (1968): Total Positivity. Stanford University Press.

11 Lehmann,E.L. (1955): Ordered families of distributions. Annals of Mathematical Statistics. 26. 399-419.

12 Marshall,A.W. (1975): Multivariate distributions with monotone hazard rate. In Reliability and Fault Tree Analysis. Eds. R.E.Barlow, J.B.Fussel and N.D.Singpurwalla. SIAM. Philadelphia.

13 Marshall,A.W. and Olkin,I. (1979): Inequalities: Theory of Majorization and its Application. Academic Press. NY.

14 Mukherjee,S.P. and Chatterjee,A. (1988): A new MIFRA class of life distributions. Calcutta Statistical Association Bulletin. 37, 67-80.

15 Mukherjee,S.P. and Chatterjee,A. (1991): On some properties of the multivariate ageing classes. Probability in the Engineering and Informational Sciences. 5. 523-534.

16 Robertson,T., Wright,F.T. and Dykstra,R.L. (1988): Order Restricted Statistical Inference. Wiley. NY.

17 Ross,S.M. (1983): Stochastic Processes. John Wiley and Sons. NY.

18 Shorack,G. and Wellner,J. (1987): Empirical Processes with Applications to Statistics. Wiley. NY.

19 Stoyan,D. (1983): Comparison Methods for Queues and other Stochastic Models. (Ed. D.J.Daley). John Wiley and Sons. NY.

20 Zahedi,H. (1985): Some new classes of multivariate survival functions. Journal of Statistical Planning and Inference. 11. 171-188.

ADVANCES IN RELIABILITY
Edited by Asit P. Basu

Detection and Modeling of Aging Properties in Lifetime Data

H. Pamme[a] and H. Kunitz[b]

[a]RWE Energie AG, Kruppstraße 5, D-4300 Essen, Germany

[b]Gesellschaft für Anlagen- und Reaktorsicherheit (GRS) mbH, Schwertnergasse 1, D-5000 Köln 1, Germany

Abstract

Parametric and non-parametric lifetime data analyses in practical applications require sensitive tools if non-monotonic aging properties ("trend changes") are to be examined. The well-known bathtub-shaped hazard rate is a special model with a trend change in aging properties over time.

The identification of trend changes in the hazard rate can be supported by graphical tools. This paper discusses the combined application of graphical tools and parametric estimation in the flexible mixed gamma distribution family to identify trend changes and model bathtub-shaped hazard rates.

1. INTRODUCTION

In literature dealing with statistical lifetime data analysis the bathtub-shaped hazard rate model is often cited as a suitable model to describe a trend change from decreasing to increasing aging behaviour. However discussions concerning lifetime data analysis mostly end with the pure theoretical introduction of this trend change model. Reasons might be difficulties in identifying these trend changes in a given data set. Additionally the familiar parametric distribution families such as the Weibull- or Gamma-distributions are not able to model trend changes or have a lack in flexibility like the log-normal distribution which can only model an inverse bathtub-shaped hazard rate with minor importance as a model in practical lifetime data analysis.

This paper describes a combined approach in identifying and modeling a bathtub-shaped hazard rate. Plots based on the total time on test statistic and the mean residual life statistic are analysed concerning their ability to indicate trend changes in a data set. A comparison of the graphical informations can emphasize the hypothesis of an underlying trend change in a data set.

Additionally the mixed gamma distribution is introduced. The mixture properties concerning parametric modeling of monotonic hazard rates and bathtub-shaped hazard rates are discussed.

The maximum likelihood estimation of the parameters of the mixed gamma distribution can provide additional support concerning a trend change hypothesis as shown in the example.

The described exploratory approach in data analysis will show advantages in practical applications as no prior information or assumptions concerning the aging properties (e.g. knowledge of position of trend change in time or proportions of populations with different aging trends as e.g. discussed by Guess, Hollander and Proschan (1986)) are required.

2. A BATHTUB-SHAPED HAZARD RATE MODEL

A bathtub-shaped hazard rate is often discussed as an appropriate model to describe especially the aging behaviour of technical components. It incorporates three specific phases of life times: the infant mortality phase (burn-in-phase) with decreasing failure rate (DFR)-behaviour, the "operational" phase with a nearly constant (exponential) hazard rate and the wear-out (aging) phase with increasing failure rate (IFR)-behaviour.

Formally the bathtub-shaped hazard rate $h(x)$ can be characterized by the condition that a local minimum $d\,h(x) / dx = 0$ and $d^2\,h(x) / dx^2 > 0$ exists on $0 < x < \infty$.

We propose a mixed gamma distribution as a flexible distribution family with probability density function (pdf) (index m to denote a mixture-pdf)

$$f_m(x) = p\,f(x;\,\alpha_1,\,\beta_1) + q\,f(x;\,\alpha_2,\,\beta_2) \qquad \text{with} \qquad (1a)$$

$$f(x;\,\alpha_i,\,\beta_i) = 1/\Gamma(\beta_i)\,(\,\alpha_i^{\,\beta_i}\,x^{\beta_i-1})\,\exp -(\alpha_i x)$$

$$p > 0, \quad q > 0, \quad p + q = 1,\ \beta_i > 0\ , \quad \alpha_i > 0\ , \quad i = 1,\,2\ .$$

The identifiability of the class of finite mixtures of two parameter gamma distributions with pdf $f(x;\,\alpha,\,\beta)$ is proven in Teicher (1963).

Defining the hazard rate of the population i of the mixture with

$h_i(x) = f_i(x) / R_i(x)$, $i = 1, 2$

the hazard rate of the mixed distribution can be written as

$h_m(x) = (p f_1(x) + q f_2(x)) / (p R_1(x) + q R_2(x))$.

A bathtub-shaped hazard rate can be modelled by a mixture-pdf with

$$f_m(x) = p f(x; \alpha_D, \beta_D) + q f(x; \alpha_I, \beta_I) \quad (1b)$$

$\beta_D < 1$, $\beta_I > 1$, $\alpha_D > 0$, $\alpha_I > 0$.

[Index D (I) to denote pdf with corresponding decreasing (increasing) hazard rate.]

It can be easily shown that

$\lim_{x \to 0} h(x) = \infty$ and $\lim_{x \to \infty} h(x) = \alpha_I$

holds for a pdf defined by (1b). Thus all bathtub-shaped hazard rates of "two component" mixed gamma distributions converge from below towards the scale parameter α_I for very large x. The proof is given in Pamme (1992).

It should be noted that the mixed gamma distribution is able to model "real" bathtub-shaped hazard rates as defined above. In literature a mixture of Weibull-distributions with shape parameters $\beta_D < 1$ and $\beta_I > 1$ is often proposed as a model for bathtub-shaped hazard rates. The interpretation of (1b) as a Weibull-mixture however would produce a hazard rate with $\lim_{x \to \infty} h(x) = 0$.

3. Graphical tools for data analysis

3.1. Some remarks

One central question in technical lifetime data analyses is very often the acceptance or rejection of the constant i.e. exponential aging model.

Especially when the sample size n is relatively small (e.g. $n < 30$) it is a difficult task to judge whether a data set "suggests" a constant hazard rate or any non-constant alternative if there is not a very clear IFR- or DFR-behaviour. Especially the discrimination between a constant and a bathtub-shaped hazard rate (e.g. with relatively flat DFR-

and IFR-phases) should generally be analysed by graphical tools.

Special test statistics to detect bathtub-shaped aging behaviour (see e.g. Bergman (1977), Aarset (1987)) are very often not able to reject the null-hypothesis of exponentiality in cases of relatively flat bathtub-shaped hazard rates or if the sample size is small (e.g. n < 30). This was shown by the authors in a simulation study using the Dhillon-distribution (see Dhillon (1981)) to generate bathtub-shaped hazard rates. So the combined use of suitable plots which anyhow indicate departure from exponential behaviour seems to be a useful way of data analysis in practical applications.

Beside the usually (and often exclusively) used plots of the empirical distribution function (plotting the empirical cumulative density function cdf over time) especially the well-known probability plots (or hazard plots) might provide first indications concerning departure from exponentiality (e.g. the Weibull probability plot). The informations in these plots however only refer to the suitability of the underlying distributional model.

3.2. TTT-Plots

Since Barlow and Campo (1975) presented their publication on the total time on test (TTT) transform of F(x)

$$H^{-1}(t) = \int_0^{F^{-1}(t)} [1 - F(x)]\, dx \qquad 0 \le t \le 1$$

the TTT-concept has proven to be a valuable graphical tool in data analysis and model identification.

We recall an important property of the TTT-transform

$$d/dt \; H^{-1}_F(t) \Big|_{t = F(x)} = 1 / h(x) \; . \tag{2}$$

With introduction of the scaled TTT-transform

$$H(t) = H^{-1}_F(t) / H^{-1}_F(1) \tag{3}$$

and its empirical "counterpart", the TTT-plot (plotting T_i/T_n against i/n), with

$$T_i = \sum_{j=1}^{i} (n-j+1)\,(x_{(j)} - x_{(j-1)}) \;, \qquad x_{(0)} = 0$$

and $x_{(1)}, x_{(2)}, \ldots, x_{(n)}$ an ordered sample of size n from X, a sensitive graphical instrument becomes available to identify departures from exponentiality and trend changes in life time data. TTT-plots are scale invariant, monotonically increasing plots in a unit square with the diagonal representing the scaled TTT-transform of the exponential distribution family. TTT-plots above (below) the diagonal in the unit square indicate increasing (decreasing) aging properties in a lifetime data set.

According to (2) and (3) hazard rates with "trend changes" (i.e. local extremes of $h(x)$ on $0 < x < \infty$) appear as turning points in the TTT-plot.

Thus for example a bathtub-shaped hazard rate is represented by a scaled TTT-transform once crossing the diagonal in the unit square from below and having exactly one turning point on $0 < t < 1$.

Thus TTT-plots are useful exploratory tools in identifying changes in aging properties and suitable distributional models. Furthermore the comparison of the graphs of TTT-plots and scaled TTT-transforms can indicate an achieved goodness of fit e.g. after parameter estimation of a lifetime distribution (see example below).

4. APPLICATION TO EMPIRICAL DATA

4.1. GRAPHICAL ANALYSIS

We illustrate the applicability of the bathtub-shaped hazard rate model on a set of data reflecting failure times of a group of centrifugal water pumps (sample size n = 32). The first data set provides the times of the first external leakage, the second data set the relative survival times after this first failure until the next external leakage occured. The data are summarized in tables 1 and 2.

Table 1
Times of the first external leakages (in hours)

666	687	1335	2044	2195	2281	2708	2764	2940	2970
2972	3004	3564	3955	4133	4230	4805	5200	5384	5766
6222	6267	6714	6794	7398	7532	7659	8696	8740	9213
9470	12213								

Table 2
Relative survival times after first failure

0.003	0.018	0.066	0.096	.0.114	0.141	0.195	0.229	0.239
0.378	0.392	0.461	0.500	0.584	0.592	0.648	0.784	0.816
0.877	0.879	0.896	1.021	1.070	1.400	1.521	1.531	1.540
1.868	2.022	2.349	2.360	2.721				

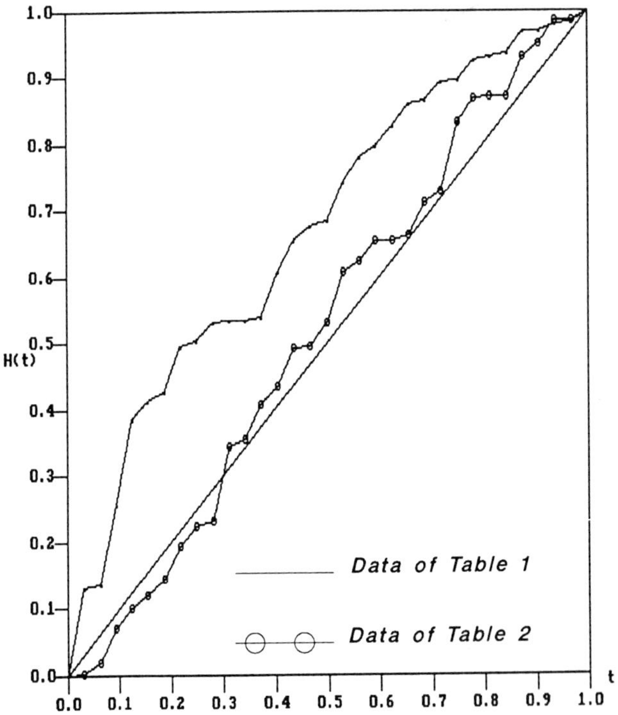

Figure 1. TTT-plots of data sets

It has to be analysed whether the life expectation after the first repair is the same as at the start of operation after the delivery from the manufacturer. From an engineering point of view it was recommended to analyse the relative survival times as the environmental conditions for each pump were slightly different which might have had already effects on the times to first leakage.

The scale-invariant TTT-plot for both data sets is shown in figure 1.

The data of the first leakages provide an IFR-trend due to the nearly concave TTT-plot (lying completely above the diagonal).

The data set of the relative survival times cross the diagonal once from below. An exponential behaviour for the survival data however seems to be an incorrect assumption as the TTT-plot significantly lies above the diagonal for $H(t) < 0.4$. Thus the crossing with the diagonal from below and the change from a convex slope to a following NBUE-behaviour might indicate a trend change from a DFR-behaviour to approximately constant aging.

The data of the first leakages can be modelled by a Weibull-distribution with cdf $F(x) = 1 - \exp(-(\alpha x)^{\beta})$. The estimated parameters are $\hat{\alpha} = 0.00017$ and $\hat{\beta} = 1.67$ (derived from a least squares regression, "^" to denote an estimator).

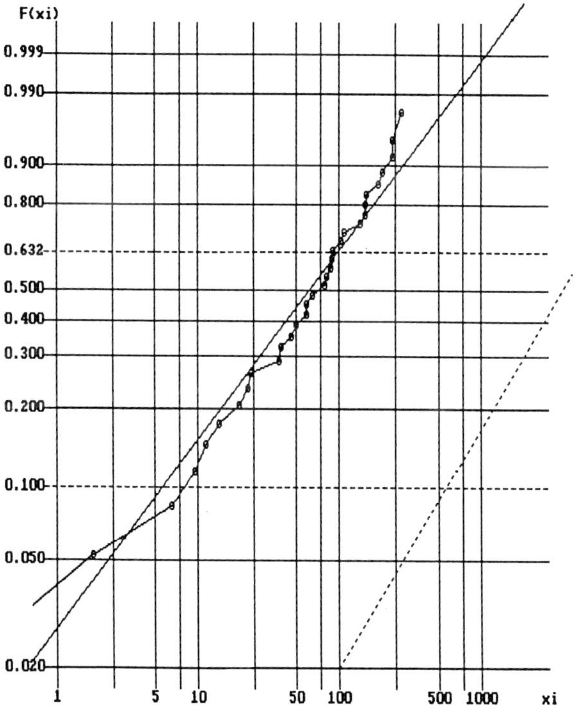

Figure 2. Weibull probability plot of relative survival times, data multiplied by 100

The use of any classical distribution such as Weibull, gamma or lognormal does not provide an acceptable fit to the relative survival data. Especially exponentiality as potential conclusion from the TTT-plot should be excluded due to graphical arguments. This is shown in figure 2 in a Weibull-probability plot.

The data set provides a convex curve. The least squares regression line crosses this curve only twice. Formally the estimation of Weibull-parameters provides $\alpha\hat{} = 1.0333$ and $\beta\hat{} = 0.79$. So the Weibull-hazard rate would be of DFR-character.

The plot of the corresponding scaled Weibull-TTT-transform (see figure 3) clearly indicates that the Weibull-DFR-estimation must be rejected as suitable aging model.

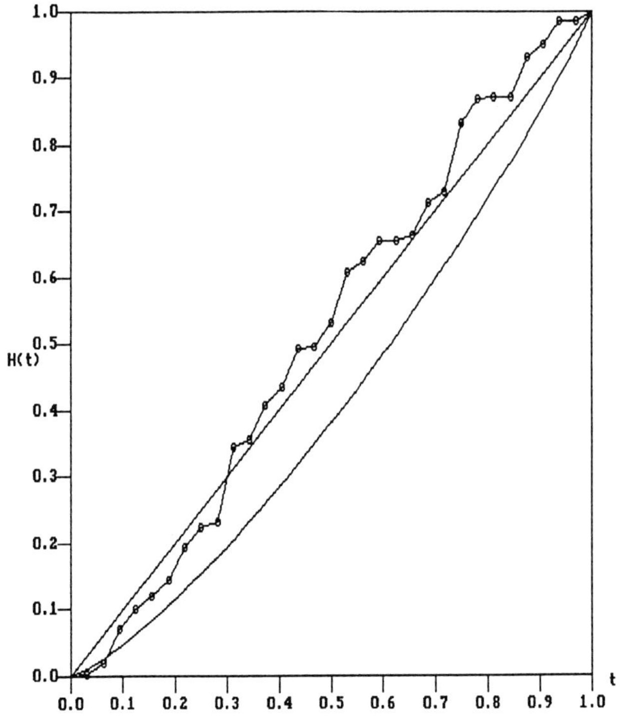

Figure 3. TTT-plot of relative survival times with Weibull-fit, shape parameter $\beta\hat{} = 0.79$

Other plots e.g. plotting the empirical mean residual life

$$\hat{m}(x_i) = \sum_{j=i+1}^{n} x_j \,/\, (n-i) - x_i \qquad (4)$$

$i = 0, 1, \ldots, n-1, \quad x(0) = 0$

also indicate the departure from exponentiality.

4.2.FIT WITH MIXED GAMMA MODEL

As "proof" to our conjectures we fit a mixed gamma distribution with pdf according (1a) to the data. The maximum likelihood estimation yields

$\hat{p} = 0.804$, $\hat{\alpha}_1 = 1.41$, $\hat{\alpha}_2 = 5.44$, $\hat{\beta}_1 = 0.93$, $\hat{\beta}_2 = 9.84$,

i.e. the best fit posesses a bathtub-shaped hazard rate.

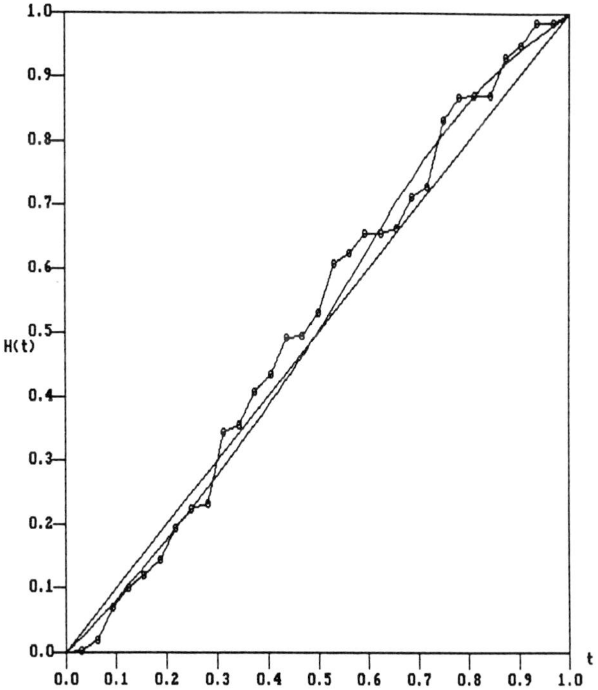

Figure 4. TTT-plot of relative survival times with mixed-gamma-fit with parameters $\hat{p}=0.804$, $\hat{\alpha}_1=1.41$, $\hat{\alpha}_2=5.44$, $\hat{\beta}_1=0.93$, $\hat{\beta}_2=9.84$

Due to the estimation of the shape parameters the pdf has a parameter structure as defined in (1b). A TTT-plot of this fit is given in figure 4.

The corresponding hazard rate in comparison with the alternative Weibull- and exponential estimation is shown in figure 5.

The estimation thus "rejects" exponentiality which might arise as a special case in the mixed gamma model with $\beta_1 \approx \beta_2 \approx 1$ and $0 < p < 1$ (see (1a)).

The expectation of mixed gamma lifetimes provides $E(x) = 0.887$ which is nearly identical to the empirically calculated expectation due to (4) with $\hat{m}(0) = 0.885$.

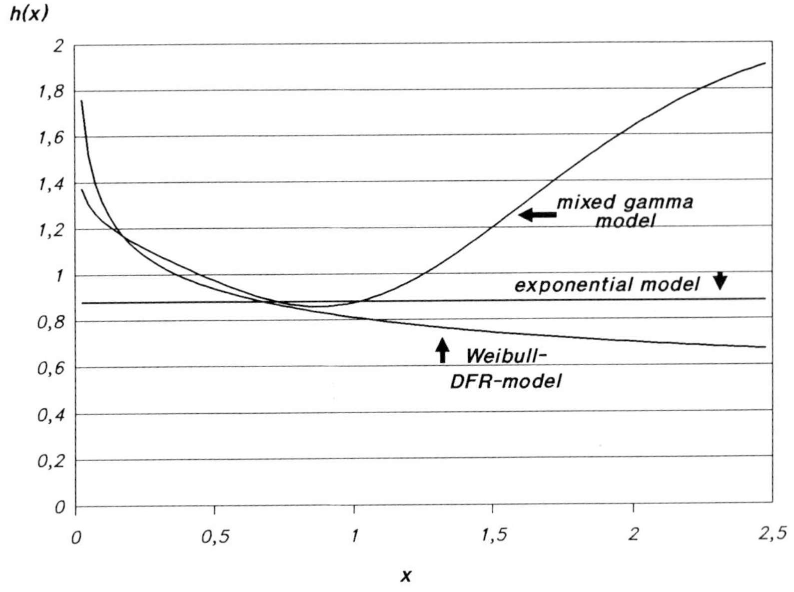

Figure 5. Alternative estimation of hazard rates for relative survival data

4.3. INTERPRETATION OF RESULTS

A potential interpretation of the revealed bathtub-shaped hazard rate from an engineering and statistical point of view is that some repairs of the leakages did not provide a state of "as good as new" for the pumps. These pumps failed in an "infant mortality phase due to bad repair" shortly after re-start of operation providing a DFR-trend of the hazard rate. Those pumps which were repaired successfully returned to their "inherent" aging behaviour which is of IFR-character. Thus their relative life expectation shows a spread around $E(x_{rel}) \approx 1$.

We have stressed the important role of graphical tools for the exploration of lifetime data. Especially the TTT-plot is a sensitive tool to detect departures from exponential aging behaviour. "Objective" parameter estimations in a flexible distribution model and "subjective" visual analyses of graphical data representations seem to be a senseful approach to analyse data sets of small size whose aging behaviour is "on average close to exponentiality".

5. LITERATURE

Aarset M.V. 1987
How to identify a bathtub hazard rate
IEEE Transactions on Reliability
Vol. R-36, No.1, pp. 106-108

Barlow R.E., Campo R. 1975
Total time on test processes and applications to failure data analysis
in: Barlow, Fussell, Singpurwalla (eds.)
Reliability and Fault Tree Analysis
SIAM, Philadelphia, pp. 451-481

Bergman B. 1977
Crossings in the total time on test plot
Scandinavian Journal of Statistics, 4, pp. 171-177

Dhillon B.S. 1981
Life Distributions
IEEE Transactions on Reliability, Vol. R-30, No.5, pp. 457-460

Guess F.M., Hollander M., Proschan F. 1986
Testing exponentiality versus a trend change in mean residual life
Annals of Statistics, Vol. 11, pp. 1388-1398

Pamme H. 1992
Beiträge zur statistischen Analyse von Alterung
Peter Lang Verlag, Frankfurt (to appear in 1992, in German)

Teicher H. 1963
Identifiability of finite mixtures
Annals of Math. Stats. 34, pp. 1265-1269

ADVANCES IN RELIABILITY
Edited by Asit P. Basu

Scheduling Inspections in Reliability

Giovanni Parmigiani

Institute of Statistics and Decision Sciences, Duke University, Durham, NC, 27706 U.S.A.

Abstract

When a system is subject to random failures that can be detected only by costly inspections, it is important to choose inspection times so that late detection costs and inspection costs are balanced optimally. This paper is a review of some recent developments in the methodology for solving inspection problems. Topics include fallible tests, replacement policies, delayed symptoms of failure, stand-by systems and Bayesian models. Areas worthy of further study are also indicated.

1. INTRODUCTION

Scheduling of inspection times appears in a variety of reliability applications. Examples include maintaining stand-by systems, replacing worn-out tools in automated manufacturing systems, designing life testing experiments, monitoring quality, planning checkups for aging aircraft and so on. In all such problems, failures can be detected only by a costly inspection, but a timely detection can reduce other costs, such as downtime cost, and the optimal schedule of inspection times has to reach a balance between the two sources of cost.

The objective of this paper is to present some recent developments in the theory of optimal scheduling of inspections and to identify areas and problems worthy of further investigation. Section 2 briefly introduces the problem and reviews the fundamental model of Barlow, Hunter and Proschan [3]. Various extensions are then considered: Section 3 deals with fallible, rather than error-free, inspection procedures; Section 4 includes an application of the results of Section 3 to inspection and replacement policies. Section 5 considers the case of delayed symptoms of failure and points out a connection with problems in the secondary prevention of chronic disease. Section 6 studies stand-by systems. Finally, Section 7 deals with relaxing the assumption of known failure distributions, that may not always be realistic. A parametric family with unknown parameter is considered instead. Uncertainty adds another facet to the problem: how does learning influence optimal schedules? The answer is obtained in the context of a Bayesian approach.

2. A BASIC MODEL FOR THE DETECTION OF SYSTEM FAILURES

Interest in optimal inspections schedules for the maintenance of stochastically deteriorating systems originated the work of Savage [23], Barlow and Hunter [2], Derman [9] and Coleman and Abrams [8]. A fundamental initial contribution is that of Barlow, Hunter and Proschan [3]. They developed a simple model capturing very effectively the fundamental trade-off involved in the choice of an optimal inspection schedule: frequent checks increase the cost of inspections but decrease the costs of late detection of failure. The set of assumptions that they introduced will be referred to as the basic model in the sequel. For ease of further reference, I summarize it as follows:

(i) The monitored system is described by a continuous time process with two states, labeled, for simplicity, safe functioning and failure;

(ii) The system starts at time 0 in the functioning state; the time to failure is a random variable Y, with known distribution F and expectation μ_y;

(iii) Information on the state of the system can be obtained only by an appropriate test;

(iv) The test is error-free, and costs a fixed amount k;

(v) Each unit of elapsed time between the failure and its detection entails a fixed cost c.

(vi) Inspections terminate upon detection of the failure.

In this context a decision, or policy, consists of a schedule of inspection times. Schedules are denoted by $\boldsymbol{\tau} = \{\tau_i\}_{i=0,1,2,\dots}$, where τ_i is time of the i-th inspection, $\tau_i > \tau_{i-1}$ and $\tau_0 = 0$. For each schedule $\boldsymbol{\tau}$, only a subset of the inspections will actually be carried out. Let the time of the last inspection performed be $X^{\boldsymbol{\tau}}$, and let $I^{\boldsymbol{\tau}}$ be the corresponding number of inspections. Then the loss $L^{\boldsymbol{\tau}}$, suffered if the schedule $\boldsymbol{\tau}$ is adopted, is:

$$L^{\boldsymbol{\tau}} = kI^{\boldsymbol{\tau}} + c(X^{\boldsymbol{\tau}} - Y).$$

The problem is in choosing the sequence $\boldsymbol{\tau}$ that minimizes the expected cost, or risk:

$$R(\boldsymbol{\tau}) \equiv E(L^{\boldsymbol{\tau}}) = kE(I^{\boldsymbol{\tau}}) + cE(X^{\boldsymbol{\tau}} - Y). \tag{1}$$

Before moving to the description of the general solution, the two following examples are helpful. First, suppose that an inspection is fixed at time T, and it is desired to schedule optimally a further inspection time τ, between 0 and T. The risk can then be rewritten as:

$$R(\tau) = c_0 - F(\tau)[k + c(T - \tau)],$$

where c_0 does not depend on τ. This simple expression captures the fundamental trade-off: increasing τ reduces $k+c(T-\tau)$, but increases $F(\tau)$. If F has a density, the optimal balance is reached at the value solving:

$$k + c(T - \tau) = \frac{F(\tau)}{f(\tau)}.$$

This pattern of solution will reappear in more complex situations.

Second, consider an exponential failure distribution. Suppose it is optimal to schedule the first inspection at time τ. If the inspection reveals a failure, the problem terminates.

Second, consider an exponential failure distribution. Suppose it is optimal to schedule the first inspection at time τ. If the inspection reveals a failure, the problem terminates. If not, the problem continues with the additional information of the observed survival. However, from the memoryless property of the exponential distribution, conditioning on the new information leaves the failure distribution unchanged. Hence it is optimal to schedule the second inspection at time 2τ. In other words, as is intuitive, a periodic schedule is optimal.

The minimization of the risk (1) requires the choice of an infinite sequence. In this regard, the question is similar to some very well studied problems in infinite horizon dynamic programming. For example, in economics, optimal growth models often take the form of minimizing:

$$R(\boldsymbol{\tau}) = \sum_{i=1}^{\infty} r^i U(\tau_i, \tau_{i+1}), \tag{2}$$

given a set of initial conditions, where U is a utility function. In such cases, methods based on solving the fundamental functional equation of dynamic programming can be applied. Alternatively, differentiation of R with respect to each of the coordinates gives a set of recursive first order conditions, known as Euler equations. These define an optimal continuation for the given initial conditions. A thorough discussion of the vast literature is in Stokey and Lucas [27].

The basic inspection model differs from minimizing (2) in one very important respect: the function U varies with i. However, as each term in the sum is still a function of only two consecutive values of the sequence, the optimality of an inspection time depends only on the value of the two neighboring times, and methods based on Euler-like conditions can still be developed.

In fact, the risk function (1) can be rewritten in terms of the individual inspection times as:

$$\begin{aligned} R(\boldsymbol{\tau}) &= \sum_{i=0}^{\infty} \int_{\tau_i}^{\tau_{i+1}} [k(i+1) + c(\tau_{i+1} - y)] dF(y) \qquad (3) \\ &= \sum_{i=0}^{\infty} (k(i+1) + c\tau_{i+1})[F(\tau_{i+1}) - F(\tau_i)] - c\mu_y. \end{aligned}$$

If F has a density f, with $f(y) > 0$ for $y > 0$, partial differentiation of (3) with respect to each element of $\boldsymbol{\tau}$ gives:

$$\tau_{i+1} - \tau_i = \frac{F(\tau_i) - F(\tau_{i-1})}{f(\tau_i)} - \frac{k}{c} \qquad i = 1, 2, \ldots \tag{4}$$

These equations define recursively the optimal inspection schedule once the optimal value of τ_1 is given. This reduces the infinite dimensional optimization to a unidimensional problem. A recurring theme throughout the rest of the discussion will be the generalization of this fundamental property to more complex situations.

The problem of determining the optimal initial inspection τ_1 was solved by Barlow, Hunter and Proschan [3] in the special case of log-concave failure densities (or Pólya density function of order 2), that is densities such that the ratio $f(y-\delta)/f(y)$ is increasing in y for fixed positive δ. Log concavity connects naturally to this problem, since the ratio $[F(y) - F(y - \delta)]/f(y)$ —appearing at the left hand side of the optimality conditions, is increasing for log-concave densities. Moreover, log-concave densities are common in applications where the failure rate is taken to be smooth and increasing.

Assuming log-concavity, Barlow, Hunter and Proschan [3] proved that the inspection intervals decrease, consistent with the fact a device aging according to such failure density wears out with time. Moreover, they showed that, when equations (4) are used to determine the continuation sequence for an arbitrary τ_1, then a value larger than optimal will eventually generate increasing inspection intervals, whereas a value smaller than optimal will eventually imply negative inspection intervals. This insures uniqueness of the optimal first inspection. Moreover, it provides a very efficient algorithm for determining the optimal τ_1 without evaluating the risk function.

3. FALLIBLE TESTS

The assumption that the inspection procedure adopted is error-free appears to be one of the most restrictive in the basic model. Optimal schedules for fallible tests when failures are exponential have been studied by Sengupta [25]. If the only possible error is that of overlooking a failure, the optimal schedule has the form: $\tau_{i+1} = \tau + \delta \cdot i$, where τ and δ solve:

$$\delta = \frac{F(\delta)}{f(\tau)} - \frac{k}{c} \qquad (5)$$

$$\beta = \frac{\lambda F(\delta)(k + c\delta)}{\lambda(k + c\delta) - cF(\delta)}. \qquad (6)$$

Also, $\tau > \delta$.

Parmigiani [18], analyzes general failure distributions in a situation where there are two tests available: one is fallible; the other is error-free, but more expensive, and it is used only when the outcome of the fallible test indicates a failure. This corresponds to common practice in many applications: if a fallible examination of an aircraft engine reveals evidence of an advanced deterioration, a more extensive work-up follows. The same applies to cancer screening tests. A further relevant generalization can be obtained by assuming that inspections take time. Let the assumptions of the basic model be modified as follows:

(iv)' There are two types of test available: one is fallible and the other is error-free. The fallible test reports the correct state with probability $\beta \in (0, 1]$ when the system is failed, and with probability $\alpha \in (0, 1]$ when it is functioning; both α and β are independent of the failure time. Each fallible tests costs k_1 whereas each error-free test costs k_2, with $k_1 < k_2$.

(iv)" At each inspection time, the fallible test is performed first; if the outcome indicates a failure, then an error-free test is performed as well.

(iv)"' Tests do not alter the state of the system, and take a random amount of time, independent of the inspection time; Z_i denotes the duration of the i-th fallible test and Z'_i the duration of the i-th error-free test, with $\mu_z = E(Z_i)$ and $\mu'_z = E(Z'_i)$. The system does not operate —and hence does not age— during inspections.

Let now $X^{\boldsymbol{\tau}}$ be the age of the system at detection of the failure (not inclusive of the time spent for inspections). The loss $L^{\boldsymbol{\tau}}$, suffered if the schedule $\boldsymbol{\tau}$ is adopted, is then:

$$L^{\boldsymbol{\tau}} = k_1 I^{\boldsymbol{\tau}} + k_2 J^{\boldsymbol{\tau}} + \left(X^{\boldsymbol{\tau}} - Y + \sum_{i=1}^{I^{\boldsymbol{\tau}}} Z_i + \sum_{i=1}^{J^{\boldsymbol{\tau}}} Z_i' \right). \tag{7}$$

Taking the expectation of the loss function in (7), and using the fact that the inspection durations are independent of the age and the state of the system at inspection, the risk $R(\boldsymbol{\tau})$ is:

$$R(\boldsymbol{\tau}) = (k_1 + \mu_z)E(I^{\boldsymbol{\tau}}) + (k_2 + \mu_z')E(J^{\boldsymbol{\tau}}) + E(X^{\boldsymbol{\tau}} - Y). \tag{8}$$

To write R explicitly as a function of inspection times, consider that, since α and β are independent of time,

$$\begin{aligned} E(I^{\boldsymbol{\tau}} | \tau_i < Y \leq \tau_{i+1}) &= i + \frac{1}{\beta} \\ E(J^{\boldsymbol{\tau}} | \tau_i < Y \leq \tau_{i+1}) &= 1 + (1-\alpha)i. \end{aligned}$$

The average increase in risk due to each inspection before the failure is $k = k_1 + \mu_z + (1-\alpha)(k_2 + \mu_z')$. The risk in (8) can be written as:

$$R(\boldsymbol{\tau}) = \sum_{i=0}^{\infty} \sum_{j=0}^{\infty} \beta(1-\beta)^j \int_{\tau_i}^{\tau_{i+1}} [ki + (k_1 + \mu_z)(j+1) + (k_2 + \mu_z') + (\tau_{j+i+1} - y)] dF(y). \tag{9}$$

The main difficulty in extending the basic model this way is that, due to the fallibility of the test, the first order optimality conditions of each inspection time depend on all other times rather than just the two neighboring. Therefore, the crucial first step consists in finding a way to obtain a solution in recursive form, thus reducing the problem to one dimension.

Consider the sequence $\boldsymbol{\gamma}$ of the expected detection times conditional on the failure occurring in the interval $(\tau_{i-1}, \tau_i]$:

$$E(X^{\boldsymbol{\tau}} | \tau_{i-1} < Y < \tau_i) = \sum_{j=0}^{\infty} \beta(1-\beta)^j \tau_{j+i} \equiv \gamma_i \qquad i = 1, 2, \ldots \tag{10}$$

Note that the sequences $\boldsymbol{\tau}$ and $\boldsymbol{\gamma}$ satisfy the relation $\gamma_i = \beta\tau_i + (1-\beta)\gamma_{i+1}$, or:

$$\tau_i = \frac{1}{\beta}(\gamma_i - (1-\beta)\gamma_{i+1}). \tag{11}$$

Applying (10) to (9) yields:

$$R(\boldsymbol{\tau}) = \sum_{i=0}^{\infty} (ki + \gamma_{i+1})[F(\tau_{i+1}) - F(\tau_i)] + k_2 + \mu_y' + (k_1 + \mu_y)\frac{1}{\beta} - \mu_y. \tag{12}$$

Assume that Y has a probability density function f, with $f(y) > 0$ for $y > 0$. Differentiating (12) with respect to τ_i, $i \geq 1$, and rearranging, gives the following set of necessary conditions for optimality:

$$\gamma_{i+1} - \gamma_i = \frac{1}{f(\tau_i)} \sum_{j=1}^{i} \beta(1-\beta)^{i-j} [F(\tau_j) - F(\tau_{j-1})] - k \qquad i = 1, 2, \ldots \tag{13}$$

Iteratively substituting the i-th condition in the $(i+1)$-st gives:

$$\begin{aligned} \gamma_2 - \gamma_1 &= \beta\frac{F(\tau_1)}{f(\tau_1)} - k & (14) \\ \gamma_{i+1} - \gamma_i &= \beta\frac{F(\tau_i) - F(\tau_{i-1})}{f(\tau_i)} + (1-\beta)\frac{f(\tau_{i-1})}{f(\tau_i)}(\gamma_i - \gamma_{i-1} + k) - k \qquad i = 2, 3, \ldots \end{aligned}$$

Suppose a starting τ_1 is fixed. Then, using (11), the top equation in (14) becomes:

$$\gamma_2 = \tau_1 + \frac{F(\tau_1)}{f(\tau_1)} - \frac{k}{\beta};$$

and can be used to compute γ_2. Now (11) can be used again to obtain γ_1 from γ_2 and τ_1. Then the two following steps can be iterated:

1. When the values of γ_i, γ_{i-1} and τ_{i-1} are known, use (11) to express the right hand side of (14) as a function of τ_{i-1}, γ_i and γ_{i+1}, and then solve for γ_{i+1};

2. Compute τ_i using (11);

In this way both $\boldsymbol{\tau}$ and $\boldsymbol{\gamma}$ can be determined from the initial τ_1. Also, at each iteration, (12) can be used to compute the terms of $R(\boldsymbol{\tau})$. Note that since $\tau_i > \tau_{i-1}$, the range for γ_{i+1} in step 1 is included in the interval $(\gamma_i, (\gamma_i - \beta\tau_{i-1})/(1-\beta))$. This may be empty, in which case the starting τ_1 cannot be optimal. The same conclusion holds when step 1 has no solution. Uniqueness of γ_{i+1} in step 1 is guaranteed if $F(y)/f(y)$ is increasing in y, a condition satisfied by many of the distributions commonly used in applications.

If the failure distribution is log-concave, the results obtained by Barlow, Hunter and Proschan for the basic model can be generalized to this context. In particular, it is possible to show that both $\gamma_{i+1} - \gamma_i$ and $\tau_{i+1} - \tau_i$ are decreasing with i, and that the optimal first inspection is unique and can be determined by binary search. Details of the proofs can be found in Parmigiani [18].

4. INSPECTION AND REPLACEMENT POLICIES WITH FALLIBLE TESTS

The assumption that the decision problem terminates upon detection of failure can be restrictive in instances such as monitoring a unit in a complex industrial process. Then, some action, like repair or replacement, usually follows the discovery of a malfunctioning. To fit this type of situation Brender [7] developed an appropriate alternative to the basic model: he assumed that upon detection of failure, a replacement occurs, at a fixed price and in a known amount of time, after which the system is new and checking resumes. The risk function is then the long run expected cost of inspection and replacement per unit of time. Brender constructed a version of the algorithm proposed by Barlow, Hunter and Proschan for the case of a log-concave density.

When tests are fallible, a key issue in designing replacement policies is deciding what strategy to follow when the test reports a failure. In most practical situations, it seems realistic to assume that the relevant trade-off involved in this decision is between the cost of replacing a unit that appears to be failed but it is not and the cost of acquiring additional information, perhaps by a more accurate —and typically more costly— test. If the device under inspection is expensive to replace, it may be convenient to adopt a two-stage inspection procedure, performing the fallible test first, and the error-free test whenever the first test reports a failure. In this way, only actually failed equipment is replaced. If instead the unit is inexpensive, it may be better to replace it upon the report of failure and ignore the possibility of a more accurate inquiry. Both types of policies are discussed in this section. I refer to the first policy as two-test policy and to the second as one-test policy. Interestingly, the property that times between inspection have to be decreasing generalizes for the two-test policy but not for the one-test policy. This is due to the cumulative effect of allowing a positive probability replacing of a good unit.

Let us consider the problem in further detail. Assume that each replacement costs k_3 and requires a random amount of time W, with known expectation μ_w. In the case of a one-test policy, if the schedule $\boldsymbol{\tau}$ is adopted, the cost suffered per unit of time between two successive replacements is:

$$C(\boldsymbol{\tau}, Y) = \frac{k_1 I^{\boldsymbol{\tau}} + c(U^{\boldsymbol{\tau}} + \sum_{i=1}^{I^{\boldsymbol{\tau}}} Z_i + W) + k_3}{X^{\boldsymbol{\tau}} + \sum_{i=1}^{I^{\boldsymbol{\tau}}} Z_i + W}, \tag{15}$$

where, $I^{\boldsymbol{\tau}}$, as before, denotes the number of tests performed, $X^{\boldsymbol{\tau}}$ the age of the device at replacement (inclusive of the time between failure and detection when applicable) and $U^{\boldsymbol{\tau}} = (X^{\boldsymbol{\tau}} - Y) I_{\{X^{\boldsymbol{\tau}} > Y\}}$ the delay in the detection of the failure; neither $X^{\boldsymbol{\tau}}$ nor $U^{\boldsymbol{\tau}}$ include the time used for inspection. A standard limiting argument and independence of the inspection durations from the age of the system allow writing the long run cost per unit of time —henceforth risk— as:

$$R(\boldsymbol{\tau}) = \frac{(k_1 + c\mu_z)E(I^{\boldsymbol{\tau}}) + cE(U^{\boldsymbol{\tau}}) + k_3 + c\mu_w}{E(X^{\boldsymbol{\tau}}) + \mu_z E(I^{\boldsymbol{\tau}}) + \mu_w} = \frac{R_N(\boldsymbol{\tau})}{R_D(\boldsymbol{\tau})}. \tag{16}$$

Note that an inspection policy will always have a risk greater than

$$r_0 = [(k_1 + c\mu_z) + k_3 + c\mu_w]/(\mu_y + \mu_z + \mu_w).$$

If $r_0 \geq c$, it is optimal not to inspect.

Moving to the two-test policy, let $J^{\boldsymbol{\tau}}$ be the number of error-free tests associated with the policy $\boldsymbol{\tau}$. The cost of a cycle between replacements is then:

$$C'(\boldsymbol{\tau}, Y) = \frac{k_1 I^{\boldsymbol{\tau}} + k_2 J^{\boldsymbol{\tau}} + c(X^{\boldsymbol{\tau}} - Y + \sum_{i=1}^{I^{\boldsymbol{\tau}}} Z_i + \sum_{i=1}^{J^{\boldsymbol{\tau}}} Z'_i + W) + k_3}{X^{\boldsymbol{\tau}} + \sum_{i=1}^{I^{\boldsymbol{\tau}}} Z_i + \sum_{i=1}^{J^{\boldsymbol{\tau}}} Z'_i + W}. \tag{17}$$

Consequently, the risk can be written as:

$$\begin{aligned} R'(\boldsymbol{\tau}) &= \frac{(k_1 + c\mu_z)E(I^{\boldsymbol{\tau}}) + cE(X^{\boldsymbol{\tau}} - Y) + (k_2 + c\mu'_z)E(J^{\boldsymbol{\tau}}) + k_3 + c\mu_w}{E(X^{\boldsymbol{\tau}}) + \mu_z E(I^{\boldsymbol{\tau}}) + \mu'_z E(J^{\boldsymbol{\tau}}) + \mu_w} \\ &= \frac{R'_N(\boldsymbol{\tau})}{R'_D(\boldsymbol{\tau})}. \end{aligned} \tag{18}$$

The decision problem consists of determining the schedules minimizing (16) and/or (18), and of comparing the two policies when both are available. A convenient way to solve these optimization problems is based on re-expressing the risk as a linear combination of the expectations involved, by means of an auxiliary variable. Then, differentiating the risk with respect to the inspection times gives recursive optimality conditions that can be used to determine the optimal continuation schedule once the first inspection is given. The optimization is then reduced to determining the values of the first inspection and of the auxiliary variable.

Starting with the two-test case, consider the function:

$$\begin{aligned}\varphi'(s,\boldsymbol{\tau}) &= R'_N(\boldsymbol{\tau}) - sR'_D(\boldsymbol{\tau}) = (k_1 + (c-s)\mu_z)E(I^{\boldsymbol{\tau}}) \\ &\quad +(k_2 + (c-s)\mu'_z)E(J^{\boldsymbol{\tau}}) + (c-s)E(X^{\boldsymbol{\tau}}) + k_3 + (c-s)\mu_w - c\mu_y.\end{aligned} \tag{19}$$

Since φ' is a linear combination of $E(I^{\boldsymbol{\tau}})$, $E(J^{\boldsymbol{\tau}})$ and $E(X^{\boldsymbol{\tau}})$, it is the risk function of an inspection problem with no replacement and fixed cost of down-time and inspection, as that studied in the previous section. Therefore all the results apply. Moreover, by a standard argument, (see Barlow and Proschan [5], p. 116) it can also be demonstrated that there exist a value s^* such that $\varphi(s^*, \boldsymbol{\tau}(s^*)) = 0$ and that the schedule $\boldsymbol{\tau}(s^*)$ minimizes (8); s^* is then the expected long run cost per unit of time associated with the optimal schedule. If $s^* > c$ it is optimal not to inspect; otherwise it is optimal to inspect according to $\boldsymbol{\tau}(s^*)$. In practice, the determination of the optimal policy is based on two separate steps: the first is an algorithm determining, for given s and starting value τ_1, the optimal continuation schedule $\tau_2, \tau_3, \dots$ and its risk; the second is the two-dimensional optimization over τ_1 and s.

A similar technique can be applied in the one-test case. However, the application of the results of the previous section is not direct, because of the possibility of replacing a unit before the actual occurrence of failure. A separate derivation is required, with the further assumption that $\alpha + \beta > 1$. As before, define:

$$\begin{aligned}\varphi(s,\boldsymbol{\tau}) &= R_N(\boldsymbol{\tau}) - sR_D(\boldsymbol{\tau}) \\ &= (k_1 + (c-s)\mu_z)E(I^{\boldsymbol{\tau}}) - sE(X^{\boldsymbol{\tau}}) + cE(U^{\boldsymbol{\tau}}) + k_3 + (c-s)\mu_w.\end{aligned} \tag{20}$$

To obtain recursive optimality conditions, it is necessary to rewrite φ as a function of $\boldsymbol{\tau}$. It is convenient to define:

$$\tilde{\delta}_i = \frac{1}{\beta}\left[(1-\alpha+\beta)\gamma_i - (\alpha\beta + (1-\alpha)(1-\beta))\gamma_{i-1}\right]. \tag{21}$$

where γ_i's are given by (10). Differentiation yields the following optimality conditions:

$$\begin{aligned}\tilde{\delta}_2 &= \beta\frac{F(\tau_1)}{f(\tau_1)} - q\frac{1-F(\tau_1)}{f(\tau_1)} - k \\ \tilde{\delta}_{i+1} &= \beta\frac{F(\tau_i)-F(\tau_{i-1})}{f(\tau_i)} - \frac{q[1-F(\tau_i)]}{f(\tau_i)} - k + \frac{1-\beta}{\alpha}\frac{f(\tau_{i-1})}{f(\tau_i)}\left(\tilde{\delta}_i + k + \frac{q[1-F(\tau_{i-1})]}{f(\tau_{i-1})}\right)\end{aligned} \tag{22}$$

where

$$k = \frac{c_1 + (c_2 - s)\mu_z}{(c_2 - s)}\left(1 - \frac{1-\alpha}{\beta}\right),$$

and

$$q = \frac{s(1-\alpha)}{(c_2 - s)}.$$

The derivation can be found in Parmigiani [19]. If $\alpha = 1$, (22) reduces to (14).

When the failure density is log concave, the times between checks for the two-test case have to be decreasing and the solution is unique. In the one-test case it is possible to show that the same applies to $\tilde{\delta}_i$, under the additional assumption that the function $g(y) = f(y)/[1 - F(y)]$ is convex. The proof is contained in [19]. This still ensures that a fast computational algorithm can be constructed. Interestingly, however, it is not true that times between inspections have to decrease, but only that they can increase at most geometrically, with ratio:

$$\rho = \frac{1 - \alpha + \beta}{\alpha\beta + (1-\alpha)(1-\beta)} < 1$$

Heuristically, to compensate for the fact that a functioning unit may be replaced, one has to schedule less frequent inspections, especially for high values of i, where the compound effect on the risk of the possible past errors is stronger. This may compensate for the increased frequency required by the aging of the machine, so that no general conclusion can be reached. As an illustration of this compound effect, consider the exponential failure distribution, for which $f(y - \Delta)/f(y)$ is constant and there is no aging. The optimal schedule for the two-test case is periodic. On the other hand, since the failure rate is constant, in the one-test case the sequence $\rho\gamma_{i+1} - \gamma_i$ has be constant. This leads to $\rho\delta_{i+1} = \delta_i$, so that times between inspection increase geometrically.

5. DELAYED SYMPTOMS OF FAILURE AND MEDICAL APPLICATIONS

In many applications, system failures become apparent without an inspection, but only after sufficient time has elapsed. For example, in a continuous manufacturing process, a failure may cause systematic defects. The first defective item, however, may not appear until a lengthy production cycle has terminated. In the meanwhile, all production is lost. Also, material failures may start as small cracks that are detectable only by an appropriate inspection procedure, but become obvious after some period of time, possibly itself random.

Sengupta [24], considered the case of delayed symptoms of failure. He extended the basic model by assuming that the failure occurs at time Y and becomes apparent at time $Y + U$, where U is a random variable independent of Y, with known distribution H and expectation μ_u. All other assumptions are maintained. In this context, it is not necessarily optimal to inspect, since the risk of not inspecting is $c\mu_u$, which is smaller than the risk associated with inspecting for large enough k. If the schedule $\boldsymbol{\tau}$ is adopted, the risk is:

$$R(\boldsymbol{\tau}) = \sum_{i=0}^{\infty} \int_{\tau_i}^{\tau_{i+1}} \left[\int_0^{\tau_{i+1}-y} (ki + cu)h(u)du + \int_{\tau_{i+1}-y}^{\infty} [(i+1)k + c(\tau_{i+1} - y)]h(u)du \right] f(y)dy.$$

Sufficient conditions for optimality become:

$$\bar{H}(\tau_{i+1}-\tau_i)+\frac{k}{c}\int_0^{\tau_{i+1}-\tau_i}\bar{H}(u)du=\int_0^{\tau_i-\tau_{i-1}}\frac{f(\tau_i-u)}{f(\tau_i)}[\bar{H}(u)-\frac{k}{c}h(u)]du \quad i\geq 1, \quad (23)$$

where $\bar{H}(u)=1-H(u)$. Sengupta assumed that $h(u)/\bar{H}(u)<c/k$, to guarantee that an optimal continuation exists. Moreover, for the case of a log-concave density for Y, he derived an elegant result concerning the stopping rule. Let:

$$\varphi(\delta,x)=\int_0^{\delta}\frac{f(x-u)}{f(x)}\left[\bar{H}(u)-\frac{k}{c}h(u)\right]du. \quad (24)$$

Then if $\lim_{x\to\infty}\varphi(x,x)\leq c\mu_u$ it is optimal not to inspect. Else, the optimal inspection schedule consists of an infinite number of inspections, as in the basic model. Moreover, he generalized the results of Barlow, Hunter and Proschan on increasing inspection intervals and on uniqueness.

Similar models arise in preventive medicine, in connection with scheduling examinations of asymptomatic patients for the early detection of chronic disease. Then the random variable U as the time spent by the patient in the asymptomatic stage of the disease. Careful treatment of the problem requires more general loss functions. Also, dependence between U and Y must also be considered.

This topic has lately gained relevance in connection with the public health effort in mass screening programs. When the screening test is to be carried out on a number of occasions during the lifetime of a person, the appropriate scheduling can play an important role. For instance, in the case of breast cancer, the questions of the age of first test and of the frequency of tests in different age groups has originated an important debate. Further discussion of this topic is in Parmigiani [21].

6. STAND-BY SYSTEMS

A frequently encountered application is the inspection of stand-by units or systems. For example, power generators in cold stand-by, used in nuclear power plants, hospitals, and so on, have to be inspected for failures. The aim of inspections, however, is preventing the catastrophic consequences of the standby failing to operate when necessary. A sensible risk function would then include the cost of inspections and the probability of catastrophe; on the other hand, no cost is associated with downtime.

To construct a model for stand-by systems, assume that the events that require the stand-by to operate are generated by a renewal process. Let the event times be $X_1, X_2, \ldots$, and the waiting times be distributed according to a cumulative distribution G, with density g. If events occur while the stand-by is still operating, no loss is suffered, and the process generating events renews. If an event occurs between the failure of the stand-by and the detection of failure, the loss is c. Typically $c>k$. Event times are observed at no cost by the decision maker. Also, when events occur, it is observed whether the stand-by is functioning.

As the process is a renewal, knowledge of the time of the latest event is a sufficient summary for the inspection decision. The optimal policy must therefore be defined as a function of the time of the latest event. I will derive the optimal policy conditional on the latest (henceforth n-th) event occurring at time x. The policy is followed until either a new event occurs and the stand-by operates (in which case the policy is updated) or a new event occurs and the stand-by does not operate (in which case the problem terminates with additional cost c) or a failure of the stand-by is detected (in which case, as before, the problem terminates at no further cost).

If $X_n = x$ and the stand-by worked at x, let $\tau_0 = x$; the risk is then:

$$R(\boldsymbol{\tau}) = \frac{1}{\bar{F}(x)} \sum_{i=0}^{\infty} \int_{\tau_i}^{\tau_{i+1}} \Big\{ c[G(\tau_{i+1} - x) - G(y - x)] + \int_0^{y-x} \rho(x+u) g(u) du \quad (25)$$
$$+ k \left[(i+1)[\bar{G}(\tau_{i+1} - x)] + \sum_{j=0}^{i} j[G(\tau_{j+1} - x) - G(\tau_j - x)] \right] \Big\} f(y) dy,$$

where $\rho(x)$ is the risk as a function of x, and $\bar{F} = 1 - F$, $\bar{G} = 1 - G$. As shown in Parmigiani [22], the optimality conditions for this case are:

$$\frac{G(\tau_{i+1} - x) - G(\tau_i - x)}{g(\tau_i - x)} + \frac{k}{c} \frac{\bar{G}(\tau_{i+1} - x)}{g(\tau_i - x)} = \frac{F(\tau_i) - F(\tau_{i-1})}{f(\tau_i)} - \frac{k}{c} \frac{\bar{F}(\tau_{i-1})}{f(\tau_i)} \quad (26)$$

for $i = 1, 2, \ldots$.

An interesting special case obtains when the events that require the stand-by to operate are generated by a Poisson process with rate μ. Then, if it is optimal to inspect, the optimality conditions give an explicit recursive solution of the form:

$$\tau_{i+1} - \tau_i = -\frac{1}{\mu} \log \left[1 - \frac{\mu(c-k)}{c+k} \frac{F(\tau_i) - F(\tau_{i-1})}{f(\tau_i)} + \frac{\mu k}{c+k} \frac{\bar{F}(\tau_i)}{f(\tau_i)} - \frac{k}{c+k} \right] \qquad i = 1, 2, \ldots$$

Assume further that F is also exponential with rate $\lambda < \mu$, and that inspecting at the optimum is preferred to not inspecting. Then the optimal schedule is periodic and the interval is the value of δ solving:

$$\delta = -\frac{1}{\mu} \log \left[1 - \frac{\mu(c-k)(e^{\lambda\delta} - 1)}{\lambda(c+k)} + \frac{\mu k}{\lambda(c+k)} - \frac{k}{c+k} \right] \qquad i = 1, 2, \ldots$$

It is easy to show that under the above assumptions the solution must be unique.

7. UNCERTAINTY ABOUT THE FAILURE DISTRIBUTION

The assumption of known failure distribution is not always met in practice. Frequently there is knowledge about the parametric family of the distribution, but there may be uncertainty about parameter values. In such cases, a natural generalization consists of assigning a prior distribution to the unknown parameters. This way a marginal failure distribution can be computed by the law of total probabilities. When the loss depends only on the waiting time to failure and the strategy adopted, as in the models studied here, the risk can be computed based on the marginal failure distribution alone, and the theory applies directly to it.

There are various aspects, however, that need special consideration in the case of unknown parameters. Of special importance is the following: when a parametric model is used, the marginal failure distribution results from a mixture. The optimal policy arising from a mixture can be very different from the policies that arise from the mixed distributions when assumed known, due to the learning taking place. For example, neither the class of densities with increasing failure rate (IFR) nor the class of log-concave densities is closed under mixture (see Barlow, Marshall and Proschan [4]). So a gamma failure distribution with parameters $A > 1$ and $B > 0$, is log-concave if A and B are known, but not necessarily so if they are not. Consequently, the theory of Barlow, Hunter and Proschan [3] may apply to every element in the family of conditional failure distributions, and not to the resulting marginal failure distribution. This provides motivation for extending the study to classes of distributions that arise as mixtures of failure distribution commonly used in practice.

In particular, the class of log-convex densities (or Sign Reverse Regular densities of order 2) has practical relevance because it contains all marginal failure densities resulting from exponential or Pareto distributions with unknown parameters. It can be shown (see [20] for details) that log-convex failure densities yield increasing interchecking times. In the case of the exponential distribution with unknown parameter this has an obvious intuitive interpretation. Suppose the optimal first inspection is at time τ_1. If the system is still working, it is known from the memoryless property that no deterioration took place; on the other hand there is new information about the value of the parameter: observing no failures in a positive interval of time will induce more optimistic expectations about the lifetime of the device, and make it optimal to wait longer before the next inspection.

i	τ_i	$\tau_i - \tau_{i-1}$	$F(\tau_i)$	$[F(\tau_i) - F(\tau_{i-1})]/[1 - F(\tau_{i-1})]$
1	2.42	2.42	0.43	
2	5.52	3.09	0.64	0.37
3	9.30	3.78	0.76	0.33
4	13.79	4.48	0.83	0.29
5	18.98	5.18	0.87	0.26
6	24.86	5.88	0.90	0.23
7	31.43	6.56	0.92	0.21
8	38.66	7.23	0.94	0.19
9	46.53	7.87	0.95	0.18
10	55.03	8.49	0.96	0.17
15	105.53	11.03	0.98	0.11
20	165.09	12.30	0.98	0.08

Table 1: Optimal policy at $k = c = 1$, $a = 1.2$ and $b = 4$.

I will illustrate this by an example. Let Y given $\Lambda = \lambda$ be exponential with mean

$1/\lambda$ and let Λ have a gamma distribution with parameters a and b. Then the marginal distribution of Y is Pareto with parameters a and b.

Consider first a situation where $k = c = 1$ and the prior parameters are $a = 1.2$ and $b = 4$. Loosely speaking, this corresponds to the observation of 1.2 failures in 4 units of time of random variables exchangeable to Y, prior to the inspection —a rather vague prior knowledge. The resulting optimal policy is in Table 1. The expected number of inspections associated with this policy is 2.97, whereas the expected detection time is 12.92. Note that if the device is still functioning after 20 inspections, the posterior predictive distribution of Y is Pareto with parameters 1.2 and 4+165.1. This represents a substantial change compared to the prior predictive distribution. The change reflects in the value of the optimal time before the next check which changes from an initial value of 2.4 to 12.3. Note, however, that the value of δ_i tends to stabilize as i gets large. The results are very sensitive to small changes in the values of the prior parameters: in particular $dR/da = 13.065$ and $dR/da = .823R$ and $dR/db = -1.589$ and $dR/db = .1R$. This is typically the case when the value of a approaches 1, leading to an improper prior predictive density.

i	τ_i	$\tau_i - \tau_{i-1}$	$F(\tau_i)$	$[F(\tau_i) - F(\tau_{i-1})]/[1 - F(\tau_{i-1})]$
1	1.46	1.46	0.61	
2	2.96	1.50	0.84	0.60
3	4.51	1.54	0.93	0.59
4	6.09	1.58	0.97	0.59
5	7.72	1.62	0.98	0.58
6	9.39	1.66	0.99	0.57
7	11.09	1.70	0.99	0.57
8	12.85	1.75	0.99	0.56
9	14.64	1.79	0.99	0.55
10	16.47	1.83	0.99	0.55
15	26.28	2.04	0.99	0.52
20	37.17	2.26	1.00	0.49

Table 2: Optimal policy at $k = c = 1$, $a = 20$ and $b = 30$.

A different situation arises when the prior distribution is based on a larger amount of knowledge. Consider the case $a = 20$, $b = 30$, roughly corresponding to the observation of 20 failures in 30 units of time for random variables exchangeable to Y, prior to the inspection. Then the optimal policy is in Table 2. The expected number of inspections is now 1.36 and the expected detection time is 1.24. The effect of the additional information on the interchecking times is still present but it is much less pronounced. Also, only 5 inspections fall within the 98th percentile, compared to the 20 of the previous case. Finally, the results are not very sensitive to changes in the values of the parameters: $dR/da = .083$ and $dR/da = .002R$, whereas $dR/db = -.002$ and $dR/db = .0006R$.

i	τ_i	$\tau_i - \tau_{i-1}$	$F(\tau_i)$	$[F(\tau_i) - F(\tau_{i-1})]/[1 - F(\tau_{i-1})]$
1	0.51	0.51	0.28	
2	1.03	0.52	0.49	0.28
3	1.56	0.52	0.63	0.28
4	2.09	0.53	0.74	0.28
5	2.63	0.53	0.81	0.28
6	3.17	0.54	0.86	0.28
7	3.71	0.54	0.90	0.27
8	4.26	0.55	0.93	0.27
9	4.82	0.55	0.94	0.27
10	5.38	0.55	0.96	0.27
15	8.25	0.58	0.99	0.26
20	11.23	0.60	0.99	0.25
25	14.34	0.63	0.99	0.24
30	17.56	0.65	0.99	0.24

Table 3: Optimal policy at $k = c = 10$, $a = 20$ and $b = 30$.

To illustrate the effect of changes in the costs, let $c = 10$. The new optimal policy is in Table 3. For this policy the expected number of inspections is 2.57 and the expected detection time is 1.24. The sensitivity of the risk is larger than in the previous case. In relative terms: $dR/da = .041R$ and $dR/db = .001R$. Since the optimal policy depends only on the ratio of costs, the same effect would have resulted from the appropriate change in k.

8. COMMENTS

In this paper I reviewed some recent developments in the theory of optimal scheduling of inspections in reliability, emphasising the essential role of recursive dynamic methods and of Bayesian methods. The problem of optimally scheduling inspections arises whenever there is the need to acquire information about a stochastic process at successive points in time, and observations are costly. Therefore, useful extensions of the methodologies reviewed here can be attempted in several directions, depending on the field of application.

Of importance in many areas would be relaxing the assumption that the outcome of an inspection is a binary response. Alternatively, the outcome may be a measure of deterioration. Then each inspection improves knowledge in two ways. First, based on the history of the process, one can update the marginal posterior distribution of the time to failure: this gives information about the speed at which the process moves toward failure. Second, the observation of the actual state gives the distance of the system from the failure. An optimal schedule will then have to learn sequentially from this information.

A second group of interesting problems —representing a significant departure from the loss functions discussed in this paper— arises when the objective of the decision problem is to keep a system as close as possible to a certain target by acting on a control variable. Then the optimal schedule has to balance the costs of frequent inspections and perhaps those associated with the use of the control variable, with the costs deriving from the process being off target. Some theoretical results regarding the optimal inspection and control of a Brownian motion, by Anderson and Friedman [1], constitute an important step in this direction.

The area of quality control is also rich of fruitful applications for inspection problems. In particular, substantial attention has been recently devoted to the design of Shewart charts with sampling frequency dependent on the current state of the process. Bather [6] gives theoretical results and identifies open problems in establishing a connection between optimal inspection and quality control.

Finally, the solution of complex scheduling problem requires progress in computational methods. As it is clear from the examples discussed in this paper, a general way of solving inspection problems is based on recursive optimality equations reducing the optimization to a small number of dimensions. This pattern appears in a variety of applications where the complexity of the problem does not allow for the elegant theory based on log-concave densities to apply. Examples include the uncertain failure distribution case and the optimal design problems for life testing, as well as applications to other areas not discussed here such as medical screening problems, where the failure rate is almost never monotone. This suggest that it is of use to develop a general approach to the numerical optimization of initial conditions in inspection problems.

Efficient algorithms for the numerical evaluation of the optimal policy can be obtained by combining the recursive dynamics with an approximate planning horizon approach, where only the horizon necessary to obtain a good accuracy is considered. The method is based on truncating the evaluation of the risk function at a large finite horizon. For fixed horizon, the exact dynamic programming solution of the finite horizon problem involves the laborious determination of the optimal number of inspections, which is finite. This is avoided by utilizing the recursive optimality condition of the infinite horizon case to evaluate the optimal continuation for a given initial inspection τ_1. This algorithm has desirable convergence properties and, in applications, it requires only a small to moderate horizon to give very accurate results.

Acknowledgements

M. H. DeGroot, J. B. Kadane and N. Singpurwalla provided useful suggestions. Work partially supported by the Statistical Center for Quality Improvement, Carnegie Mellon University, Pittsburgh.

REFERENCES

1. ANDERSON, R. F. and A. FRIEDMAN: (1977) "Optimal Inspections in a Stochastic Control Problem Problem with Costly Observations," *Mathematics of Operations Research* **2** 155–190.

2. BARLOW, R. E. and L.C. HUNTER: (1960) "Optimum Checking procedures," *Proceedings of the Seventh National Symposium on Reliability and Quality Control* **9** 73.
3. BARLOW, R. E., L.C. HUNTER and F. PROSCHAN: (1963) "Optimum Checking procedures," *Journal of the Society for Industrial and Applied Mathematics* **4** 1078–1095.
4. BARLOW, R. E., A. W. MARSHALL and F. PROSCHAN: (1963) "Properties of Probability Distributions with Monotone Hazard Rate," *Annals of Mathematical Statistics* **34** 375–389.
5. BARLOW, R. E. and F. PROSCHAN: (1965) *Mathematical Theory of Reliability.* Wiley, New York.
6. BATHER J.A.: (1976) "A Control Chart Model and a Generalised Stopping Problem for Brownian Motion," *Mathematics of Operations Research* **1** 209–224.
7. BRENDER, D. M.: (1963) "Surveillance Methods for Recurrent Events," *IBM Watson Research Center Report* .
8. COLEMAN J. and I. ABRAMS: (1962) "Mathematical Models for Operational Readiness," *Journal of the Operational Research Society* **1** 126–139.
9. DERMAN, C.: (1961) "On Minimax Surveillance Schedules," *Naval Research Logistics Quarterly* **8** 415–419.
10. KAIO, N. and S. OSAKI: (1984) "Some Remarks on Optimum Inspection Policies," *IEEE Transactions on Reliability* **R-33, 4** 277–279.
11. KAWAI, H.: (1984) "An Optimal Inspection and Replacement Policy of a Markovian Deterioration System," *Stochastic Models in Reliability Theory* (S. Osaki and Y. Hatoyama, ed.) 177–186. Springer-Verlag, Berlin.
12. LUSS, H.: (1976) "Maintenance Policies when Deterioration Can Be observed by Inspection," *Operations Research* **24** 359–366.
13. MUNFORD, A. G. and A. K. SHAHANI: (1972) "A Nearly Optimal Inspection Policy," *Journal of the Operational Research Society* **23** 851–853.
14. MUNFORD, A. G. and A. K. SHAHANI: (1973) "An Inspection Problem for the Weibull Case," *Journal of the Operational Research Society* **24** 453–458.
15. NAKAGAWA, T.: (1980) "Optimum Inspection Policies for a Standby Unit," *Journal of the Operational Research Society of Japan* **23** 13–26.
16. NAKAGAWA, T. and K. YASUI: (1980) "Approximate Calculations of Optimal Inspection Times," *Journal of the Operational Research Society* **31** 851–853.
17. OHNISHI, M. and H. MINE: (1984) "An Optimal Inspection and Replacement Policy under Incomplete State Information: Average Cost Criterion," *Stochastic Models in Reliability Theory* (S. Osaki and Y. Hatoyama, ed.) 187–197. Springer-Verlag, Berlin.
18. PARMIGIANI G.: (1990) "Optimal Scheduling of Fallible Inspections," *SCQI Technical Report n. 4, CMU* Pittsburgh, PA.
19. PARMIGIANI G.: (1990) "An Optimal Inspection and Replacement Plan with Time-dependent Failures and Fallible Tests," *SCQI Technical Report n. 6, CMU* Pittsburgh, PA.

20. PARMIGIANI G.: (1991) "Inspecting for Failures While Learning About the Failure Rate," *ISDS Discussion Paper*, Duke U., Durham, NC.
21. PARMIGIANI G.: (1992) "Optimal Screening Ages," *JASA*, Forthcoming.
22. PARMIGIANI G.: (1992) "Inspection and Replacement of Stand-By Systems," *ISDS Discussion Paper*, Duke U., Durham, NC.
23. SAVAGE, I. R.: (1956) "Cycling," *Naval Research Logistics Quarterly* **3** 163–175.
24. SENGUPTA, B.: (1980) "Inspection Procedures when Failure Symptoms are Delayed," *Operations Research* **28** 768–776.
25. SENGUPTA, B.: (1982) "An Exponential Riddle," *Journal of Applied Probability* **19** 737–740.
26. SHAHANI, A. K. and D. M. CREASE: (1977) "Towards Models of Screening for Early Detection of Disease," *Advances in Applied Probability* **9** 665–680.
27. STOKEY N. L. and R. E. LUCAS: (1989) *Recursive Methods in Economic Dynamic.* Harvard University, Cambridge.
28. THOMAS, L. C., P. A. JACOBS and P. GAVER: (1987) "Optimal Inspection Policies for Standby Systems," *Communications in Statistics: Stochastic Models* **3** 259–273.
29. WEISS, G. H.: (1963) "Optimal Periodic Inspection Programs for Randomly Failing Equipment," *Journal of Research of the National Bureau of Standard* **67B, 4** 223–228.
30. ZUCKERMAN, D.: (1980) "Inspection and Replacement Policies," *Journal of Applied Probability* **17** 168–177.

ADVANCES IN RELIABILITY
Edited by Asit P. Basu

A Bayesian perspective on the design of accelerated life tests

NICHOLAS G. POLSON

University of Chicago

SUMMARY

A Bayesian approach to optimal design issues in accelerated life tests is outlined. The approach focuses on the use of utility functions based on the Shannon information of the posterior distribution of the parameter of interest as a pragmatic mechanism for generating useful design criteria. Within the Bayesian hierarchical linear model we show how this approach can be used to characterise previous design criteria. Several issues of design in accelerated life tests are considered in this framework; for example, censoring and step-stress.

Some key words: Bayesian inference, Design, Shannon information, Accelerated life tests, Decision theory, Nonlinear models.

1. INTRODUCTION

The objective of this paper is to address issues arising in the design of accelerated life tests from a Bayesian perspective. The approach adopts utility functions motivated by the Shannon information of the posterior distribution for the quantity of interest. In many applications of life testing, the test process requires an unacceptably long time period for its completion if the test is carried out at the use stress level, x_u. An accelerated life test provides information about the lifetime, T_u, at x_u by testing units at higher than usual levels of stress, thus shortening the time to completion, and then extrapolating to the use stress level. See Chernoff (1962), Mann, Shafer and Singpurwalla (1974) and Lawless (1982) for statistical analysis and design of life test problems. In an accelerated life test, the lifetime T of an item is assumed to have distribution $F(t|\theta)$, where the unknown parameter follows a physical law of the form $\theta = f(x, \beta)$. Here x is the stress at which the item is tested and β is a vector of parameters describing the relationship between θ and x. Typical error distributions are: the lognormal, Weibull, extreme value and log-logistic, and for the physical law: the power, Arrhenius and Erying relationships (see, Mann, Shafer and Singpurwalla, 1974, Sethuraman and Singpurwalla, 1982). Of course, care must be exhibited when extrapolating any statistical model (for example, when predicting T_u) as the resulting distribution will be highly model dependent. We assume throughout that the experimenter can test in the region

$[x_l, x_h]$ (where $x_u < x_l < x_h$ are prespecified) and that the physical law relating stress to lifetime holds throughout the region $[x_u, x_h]$. Testing below x_l takes too long and above x_h the model is in doubt. The experimenter has n items to tests and k stresses will be chosen with n_j items placed on test at the jth stress x_j. Common design problems include the choice of number of levels k, the values of the stresses x_j, the amount of censoring at each level before stopping the tests.

Lindley (1956) proposed the use of Shannon information gain between the prior and the posterior as a tool in Bayesian inference. In the context of design we will see that this is equivalent to maximising the negative expected posterior entropy. Stone (1959a) and Smith and Verdinelli (1980) discuss the application of Shannon information to the linear Bayes hierarchical model. There are many other applications of Shannon information to related design problems; for example, to sample size calculations (Polson, 1990) and to optimal allocation problems (Brooks, 1980, 1987).

The paper is outlined as follows. Section 2 discusses the Bayesian approach to design and the interpretation of Shannon information as an expected utility from the decision problem of reporting beliefs. Section 3 discusses different design strategies in accelerated life testing. A hierarchical Bayesian model is proposed for the design and analysis of accelerated life tests Within this framework, it is shown how the classical design criteria of $D-$ and $A-$ optimality are special cases of the Shannon information approach. The applications of Shannon information to censored data, step stress problems and nonlinear models are discussed.

We will not discuss computational issues involved in constructing optimal designs. For a general discussion, based on Fréchet derivatives, see Whittle (1973) and Chaloner and Larntz (1989). The latter derives Bayes designs for logistic regression. It is in the context of nonlinear models where Bayes designs, which incorporate the use of prior information, can differ dramatically from the classical approach of using locally optimal designs. We illustrate this with an example in section 3. With specific regard to Bayesian design of accelerated life tests see Verdinelli, Polson and Singpurwalla (1990), Chaloner and Larntz (1990) and Menzefricke (1991).

2. Bayesian design and Shannon information

Let θ denote the parameter of interest. The Bayesian approach to statistical modelling and design requires three ingredients: the specification of a utility function $U(\theta, d)$ (representing the experimenter's gain if he selects decision d when θ is the state of nature); the likelihood function (denoted by $f(Y|\theta, X)$, corresponding to the joint beliefs about the vector of data Y given a specified value of θ and design X); and the experimenter's *a priori* beliefs about θ (denoted by $p(\theta)$). All of these specifications are subjective and contribute to the calculation of the optimal design in the following way: first, for a specified design X, the optimal decision d (in light of data Y) attains $\sup_{d \in D} E_{\theta|Y,X}\left(U(\theta, d)\right)$ where D denotes the space of decisions and $E_{\theta|Y,X}$ denotes expectation with respect to the posterior $p(\theta|Y, X)$. Secondly, the preposterior expected utility, under a design X, of performing the experiment and then acting optimally, is given by

$$E_{Y|X}\left(\sup_{d \in D} E_{\theta|Y,X}\left(U(\theta, d)\right)\right) \tag{1}$$

where $E_{Y|X}$ denotes expectation with respect to $p(Y|X) = \int f(Y|\theta, X)p(\theta)d\theta$. The experimenter selects the design X optimally by maximising his expected utility.

In this paper we will consider a particular decision problem. Rather than adopting the usual decision problem of estimation of parameters under quadratic loss, we will quantify the aim of the analysis as that of reporting beliefs about a random variable of interest. Here the decision space D is given by $\mathcal{P}$, the space of probability distributions. For a suitable utility function for reporting probability densities we adopt the logarithmic function, that is, we assume that $U(\theta, d) = \log d(\theta)$ where $d(\cdot) \in \mathcal{P}$. Then the experimenter is honest, in the sense that he reports his true beliefs, and the optimal decision d is given by $p(\theta|Y, X)$ with associated utility

$$\sup_{d \in D} E_{\theta|Y,X}\left(U(\theta, d)\right) = \int p(\theta|Y, X) \log p(\theta|Y, X) d\theta \; . \tag{2}$$

Therefore, under a logarithmic utility, the design criterion (1) becomes the maximisation of the negative expected posterior entropy, that is to select X via

$$\max_X E_{Y|X}\left(\int p(\theta|Y, X) \log p(\theta|Y, X) d\theta\right) \; . \tag{3}$$

As an aside, we note that even though this functional is not invariant with respect to 1-1 reparameterisations, the optimal design is invariant. This can be shown as follows, first, subtract the prior entropy from the design criterion (3). This does not affect the design and the criterion then becomes that of maximising (with respect to X) the functional

$$E_{Y|X}\left(E_{\theta|Y,X}\left(\log \frac{p(\theta|Y, X)}{p(\theta)}\right)\right) \; .$$

Secondly, note that this is precisely the Shannon information gain between the prior $p(\theta)$ and the posterior $p(\theta|Y, X)$, the measure originally proposed by Lindley (1956). This measure is invariant with respect to 1-1 reparameterisations and hence the design obtained by (3) is also invariant.

3. Design issues in accelerated life tests

We now consider four design scenarios in accelerated life testing: a Bayesian hierarchical model for uncensored data (Verdinelli, Polson and Singpurwalla, 1990); modelling censored data (Escobar and Meeker, 1986); step stress experiments (DeGroot and Goel, 1986, Shaked and Singpurwalla, 1983) and binary data (Abdelbasit and Plackett, 1983).

3.1 Hierarchical Model

Suppose that the experimenter will place n_i items at stress level x_i, where i indexes the k subexperiments. It is common to suppose that after a suitable transformation of the lifetime T_{ij} the model follows a linear relationship of the form

$$Y_{ij} = \mu_i + \sigma_i \epsilon_{ij} \tag{4}$$

where Y_{ij} denotes the transformed lifetime, $\mu_i = f(x_i)^T\beta$ where $f(x_i)^T = (f_1(x_i), \ldots, f_k(x_i))^T$ for given functions f_j and where ϵ_{ij} has a specified distribution $F(\cdot)$. Therefore we have a linear model

$$Y = X\beta + \epsilon, \tag{5}$$

for a suitably defined design matrix X and multivariate error, ϵ. For the purpose of illustration, we suppose that lifetimes are lognormal, corresponding to $\epsilon \sim N(0, \Sigma)$, and where $\Sigma = diag(\sigma_1^2 I_{n_1}, \ldots, \sigma_k^2 I_{n_k})$ where σ_j^2 is known and I_{n_j} is the $n_j \times n_j$ identity matrix. To complete the Bayesian specification we assume that *a priori*

$$\beta \mid \beta_0, \Sigma_0 \sim N(\beta_0, \Sigma_0), \tag{6}$$

where β_0, Σ_0 are specified. It is well known (Lindley and Smith, 1972) that, under this hierarchical Bayes model, the posterior distribution of $\beta|\mathbf{y}, X$ is given by

$$\beta \mid \boldsymbol{y}, X \sim N(Bb, B), \tag{7}$$

where $B = (X^T\Sigma^{-1}X + \Sigma_0^{-1})^{-1}$ and $b = (X^T\Sigma^{-1}\boldsymbol{y} + \Sigma_0^{-1}\boldsymbol{\beta}_0)$. The marginal distribution depends on X, β_0, Σ_0 and Σ, and is given by

$$\mathbf{y} \mid X, \Sigma \sim N(X\beta_0, X\Sigma_0X^T + \Sigma)\,. \tag{8}$$

These distributions allow the experimenter to compute the preposterior expected utility (3). It can be shown that the negative expected posterior entropy of $p(\beta|\mathbf{y}, X)$ is an increasing function of

$$|(X^TX)^{-1}\sigma^{-2} + \Sigma_0^{-1}|^{-1} \tag{9}$$

where $|\cdot|$ denotes determinant. Stone (1959a) contains results pertaining to the number of support points required to maximise this functional with respect to X. If β is a k dimensional parameter of interest, then the experimenter needs at most $\frac{1}{2}k(k+1)$ support points for the design. This Bayesian result parallels that of the classical development of optimal design in Kiefer and Wolfowitz (1959), Elfving (1952). See Chaloner (1984) for results on Bayesian design of linear models. We now show how the classical criteria of $D-$ and $A-$ optimality can be obtained as special cases of the above scenario. See, also Stone (1959b).

First, consider the case where *a priori* beliefs for the parameters are weak, that is $\Sigma_0^{-1} \to 0$. We see from (9) that the criterion of maximising negative expected posterior entropy is equivalent to $D-$optimality, that is

$$\max_X |X^TX|\,.$$

Secondly, consider the case when the information in the experiment is weak for the parameters (possibly due to dominating prior information), in the sense that $\Sigma_0(X^TX)^{-1}\sigma^{-2}$ is small, we can expand (9) using the matrix result:

$$|I + An^{-\frac{1}{2}}| = 1 + n^{-\frac{1}{2}}tr(A) + \frac{1}{2}n^{-1}\left((tr(A))^2 - tr(A^2)\right) + o(n^{-1})$$

for any matrix A as $n \to \infty$. Therefore we have,

$$|(X^TX)^{-1}\sigma^{-2} + \Sigma_0^{-1}|^{-1} \approx |\Sigma_0|(1 - tr(\Sigma_0(X^TX)^{-1}\sigma^{-2})$$

and the Bayesian design criterion becomes that of maximising

$$-|\Sigma_0|tr(\Sigma_0(X^TX)^{-1}\sigma^{-2})$$

In the case where $\Sigma_0 = I$, this becomes the classical criterion of A−optimality.

Therefore, the hierarchical Bayesian model, together with utility functions based on Shannon information, provides a unified perspective on possible design strategies.

3.2 Censoring

The simplest form of censoring (type I) occurs when the experimenter terminates the experiment after a prespecified length of time. A more common situation in accelerated life tests occurs when the censoring mechanism (type II censoring) is performed as follows: items are subjected to k levels of stresses, n_j items are placed at stress x_j in subexperiment j, and the experimenter decides to stop when he observes that a prespecified number m_j of the n_j items have failed.

It is common to assume that after a suitable transformation, that there is a linear structure relating the (transformed) failure times $Y_j = (Y_{1j}, \ldots, Y_{n_j j})$ to the underlying stresses, that is

$$Y_j = x_j^T\beta + \epsilon_j$$

where ϵ_j is an error distribution, $F(\cdot)$, with zero mean and (unknown) scale σ (assumed to be constant over subexperiments). Under this design all the information in subexperiment j about β is contained in the first m_j order statistics (from the sample of n_j on test). We can deduce that these m_j uncensored observations also follow a linear model of the form

$$Y_j = (1_j x_j^t)\beta + \sigma Z_j + \sigma\nu_j$$

where 1_j is a column vector of ones, Z_j is the vector of expected values of the first m_j order statistics in a sample of size n_j from distribution F and the error term ν_j has mean zero and variance-covariance matrix V_j corresponding to the first m_j order statistics in a sample of size n_j from distribution F. Again we can write this as a linear model

$$Y = X\beta + \sigma Z + \sigma\nu$$

where ϵ has mean zero and covariance matrix $V = diag(V_1, \ldots, V_k)$. In the normal linear model, the Shannon information criterion reduces to the minimisation of the logarithm of the determinant of the posterior variance. Mimicing this we see that a reasonable design criterion is to maximise the functional

$$|\Sigma_0^{-1} + \sigma^{-2}X^TV^{-1}X|\,.$$

If prediction is the aim of the analysis and the predictive distribution $p(Y_u|\mathbf{y}, X)$ is reported, where $Y_u = x_u^T\beta + \epsilon$, then a reasonable design criterion would be to minimise the functional

$$x_u^T(\Sigma_0^{-1} + \sigma^{-2}X^TV^{-1}X)^{-1}x_u = tr((\Sigma_0^{-1} + \sigma^{-2}X^TV^{-1}X)^{-1}x_u x_u^T)$$

This is equivalent to the $c-$optimality criterion of Chaloner (1984) and the design criterion used by Menzefricke (1991). The optimal design consists of determining: the number of distinct accelerated levels and their stress values x_j; the degree of censoring at each level, that is $\lambda_j = m_j/n_j$. Therefore, the design criterion is maximised with respect to these variables under a cost constraint of the form $\sum_{i=1}^{k} c_j n_j \leq c$ where c_j is the unit cost of taking an observation at stress level x_j and c is an overall cost parameter.

3.3 Step Stress

Miller and Nelson (1983), DeGroot and Goel (1986), LuValle, Welsher and Svoboda (1988) consider the design and analysis of step-stress accelerated life tests. Here the experimenter subjects items to a standard test enviroment and then if item i has not failed by some prespecified time x_i, then it is put under a higher level of stress and the test is continued. Let T_i denote the would be lifetime of item i in the standard enviroment, then the total lifetime Y_i under this step-stress experiment is given by

$$Y_i = \begin{cases} T_i & T_i \leq x_i \\ x_i + \alpha(T_i - x_i) & T_i > x_i \end{cases} .$$

The x_i's are design parameters, called tampering points, and α is called the tampering coefficient. The experimenter can control $X = (x_1, \ldots, x_n)$ and he wishes to do so optimally.

Suppose that lifetimes follow an exponential model that has known mean θ_0. Let α be the parameter of interest and let $p(\alpha)$ be inverse Gamma with known hyperparameters, denoted by $IGa(\alpha_0, \beta_0)$. DeGroot and Goel (1986) show that the posterior for $\beta = 1/\alpha$ is given by

$$\beta|\mathbf{y}, X \sim Ga\left(\alpha_0 + m, \beta_0 + \sum_{i \in A}(Y_i - x_i)\right) \tag{10}$$

where m denotes the number of tampered observations with indices lying in the set A. The design based on the Shannon information criterion can sometimes be analytically intractable. Here, the optimal design criterion (3) is given by

$$\max_X E_{\mathbf{y}|X}\left(\int p(\beta|\mathbf{y}, X) \log p(\beta|\mathbf{y}, X) d\beta\right) .$$

Now, for a Gamma distribution $Ga(\alpha, \lambda)$, the entropy is given by

$$\log \lambda - \log \Gamma(\alpha) - \alpha + (\alpha - 1)\frac{d}{d\alpha} \log \Gamma(\alpha)$$

where $\Gamma(\alpha)$ is the gamma function. Therefore, the entropy of the posterior (10) is a complicated functional of m and Y_i, moreover, the preposterior expected utility is required. Therefore, this example, under a logarithmic utility, does not lend itself to a simple analytical solution. For designs based on quadratic loss functions, see DeGroot and Goel (1986).

3.4 Binary data and Nonlinear Models

A standard statistical problem in life testing, especially in dose-response experiments, occurs when failures are recorded on the n items on test where the probability of failure has a distribution dependent on the level of stress (dose).

First, we note a useful asymptotic approximation to the design criterion (3) for a nonlinear model. Consider the model $y_i = \eta(x_i, \theta) + \epsilon_i$. Here $\eta(\cdot, \cdot)$ is a given response surface, x_i are design points and ϵ_i are independent and identically distributed with density $f(\cdot)$. Let X denote the set of design points, $\eta_i = \eta(x_i, \theta)$ and let $I_f = \int (f')^2/f$ be Fisher information. Let $I_i(\theta)$ denote Fisher information for the ith item, then $I_i(\theta) = I_f \frac{\partial \eta_i}{\partial \theta}^T \frac{\partial \eta_i}{\partial \theta}$. Let $I_X(\theta) = \sum_{i=1}^n I_i(\theta)$ denote the full Fisher information. Under suitable regularity conditions, Polson (1990) shows that as $n \to \infty$,

$$E_{Y|X}\left(\int p(\theta|Y,X)\log p(\theta|Y,X)d\theta\right) = E_\theta\left(\log|\sum_{i=1}^n \frac{\partial \eta_i}{\partial \theta}^T \frac{\partial \eta_i}{\partial \theta}|^{\frac{1}{2}}\right) + \frac{k}{2}\log\left(\frac{I_f}{2\pi e}\right) + o(1) \tag{11}$$

where E_θ denotes expectation with respect to the prior $p(\theta)$. Therefore, the design criterion of maximising the negative expected posterior entropy is approximately equivalent to the maximisation of the expected logarithm of the determinant of Fisher information, where the expectation is taken with respect to the prior beliefs about θ. For an application of this criterion to designing the barriers of a random walk, see Parmigiani and Polson (1990). We now consider an example of the latter criterion in the context of life testing.

Suppose that the probability that an item fails at stress x is $F(x;\theta)$ for a specified distribution $F(\cdot)$. The experimenter places n_i items on test at the ith stress level x_i and observes that Y_i fail. We assume that $Y_1, \ldots, Y_k$ are independent and are distributed as $Y_i \sim Bin(n_i, F(x_i;\theta))$. For the purpose of illustration, consider the exponential model,

$$F(x_i;\theta) = 1 - \exp(-\beta x_i) \ x_i > 0 \ .$$

Now, suppose that the experimenter places all n items at stress x. Under design X, the Fisher information about $\log \beta$, denoted by $I_X(\log \beta)$, is given by

$$I_X(\log \beta) = \frac{n(\beta x)^2}{e^{x\beta} - 1} \ . \tag{12}$$

Using the design criterion based on the expected logarithm of Fisher information, and differentiating, we can that the optimal design stress x^* satisfies

$$E_\beta\left(\frac{\beta e^{x^*\beta}}{e^{x^*\beta} - 1}\right) = \frac{2}{x^*}$$

where E_β denotes expectation with respect to the *a priori* beliefs $p(\beta)$. A simple approximation, involving only the *a priori* mean of β, is given by $x^* = 2/E(\beta)$. This solution is to be contrasted with the classical approach that first notes that (12) is maximised when $\beta x^* \approx 1.59$ and then selects $x^* = 1.59/\beta_0$ where β_0 is an initial estimate for β (see Abdelbasit and Plackett, 1983). It is interesting to note that Fisher (1966, p.130) avoids such *ad hoc* procedures by proposing the use of an auxiliary experiment in order to determine a fiducial probability distribution for the parameter β. Then the design is selected to maximise the expected Fisher information, where the expectation is taken with respect to the fiducial distribution.

Finally, we note that in certain instances the experimenter can reduce the dependence of the expected utility on the prior by carefully selecting the design points (maybe after a reparameterisation). This clearly has application to generating robust designs. For example, in the above scenario, if instead of experimenting at one stress level x, the experimenter places n items geometrically at design points $X = \{ca^{-t}|t = 0, 1, \ldots\}$ where $c, a > 0$, then the information about $\log\beta$ takes the form (Abdelbasit and Plackett, 1983),

$$I_X(\log\beta) = n\sum_{t=0}^{\infty}\frac{(\beta ca^{-t})^2}{e^{\beta ca^{-t}} - 1} .$$

When a is close to one, Fisher (1966) derives the approximation $I_X(\log\beta) \approx n\pi^2/6\log a$ and thus the negative expected posterior entropy is approximately independent of the *a priori* specification of $p(\beta)$.

4. Discussion

The principle of maximum expected utility provides a formal approach to the design of experiments. The optimal design, therefore, is a functional of the assumed utility, model and prior specifications. As a mechanism for generating useful design criteria, a logarithmic utility for the decision problem of reporting a posterior for the random variable of interest is proposed. Many life test problems can, after suitable transformation, be modelled within the hierarchical Bayes linear model and in this case the logarithmic utility provides a unifying perspective on D- and A- optimal designs. For a nonlinear model, the expected utility (that is, the negative expected posterior entropy) is asymptotically equivalent to the expected logarithm of the determinant of Fisher information. This was applied to the design of life tests that result in binary data and the Bayesian and classical approaches were contrasted. Applications to censored data and step stress life tests were also explored. Clearly, there are many other directions to explore including; computational issues, elicitation of costs, non-normal errors and accounting for unknown variances.

References

[1] Abdelbasit, K.M. and Plackett, R.L. (1983). Experimental Design for Binary Data. *J. Amer. Stat. Assoc.*, **78**, 90-98.

[2] Brooks, R.J. (1980). On the relative efficiency of two-paired experiments. *J.R. Statist. Soc.*, **B**, 42, 186-191.

[3] Brooks, R.J. (1987). Optimal allocation for Bayesian inference about an odds ratio. *Biometrika*, 74, 196-199.

[4] CHALONER, K. (1984). Optimal Bayesian Experimental Designs for linear models. *Ann. Statist.*, 12, 283-300.

[5] CHALONER, K. and LARNTZ, K. (1989). Optimal Bayesian design applied to logistic regression experiments. *J. Stat. Plan. and Inf.*, 21,191-208.

[6] CHALONER, K. and LARNTZ, K. (1990). Bayesian design for accelerated life testing. *Technical Report, University of Minnesota.*

[7] CHERNOFF, H. (1962). Optimal accelerated life designs for estimation. *Technometrics*, **4**, 381-408.

[8] ESCOBAR, L.A. and MEEKER, W.Q. (1986). Planning accelerated life tests with type II censored data. *J. Stat. Comp. Simul.*, **23**, 273-297.

[9] ELFVING, G. (1952). Optimum allocation in linear regression theory. *Ann. Math. Stats.*, 23, 255-262.

[10] FISHER, R.A. (1966) *The Design of Experiments.* (8th ed.), Edinburgh: Oliver & Boyd.

[11] KIEFER, J. and WOLFOWITZ, J. (1959). Optimum designs in regression problems. *Ann. Math. Statist.*, 30, 271-294.

[12] LAWLESS, J.F. (1982). *Statistical Models and Methods for lifetime data.* Wiley, New York.

[13] LINDLEY, D.V. (1956). On the measure of information provided by an experiment. *Ann. Statist.*, **27**, 986-1005.

[14] LINDLEY, D.V. (1972). *Bayesian Statistics: A Review.* SIAM, Philadelphia.

[15] LUVALLE, M.J., WELSHER, T.L and SVOBODA, K. (1988). Acceleration transforms & statistical kinetic models. *J. Stat. Phys*, **52**, 311-320.

[16] MANN, N.R., SCHAFER, R.E. and SINGPURWALLA, N.D. (1974). *Methods for statistical analysis of reliability and life data.* John Wiley and Sons, Inc.

[17] MEEKER, W.Q. (1984). A companion of accelerated life test plans for Weibull and lognormal distributions and type I censored data. *Technometrics*, **26**, 157-171.

[18] MENZEFRICKE, U. (1991). Designing accelerated life tests when there is type II censoring. Technical Report, University of Toronto.

[19] MILLER, R. and NELSON, W. (1983). Optimum simple step stress plans for accelerated life testing. *I.E.E.E. Reliability*, **32**, 59-65.

[20] PARMIGIANI, G. and POLSON, N.G. (1990). Bayesian design for random walk barriers. To appear *Bayesian Statistics 4*.

[21] POLSON, N.G. (1990). On the expected amount of information from a nonlinear model. To appear *J.R. Statist. Soc. B.*

[22] SETHURAMAN, J.S. and SINGPURWALLA, N.D. (1982). Testing of hypotheses for distributions in accelerated testing. *J. Amer. Stat. Assoc.*, **77**, 204-208.

[23] SHAKED, M. and SINGPURWALLA, N.D. (1983). Inference from step-stress accelerated tests. *J. Stat. Plan. and Inf.*, **7**, 295-306.

[24] SMITH, A.F.M. and VERDINELLI, I. (1980). A note on Bayes designs for inference using a hierarchical linear model. *Biometrika*, **67**, 613-619.

[25] STONE, M. (1959a). Application of a measure of information to the design and comparison of regression experiments. *Ann. Statist.*, **21**, 55-70.

[26] STONE, M. (1959b). Discussion of Kiefer, *J.R. Statist. Soc. B*, **21**, 313-315.

[27] WHITTLE, P. (1973). Some general points in the theory of optimum experimental design. *J.R. Statist. Soc. B*, **35**, 123-130.

[28] VERDINELLI, I., POLSON, N.G. and SINGPURWALLA, N.D. (1990). Shannon information, Bayesian design and accelerated life testing. *SCQI Technical Report no. 13, Carnegie Mellon University.*

ADVANCES IN RELIABILITY
Edited by Asit P. Basu

A Piecewise Exponential Model for Reliability Growth and Associated Inferences

Ananda Sen and Gouri K. Bhattacharyya

Department of Statistics, University of Wisconsin, Madison, WI 53706

Abstract

Nonhomogeneous Poisson process (NHPP) with Weibull intensity has been widely used in modeling reliability growth (RG), and some elegant results are available for statistical inferences. However, a key feature of this model, namely, the intensity changing continuously over time regardless of the failure history, is not physically meaningful if fixes or design changes for a system improvement take place only after the observation of failures. For a test-fix-retest setting, we propose a simple stochastic model of RG, called a piecewise exponential (PEXP) model, which assumes that, after the $(i-1)$st failure, a homogeneous Poisson process with intensity $\lambda_i = (\mu/\delta)i^{1-\delta}$, $1 \leq \delta$, $0 < \mu$, governs the event of the next failure. The step intensity with this particular parameterization provides an alternative to the NHPP as a stochastic formulation of the Duane plot. By an analogy with the NHPP, closed-form estimators of the model parameters are constructed and are compared to the maximum likelihood estimators (MLE) in terms of asymptotic efficiency as well as finite sample simulation. The development of asymptotic properties of the MLE's involves some modifications of the standard arguments due to the singularity of the covariance matrix. The proposed model is applied to two data sets which were previously analyzed using the NHPP model.

1. Introduction

Modeling reliability growth has received considerable attention in the statistical and engineering literature over the past three decades. At the initial stage of any production involving complex systems, prototypes are put into life test under a development testing program, corrective actions or design changes are made when failures occur, and the modified system is tested again. As this test-redesign-retest sequence contributes to an improvement in the system performance, failure data become increasingly sparse at the later stages of testing making it more difficult to assess the current reliability. A reliability growth (RG) model provides a structure through which the failure data from the current as well as previous stages of testing could be analyzed in an integrated way in order to make inferences on the current system reliability.

A major thrust to RG modeling rose from certain empirical findings of Duane(1964) from examination of the failure data of a variety of systems such as complex hydromechanical devices, aircraft generators and jet engines in the course of their development. When plotted on a log-log scale, the cumulative number of failures was typically found to produce a linear pattern of relationship with the cumulative operating time. This phenomenon, later came to be known as the "Duane postulate", was given a concrete stochastic basis by Crow(1974) who assumed that the failures during the development stage of a new system follow a nonhomogeneous Poisson process (NHPP) with an intensity function $\lambda(t)$ of the form $\mu\delta t^{\delta-1}$. The corresponding cumulative failure rate $\Lambda(t) = \mu t^{\delta}$ is linear on a log-log scale. Incidentally, an NHPP formulation was proposed by Ascher(1968) in modeling the reliability change of a *bad-as-old system*, and was later used by Bassin(1969, 1973) with the Weibull intensity to obtain optimal overhaul intervals for various machines. A large body of literature has evolved thereafter in the areas of statistical inferences as well as applications of the NHPP model in RG analysis (see, for instance, Crow(1974), Bell and Midouski(1976), Finkelstein(1976), Lee and Lee(1978), Bain and Engelhardt(1980, 1982), Durr(1980), Lee(1980), Higgins and Tsokos(1981), Crow(1982), Rigdon and Basu(1988), Guida, Calabria and Pulcini(1989)).

For single-mission systems such as missiles or torpedos, the test results are binary in nature as opposed to time-to-failure in a continuous-time framework. The NHPP model in the continuous-time case has yielded a natural counterpart for the discrete case, which is called a Logarithmic Growth model or a nonhomogeneous binomial (NHB) model. Some estimation procedures were suggested by Finkelstein(1983), and asymptotic properties were studied by Bhattacharyya, Fries and Johnson(1989) and Bhattacharyya and Ghosh(1991).

In the continuous-time case, the NHPP has gained vast popularity due to its empirical fit to a variety of data vis-a-vis its conformity to the Duane learning curve, and availability of elegant distributional results concerning statistical inferences. While empirical fit, nice mathematical properties and tractability of statistical inferences are very important aspects of a stochastic model, a clear physical interpretation is also of paramount importance. Duane(1964) indicates that a learning curve is used to monitor developmental progress and plan for reliability improvement. One point of concern with the NHPP model is its continually changing intensity function irrespective of the failure history, which is in direct conflict with the concept of the Duane learning curve as well as the conceptual framework of a test-redesign-retest course of development testing. If system improvement is assumed to be effected only after a failure is observed, a realistic model should be flexible enough to incorporate a change in the failure rate at the occurrences of failures. Thompson(1988) expresses the same concern by saying that "$\cdots$ some provision needs to be present for altering the process of failures when modifications or corrective actions are applied to the system".

This is however, not to imply that the NHPP is inappropriate in all cases. Much depends on whether failure occurrences and fixes or design changes are synchronized in the real operational setting. Even if they are, an NHPP can be thought of as an approximation or "idealization" of a step-intensity model, as mentioned in Benton and Crow(1989). Our goal here is to formulate a model that avoids the approximation and also to make a comparative study of the two approaches of modeling. The discrete model which assumes that the probability of a failure decreases from stage to stage irrespective of the outcome of a trial, also needs to be modified for an operational setting where no design changes are made until a failure is observed. The continuous-time model described in this paper yields a discrete analog which handles this concern in a physically meaningful manner.

A continuous-time RG model, called "Piecewise Exponential" (PEXP), is formulated in the next section as a stochastic version of the Duane learning curve. In Section 3 we describe estimation procedures for our model and derive large sample inference results. In Section 4 some simulation results are reported along with a comparative study of the different estimation procedures. Finally, in the same section, we fit the PEXP model to two data sets which were previously analyzed under the assumption of NHPP, and compare the results.

2. The PEXP and related models

We incorporate the philosophy of learning curve into building a simple model for reliability growth. At the initial stage of testing, we consider a unit to have a constant failure intensity λ_1. At the first failure, fixes or design changes are made, thereby decreasing the rate of failure to a constant λ_2, and the modified unit is tested again. In this process we consider observations until n failures (failure truncated scheme). Then the data would consist of n ordered failure times $0 < T_1 < T_2 < \ldots < T_n$. A constant rate of failure prevailing at each stage of testing and the rate decreasing at each failure (after the corrective actions are taken) would amount to the assumption that the inter-failure times $T_i - T_{i-1}$ are independent exponential random variables with means $1/\lambda_i$, $i = 1, \ldots, n$. As for the pattern of change of failure rate in the successive stages, we assume the parameterization

$$\frac{1}{\lambda_i} = (\delta/\mu) i^{\delta-1}, \quad \mu > 0, \ \delta \geq 1 \tag{2.1}$$

and call the resulting model *Piecewise Exponential*, abbreviated as PEXP. Here λ_i is parameterized in a way that makes the model a stochastic version of the *Duane postulate*. To this end, we recall that the NHPP intensity $\lambda(t) = \mu\delta t^{\delta-1}$ was already formulated as a reflection of the Duane curve. Therefore, it would be appropriate to have a structure for λ_i that would bring it in line with the NHPP intensity.

To pursue this idea of "parameter matching" we observe that the cumulative failure rate at the nth failure time, $\int_0^{T_n} \lambda(t)\, dt$, equals μT_n^δ and $\sum_{i=1}^n \lambda_i(T_i - T_{i-1})$ for NHPP and PEXP, respectively. Since both quantities equal n in expectation, they form a common basis for matching the two models. Equating μT_n^δ to its expectation n amounts to setting the correspondence of T_n with $(n/\mu)^{1/\delta} = n^{\delta'}/\mu'$, where $\mu' = \mu^{1/\delta}$and $\delta' = 1/\delta$. Consequently, $T_i - T_{i-1}$ would correspond to $[i^{\delta'} - (i-1)^{\delta'}]/\mu'$. Finally, we replace the random variable $T_i - T_{i-1}$ by its expectation under PEXP model, and arrive at the relation

$$\frac{1}{\lambda_i} = \frac{i^{\delta'} - (i-1)^{\delta'}}{\mu'}.$$

For large i, the right hand side can be approximated by $(\delta'/\mu')i^{\delta'-1}$ which yields (2.1). Alternatively, we can directly motivate (2.1) as a model of "logarithmic growth" of the failure rate. Henceforth, for clarity and notational convenience, we shall drop the primes attached to the parameters.

Discrete Analog of PEXP

Concerning the test-redesign-retest cycle of development program for single-mission systems, a discrete analog of the PEXP is readily apparent. At each stage (configuration), independent trials are repeated until a failure is observed. Fixes or design changes are then made, and the modified system is tested again according to the same inverse sampling scheme. Then N_i, the *number of trials to the first failure under the ith configuration*, can be modeled as a geometric (q_i) random variable where q_i denotes the system failure probability. As for a reasonable structure for q_i that mimics the Duane postulate, we consider each trial to take a unit amount of time. Then the number of trials N_i between consecutive failures in the discrete case would correspond to the *inter-failure times* in the continuous case for which the PEXP model is appropriate. Thus taking q_i to be of the same form as the λ_i for PEXP, we arrive at a *Piecewise Geometric* model with logarithmic growth. Specifically, the failure probability at the *ith* configuration is given by

$$q_i = (\mu/\delta)i^{1-\delta}, \quad 0 < \mu < 1, \quad \delta \geq 1 \quad i = 1, \ldots, n.$$

An alternative parameterization, namely, $q_i = \mu\delta^i$, yields the discrete RG model due to Dubman and Sherman(1969).

A Generalization of the PEXP

The PEXP model assumes that the failure rate remains constant between failures. A natural generalization of the model would be to incorporate into the intensity function another component that changes continually with time. The role of the second component would be to account for such factors as wear out or other contributors to failure that are not affected by design changes. Essentially, we assume that the failure process has two components : one relates to the step intensity indicating reliability growth following an intervention, while the other pertains to reliability changes (i.e. degradation) not associated with the intervention.

To formalize this idea in a concrete physical setting, we consider observing the failure time of a two-component series system where the components are subject to two kinds of failures. As for the mechanism of the test-analyze-fix program, we assume that every time the system fails, fixes or design changes are made to component A while component B is replaced by a good-as-new unit (see Ascher(1968)). Consequently, component A undergoes reliability growth, while the clock for the failure process of B is reset to zero at each system failure. If T_{1i} and T_{2i} denote the lifetimes of A and B, respectively, at the i-th stage of the development program, then the system inter-failure time $T_i - T_{i-1}$ equals $min(T_{1i}, T_{2i})$, $i = 1, \ldots, n$. We further assume the following :

1. For all i, T_{1i} and T_{2i} are independent (independence of the components).
2. T_{1i}'s are independent exponential random variables with parameters λ_i (step-changing pattern) while T_{2i}'s are i.i.d. copies from a Weibull distribution W(λ, β).

These amount to the assumption that the successive failures arise from a composite intensity function

$$\lambda(t) = \lambda_{N(t)} + \lambda\beta t^{\beta-1},$$

where N(t) stands for the number of system failures in the time interval [0,t). Bain and Engelhardt (1991, p. 385) discusses a model where the failure rate is a polynomial in t. Note that our formulation is a generalization of the polynomial hazard function model in that it allows the constant term to be a function of the cumulative number of failures, and the power of t to be a positive real number. This model yields the PEXP as a limiting case when $\lambda \longrightarrow 0$ (or when only component A is present). In the rest of this paper we confine our investigation to the PEXP which itself is physically meaningful and serves as a simple alternative to NHPP for reliability growth modeling.

3. Parameter Estimation for PEXP

In this section we develop estimation procedures for the parameters of PEXP. We first study maximum likelihood estimation (MLE) which requires iterative solutions and also gives rise to an interesting non-standard situation of asymptotic theory. This will be followed by the development of an alternative simple estimation procedure which is motivated from the connection between the PEXP and the NHPP models.

3.1 The MLE and its Asymptotics

The likelihood of the failure times $T_1 < T_2 < \ldots < T_n$ under a failure-truncated sampling scheme can be written by using the fact that $Y_i \equiv T_i - T_{i-1}$ are independent exponential random variables with means $1/\lambda_i = (\delta/\mu)i^{\delta-1}$, $i = 1, \ldots n$. For simplicity of exposition, we write the mean in the log-linear form (exp($\boldsymbol{\beta}'\boldsymbol{x}_i$)) where $\boldsymbol{\beta}' = (\beta_1, \beta_2)$, $\beta_1 = log(\delta/\mu)$, $\beta_2 = \delta - 1$ and $\boldsymbol{x}_i' = (1, logi)$. The log-likelihood is then:

$$\log L = -\boldsymbol{\beta}'\sum_{i=1}^{n}\boldsymbol{x}_i - \sum_{i=1}^{n} Y_i \, exp(-\boldsymbol{\beta}'\boldsymbol{x}_i)$$

and we have

$$\boldsymbol{\psi}_n(\boldsymbol{\beta}) \equiv \frac{\partial logL}{\partial \boldsymbol{\beta}} = -\sum_{i=1}^{n}\boldsymbol{x}_i + \sum_{i=1}^{n} Y_i \, exp(-\boldsymbol{\beta}'\boldsymbol{x}_i)\boldsymbol{x}_i \tag{3.1}$$

$$\boldsymbol{A}_n(\boldsymbol{\beta}) \equiv -\frac{\partial^2 logL}{\partial\boldsymbol{\beta}\partial\boldsymbol{\beta}'} = \Sigma_{i=1}^n Y_i\, exp(-\boldsymbol{\beta}'\boldsymbol{x}_i)\boldsymbol{x}_i\boldsymbol{x}_i' \tag{3.2}$$

Expression (3.2) shows that $\boldsymbol{A}_n(\boldsymbol{\beta})$ is positive definite so $\boldsymbol{\psi}_n(\boldsymbol{\beta})$ is strictly concave. However, the likelihood equations $\boldsymbol{\psi}_n(\boldsymbol{\beta}) = \boldsymbol{0}$ do not seem to have a closed form solution. Numerical solutions can be obtained through standard iterative methods such as the Newton-Raphson or the scoring method. We proceed to derive the large sample properties of the maximum likelihood estimates. In our subsequent discussion all limits will be taken as $n \longrightarrow \infty$ unless otherwise mentioned. Also, the symbol $\sim$ placed between two functions of n will indicate that the ratio of the two tends to 1 as $n \longrightarrow \infty$. Denoting the true parameter point in the interior of the parameter space by $\boldsymbol{\beta}_0$, we define

$$\boldsymbol{\psi}_n = \boldsymbol{\psi}_n(\boldsymbol{\beta}_0), \quad \boldsymbol{A}_n = \boldsymbol{A}_n(\boldsymbol{\beta}_0),$$

$$Y_{i0} = Y_i\, exp(-\boldsymbol{\beta}_0'\boldsymbol{x}_i), \quad e_i = Y_{i0} - 1, \quad i = 1, 2, \ldots, n$$

and note that e_i's are i.i.d random variables with zero mean and unit variance. Using (3.1) and (3.2) we can then express $\boldsymbol{\psi}_n$ and $\boldsymbol{A}_n$ in terms of e_i's as :

$$\boldsymbol{\psi}_n \equiv (l_{1n},\, l_{2n})' = \sum_{i=1}^n e_i\boldsymbol{x}_i,$$

$$\boldsymbol{A}_n = \sum_{i=1}^n \boldsymbol{x}_i\boldsymbol{x}_i' + \sum_{i=1}^n e_i\boldsymbol{x}_i\boldsymbol{x}_i'.$$

These in our special case, have the components

$$l_{1n} = \sum_{i=1}^n e_i, \quad l_{2n} = \sum_{i=1}^n e_i \log i,$$

$$a_{11} = \sum_{i=1}^n e_i + n, \quad a_{12} = \sum_{i=1}^n e_i \log i + \sum_{i=1}^n \log i,$$

$$a_{22} = \sum_{i=1}^n e_i(\log i)^2 + \sum_{i=1}^n (\log i)^2. \tag{3.3}$$

We state a general asymptotic result for the sum of powers of $\log i$ which will be repeatedly used in the sequel. The proof is easy and hence omitted.

Lemma 3.1 *For all fixed nonnegative integer* k,
$n^{-1}(\log n)^{-k}\sum_{i=1}^n(\log i)^k = 1 - k(\log n)^{-1}\epsilon_{kn}$ *where* ϵ_{kn} *converges to 1 as* $n \longrightarrow \infty$.

Let

$$U_k = n^{-1/2}(\log n)^{-k}\sum_{i=1}^{n} e_i(\log i)^k, \quad k = 0,1,2, \tag{3.4}$$

$$Z'_n = (Z_{1n},\ Z_{2n}) = (n^{-1/2}l_{1n},\ n^{-1/2}(\log n)^{-1}l_{2n}). \tag{3.5}$$

Lemma 3.2 *Asymptotically* $\boldsymbol{Z}_n$ *is bivariate normal* $N_2(\boldsymbol{0}, \boldsymbol{\Sigma})$, *where* $\boldsymbol{\Sigma} = \boldsymbol{11}'$.

Proof : From (3.4) and (3.5) identify Z_{1n} and Z_{2n} to be U_0 and U_1, respectively. By the central limit theorem the asymptotic distribution of U_0 is standard normal. Also, $U_0 - U_1 = o_p(1)$ because $E[U_0 - U_1] = 0$ and $Var[U_0 - U_1]$ converges to zero by virtue of Lemma 3.1. The stated result then follows. //

The singularity of $\boldsymbol{\Sigma}$ poses a problem in doing the usual Taylor series expansion proof for the MLE's. This situation is very similar to one encountered by Bhattacharyya and Ghosh(1991) in the context of a nonhomogeneous binomial model. In order to use their line of argument we will show that although the scaled matrix of second derivatives of the log likelihood is asymptotically singular, the probability that it is positive definite tends to 1. To this end let us denote

$$\boldsymbol{C}_n(\boldsymbol{\beta}) = n^{-1}\begin{bmatrix} a_{11}(\boldsymbol{\beta}) & a_{12}(\boldsymbol{\beta})/(\log n) \\ a_{12}(\boldsymbol{\beta})/(\log n) & a_{22}(\boldsymbol{\beta})/(\log n)^2 \end{bmatrix},$$

$$\boldsymbol{C}_n = \boldsymbol{C}_n(\boldsymbol{\beta}_0) = (c_{ij}), \quad d_n = |\,\boldsymbol{C}_n\,|\,. \tag{3.6}$$

Lemma 3.3 (i) $\boldsymbol{C}_n \xrightarrow{P} \boldsymbol{\Sigma}$, (ii) $(\log n)^2 d_n \xrightarrow{P} 1$.

Proof : (i) From (3.3) and (3.6) we obtain

$$c_{11} = n^{-1/2}U_0 + 1,$$

$$c_{12} = n^{-1/2}U_1 + (n\log n)^{-1}\sum_{i=1}^{n}\log i\ ,$$

$$c_{22} = n^{-1/2}U_2 + n^{-1}(\log n)^{-2}\sum_{i=1}^{n}(\log i)^2\ . \tag{3.7}$$

Hence (i) follows from Lemma 3.1 and the fact that $U_k = O_p(1)$.
(ii) Note that $d_n = (n\log n)^{-2}(a_{11}a_{22} - a_{12}^2)$, where (using (3.3) and (3.4))

$$a_{11} = n(1 + n^{-1/2}U_0) \ ,$$

$$a_{12} = n(\log n)[(n \log n)^{-1}\sum_{i=1}^{n} \log i + n^{-1/2}U_1] \ ,$$

$$a_{22} = n(\log n)^2[n^{-1}(\log n)^{-2}\sum_{i=1}^{n}(\log i)^2 + n^{-1/2}U_2] \ .$$

Using these expressions, $(\log n)^2 d_n$ equals

$$\left(1 + n^{-1/2}U_0\right)\left[n^{-1}\sum_{i=1}^{n}(\log i)^2 + n^{-1/2}(\log n)^2 U_2\right] - \left[n^{-1}\sum_{i=1}^{n}(\log i) + n^{-1/2}(\log n)U_1\right]^2$$

$$= \left\{n^{-1}\sum_{i=1}^{n}(\log i)^2 - n^{-2}\left(\textstyle\sum_{i=1}^{n}\log i\right)^2\right\} + n^{-1/2}U_0\left[n^{-1}\sum_{i=1}^{n}(\log i)^2\right] + n^{-1/2}(\log n)^2 U_2$$

$$- 2n^{-1/2}(\log n)U_1[n^{-1}\sum_{i=1}^{n}\log i] + n^{-1}(\log n)^2 U_0 U_2 - n^{-1}(\log n)^2 U_1^2$$

$$= n^{-1}\textstyle\sum_{i=1}^{n}(\log i)^2 - \left(n^{-1}\displaystyle\sum_{i=1}^{n}\log i\right)^2 + o_p(1). \tag{3.8}$$

The last equality follows by observing that $\sum_{i=1}^{n}(\log i)^k \sim n(\log n)^k$ and $n^{-1/2}(\log n)^k U_j = o_p(1)$. Using the identity $n^{-1}\sum_{i=1}^{n} h_i^2 - \bar{h}^2 = n^{-1}\sum_{i=1}^{n} h_i^{*2} - \bar{h}^{*2}$ with $h_i^* = h_i - h_n$, the first two terms in the right hand side of (3.8) can be written as

$$\frac{1}{n}\sum_{i=1}^{n}[log(i/n)]^2 - \left[\frac{1}{n}\sum_{i=1}^{n}\log(i/n)\right]^2$$

which converges to $\int_0^1(\log u)^2\, du - \left[\int_0^1(\log u)\, du\right]^2 = 1.$ //
Part(ii) of Lemma 3.3 yields the crucial result that $P[d_n > 0] = P[(\log n)^2 d_n > 0] \longrightarrow 1$. Therefore, defining the set $G_n = \{d_n \neq 0\}$, we form a perturbed inverse of $\boldsymbol{C}_n$ as

$$\boldsymbol{F}_n = \boldsymbol{C}_n^{-1}I(G_n) + \boldsymbol{I}_2 I(G_n^c) \tag{3.9}$$

where I denotes the the indicator function and $\boldsymbol{I}_2$ the 2×2 identity matrix. Assume that $\boldsymbol{\psi}_n(\boldsymbol{\beta}) = \boldsymbol{0}$ has a solution $\widehat{\boldsymbol{\beta}}_n = (\widehat{\beta}_{1n}, \widehat{\beta}_{2n})$. The appropriate neighborhood of $\boldsymbol{\beta}_0$ in which the solution exists is specified in Lemma 3.5. A Taylor expansion of $\boldsymbol{\psi}_n(\widehat{\boldsymbol{\beta}}_n) = \boldsymbol{0}$ around $\boldsymbol{\beta}_0$ yields

$$\boldsymbol{\psi}_n(\boldsymbol{\beta}_0) = \boldsymbol{A}_n(\boldsymbol{\zeta}_n)(\widehat{\boldsymbol{\beta}}_n - \boldsymbol{\beta}_0) \tag{3.10}$$

where $\boldsymbol{\zeta}_n$ is on the line segment joining $\widehat{\boldsymbol{\beta}}_n$ and $\boldsymbol{\beta}_0$.

Defining,

$$W_{1n} = n^{1/2}(\log n)^{-1}(\widehat{\beta}_{1n} - \beta_{10}), \quad W_{2n} = n^{1/2}(\widehat{\beta}_{2n} - \beta_{20}), \quad \boldsymbol{W}_n = (W_{1n}, W_{2n})',$$

we observe from (3.5) and (3.10) that

$$\boldsymbol{Z}_n = (\log n)\boldsymbol{C}_n(\boldsymbol{\zeta}_n)\boldsymbol{W}_n \ , \tag{3.11}$$

$$\boldsymbol{K}_n \equiv (\log n)^{-1}\boldsymbol{F}_n\boldsymbol{Z}_n = \boldsymbol{F}_n\boldsymbol{C}_n(\boldsymbol{\zeta}_n)\boldsymbol{W}_n \ . \tag{3.12}$$

Lemma 3.4 *Asymptotically,* $\boldsymbol{K}_n$ *is bivariate* (singular) *normal* $N_2(\boldsymbol{0}, \boldsymbol{\Sigma}_1)$ *with*

$$\boldsymbol{\Sigma}_1 = \begin{bmatrix} 1 & -1 \\ -1 & 1 \end{bmatrix}.$$

Proof : We observe that on the set G_n

$$\boldsymbol{F}_n\boldsymbol{Z}_n = \frac{1}{d_n}\begin{bmatrix} c_{22} & -c_{12} \\ -c_{12} & c_{11} \end{bmatrix}\begin{pmatrix} U_0 \\ U_1 \end{pmatrix}$$

where, using (3.7) and Lemma 3.1, we write

$$\begin{aligned} c_{11} &= n^{-1/2}U_0 + 1, \\ c_{12} &= n^{-1/2}U_1 + 1 - (\log n)^{-1}\epsilon_{1n}, \\ c_{22} &= n^{-1/2}U_2 + 1 - 2(\log n)^{-1}\epsilon_{2n}. \end{aligned}$$

Denoting $\boldsymbol{K}_n^* = d_n(\log n)^2\boldsymbol{K}_n$, we have

$$\begin{aligned} K_{1n}^* &= (logn)(c_{22}U_0 - c_{12}U_1) \\ &= (logn)\left[\left(n^{-1/2}U_2 + 1 - 2(logn)^{-1}\epsilon_{2n}\right)U_0 - \left(n^{-1/2}U_1 + 1 - (logn)^{-1}\epsilon_{1n}\right)U_1\right] \\ &= (logn)(U_0 - U_1) - U_0 + o_p(1). \end{aligned}$$

The last equality follows from the fact that (U_0, U_1) is asymptotically distributed as $N_2(\boldsymbol{0}, \boldsymbol{11}')$. By similar steps, it follows that

$$\begin{aligned} K_{2n}^* &= (\log n)(U_1 - U_0) + U_0 + o_p(1) \\ &= -K_{1n}^* + o_p(1). \end{aligned}$$

By characteristic function argument for exponential random variables it can be shown that $(\log n)(U_0 - U_1) - U_0$ has an asymptotic standard normal distribution. The stated result then follows by observing that $(\log n)^2 d_n$ converges to 1 in probability (by part (ii) of Lemma 3.3) and the fact that on G_n^c, $\boldsymbol{K}_n = (logn)^{-1}\boldsymbol{Z}_n = o_p(1)$ (by Lemma 3.2). //

Lemma 3.5 *Define a sequence of neighborhoods of* β_0 *by :*

$$M_n(\boldsymbol{\beta}_0) = \{(\beta_1, \beta_2) : \beta_1 = \beta_{10} + \tau_1(\log n)n^{-\gamma}, \quad \beta_2 = \beta_{20} + \tau_2 n^{-\gamma}, \quad \|\tau\| \leq h\},$$

where τ_1, τ_2, γ *and* h *are fixed numbers with* $0 < \gamma < 1/2$, *and* $0 < h < \infty$. *Then* $\boldsymbol{F}_n[\boldsymbol{C}_n(\boldsymbol{\beta}) - \boldsymbol{C}_n] \xrightarrow{P} \boldsymbol{0}$ *uniformly in* $\boldsymbol{\beta} \in M_n(\boldsymbol{\beta}_0)$.

The proof rests on showing the uniform convergence of certain exponential functions that arise in the expressions for the mean of the elements in $C_n(\boldsymbol{\beta})$. The details are outlined in the Appendix.

The main results concerning the existence and asymptotic normality of a consistent sequence of roots of the likelihood equations are stated in the next two theorems.

Theorem 3.1 (Existence) *With probability tending to 1 as* $n \longrightarrow \infty$, *there exists a sequence of roots* $\widehat{\boldsymbol{\beta}}_n \in M_n(\boldsymbol{\beta}_0)$ *of the likelihood equations. Furthermore, such* $\widehat{\boldsymbol{\beta}}_n$*'s correspond to local maxima of the likelihood function.*

Theorem 3.2 (Asymptotic Normality) $\boldsymbol{W}_n$ *is asymptotically bivariate* (singular) *normal* $N_2(\boldsymbol{0}, \boldsymbol{\Sigma}_1)$.

The proof of Theorem 3.1 follows along the same lines of the proof of a corresponding result in Bhattacharyya and Ghosh(1991) and is outlined in the Appendix. As for the proof of Theorem 3.2 note that

$$\boldsymbol{F}_n\boldsymbol{C}_n = \boldsymbol{I}_2 I(G_n) + \boldsymbol{C}_n(\boldsymbol{\beta}_0)I(G_n^c).$$

By Lemma 3.3 we have, $P(G_n) \longrightarrow 1$ and $\boldsymbol{C}_n \xrightarrow{P} \boldsymbol{\Sigma}$. We can then conclude

$$\boldsymbol{F}_n\boldsymbol{C}_n \xrightarrow{P} \boldsymbol{I}_2. \tag{3.13}$$

Equation (3.12) gives

$$\boldsymbol{K}_n = \boldsymbol{F}_n\boldsymbol{C}_n(\boldsymbol{\zeta}_n)\boldsymbol{W}_n = \boldsymbol{F}_n[\boldsymbol{C}_n(\boldsymbol{\zeta}_n) - \boldsymbol{C}_n(\boldsymbol{\beta}_0) + \boldsymbol{C}_n(\boldsymbol{\beta}_0)]\boldsymbol{W}_n.$$

For $\boldsymbol{\zeta}_n \in M_n(\boldsymbol{\beta}_0)$, Lemmas 3.4-3.5 along with equation(3.13) then yields the result.

Consequences

(i) Noting the parameter relations $\beta_1 = log(\delta/\mu)$, $\beta_2 = \delta - 1$, we can translate the results of Theorem 3.2 (via delta method and Slutsky's theorem) in terms of the original parameters as :

$$\sqrt{n}(\hat{\delta} - \delta_0) \xrightarrow{d} N(0,1), \quad \sqrt{n}(\log n)^{-1}(\hat{\mu} - \mu_0) \xrightarrow{d} N(0, \mu_0^2).$$

(ii) The *current system reliability* under the PEXP model is a 1-1 function of the current intensity of failure. The intensity λ_n at the n-th failure can be expressed in terms of the parameters β_1 and β_2 as :

$$-log\ \lambda_n \ = \ \beta_1 \ + \ \beta_2 \log n.$$

An estimate of this can be obtained by replacing the parameters by their MLE's. By virtue of Lemma 3.5 and Lemma 3.3 we have $\boldsymbol{C}_n(\boldsymbol{\zeta}_n) \xrightarrow{P} \boldsymbol{\Sigma}$ uniformly in the neighborhood $M_n(\boldsymbol{\beta}_0)$. Also, Z_{1n} has an asymptotic standard normal distribution by Lemma 3.2. Using these two facts in (3.11), we deduce

$$\sqrt{n}(\log\ \hat{\lambda}_n \ - \ \log\ \lambda_n) \ = \ -(\log n)(W_{1n} \ + \ W_{2n}) \ \xrightarrow{d} \ N(0,1).$$

3.2 Simple Estimators — Weibull Process Analog

We have noted in Section 3.1 that the MLE's for the parameters in the PEXP are not available in closed form. Also, the simulations described in Section 4.1 indicate that for small sample sizes, the MLE of μ often falls far off the true parameter value. Here we construct an alternative set of estimators by exploiting the link between the PEXP and NHPP models. To this end, we note the following correspondence between parameters of the two models :

NHPP	PEXP
$\mu^{1/\delta}$	μ
$1/\delta$	δ

The fact that the MLE's under the NHPP model are given by $\hat{\delta} \ = \ n/\sum_{i=1}^{n} \log(T_n/T_i)$ and $\hat{\mu} \ = \ n/T_n^{\hat{\delta}}$ then motivates the following set of estimators for the PEXP :

$$\delta^* \ = \ \frac{1}{n} \textstyle\sum_{i=1}^{n} log(T_n/T_i), \qquad \mu^* \ = \ \frac{n^{\delta^*}}{T_n}. \tag{3.14}$$

Since these estimators stem from the NHPP model with Weibull intensity, we call them Weibull process analog estimators (WPAE). The main advantage of (3.14) over the MLE's is the simple closed form expressions of the estimators. Simulations demonstrate that μ^* behaves better than the MLE $\hat{\mu}$ for small sample sizes.

If the true model is PEXP and one wrongly assumes the model to be NHPP, then the WPAE's constitute a set of estimators for the "misspecified" model. From this point of view of misspecificaton, it is worth comparing the properties of the WPAE's with those of the MLE's.

We denote the true parameter values by μ_0 and δ_0 and define,

$$W_{1n}^* \ = \ n^{1/2}(\log n)^{-1}(\mu^* - \mu_0), \ \ W_{2n}^* \ = \ n^{1/2}(\delta^* - \delta_0), \ \ \boldsymbol{W}_n^* = (W_{1n}^*, \ W_{2n}^*)'. \tag{3.15}$$

The following theorem shows that the WPAE's are consistent and asymptotically normal (CAN) estimators of the parameters under PEXP.

Theorem 3.3 *Asymptotically, $\mathbf{W}_n^*$ is bivariate normal $N_2(\mathbf{0}, \mathbf{\Sigma}_1^*)$ where*

$$\mathbf{\Sigma}_1^* = \delta_0^2/(2\delta_0 - 1)\begin{bmatrix} \mu_0^2 & \mu_0 \\ \mu_0 & 1 \end{bmatrix}.$$

To prepare the groundwork for proving the theorem we define

$$X_i = \frac{\mu_0 T_i}{i^{\delta_0}}, \quad i = 1, \ldots, n, \quad \overline{X}_n = n^{-1}\sum_{i=1}^{n} X_i$$

and express the WPAE's in terms of these random variables as :

$$\delta^* = \log(T_n/n^{\delta_0}) - n^{-1}\sum_{i=1}^{n}\log(T_i/i^{\delta_0}) + \delta_0 n^{-1}\sum_{i=1}^{n}\log(n/i)$$

$$= n^{-1}\sum_{i=1}^{n}\log(X_n/X_i) + \delta_0[\log n - (\log n!)/n] , \tag{3.16}$$

$$\log\mu^* - \log\mu_0 = \log(n^{\delta_0}/\mu_0 T_n) + (\log n)(\delta^* - \delta_0)$$

$$= (\log n)(\delta^* - \delta_0) - \log X_n . \tag{3.17}$$

The next lemma provides some results concerning the random variables X_i which will be used in proving our main results.

Lemma 3.6 $(a) n^{1/2}(X_n - 1) \xrightarrow{d} N(0, \delta_0^2/(2\delta_0 - 1))$.
$(b) n^{1/2}(X_n - \overline{X}_n) \xrightarrow{d} N(0, \delta_0^2/(2\delta_0 - 1))$.
$(c) n^{-1/2}\sum_{i=1}^{n} log(X_n/X_i) = n^{1/2}(X_n - \overline{X}_n) + o_p(1)$.

Proof : (a) Noting that X_n is a linear function of the independent exponential random variables $Y_i \equiv T_i - T_{i-1},\ i = 1, \ldots, n$, the result follows from an application of the Lindeberg-Feller central limit theorem.
(b) We first write $\overline{X}_n - X_n$ as a linear function of Y_i's. Using the fact : $\sum_{i=1}^{n}\sum_{j=1}^{i} a_i b_j = \sum_{i=1}^{n}\sum_{j=i}^{n} b_i a_j$, we write

$$\sum_{i=1}^{n} X_i = \mu_0 \sum_{i=1}^{n} i^{-\delta_0}\sum_{j=1}^{i} Y_j$$

$$= \mu_0 \sum_{i=1}^{n} Y_i \left(\sum_{j=i}^{n} j^{-\delta_0}\right).$$

Also, $nX_n = \mu_0 n^{1-\delta_0}\sum_{i=1}^{n} Y_i = \mu_0 n^{1-\delta_0}\sum_{i=1}^{n} Y_i \sum_{j=i}^{n}(n - i + 1)^{-1}$.
Thus, we have

$$S_n \equiv n^{-1/2}\sum_{i=1}^{n}(X_i - X_n)$$
$$= \mu_0 n^{-1/2}\sum_{i=1}^{n} Y_i\, d_{in},$$

where

$$d_{in} = \sum_{j=i}^{n}\left(\frac{1}{j^{\delta_0}} - \frac{n}{n^{\delta_0}(n-i+1)}\right).$$

By using the relations

$$\delta(j+1)^{\delta-1} \geq (j+1)^{\delta} - j^{\delta} \geq \delta j^{\delta-1}, \tag{3.18}$$

we obtain bounds for $E(Y_j)$ and correspondingly for $E(S_n)$ as :

$$-\delta_0 n^{-1/2}\sum_{i=1}^{n} i^{-\delta_0}(i+1)^{\delta_0-1} \leq E(S_n) \leq \delta_0 n^{1/2-\delta_0}(n+1)^{\delta_0-1}.$$

Since the lower bound is $O(n^{-1/2}\log n)$ and the upper bound is $O(n^{-1/2})$, we have $E(S_n) \longrightarrow 0$. Also, from the expression of S_n we readily obtain

$$s_n^2 \equiv Var(S_n) = \delta_0^2 n^{2\delta_0-3}\sum_{i=1}^{n}(i/n)^{2\delta_0-2}\, d_{in}^2$$
$$\mu_0^4 n^{-2}\sum_{i=1}^{n} E(Y_i)^4\, d_{in}^4 = 24\delta_0^4 n^{4\delta_0-6}\sum_{i=1}^{n}(i/n)^{4\delta_0-4}\, d_{in}^4.$$

Setting a correspondence of d_{in} with a Riemann sum, we observe that as $n \longrightarrow \infty$,

$$s_n^2 \longrightarrow \delta_0^2 \int_0^1 u^{2\delta_0-2}\left\{\int_u^1\left(\frac{1}{v^{\delta_0}} - \frac{1}{1-u}\right)dv\right\}^2 du = \delta_0^2/(2\delta_0 - 1)$$
$$\mu_0^4 n^{-2}\sum_{i=1}^{n} E(Y_i)^4\, d_{in}^4 \sim n^{-1}24\delta_0^4 \int_0^1 u^{4\delta_0-4}\left\{\int_u^1\left(\frac{1}{v^{\delta_0}} - \frac{1}{1-u}\right)dv\right\}^4 du \longrightarrow 0.$$

These facts in conjunction with the result that $E(S_n) \longrightarrow 0$, enable us to use the central limit theorem to conclude part(b) of the lemma.

(c) Let

$$U_{in} \equiv log(X_n/X_i) - (X_n - X_i).$$

Since $(x-1)/x \leq \log x \leq x-1$ for $x > 0$, we have

$$\left(\frac{1}{X_n} - 1\right) n^{-1/2}\sum_{i=1}^{n}(X_n - X_i) \le n^{-1/2}\sum_{i=1}^{n} U_{in} \le n^{-1/2}\sum_{i=1}^{n}(X_n - X_i)\left(\frac{1}{X_i} - 1\right) \quad (3.19)$$

We use Slutsky's theorem in conjunction with the results in parts (a) and (b) to conclude that the lower bound in equation (3.19) is $o_p(1)$. The upper bound equals

$$n^{-1/2}\sum_{i=1}^{n}[(X_n - 1) - (X_i - 1)]\left(\frac{1}{X_i} - 1\right)$$

$$= n^{1/2}(X_n - 1)\left\{n^{-1}\sum_{i=1}^{n}\left(\frac{1}{X_i} - 1\right)\right\} - n^{-1/2}\sum_{i=1}^{n}(X_i - 1)\left(\frac{1}{X_i} - 1\right).$$

Denote the first and second terms on the right hand side by B_1 and B_2, respectively. By part(a) of the lemma, we have $B_1 = o_p(1)$. The proof is completed once we establish that $B_2 = o_p(1)$. Now, by the Cauchy-Schwarz inequality,

$$B_2^2 \le \sum_{i=1}^{n}(X_i - 1)^2\left\{n^{-1}\sum_{i=1}^{n}\left(\frac{1}{X_i} - 1\right)^2\right\}.$$

Again part(a) of the lemma implies that $n^{-1}\sum_{i=1}^{n}(1/X_i - 1)^2 = o_p(1)$. To show that $\sum_{i=1}^{n}(X_i - 1)^2 = O_p(1)$ we note that

$$\begin{aligned}\sum_{i=1}^{n} Var(X_i) &= \delta_0^2\sum_{i=1}^{n}(1/i^2)\sum_{j=1}^{i}(j/i)^{2\delta_0-2}\\ &= \delta_0^2 n^{-1}\sum_{i=1}^{n}\frac{1}{(i/n)}\frac{1}{i}\sum_{j=1}^{i}(j/i)^{2\delta_0-2}\\ &\longrightarrow \delta_0^2\int_0^1\frac{1}{u}\int_0^u v^{2\delta_0-2}\,dv\,du = \delta_0^2/(2\delta_0-1)^2\end{aligned}$$

Using equation (3.18) we also have,

$$0 \le \textstyle\sum_{i=1}^{n}\{E(X_i) - 1\}^2 \le \sum_{i=1}^{n} i^{-2\delta_0}\{\delta_0^2(i+1)^{2\delta_0-2} - 2\delta_0(i+1)^{\delta_0-1} + 1\}.$$

Since $\delta_0 > 1$, all the terms in the sum on the right hand side converge to finite numbers. Thus, $\sum_{i=1}^{n}(X_i - 1)^2$ is bounded in expectation and hence is $O_p(1)$ which implies $B_2 = o_p(1)$. //

Proof of Theorem 3.3

From (3.15) and (3.16), we have

$$\begin{aligned}W_{2n}^* &= n^{-1/2}\sum_{i=1}^{n}log(X_n/X_i) + n^{1/2}\delta_0\,[logn - (logn!)/n - 1]\\ &= n^{1/2}(X_n - \overline{X}_n) + n^{1/2}\delta_0\,[logn - (logn!)/n - 1] + o_p(1)\\ &= n^{1/2}(X_n - \overline{X}_n) + o_p(1).\end{aligned}$$

The second equality follows from lemma 3.6(c) and the last equality obtains from Stirling's formula,

$$\log n \; - \; (\log n!)/n \; - \; 1 \; = \; -(\log n)/2n \; + \; O(1/n).$$

Then Lemma 3.6(b) entails that W^*_{2n} is asymptotically distributed as $N(0,\ \delta_0^2/(2\delta_0 - 1))$. Finally, from (3.17) we have,

$$n^{1/2}(\log n)^{-1}(\log \mu^* \; - \; \log \mu_0) \;\; = \;\; n^{1/2}(\delta^* \; - \; \delta_0) \; - \; n^{1/2}(\log n)^{-1} \; \log X_n \; .$$

Since Lemma 3.6(a) along with an application of the delta method yields the fact that $n^{1/2}(logn)^{-1}\ logX_n$ converges to zero in probability, we obtain

$$W^*_{1n} \; = \; \mu_0 W^*_{2n} \; + \; o_p(1)$$

which completes the proof. //

Calculation of ARE's

In view of the asymptotic results derived in this section, we summarize the comparative features between the MLE's and the WPAE's.

- The rates of convergence for both sets of estimators $(\hat{\mu}, \hat{\delta})$ and (μ^*, δ^*) are identical.
- Both sets of estimators (properly scaled) have an asymptotically singular normal distribution.
- The WPAE's incur a loss of asymptotic efficiency compared to the MLE's. In fact, for both μ and δ the

$$\text{ARE (WPAE : MLE)} \; = \; \frac{2\delta_0 \; - \; 1}{\delta_0^2} \; < \; 1.$$

The ARE decreases with an increase in δ, becomes close to 1 as δ gets close to 1. For large δ the ARE tends to 0. Also note that the ARE does not depend on the parameter μ.

4. Simulation and applications

Summary of simulation results

Monte Carlo simulation techniques were employed to study the performances of the maximum likelihood and the Weibull process analog estimators in both small and large samples. Three pairs of (μ, δ) values, (0.5,1.5), (1.0,2.0) and (2.5,2.5) were used for the study, and for each case 100 realizations of the MLE's and the WPAE's were obtained with the sample sizes n=10, 25, 50 and 100.

Exponential random variables were generated using the inverse cdf transformation on uniform(0,1) random numbers obtained from the *Uniform Random Number Generator* (UNI) residing in the Fortran Library **CMLIB**. The MLE $\hat{\delta}$ was obtained through a single-variable Newton-Raphson iteration procedure using the WPAE δ^* as the initial value. The MLE $\hat{\mu}$ was then calculated from the relation

$\hat{\mu} = n\hat{\delta}/\sum_{i=1}^{n} i^{1-\hat{\delta}} Y_i$

where Y_i's represent the generated values of the inter-failure times which are independent exponential random variables under the PEXP model. Table 1 gives the estimated bias and mean squared error of the MLE's as well as the WPAE's.

The MLE for μ has a tendency to overestimate as is evidenced from positive bias in all cases in Table 1. Also, it shows a substantial variability especially for small sample sizes (*e.g. n=10*). By contrast, the MLE for δ appears to be quite stable. Although it exhibits a positive bias in most cases, the magnitudes of the bias as well as the MSE's are substantially smaller compared to those for $\hat{\mu}$. The WPAE's for both the parameters show a tendency of underestimation in almost all cases. With respect to the MSE's the performances of the estimates of δ are comparable (with the MLE behaving slightly better for larger δ values), while the WPAE of μ performs better than the corresponding MLE in all cases. Even for n as large as 50 or 100, the finite-sample efficiency of μ^* relative to $\hat{\mu}$ as measured by the ratio (estimated) MSE($\hat{\mu}$)/MSE(μ^*) is quite different from the value of the asymptotic relative efficiency $(2\delta - 1)/\delta^2$. Specifically, the ARE values are 0.88, 0.75, and 0.64 for the three cases $\delta = 1.5$, 2.0, and 2.5 respectively, while the corresponding finite-sample relative efficiencies are found to be 2.04, 3.068, and 2.89 for n=50, and 2.10, 1.49, and 1.62 for n=100.

Table 1 :

Comparative Study of MLE and WPAE

Sample Size (n)	Maximum Likelihood				Weibull Process Analog			
	Bias($\hat{\mu}$)	MSE($\hat{\mu}$)	Bias($\hat{\delta}$)	MSE($\hat{\delta}$)	Bias(μ^*)	MSE(μ^*)	Bias(δ^*)	MSE(δ^*)
	$\mu_0 = 0.5,\ \delta_0 = 1.5$							
10	0.500	8.718	0.147	0.242	−0.049	0.341	−0.233	0.162
25	0.107	0.218	−0.029	0.056	−0.171	0.067	−0.185	0.071
50	0.127	0.167	0.010	0.028	−0.084	0.082	−0.089	0.032
100	0.076	0.082	0.008	0.011	−0.075	0.039	−0.053	0.012
	$\mu_0 = 1.0,\ \delta_0 = 2.0$							
10	2.143	64.438	0.160	0.250	−0.476	0.505	−0.396	0.271
25	0.567	4.1006	0.006	0.073	−0.324	1.033	−0.234	0.115
50	0.328	1.031	0.011	0.034	−0.282	0.336	−0.135	0.049
100	0.175	0.345	0.013	0.013	−0.196	0.232	−0.070	0.018
	$\mu_0 = 2.5,\ \delta_0 = 2.5$							
10	1.970	38.785	0.067	0.180	−1.721	3.589	−0.583	0.453
25	0.83	6.391	0.005	0.059	−1.356	2.481	−0.306	0.147
50	0.789	5.698	0.019	0.030	−0.955	1.971	−0.172	0.062
100	0.406	2.181	0.005	0.012	−0.619	1.345	−0.091	0.023

For practical applications of the asymptotic results, it is important to examine how the normal approximation improves with increasing sample sizes. An investigation in that direction is made through the normal scores plots of the estimates. Plots for both $\hat{\delta}$ and δ^* (Figure 1) indicate a fairly linear pattern in all cases of n, small or large. However, the corresponding plots for $\hat{\mu}$ and μ^* (Figure 2) show a substantial departure from a straight line pattern, which persists even for n as large as 50. This appears to be due to a considerable fluctuation in the estimated μ values and also the slow rate $(\sqrt{n}/\log n)$ of convergence. However, logarithmic transformation on the estimates

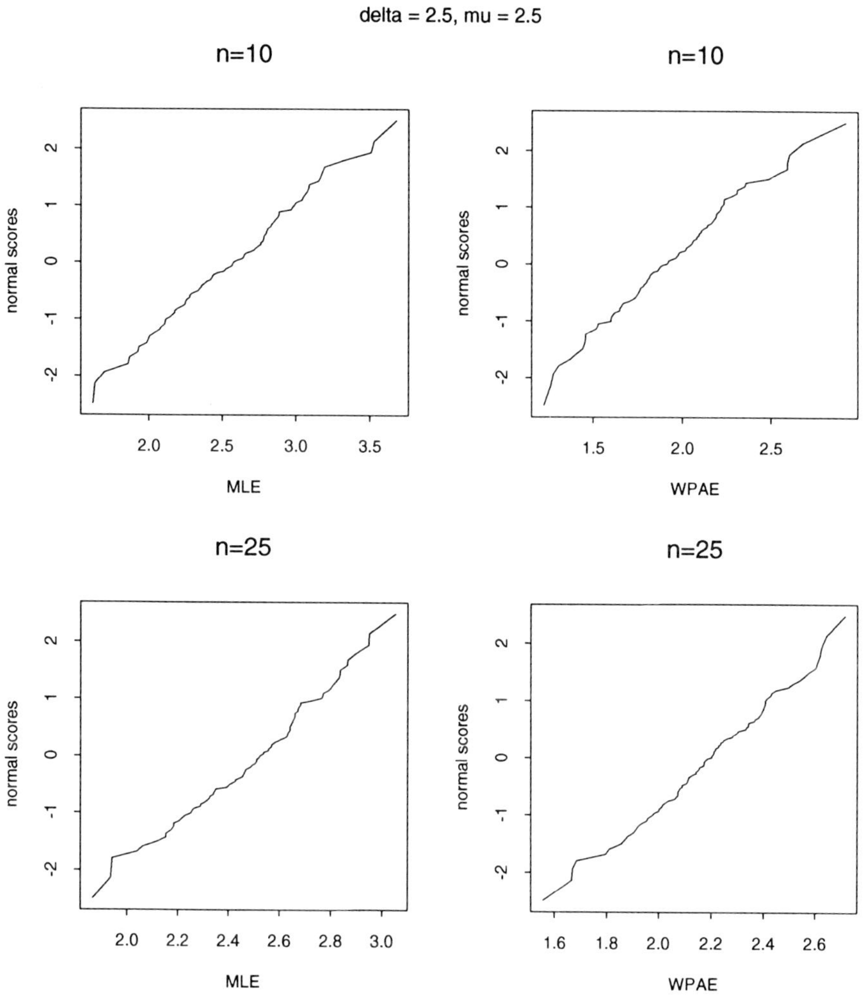

Figure 1: Normal Scores plots for $\hat{\delta}$ and δ^*

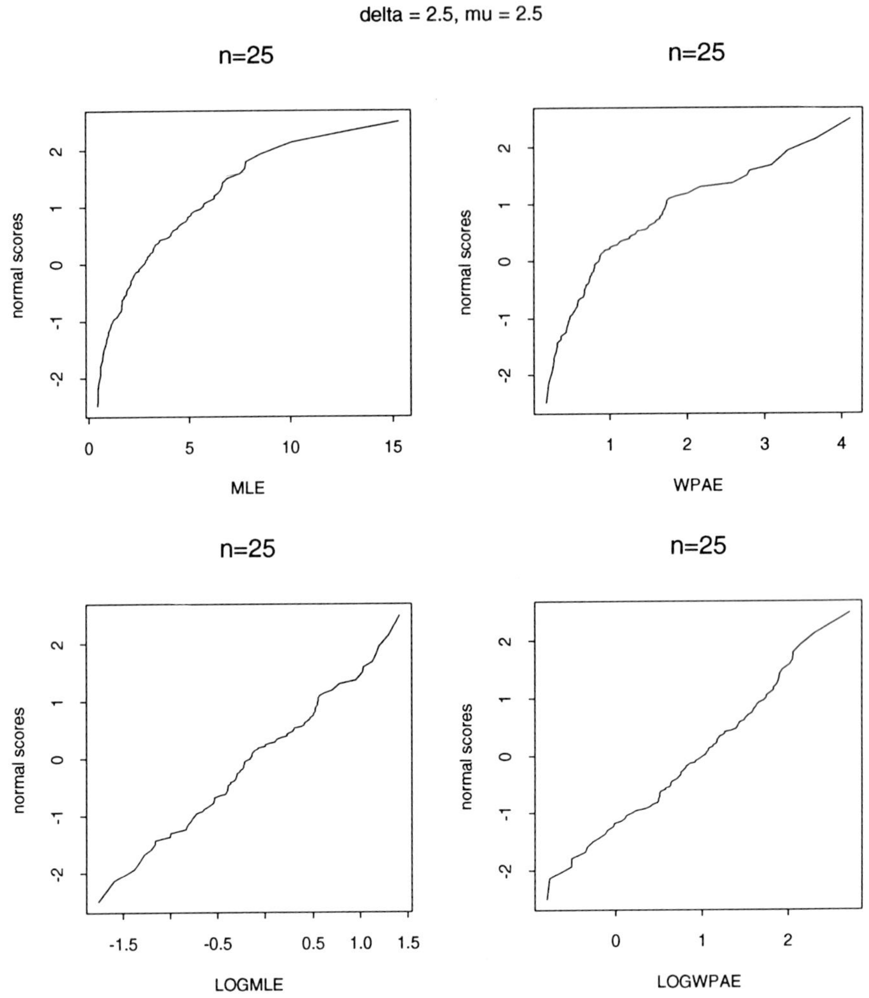

Figure 2: Normal Scores plots for the estimates of μ and their logarithms

of μ is found to stabilize their variations substantially, and the agreement with the normal scores is also considerably improved. Figure 2 exhibits these features for the case n=25. This indicates that when setting large-sample confidence interval for μ, one should first construct a confidence interval for $\log \mu$ using the asymptotic normality result and then transform the result to μ. Specifically, a $100(1-\alpha)\%$ confidence interval for μ constructed in this process would be of the form

$$\hat{\mu}\left[n^{\pm z_{\alpha/2}/\sqrt{n}}\right]. \tag{4.1}$$

Applications

Here we fit the PEXP and apply the inference results of Section 3 to two data sets which were previously analyzed under the NHPP model. We also employ some graphical checks for model adequacy, and compare the inference results between the two models, especially with regard to estimating the current system reliability.

Example 1 **Tank failure data** For a tank system, the number of miles accumulated was recorded for the first 25 failures : 1, 57, 252, 310, 485, 693, 720, 727, 779, 1028, 1561, 1766, 1793, 1938, 2030, 2065, 2289, 2423, 2560, 3086, 3458, 3626, 4252 and 4582 (Source : Military Handbook 189 (1981, pp 111)). The objective of the study was to assess the extent to which parts improvement and other design changes reduced the intensity of failure. From the PEXP fit, the maximum likelihood estimates for the parameters are found to be $\hat{\mu} = 0.04699$ and $\hat{\delta} = 1.6614$. A large sample 95% confidence interval for δ is given by $\hat{\delta} \pm 1.96/\sqrt{n} = [1.27, 2.05]$, and it indicates reliability growth. Using (4.1), a 95% confidence nterval for μ is found to be [0.17, 1.73]. The corresponding WPAE's (or the MLE's for the misspecified model) are found to be $\mu^* = 0.03027$, $\delta^* = 1.5323$, with the associated large sample 95% confidence intervals [1.11, 1.95] and [.008, .116] for δ and μ, respectively.

To develop a graphical model checking procedure we define the residuals

$$\hat{e}_i = \frac{\hat{\mu} Y_i}{\hat{\delta}_i^{\hat{\delta}-1}}, \qquad i = 1, \ldots, 25$$

where Y_i denotes the miles between the ith and $(i-1)st$ failure. Note that these residuals correspond to the standard exponential variates $e_i = (\mu/\delta)Y_i/(i^{\delta-1})$. Figure 3(a) shows a plot of the points $(\hat{e}_{(i)}, \alpha_i)$ where $\alpha_i = \sum_{i=1}^{25}(25-j+1)^{-1}$ is the expected value of the ith standard exponential order statistic in a sample of size 25 and $\hat{e}_{(i)}$ denotes the ordered residuals. The points lie roughly along a straight line with unit slope – a pattern that supports the assumption of exponentiality and hence the PEXP model. Note that $\hat{e}_i$ is of the scale free form $a_i(\hat{\delta})/\overline{a}(\hat{\delta})$, where $a_i(\hat{\delta}) = Y_i i^{1-\hat{\delta}}$. For a corresponding graphical check for the NHPP fit, we define the residuals to be equal to $\hat{\delta}\log(t_n/t_i)$, $i = 1, \ldots, n-1$, t_i being the accumulated mileage at the ith failure and $\hat{\delta}$ being the maximum likelihood estimate of δ under the NHPP model. Note that these residuals correspond to a set of order statistics of size $n-1$ from the standard exponential and are also of the scale free form $b_i/\overline{b}$, where $b_i = \log(t_n/t_i)$. Figure 3(b) shows a plot of $(\tilde{e}_{(i)}, \tilde{\alpha}_i)$, where $\tilde{e}_{(i)}$ denotes the ordered residuals and $\tilde{\alpha}_i$ are the exponential scores in a sample of size 24.

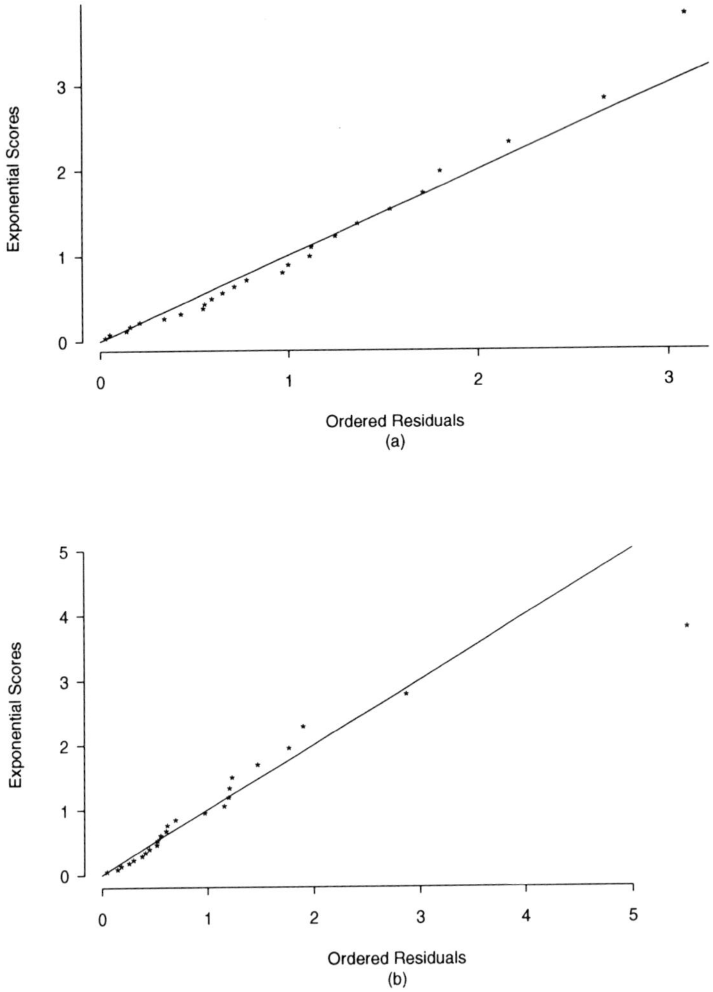

Figure 3: Residual checks for the tank data with (a) PEXP fit, (b) NHPP fit

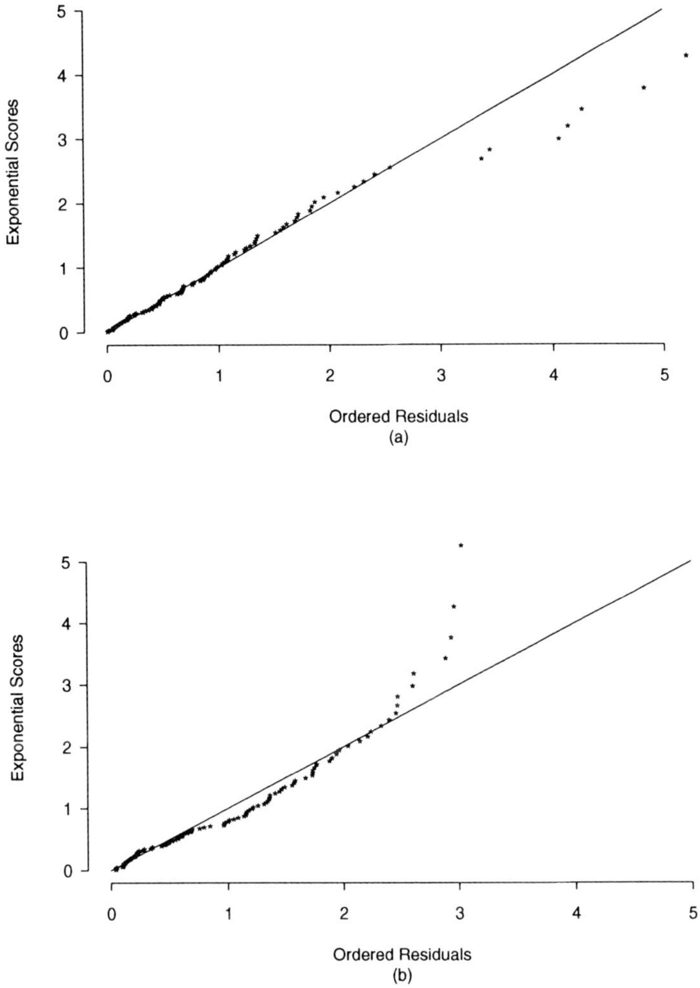

Figure 4: Residual checks for the mine data with (a) PEXP fit, (b) NHPP fit

As mentioned in Section 3, one aspect of practical importance in RG analysis is the estimation of the current reliability of the system, as measured by the reciprocal of the current value of the intensity. For our situation, this is also the common ground for comparing the two models. For the present data "time" is identified with "miles" and the current "mean time between failures" (MTBF), defined as the reciprocal of the current intensity of failure, is 297.21 and 292.78 for the PEXP and the NHPP fits respectively. An associated approximate 95% confidence interval for the MTBF for the PEXP is provided by the formula $\hat{\lambda}_n[1 \pm 1.96/\sqrt{n}]^{-1}$ and is calculated as : [213.51, 488.83]. The corresponding large sample confidence interval for the NHPP is computed as [188.36, 657].

Example 2 **Mine explosion data** Maguire, Pearson and Wynn (1952) provide the data of the number of days between mine explosions in Great Britain involving more than 10 men killed between December, 1875 and May, 1951. We want to see to what extent the safety regulations and other necessary precautions decrease the intensity of the accidents. For a graphical check for the PEXP fit, the residuals are plotted against the exponential scores (Figure 4(a)). Most of the points lie around the straight line with unit slope thereby ensuring a reasonable fit. By contrast, the corresponding residual plot for the NHPP (Figure 4(b)) shows more departure from the line. It is, however, evident from both the plots that a different model may be needed for the latter part of the data. The MLE's for the parameters are $\hat{\mu} = 0.08885$ and $\hat{\delta} = 1.6512$ with associated approximate confidence intervals [0.036, 0.214] and [1.46, 1.84] for μ and δ, respectively, thereby indicating reliability growth. The WPAE's are $\mu^* = 0.02719$ and $\delta^* = 1.40068$. The associated 95% large sample confidence intervals are [0.036, 0.22] and [1.20, 1.60] respectively. The estimate of the current MTBF is 394.38 days with an associated approximate 95% confidence interval [332.04, 485.53]. For the NHPP fit which is demonstrated by Crow (1974), the current MTBF is estimated as 337.51 days with the associated large sample 95% confidence interval [266.71, 459.51].

5. Appendix

Proof of Lemma 3.5: Observe that $\boldsymbol{F}_n$ involves the fixed point $\boldsymbol{\beta}_0$ and is $O_p((\log n)^2)$. By virtue of expressions (3.4) and (3.7), it then suffices to show that

$$T_{n,k} = (\log n)^2 n^{-1} (\log n)^{-k} \sum_{i=1}^{n} (\log i)^k [e_i(\beta) - e_i]$$

converges in probability to 0 uniformly in $\boldsymbol{\beta} \in M_n(\boldsymbol{\beta}_0)$ for k = 0,1,2 where

$$e_i(\boldsymbol{\beta}) = Y_i exp(-\boldsymbol{\beta}'\boldsymbol{x}_i) - 1, \quad e_i = e_i(\boldsymbol{\beta}_0), \qquad i = 1, 2, \ldots, n.$$

The result would follow by an application of the Markov inequality once we show that $E_{\boldsymbol{\beta}_0}(| T_{n,k} |) \longrightarrow 0$ uniformly in $\boldsymbol{\beta} \in M_n(\boldsymbol{\beta}_0)$. By the triangle inequality, we have

$$E_{\boldsymbol{\beta}_0}(| T_{n,k} |) \leq (\log n)^2 n^{-1} (\log n)^{-k} \sum_{i=1}^{n} (\log i)^k \mid 1 - exp\{-(\boldsymbol{\beta} - \boldsymbol{\beta}_0)'\boldsymbol{x}_i\} \mid.$$

For any real number x, we note

$$\begin{aligned} |\, 1 - exp(-x)\, | &\leq x, \qquad \text{for } x > 0 \\ &\leq |\, x\, |\, exp(-x), \qquad for\ x < 0 \end{aligned} .$$

For $\boldsymbol{\beta}$ *in* $M_n(\boldsymbol{\beta}_0)$, we have

$$|\, (\boldsymbol{\beta} - \boldsymbol{\beta}_0)'\boldsymbol{x}_i\, | \leq 2hn^{-\gamma}(\log n + \log i).$$

Thus, we arrive at the inequality

$$E_{\boldsymbol{\beta}_0}(|\, T_{n,k}\, |) \leq 4hn^{-\gamma}(\log n)^3 \exp(4hn^{-\gamma}\log n) .$$

Since $\gamma > 0$ and $(\log n)^m n^{-\gamma} \longrightarrow 0$ for any fixed nonnegative integer m, we have the required uniform convergence for $E_{\boldsymbol{\beta}_0}(|\, T_{n,k}\, |)$. //

Proof of Theorem 3.1 : For $\boldsymbol{\beta} \in M_n(\boldsymbol{\beta}_0)$ we have $\boldsymbol{\beta} - \boldsymbol{\beta}_0 = n^{-\gamma}(\tau_1 \log n,\ \tau_2)'$. Viewing $\boldsymbol{\beta}$ as a function of $\boldsymbol{\tau}$, a Taylor expansion of $\boldsymbol{\psi}_n(\boldsymbol{\beta})$ around $\boldsymbol{\beta}_0$ yields

$$\begin{aligned} \lambda_{1n}(\boldsymbol{\tau}) &\equiv l_{1n}(\boldsymbol{\beta}) = l_{1n} - n^{-\gamma}(a_{11}(\boldsymbol{\zeta})\tau_1 logn + a_{12}(\boldsymbol{\zeta})\tau_2), \\ \lambda_{2n}(\boldsymbol{\tau}) &\equiv (\log n)^{-1} l_{2n}(\boldsymbol{\beta}) = \log n^{-1} l_{2n} - n^{-\gamma}(a_{21}(\boldsymbol{\zeta})\tau_1 + a_{22}(\boldsymbol{\zeta})(\log n)^{-1}\tau_2) \end{aligned} .$$

where $(l_{1n},\ l_{2n})'$ and $a_{ij}(\boldsymbol{\zeta})$ are as defined in (3.3) and (3.2) respectively, with $\boldsymbol{\zeta}$ being a point on the line segment joining $\boldsymbol{\beta}$ and $\boldsymbol{\beta}_0$. Denoting $\boldsymbol{\lambda}_n(\boldsymbol{\tau}) = (\lambda_{1n}(\boldsymbol{\tau}), \lambda_{2n}(\boldsymbol{\tau}))$, and referring to the definitions of Z_n and $C_n(\boldsymbol{\beta})$ in (3.5) and (3.6), it follows that

$$\boldsymbol{\lambda}_n(\boldsymbol{\tau}) = n^{1/2}\boldsymbol{Z}_n - (\log n)n^{1-\gamma}\boldsymbol{C}_n(\boldsymbol{\zeta})\boldsymbol{\tau}. \tag{5.1}$$

Define

$$\boldsymbol{g}_n(\boldsymbol{\tau}) = (n^{1-\gamma}\log n)^{-1}\boldsymbol{F}_n\boldsymbol{\lambda}_n(\boldsymbol{\tau}), \quad \boldsymbol{g}_{n0} = (n^{1/2-\gamma}\log n)^{-1}\boldsymbol{F}_n\boldsymbol{Z}_n.$$

Premultiplying both sides of (5.1) by $\boldsymbol{\tau}'\boldsymbol{F}_n(n^{1/2-\gamma}\log n)^{-1}$, we obtain the relation

$$\begin{aligned} \boldsymbol{\tau}'\boldsymbol{g}_n(\boldsymbol{\tau}) &= -\boldsymbol{\tau}'\boldsymbol{\tau} + \boldsymbol{\tau}'\boldsymbol{g}_{n0} + \boldsymbol{\tau}'[\boldsymbol{I}_2 - \boldsymbol{F}_n\boldsymbol{C}_n(\boldsymbol{\tau})]\boldsymbol{\tau} \\ &= -\tau'\boldsymbol{\tau} + o_p(1) \end{aligned} \tag{5.2}$$

where the last equality follows from Lemma 3.5, the fact that $\gamma < 1/2$, and $(\log n)^{-1}\boldsymbol{F}_n\boldsymbol{Z}_n = O_p(1)$ by Lemma 3.4.

Result (5.2) implies that, given an $\epsilon > 0$, there exists $n_0 = n_0(\epsilon, h)$ such that

$$P\left(\sup_{||\tau||=h} \boldsymbol{\tau}'\boldsymbol{g}_n(\boldsymbol{\tau}) < 0\right) \geq 1 - \epsilon \qquad \forall\, n > n_0.$$

According to a version of Brouwer's fixed point theorem (see Smith(1985)), we have that $\boldsymbol{g}_n(\hat{\boldsymbol{\tau}}) = 0$ for some $\hat{\boldsymbol{\tau}} < h$. Thus, for all $n > n_0$, the probability is atleast

$1 - \epsilon$ that a $\hat{\boldsymbol{\tau}}_n$ exists that satisfies $\boldsymbol{g}_n(\hat{\boldsymbol{\tau}}_n) = 0$ and $\|\hat{\boldsymbol{\tau}}_n\| < h$. The corresponding $\hat{\boldsymbol{\beta}}_n = \boldsymbol{\beta}_0 + n^{-\gamma}(\hat{\tau}_{1n} \log n, \hat{\tau}_{2n})$ meets the requirements of the theorem. //

References

Ascher, H. E. (1968), "Evaluation of Repairable System Reliability Using the Bad-as-Old Concept," *IEEE Transactions on Reliability,* R-17, No.2, 103-110.

Bain, L.J. and Engelhardt, M. (1991), *Statistical Analysis of Reliability and Life-Testing Models,* 2nd Ed. Marcel Dekker, NY.

—— (1980), "Inferences on the Parameters and Current System Reliability for a Time Truncated Weibull Process," *Technometrics,* 22, No.3, 421-426.

—— (1982), "Sequential Probability Ratio Tests for the Shape Parameter of a NHPP," *IEEE Transactions on Reliability,* R-31, No.1, 79-83.

Bassin, W.M. (1969), "Increasing Hazard Functions and Overhaul Policy," *ARMS, IEEE-69,* C 8-R, 173-180.

—— (1973), "A Bayesian Optimal Overhaul Interval Model for the Weibull Restoration Process Case," *Journal of the American Statistical Association,* 68, No.343, 575-578.

Bell, R. and Mioduski, R. (1976), "Extension of Life of US Army Trucks," *ARMS, IEEE-76, CHO-1044-7RQC,* 200-205.

Benton, A.W. and Crow, L.H. (1989), "Integrated Reliability Growth Testing," *Proceedings of Annual Reliability and Maintainability Symposium,* 160-166.

Bhattacharyya, G.K., Fries, A. and Johnson, R.A. (1989), "Properties of Continuous Analog Estimators for a Discrete Reliability Growth Model," *IEEE Transactions on Reliability,* 38, No. 3, 373-378.

Bhattacharyya, G.K. and Ghosh, J.K. (1991), "Asymptotic Properties of estimators in a binomial reliability growth model and its continuous-time analog," *Journal of Statistical Planning and Inference,* 29, 43-53.

Crow, L.H. (1974), "Reliability Analysis for Complex Repairable Systems," *Reliability and Biometry,* ed. by F. Proschan and R.J. Serfling, 379-410.

—— (1982), "Confidence Interval Procedures for the Weibull Process with Applications to Reliability Growth," *Technometrics,* 24, No.1, 67-72.

Duane, J.T. (1964), "Learning Curve Approach to Reliability Monitoring," *IEEE transactions on Aerospace,* 2, No.2, 563-566.

Dubman, M. and Sherman, B. (1969), "Estimation of Parameters in a Transient Markov Chain Arising in a Reliability Growth Model," *The Annals of Mathematical Statistics,* 40, No.5, 1542-1556.

Durr, A.C. (1980), "Operational Repairable Equipments and the Duane Model," *Proceedings of Second International Conference on Reliability and Maintainability, CNET, France,* 189-193.

Finkelstein, J.M. (1976), "Confidence Bounds on the Parameters of the Weibull Process," *Technometrics,*18, No.1, 115-117.

Guida, M., Calabria, R. and Pulcini, G. (1989), "Bayes Inference for a NHPP with Power Intensity Law," *IEEE Transactions on Reliability,* 38, No.5, 603-609.

Higgins, J.J. and Tsokos, C.P. (1981), "A Quasi-Bayes Estimate of the Failure Intensity of a Reliability Growth Model," *IEEE Transactions on Reliability,* R-30, No.5, 471-475.

Lee, L. and Lee, S.K. (1978), "Some Results on Inference for the Weibull Process," *Technometrics,* 20, No.1, 41-45.

Lee, L. (1980), "Testing Adequacy of the Weibull and Log Linear Rate Models for a Poisson Process," *Technometrics,* 22, No.2,195-199.

—— (1980), "Comparing Rates of Several Independent Weibull Processes," *Technometrics,* 22, No.3, 427-430.

Mil-HDBK-189 (1981).

Rigdon, S.E. and Basu, A.P. (1988), "Estimating the Intensity Function of a Weibull Process at the Current Time: Failure Truncated Case," *Journal of Statistical Computation and Simulation,* 30, 17-38.

Smith, R.L. (1985), "Maximum Likelihood Estimation in a class of Nonregular cases," *Biometrika,* 72, 67-90.

Thompson, W.A., Jr. (1988), *Point Process Models with Applications to Safety and Reliability,* Chapman and Hall, NY.

ADVANCES IN RELIABILITY
Edited by Asit P. Basu

Approximate Confidence Intervals for the Difference in Means of Two Gamma Distributions

Wei-Kei Shiue[a] and Lee J. Bain[b]

[a]Department of Mathematics & Statistics, Southern Illinois University-Edwardsville

[b]Department of Mathematics & Statistics University of Missouri-Rolla

ABSTRACT

Two methods for setting approximate confidence intervals on the difference of expected mean squares in random effects experimental design models were modified to apply to the problem of setting confidence intervals on the difference of means of two gamma distributions. Most approximate methods in the experimental design area are designed to apply to positive components of variance, but since the above methods were valid for differences of unrestricted sign, they can be extended directly to the case of differences in exponential means. It is verified by a simulation study that the methods can also be extended to the gamma case, where the shape parameters are estimated.

1. INTRODUCTION

An important problem in the application of random or mixed experimental design models is to determine confidence intervals on linear combinations of variance components. For example, consider the one factor components-of-variance model

$$Y_{ij} = \mu + A_i + E_{ij};\ i = 1,\cdots,I,\ j = 1,\cdots,J,$$

where A_i and E_{ij} are jointly independent normal random variables with zero means, and variances σ_A^2 and σ_E^2, respectively. Denoting the among group mean square as S_1^2 and the within group mean square by S_2^2, then $E(S_1^2) = \theta_1 = J\sigma_A^2 + \sigma_E^2$, $E(S_2^2) = \theta_2 = \sigma_E^2$, and the among group variance component, σ_A^2, is the difference,

$$\gamma = \sigma_A^2 = (\theta_1-\theta_2)/J.$$

Also,

$$\frac{r_i S_i^2}{\theta_i} \sim \chi^2(r_i), \text{ where } r_1 = I-1,\ r_2 = I(j-1).$$

Much work has been devoted to developing approximate methods for setting confidence intervals on γ, where it is assumed $\gamma > 0$. Boardman (1974) compares nine such methods. Howe (1974) used the Cornish-Fisher expansion technique to obtain approximate confidence intervals for $\gamma \geq 0$. LU, Graybill and Burdick (1988) reviews previous results and extends Howe's results to the case where the sign of γ is unrestricted. Samaranayake and Bain (1988), also consider the unrestricted sign case, with an application to the difference in exponential means problem. Ting, et al (1990) studies the method based on Howe's result, and a second modified method, and they discuss the general problem of setting a confidence interval on $\delta-\delta^*$, where $\delta = \sum_{i=1}^{A} C_i\theta_i$, $\delta^* = \sum_{i=A=1}^{B} C_i\theta_i$, $C_i \geq 0$, and the sign of the difference is unrestricted. These results can be applied directly to the difference in exponential means problem. Suppose $x_{i1},\cdots,x_{in_i}$ represents a random sample of size n_i from an exponential distribution with mean θ_i, $i = 1,2$. Then

$$\frac{2n_i\bar{X}_i}{\theta_i} \sim \chi^2(2n_i),$$

so the approximate confidence intervals for $\theta_1-\theta_2$ developed for the components-of-variance problem (with unrestricted sign) applies directly to the exponential case by replacing S_i^2 by $\bar{X}_i$ and $r_i = 2n_i$. Our primary purpose is to consider the problem of confidence intervals for the difference in means of two gamma distributions.

2. THE GAMMA CASE

Suppose $X_1,\cdots,X_m$ and $Y_1,\cdots,Y_n$ represent random samples from two gamma distributions with shape parameters κ_1 and κ_2 and scale parameters β_1 and β_2 respectively. It is known that

$$\frac{2m\kappa_1\bar{X}}{\mu_1} \sim \chi^2(2m\kappa_1), \quad \frac{2n\kappa_2\bar{Y}}{\mu_2} \sim \chi^2(2n\kappa_2),$$

where $\mu_1 = \kappa_1\beta_1$ and $\mu_2 = \kappa_2\beta_2$ are the respective means. Approximate confidence limits for $\mu_1-\mu_2$ may also be obtained directly for this case if κ_1 and κ_2 are known. The two methods discussed by Ting, et al, provide two-sided 1-α level confidence limits for $\mu_1-\mu_2$, (L,U), which may be stated as follows for this case.

METHOD 1.

$$L_0 = \begin{cases} \bar{x}-\bar{y} - [\bar{x}^2[(1-1/F_1)^2 - (L_2/F_1)^2] + \bar{y}^2L_2^2]^{1/2}, & \text{if } \bar{x}/\bar{y} \le F_1 \\ \bar{x}-\bar{y} - [\bar{x}^2L_1^2 + \bar{y}^2[(F_1-1)^2 - F_1^2L_1^2]]^{1/2}, & \text{if } \bar{x}/\bar{y} > F_1. \end{cases}$$

$$U_0 = \begin{cases} \bar{x}-\bar{y} + [\bar{x}^2[(1/F_2-1)^2 - (H_2/F_2)^2] + \bar{y}^2H_2^2]^{1/2}, & \text{if } \bar{x}/\bar{y} \le F_2 \\ \bar{x}-\bar{y} + [\bar{x}^2H_1^2 + \bar{y}^2[(1-F_2)^2 - F_2^2H_1^2]]^{1/2}, & \bar{x}/\bar{y} > F_2, \end{cases}$$

where $F_1 = F(\alpha/2;\ n_1,n_2)$, $F_2 = F(1-\alpha/2;\ n_1,n_2)$,

$L_1 = 1 - 1/F(\alpha/2;\ n_1,\infty)$, $L_2 = 1/F(1-\alpha/2;\ n_2,\infty) - 1$

$H_1 = 1/F(1-\alpha/2;\ n_1,\infty) - 1$, $H_2 = 1-1/F(\alpha/2;\ n_2,\infty)$,

and $F(\gamma;n_1,n_2)$ denotes the upper γ percentile of the F-distribution with n_1 and n_2 degrees of freedom. Note that $F(\gamma;n,\infty) = \chi^2(\gamma;n)/n$.

The second method includes a cross-product term and the two sided 1-α level confidence interval, (L,U), may be stated as follows.

METHOD 2.

$$L_c = \bar{x} - \bar{y} - [L_1^2\bar{x}^2 + L_2^2\bar{y}^2 + L_{12}\bar{x}\,\bar{y}]^{1/2}$$

$$U_c = \bar{x} - \bar{y} + [H_1^2\bar{x}^2 + H_2^2\bar{y}^2 + H_{12}\bar{x}\,\bar{y}]^{1/2},$$

where

$$L_{12} = [(F_1-1)^2 - L_1^2F_1^2 - L_2^2]/F_1,$$

and

$$H_{12} = [(1-F_2)^2 - H_1^2F_2^2 - H_2^2]/F_2.$$

As suggested earlier, these results give approximate $1-\alpha$ level confidence intervals for $\mu_1-\mu_2$ in the gamma case if $n_1 = 2m\kappa_1$ and $n_2 = 2n\kappa_2$. The actual probability levels achieved by the intervals depend on $\delta = \mu_1/(\mu_1+\mu_2)$, and the intervals were constructed to be exact for $\delta = 0$, $\delta = 1/2$ $(\mu_1=\mu_2)$ and $\delta = 1$, and they were verified to be approximately correct otherwise.

Of course, in practice the values of κ_1 and κ_2 would be unknown. A natural approach is to consider the possibility of replacing κ_1 and κ_2 with sample estimates, and this seems particularly appealing in this case since n_1 and n_2, and thus κ_1 and κ_2, do not appear directly in the formulas, but only in the degrees of freedom in the percentiles. Consequently, we considered the above intervals with n_1 and n_2 replaced by $2m\hat{\kappa}_1$ and $2n\hat{\kappa}_2$, where the $\hat{\kappa}_i$ denote the approximately unbiased maximum likelihood estimators, e.g., $\hat{\kappa}_1 = \frac{m-3}{m}\tilde{\kappa}_1 + \frac{2}{3m}$, where $\tilde{\kappa}$ denotes the usual maximum likelihood estimator. In the simulation study mentioned below, the closed form approximation of $\tilde{\kappa}$ given by Greenwood and Duran (1960) was used.

3. SIMULATION RESULTS AND CONCLUSIONS

If estimates for κ_1 and κ_2 are used then the actual confidence levels for the approximate intervals not only depend on δ, but also on the values of κ_1 and κ_2. A simulation study was conducted to determine the actual coverages achieved by the approximate intervals, with the κ_i estimated by $\hat{\kappa}_i$, for values of $\delta = 1/4, 1/2, 3/4$, $\kappa_1 = .5,1,2$, $\kappa_2 = \kappa_1(\beta_1/\beta_2)(1/\delta-1)$, and a number of combinations of sample sizes and confidence levels as shown in Table 1.

For the simulation β_1 and β_2 were taken to be one, since the distribution of the $\hat{\kappa}_i$ does not depend on the β_i, and various combinations of κ_1 and κ_2 were obtained by varying δ. In terms of $n_1=2m\kappa_1$, and $n_2=2n\kappa_2$, changing β_1/β_2 or the κ_i are comparable to changing m and n. So table 1 should provide a representative overview of the general suitability of the methods, even though there are a large number of different factors to consider. It appears both methods give very good results, and they should be quite suitable for use in computing confidence intervals on the differences of means from gamma distributions. The average length of the intervals was also computed for both methods, but there was found to be very little difference in the two methods, and these results are not reported. Method 2 has the advantage of being somewhat simpler to compute.

Note that these results can also be used to obtain confidence intervals on $c_1\mu_1 - c_2\mu_2$ by simply replacing $\bar{x}$ with $c_1\bar{x}$ and $\bar{y}$ with $c_2\bar{y}$.

In the simulation study the random samples were computed using IMSL subroutine RNGAM, and the results are based on 1000 samples in each case.

TABLE 1

Actual Probability coverage of approximate confidence Intervals given by Method 2(M2) and Method 1(M1)

$1-\alpha = .90$

m	n	δ	.25			.50			.75		
		κ_1	.5	1	2	.5	1	2	.5	1	2
10	20	M2	.916	.907	.911	.910	.911	.900	.897	.906	.901
		M1	.907	.924	.903	.900	.888	.908	.897	.905	.915
20	10	M2	.913	.889	.927	.918	.901	.911	.899	.909	.909
		M1	.912	.902	.913	.906	.903	.927	.903	.911	.903
10	40	M2	.904	.912	.919	.901	.901	.916	.891	.893	.903
		M1	.899	.899	.906	.907	.902	.917	.907	.897	.908
40	10	M2	.902	.925	.889	.904	.901	.898	.899	.894	.902
		M1	.895	.920	.908	.897	.902	.920	.895	.877	.894
20	40	M2	.895	.906	.893	.900	.920	.912	.877	.894	.900
		M1	.920	.902	.898	.900	.908	.911	.906	.900	.907
40	20	M2	.915	.908	.906	.897	.896	.905	.903	.893	.904
		M1	.904	.894	.897	.891	.899	.914	.900	.894	.888

1-α = .95

10	20	M2	.956	.954	.952	.945	.959	.941	.946	.949	.956
		M1	.952	.960	.957	.946	.931	.955	.945	.951	.951
20	10	M2	.962	.947	.958	.965	.947	.955	.950	.955	.957
		M1	.952	.953	.944	.952	.944	.965	.959	.955	.950
10	40	M2	.945	.960	.952	.948	.944	.961	.939	.943	.953
		M1	.942	.942	.958	.945	.947	.962	.958	.946	.953
40	10	M2	.951	.955	.946	.956	.953	.957	.945	.948	.955
		M1	.937	.968	.945	.948	.961	.954	.937	.932	.941
20	40	M2	.943	.961	.941	.943	.961	.954	.929	.944	.952
		M1	.960	.953	.948	.946	.956	.957	.953	.950	.950
40	20	M2	.949	.953	.951	.952	.937	.951	.953	.940	.947
		M1	.944	.943	.944	.944	.955	.958	.954	.949	.940

1-α = .975

10	20	M2	.974	.974	.975	.976	.983	.967	.975	.975	.974
		M1	.971	.975	.979	.977	.961	.972	.973	.972	.975
20	10	M2	.979	.968	.980	.981	.976	.973	.968	.974	.976
		M1	.976	.976	.973	.974	.971	.981	.977	.979	.975
10	40	M2	.973	.982	.980	.970	.971	.975	.970	.973	.975
		M1	.973	.968	.982	.973	.972	.978	.977	.963	.971
40	10	M2	.973	.971	.967	.975	.969	.972	.976	.973	.976
		M1	.962	.983	.970	.966	.980	.973	.967	.964	.970
20	40	M2	.972	.979	.969	.973	.983	.976	.966	.973	.973
		M1	.975	.976	.977	.975	.979	.979	.976	.971	.973
40	20	M2	.977	.980	.973	.975	.965	.975	.979	.963	.970
		M1	.974	.975	.971	.967	.974	.981	.978	.977	.976

1-α = .99

10	20	M2	.988	.990	.985	.989	.991	.985	.988	.990	.990
		M1	.988	.991	.989	.990	.977	.985	.987	.985	.987
20	10	M2	.991	.986	.987	.991	.990	.987	.982	.990	.990
		M1	.984	.988	.991	.988	.991	.993	.988	.989	.986
10	40	M2	.985	.991	.991	.991	.986	.991	.987	.987	.985
		M1	.988	.984	.996	.986	.991	.990	.989	.978	.987
40	10	M2	.992	.985	.983	.990	.987	.991	.989	.987	.989
		M1	.982	.991	.982	.990	.991	.984	.982	.986	.988
20	40	M2	.985	.993	.989	.990	.996	.990	.985	.989	.990
		M1	.991	.990	.993	.984	.989	.994	.990	.987	.990
40	20	M2	.988	.988	.991	.989	.985	.990	.997	.980	.992
		M1	.984	.988	.988	.989	.982	.993	.995	.992	.992

REFERENCES

Boardman, T. J. (1974). Confidence intervals for variance components - A Comparative Monte-Carlo study. Biometrics 30, 251-262.
Graybill, F. A. (1976). Theory and Application of Linear Model. Wadsworth, California.
Greenwood, J. A. and Durand, D. (1960). Aids for fitting the gamma distribution by maximum likelihood. Technometrics 2, 55-65.
Howe, W. G. (1974). Approximate confidence limits on the mean X+Y where X and Y are two tabled independent random variables, J. Amer. Statist. Assoc. 69, 789-794.
Lu, T. -F.C., Graybill, F. A. and Burdick, R. K. (1988). Confidence intervals on a difference of expected mean squares. J. Statist. planning and Inference. 18, 35-43.
Ting, N., Burdick, R. K., Graybill, F. A., Jeyaratnam, S., and Lu, T.-F.C. (1990). Confidence Intervals on Linear Combinations of variance components that are unrestricted in sign. J. Statist. Comput. Simul., 35, 135-143.
Samaranayake, V. A., and Bain, L. J. (1988). A confidence interval for treatment component-of-variance with applications to differences in means of two exponential distributions, J. Statist. Comput. Simul. 29, 317-332.

ADVANCES IN RELIABILITY
Edited by Asit P. Basu

Some basic contributions to the theory of comparative life testing experiments [1]

Jagdish N. Srivastava

Statistics Department, Colorado State University, Fort Collins, Colorado 80523

Abstract

Suppose we wish to compare m different brands of a machine under the proportional hazards model. In this paper, we briefly discuss the properties of four censoring schemes. Two of these schemes correspond to classical censoring of Type I and II, and the other two are modified versions of the same. The first two schemes involve independent experiments on each brand, while the others lead to a combined experiment. It is seen that one of the new schemes is the most appealing.

1. INTRODUCTION

Consider m different brands of a particular machine. For the ith brand ($i = 1, ..., m$), let y_i denote the lifetime and $\psi_i(t)$ the survival function, where t denotes time. We will assume that

$$\psi_i(t) = e^{-\lambda_i \theta_i(t)}; \qquad i = 1, ..., m \tag{1.1}$$

where λ_i is a positive real number, and $\theta_i(t)$ is a nondecreasing function of t, such that

$$\theta_i(0) = 0, \quad \theta_i(\infty) = \infty, \quad \text{for all } i. \tag{1.2}$$

It is clear that the density function of y_i is given by

$$f_{y_i}(t) = \ -\psi_i'(t) = \lambda\theta_i'(t)e^{-\lambda_i\theta_i(t)}, \tag{1.3}$$

where prime ($'$) denotes differential coefficient. Also then, $\Lambda_i(t)$, the hazard rate for y_i is given by

$$\Lambda_i(t) = \lambda_i\theta_i'(t), \quad \text{for all } i. \tag{1.4}$$

[1]This work was supported by AFOSR grant #91-0031

A random variable y which obeys a survival function of the form at (1.1) will be said to have the generalized Weibull distribution (W_0); its special case, when θ_i is of the form

$$\theta_i(t) = t_i^{\beta}, \tag{1.5}$$

is the well known Weibull distribution (W). The Weibull distribution arises quite often in real life. However, the W_0-distribution is the most general possibility. As a special case, consider the situation where a machine had several components connected in series, and the lifetime of the different components were mutually independent, and obeyed a Weibull survival function. For such a situation, the special case of W_0 will be denoted by W_1, and $\theta_i(t)$ will be of the form

$$\theta_i(t) = \sum_{i=1}^{k} \lambda_i t^{\beta_i}, \tag{1.6}$$

where α_i and β_i are suitable positive real numbers.

In this paper, we shall consider the special situation where the θ_i are independent of i, i.e.

$$\theta_1(t) = \theta_2(t) = \cdots = \theta_m(t) = \theta(t), \text{ say.} \tag{1.7}$$

It is clear that this case corresponds to the well known "proportional hazards" model.

Under proportional hazards, the problem of comparing the m different brands reduces to the problem to comparing $\lambda_1, \lambda_2, .., \lambda_m$. To be able to compare the λ's, experimentation is to be performed. In this paper, we shall summarize some classical ways of experimentation, and some recently developed techniques. One of the techniques, which appears to be better than others will be especially explained. Most of the results summarized in this paper are contained in the references at the end.

2. CLASSICAL AND NEW CENSORING SCHEMES

Basically, the classical techniques involve doing independent experiments on each brand, whereas the modifications consist of doing a kind of a combined experiment. We now explain these.

One of the main problems in life testing experiments is that the experimentation may take a long time. In other words, if we start with a certain number of machines of any brand, and wait to study the life time of each of these machines, then we have to wait until all the machines fail, which may involve too much time. Because of this, two classical techniques of shortening the experiment are popular. These are called censoring schemes. Censoring of Type I (C1) consists of stopping the experiment at a predetermined time. Type II censoring (C2) consists of stopping the experiment as soon as a given number of machines has failed. Clearly, under C1, it is possible that no machine may fail during the given time period. Similarly, under C2, it is possible that one may have to wait indefinitely for the given number of machines to fail.

In this paper, we shall assume that we are in a state of total ignorance concerning the relative values of the λ's. Thus, we shall treat each brand in the "same" manner. For example, under C1 we shall assume that we start with u machines of each of the m types, and watch the failure of machines of different brands, until a predetermined time T has passed, so that u and T are same for each brand.

Under C2, we shall assume that we start with u machines of each brand, and stop the experiment for the ith brand ($i = 1, ..., m$), when G_0 machines of that brand have failed. Here again, we notice that we are treating the brands the same way in the sense that we are starting with the same number of machines for each brand and stopping the experiment for any brand after the same predetermined number of machines G_0 has failed.

Each of these censoring schemes is a "design" in the sense that it describes how to do the experiment. We now describe the modified schemes, which are called self relocating designs (SRD). Basically, the idea behind an SRD is as follows. As under the classical case, we start with the same number u of machines of each type. However, here, at all times during the experiment we maintain the same number of competing machines of each brand. This is done as follows. If we have u machines of each type, we have a total of um machines. This set of um machines is divided into u subsets, such that in each subset, there is exactly one machine of each brand. These subsets are made at the start of the experiment. We start the experiment with the um machines of the u subsets at time $t_0(=0)$. We wait until the first failure occurs. Instantaneously after the first failure, the $(m-1)$ machines which are in the set to which the failed machine belongs, but which have not failed, are censored (i.e., they are removed from the experiment). Thus, instantaneously after the first failure, there are exactly $u-1$ machines of each brand still continuing in the experiment. The experiment is continued until the second failure. Instantaneously after the second failure, we select the subset which contains the machine which was the second failure and censor the $(m-1)$ machines which are in this set and which have not failed. Thus, after the second failure, we are left with $(u-2)$ machines of each type still continuing in the experiment. This process is continued until a specified number G_0 of failures has occurred (corresponding to Type II censoring) or until a specified time period T has passed (corresponding to censoring of Type I).

In contrast to this, the censoring procedure used in the earlier papers is as follows. We start with um machines as before. At the time of the first failure, we randomly censor one machine of each brand other than the one to which the failed machine belongs. Thus, instantaneously after the first failure we have a total of $(u-1)m$ machines continuing in the experiment. As before, the experiment is continued until the second failure, at which time we randomly censor one machine belonging to each brand to which the newly failed machine does not belong. The experiment is continued in the same fashion until a specified number of machines has failed, or until a specified time period has elapsed.

It has been shown in Srivastava and Khodadadi (1991) that the two procedures presented above have identical statistical properties. Because of simplicity, in this paper, we shall assume that we are following the first procedure.

The SRD obtained by modifying C1 and C2 as above will be denoted respectively by SRD11, and SRD21. In Srivastava (1986), a different design was called SRD11; this design (which was defined only for the special case of exponential distributions) has properties similar to SRD21, and will not be considered here. Also, the designs that we have defined under procedure C1 and C2 above will respectively be denoted by C11 and C21. To explain the notation, consider C11. The second "1" in C11 refers to one particular kind of design under C1 (in the sense that we could have other designs under C1 which are different from C11). For instance, under C1 we could have a design in which we start with unequal number of machines for the different brands, or start the experiment at different times for different brands, or proceed in some other manner. Similar remarks hold for the notation for other procedures.

In this paper, we shall present a comparative discussion of these censoring schemes. The scheme SRD21 will be discussed in detail, since it is the most appealing.

3. VARIANCE MATRICES

Let $\underline{\lambda} = (\lambda_1, ..., \lambda_m)'$, denote the vector of parameters. Let $\hat{\lambda}_{c1}, \hat{\lambda}_{c2}, \hat{\lambda}_{s1}$, and $\hat{\lambda}_{s2}$ respectively denote the maximum likelihood estimates (m.l.e) of $\underline{\lambda}$ under C11, C21, SRD11, and SRD21. Also, let V_{c1}, V_{c2}, V_{s1}, and V_{s2} respectively denote the asymptotic variance (ASVAR) matrices under the four cases. We now discuss these matrices. Define

$$
\begin{aligned}
\lambda &= \lambda_1 + \lambda_2 + .. + \lambda_m \\
c &= 1 - e^{-\lambda\theta(T)} \\
c_i &= 1 - e^{-\lambda_i\theta(T^*)},
\end{aligned}
\tag{3.1}
$$

where T^* is the period of time for which each of the m experiments under C11 are run. Now, recall that under SRD21, the number of failures which we shall observe is fixed to be G. Similarly, for each of the m experiments under C21, the number of failures which we will observe is fixed to be G_0. However, in the case of C11 and SRD11, the experiment is going to be run for a fixed period of time, respectively T and T^*. Thus, in the case of C11 and SRD11, the number of failures that we will observe will be random variables. Let H and $H_i (i = 1, ..., m)$ respectively denote the number of failures observed for SRD11, and for the ith experiment under C11. Now it turns out that we have

$$
\begin{aligned}
E(H) &= uc \\
E(H_i) &= uc_i, \quad \text{for } i = 1, ..., m,
\end{aligned}
\tag{3.2}
$$

In order to be able to compare the four approaches, namely, C11, C12, SRD11, and SRD21, we need to put them on the same footing with respect to the number of failures. Thus, for this purpose, we require that

$$
G = mG_0 = E(H) = \sum_{i=1}^{m} E(H_i),
\tag{3.3}
$$

so that, from 3.2, we get

$$G = mG_0 = uc = u\sum_{i=1}^{m} c_i \tag{3.4}$$

Thus, condition (3.4) insures that we will observe the same number G of failures under SRD21 and C21 and furthermore, we will also expect to observe this number of failures under SRD11 and C11. In most situations, the number of failures constitutes the principal cost of experimentation, since the testing is destructive. However, in other situations, the experimental units may not be expensive and/or, there may be other avenues of expense. In this case, one may wish to consider the total cost associated with the experiment more elaborately. A brief discussion of this will be presented later on.

It turns out that

$$\begin{aligned}
V_{c1} &= (1/u) \text{ diag } (\lambda_1^2/c_1, ..., \lambda_m^2/c_m) \\
V_{c2} &= (m/G) \text{ diag } (\lambda_1^2, ..., \lambda_m^2) \\
V_{s1} &= (\lambda/uc) \text{ diag } (\lambda_1, ..\lambda_m) = (\bar{\lambda}m/G) \text{ diag } (\lambda_1, ..., \lambda_m) \\
V_{s2} &= (\lambda/G) \text{ diag } (\lambda_1, ..., \lambda_m) = \bar{\lambda}m/G) \text{ diag } (\lambda_1, ..., \lambda_m),
\end{aligned} \tag{3.5}$$

where the second expression for V_{s1} and V_{s2} are obtained using (3.4), and where $\bar{\lambda}$ is the arithmetic mean of $\lambda_1, ..., \lambda_m$. We compare the traces of these ASVAR matrices. It is well known that these traces are proportional to the average variance of the estimates all normalized linear combinations of the parameters. We have

$$\begin{aligned}
trV_{c1} &= (1/u)\sum_{i=1}^{m}(\lambda_i^2/c_i) = (m/G)\left[(\sum_{i=1}^{m} c_i/m)\{\sum_{i=1}^{m}(\lambda_i^2/c_i)\}\right] \\
trV_{c2} &= (m/G)\left[\sum_{i=1}^{m}\lambda_i^2\right] \\
trV_{s1} &= trV_{s2} = m^2\bar{\lambda}^2/G = (m/G)[m\bar{\lambda}^2].
\end{aligned} \tag{3.6}$$

Because of the Cauchy-Schwartz inequality, and since variance is always nonnegative, we have

$$\left(\sum_{i-1}^{m} c_i/m\right)\left\{\sum_{i-1}^{m}\lambda_i^2/c_i\right\} \geq \left\{\sum_{i=1}^{m}\sqrt{\frac{c_i}{m}}\cdot\frac{\lambda_i}{\sqrt{c_i}}\right\}^2 = m\bar{\lambda}^2$$

$$\sum_{i=1}^{m}\lambda_i^2 \geq m\bar{\lambda}^2. \tag{3.7}$$

This leads to the following important result obtained in the earlier papers.

THEOREM 3.1

For the ASVAR matrices of the estimators of $\bar{\lambda}$ under the four approaches, we have

$$trV_{s1} = trV_{s2} \leq trV_{c1}, trV_{c2}, \tag{3.8}$$

provided (3.4) holds.

The above assures the superiority of the SRD procedures compared to the classical ones with respect to the trace of ASVAR matrices.

4. TOTAL EXPECTED TIME UNDER EXPERIMENT

Besides the total number of machines that fail in an experiment, another important contribution to the cost of experiment comes from the total time for which the machines are experimented upon. To make ideas clear, let us consider the case of C21 with only one brand of machines. Suppose that we start with u machines, and wait until a total G_0 machines have failed. Let $t_i (i = 1, ..., G_0)$ denote the time of failure of the ith machine. Then the total time for which machines have been experimented upon, or the total time under experimentation (TTUE) is given by

$$TTUE = t_1 + t_2 + \cdot + t_{G_0 - 1} + (u - G_0 + 1) t_{G_0}. \tag{4.1}$$

since, clearly, $(u - G_0 + 1)$ machines were in operation until time t_{G_0}, and out of the remaining $(G_0 - 1)$ machines, one each failed at respective times $t_1, t_2, ..., t_{G_0 - 1}$. Note that the TTUE is a random variable. Consider the expected value of TTUE, or in other words, the total expected time under experiment (TETUE). To obtain the TETUE, the joint distribution of $t_1, t_2, ..., t_{G_0}$, has been found, from which the expected value of TTUE is determined. We now provide expressions for the same.

Let $T_{c1a}, T_{c2a}, T_{s1a}$ and T_{s2a} respectively denote the TTUE under the case of C11, C21, SRD11, and SRD21, and let T_{c1}, T_{c2}, T_{s1}, and T_{s2} respectively denote the corresponding values for the TETUE. We then obviously have

$$\begin{aligned} T_{c1a} &= \textstyle\sum_{i=1}^{m} \left[t_{c1i}(1) + \cdots + t_{c1i}(H_i - 1) + (u - H_i + 1) t_{c1i}(H_i) \right] \\ T_{c2a} &= \textstyle\sum_{i=1}^{m} \left[t_{c2i}(1) + \cdots + t_{c2i}(G_0 - 1) + (u - G_0 + 1) t_{c2i}(G_0) \right] \\ T_{s1a} &= m \left[t_{s1}(1) + \cdots + t_{s1}(H - 1) + (u - H + 1) t_{s1}(H) \right] \\ T_{s2a} &= m \left[t_{s2}(1) + \cdots + t_{s2}(G - 1) + (u - G + 1) t_{s2}(G) \right], \end{aligned} \tag{4.2}$$

where $t_{c1i}(g)(g = 1, ..., H_i)$ denotes the failure time of the gth machine in the ith experiment $(i = 1, ..., m)$, and H_i is the total number of machines that fail during the time period T^* under C11 in the ith subexperiment. Also, $t_{c2i}(g)(g = 1, ... G_0)$ denotes the time of failure of the gth machine in the ith $(i = 1, ..., m)$ subexperiment under C2, and furthermore $t_{s1}(g)(g = 1, ... H)$ and $t_{s2}(g)(g = 1, ... G)$ respectively denote the time of failure of the gth machine under SRD11 and SRD21. Note that the H_i, and H are random variables.

We now present the values of the TETUE, when the underlying distribution is Weibull, so that the function θ is of the form

$$\theta(t) = t^{\beta} \tag{4.3}$$

where β is a positive real number. We shall first consider T_{s1} and T_{s2}. Define (for real numbers u, and positive integers k, with $u \geq k$)

$$\zeta(u,k,\beta) = \frac{\Gamma(1+u)\Gamma(1+1/\beta)}{\Gamma(1+u-k)} \sum_{h=0}^{k-1} \left\{ \frac{(-1)^h}{h!(k-1-h)!(u-k+1+h)^{1+1/\beta}} \right\}$$

$$\varepsilon(u,k,\beta) = \sum_{j=1}^{k} \zeta(u,j,\beta) + (u-k)\zeta(u,k,\beta) \tag{4.4}$$

Then, it can be shown that

$$\begin{aligned} T_{s2} &= m\lambda^{-1/\beta}\varepsilon(u,G,\beta) = T_1(u,G,m,\lambda,\beta), \text{ say} \\ T_{c2} &= \left[\sum_{i=1}^{m}(\lambda_i)^{-1/\beta}\right] \varepsilon(u,G_0,\beta) = \sum_{i=1}^{m} T_1(u,G_0,1,\lambda_i,\beta) \end{aligned} \tag{4.5}$$

We next consider T_{s1} and T_{c1}. We have

$$T_{s1} = muTe^{-\lambda\theta(t)} + m\sum_{g=1}^{u} \delta_g Q_g \tag{4.6}$$

where

$$\begin{aligned} \delta_g &= e^{-\lambda\theta(T)(u-g)} \binom{u}{g} \\ Q_g &= \sum_{j=1}^{g} \left[\frac{g!R_{gj}}{(j-1)!(g-j)!}\right] \end{aligned} \tag{4.7}$$

where

$$\begin{aligned} R_{gj} &= \sum_{k-0}^{j-1} \sum_{s=0}^{g-j} \binom{j-1}{k} \binom{g-j}{s} (-1)^{k+g-j+s} e^{-(g-j-s)\lambda T^{\beta}} \lambda^{-1/\beta}(k+s+1)^{-1/\beta} L_{ks} \\ L_{ks} &= \int_0^{(k+s+1)\lambda T^{\beta}} y^{1/\beta} e^{-y} dy. \end{aligned} \tag{4.8}$$

Let $T_{s1} = T_{s1}(m,T,u,\beta;\lambda)$ (4.9)

Then, we have

$$T_{c1} = \sum_{i=1}^{m} T_{s1}(1, T^*, u, \beta; \lambda_i) \tag{4.10}$$

The expressions for the TETUE are in general quite complicated, this being particularly so for the two cases under Type I censoring. This is a reflection of the fact that in general the theory of Type I censoring tends to be much complicated and intractable than that of Type II.

5. COMPARISON OF CENSORING SCHEMES

We now discuss the different schemes from various points of view. In Srivastava (1986), a numerical study of the TETUE was made for SRD21 and C21. For this purpose four sets of values of m, λ's and β's were selected. These values are displayed below:

$$\begin{aligned}
(I) \quad m &= 3, \quad \underset{\sim}{\lambda} = (1, 4, 7)' \\
(II) \quad m &= 3, \quad \underset{\sim}{\lambda} = (0.5, 2.5, 4.5)' \\
(III) \quad m &= 4, \quad \underset{\sim}{\lambda} = (0.25, 0.50, 0.75, 1.00)' \\
(IV) \quad m &= 4, \quad \underset{\sim}{\lambda} = (0.03, 0.06, 0.09, 0.12, 0.15, 0.18)'
\end{aligned} \tag{5.1}$$

For of each of these sets of values of m and λ, the following values of u, g, and β's were tried:

$$\begin{aligned}
\beta &= 2.0, 1.75, 1.50, 1.00, 0.75, 0.50 \\
(u, G) &= (12, 12), (24, 24), (24, 12)
\end{aligned} \tag{5.2}$$

The same set of values of the above parameters was used for the study of SRD11 and C11. It is seen that in two of the chosen values of the pair (u, g), we have $u = g$. For this case, some simple and very important results hold. We present them below.

<u>THEOREM 5.1</u> (a) When $u = G$, the following results hold:

$$\begin{aligned}
T_{s1} = T_{s2} &= mG \textstyle\int_{x=o}^{\infty} \lambda x \theta'(x) e^{-\lambda\theta(x)} dx \\
&= mG[\Gamma(1 + 1/\beta)]\lambda^{-1/\beta}, \quad \text{for Weibull case}
\end{aligned} \tag{5.3}$$

(b) For the case of exponential distribution, i.e. when $\beta = 1$ and for all $u \geq G$, we have

$$\begin{aligned}
T_{s2} &= mG/\lambda, \\
T_{c2} &= (G/m)\sum_{i=1}^{m} \lambda_i^{-1}
\end{aligned}$$

Table 1

Value of TETUE for various cases

(u,G)	Parameter Set	Scheme	Value of β					
			2.00	1.50	1.25	1.00	0.75	0.50
(24,24)	I	SRD21	18.420	12.400	9.186	6.000	3.120	1.000
		C21	24.950	18.605	14.986	11.143	7.244	3.361
		C11	20.954	14.120	10.333	6.497	3.025	0.669
	II	SRD21	23.300	16.964	13.379	9.600	5.840	2.560
		C21	33.454	27.818	24.530	20.978	17.338	14.113
		C11	26.666	19.475	15.200	10.528	5.756	1.799
	III	SDR21	53.808	47.050	42.958	38.400	33.688	30.720
		C21	64.252	58.216	54.397	50.000	45.368	43.036
		C11	60.218	52.164	46.620	39.512	30.174	17.844
	IV	SRD21	160.782	176.892	194.096	228.570	317.468	725.624
		C21	197.429	228.982	261.498	326.667	502.331	1443.760
		C11	180.743	196.484	210.393	233.594	279.162	402.211
(24,12)	IV	SRD21	121.538	118.047	116.096	114.286	113.588	118.731
		C21	136.374	141.302	148.015	163.333	206.889	409.279
		C11	128.037	123.586	120.160	115.400	108.085	95.219
		SRD11	122.349	118.774	116.774	114.285	112.050	111.329

The above theorem gives some general results on the censoring schemes. Numerical comparisons of these schemes have also been made. Table 1 gives some of the results from the previous papers. The results concerning SRD21 and C21 are from Srivastava (1986), and those concerning C11 and SRD11 are from Srivastava and Khodadadi (1991), and Khodadadi (1990).

In Table 1, we have considered two values for the pair (u, G), namely, $(24, 24)$ and $(24, 12)$. Of these, we have given the values of the TETUE only for the case $(24, 24)$ for each of the four parameter sets. The ratio of the value of the TETUE for the pair $(24, 12)$ to that for the pair $(24, 24)$ is roughly the same for the four parameter sets. Therefore, only for the last parameter set, the values for both of these pairs have been given. In the papers cited above, the pair (12,12) was also studied. We have ignored this pair in Table 1. The reason is that the values for the case $(24, 24)$ are respectively double of the values for the pair $(12, 12)$ for the three censoring schemes SRD21, SRD11, and C11. Only for C21, the values are not double. However, C21 does not fare too well in comparison with the other schemes, and therefore the fact that for this case the dependence is of a different kind was not considered a significant fact. We have omitted the value for SRD11 for the case when $u = G$, in view of the last theorem, since the values are equal to the corresponding values for SRD21.

As β increases the value of TETUE decreases in all cases except for the case IV when $u = G$. This fact is very easily seen for SRD21 and SRD11 from Theorem 4.1. For parameter sets I, II and III, the value of $\lambda(= \sum_l^m \lambda_i)$ is greater than 1. However, for case IV, it is less than 1. Since $\Gamma(1+1/\beta)\lambda^{-1/\beta}$ is a decreasing function of β when $\beta \geq 1$, and an increasing function of β otherwise, the trend under discussion would follow.

One also observes that except C21, the other schemes have TETUE proportional to u when $u = G$. Again this is reflected in Theorem 4.1.

Comparing the four schemes with one other, it is clear that when $\beta \geq 1$, SRD21 is the winner. When $\beta < 1$, C11 is the winner in most cases except a few which correspond to $(u, G) = (24, 12)$, and where SRD11 wins. Since the hazard function for the Weibull case is given by $\Lambda(t) = \lambda\beta t^{\beta-1}$, it is clear that when $\beta < 1$, the hazard decreases as time increases. This can happen sometimes. For example, when children approach adulthood, health risks may decrease. However, in most cases, the hazard function would increase. For many machines the hazard functions is of the bathtub type, i.e., it first decreases, and then it increases. This is where the machine has to go through an initial break-through period, during which if there is something greatly wrong with it, it may fail. After this period, the machine's hazard rate decreases for a while and then increases. Thus, the case $\beta \geq 1$ is much more important, and correspondingly SRD21 is the winner among the four schemes under consideration. On the other hand, if there are strong reasons to believe that a machine or a subject will have a decreasing hazard rate for a while, then the usefulness of C11 should be investigated taking into account the associated variances. But, even for these cases, the risk will eventually increase. Thus, it will be necessary to study the function θ more closely. Notice that a bathtub type of hazard function may be obtained by taking

$$\Lambda(t) = \mu_1\delta_1 t^{\delta_1 - 1} + \mu_2\delta_2 t^{\delta_2 - 1}, \quad \text{with} \quad \delta_1 < 1, \delta_2 > 1 \tag{5.4}$$

where μ_1 and μ_2 are appropriately chosen. For such a function, using Theorem 4.1, the TETUE for SRD21 (when $u = G$) can be easily computed. It can then be compared with the value of TETUE for C11, and the winning scheme should be chosen. However, because SRD21 appears to be very good with respect to the ASVAR matrices, it is likely that SRD21 will turn out to be the winner.

It should be remarked here that the ASVAR matrices have been compared only with respect to A–optimality. Consider SRD21 and C21 under D-optimality. Since the ASVAR matrices are diagonal, we compare the geometric means of the diagonal elements of the ASVAR matrices. These turn out to be respectively $(m/G)\overline{\lambda}_g\overline{\lambda}_a$ and $(m/G)\overline{\lambda}_g^2$, where $\overline{\lambda}_g$ and $\overline{\lambda}_a$ respectively are the geometric and the arithmetic means of $\lambda_1, \lambda_2, ..., \lambda_m$.

Since $\overline{\lambda}_a \geq \overline{\lambda}_g$ (with equality holding if and only if all the λ's are equal), it follows that C21 is D−optimal relative to SRD21. Thus we reach the interesting phenomenon that one scheme is A-optimal and the other is D-optimal.

The question arises as to which scheme should be used assuming that all other considerations are neglected. This question has been studied in Srivastava and Siddiqui (1986). Some interesting situations where SRD21 is better than C21 include the following: (1) when we wish to estimate $\overline{\lambda}_a$, (2) when we wish to estimate λ_i and $\lambda_i > \lambda_a$, (3) when we wish to estimate a linear function $\sum_{i=1}^{m} c_i\lambda_i$, and the λ's are unknown and $\underline{\lambda}$ has a non-degenerate prior distribution, (in which case the expected value of the variance under SRD21 is less than that under C21), (4) when we wish to estimate $\theta = \theta_1 - \theta_2$, and $\theta_1 = (1/k)(\sum_1^k \lambda_i), \theta_2 = (1/(m-k))(\sum_{k+1}^{m} \lambda_i)$, and (θ_1/θ_2) is either larger than $\max(1, k^2/((m-k)^2)$ or less than $\min(1, k^2/(m-k)^2)$. Under cases (2) and (4), the scheme C21 is better if the conditions stated for SRD21 being better do not hold.

In this paper, so far we have considered two important components of the total cost of the experiment. These are the total number of failures, and the total expected time under experiment. However, for the choice of an appropriate censoring scheme for conducting the experiment, it is important to look at all associated costs. In other words, an appropriate cost function should be formulated.

Let C_0 denote the overhead cost which may consist of administrative, consultancy, and data analysis costs. Let C_1 be the cost incurred due to the failure of a particular machine; we are assuming that all machines have the same cost. If costs differ, for example, from brand to brand, an appropriate cost function should be used. Let C_2 be the cost of running a machine for one unit of time for the experiment. Let C_3 be the cost associated with observing the whole experiment for one unit of time; this could include, for example, the observing of failure times of different machines. Let $\gamma(n)$ be the cost associated for the procurement, storage, and handling of n machines for the purposes of conducting the experiment. We should not lose generality in assuming that γ is an increasing function. However, it is not necessarily linear. It may even be discontinuous, or remain constant for certain values of n. If C denotes the total cost, then we have

$$C = C_0 + C_1G + C_2T_0 + C_3T_0^* + \gamma(\mu), \tag{5.5}$$

where G denotes the expected number of failures, T_0 the total expected time under experiment, T_0^* the length for which the experiment is continued, and where we assume that we use mu machines in the experiment. Since C_0 and mu are the same for competing censoring schemes, it is clear that the comparative costs will be functions of G, T_0, and T_0^*. Thus, the values of C_1, C_2, C_3 will play an important role in choosing the appropriate censoring scheme. Among these, we believe that C_3 would in general be much smaller relative to C_1 and C_2. This is the reason we emphasize the values of G, and T_0 in our general discussion. On the other hand, if C_3 turns out to be relatively large, the value of T_0^* will become important. The value of T_0^* has not been discussed in this paper for the sake of brevity, but a discussion of the same will be found in the references cited, where the results of numerical computations have been given for the same set of values as in Table 1 of this paper. These computations show that C11 is the winner, followed by

SRD21. Thus, indeed, if C_3 is quite large relative to C_1 and C_2, C11 may very well turn out to be the scheme of choice from the cost view point.

What about the choice of u? This has been briefly studied in Srivastava (1987). The important point in this connection is that as u increases the value of TETUE decreases for all censoring schemes. Thus, for any scheme, T_0 may decrease while, $\gamma(mu)$ would increase. This would in general lead to a decision on an optimal value on u. However, the author feels that if u is too large, then the experiment may end too early, and the whole nature of the function $\theta(t)$ may not come into play.

Notice that C11 and C21 allow us the choice of having $G_0 > u$. This will become important if $\gamma(mu)$ is large relative to C_1, C_2 and C_3 for fixed mu. In this case, these schemes may become more attractive relative to SRD21. For such situations, some other SRD's (which may be better) will be discussed elsewhere.

As mentioned earlier, it is expected that C_1 and C_2 will usually be the large components of C_0. Also, we usually expect $\Lambda(t)$ to increase with t. Because of all this, and because of its mathematical tractability, SRD21 appears to be very promising. In view of this, we present some further results concerning SRD21.

6. FURTHER REMARKS ON SRD21

Clearly, the data set is of the form (i_g, t_g) for $g = 1, ..., G$, where we start with u machines of each type, and the experiment is continued under the SRD21 until $G(\leq u)$ failures are observed, and where

$$\begin{aligned} i_g &= \text{brand of machine that failed at time } t_g, \\ b_g &= \text{time when } g\text{th failure occurred,} \end{aligned} \tag{6.1}$$

it being presumed that the experiment is started at time $t_0(=0)$. Furthermore, for $j = 1, ..., m$, and $g = 1, ..., G$, let

$$\begin{aligned} b_{jg} &= 1, \text{ if } i_g = j \\ &= 0, \text{ otherwise} \\ b_j &= \sum_{g=1}^{G} b_{jg}, \end{aligned} \tag{6.2}$$

so that b_j denotes the total number of times a machine of brand j failed in the whole experiment. Then, we have the following results in the general case.

<u>TEOREM 6.1</u> (a) The likelihood of the data set $(i_g, t_g)(g = 1, ..., G)$ is given by

$$l_g = \frac{u!}{(u-G)!} \left(\Pi_{i=1}^{m} \lambda_i^{b_i}\right) \left(\Pi_{g=1}^{g} \theta'_{i_g}(t_g)\right) \exp\left(-\sum_{i=1}^{m} \lambda_i \theta_i(t)\right) \tag{6.3}$$

(b) The maximum likelihood estimate of λ_i is $\hat{\lambda}_i$ where

$$\hat{\lambda}_i - b_i\theta_i(\underline{t}); \text{ for } i = 1, ..., m \tag{6.4}$$

where

$$\theta_i(\underline{t}) = \theta_i(t_1) + ... + \theta_i(t_{G-1}) + (u - G + 1)\theta_i(t_G) \tag{6.5}$$

(c) Let $\zeta(t_1, t_2, ..., t_G)$ denote the joint density of the failure time $t_1, ...t_g$. Then, we have

$$\begin{aligned} \zeta(t_1, ...t_G) &= \tfrac{u!}{(u-G)!}\Pi_{g=1}^{G}\left[\textstyle\sum_{i=1}^{m}\lambda_i\theta_i(t_g)\right]\exp\left(-\textstyle\sum_{i=1}^{m}\lambda_i\theta_i(t)\right), \\ &\qquad \text{for } 0 \le t_1 \le t_2 \le \cdots \le t_G; \\ &= 0, \quad \text{otherwise,} \end{aligned} \tag{6.6}$$

(d) When $\theta_i = \theta$, for all i, then we have

$$E(t_G) = \lambda \binom{u}{g-k} (u - G + 1) \int_{x=o}^{\infty} x\theta'(x) e^{-\lambda(u-G+1)\theta(x)} \{1 - e^{\lambda\theta(x)}\}^{G-1} dx \tag{6.7}$$

The above results would be useful in the actual usage of SRD21.

7. REFERENCES

1 Srivastava, J.N. and Khodadadi, A. Studies on modified censoring of Type I for comparative experiments under the proportional hazards model (1991). (Unpublished).

2 Srivastava, J.N. Experimental design for the assessment of reliability. In: A.P. Basu Ed., In: Proceedings of the International Symposium on Reliability and Quality Control (A.P. Basu, ed.) North-Holland Amsterdam, (1986).

3 Khodadadi, A. Studies on a general distribution, and testing procedures in lifetesting, (1990). (Ph.D. dissertation, Colorado State University).

4 Srivastava, J.N. and Siddiqui, M.M. When A- and D-optimality compete. Commun. Statist. - Theory Meth., 16 (1987).

5 Lawless, J.F. Statistical Models and Methods for Lifetime Data, John Wiley and Sons, New York, (1982).

6 Mann, N.R., Schafer, R.E. and Singpurwalla, N., Methods for Statistical Analysis of Reliability and Life Data, John Wiley and Sons, New York, (1974).

7 McCool, J.I. J. Analysis of single classification experiments based on censored samples from the two-parameter Weibull distribution. Statist. Plann. Inference, 3, (1979).

8 Myers, R., et al. Reliability Engineering for Electronic Systems, John Wiley and Sons, New York, (1964).

9 Pieruschka, E. Principles of Reliability, Prentice-Hall, Englewood Cliffs, N.J., (1963).

10 Srivastava, J.N. (1987). More efficient and less time consuming censoring designs for life testing. J. Statist. Plann. Inference., 16 (1987).

ADVANCES IN RELIABILITY
Edited by Asit P. Basu

Recent Developments in Bayesian Sequential Reliability Demonstration Tests [1]

Dongchu Sun[a] and James O. Berger[b]

[a]Department of Statistics, University of Missouri–Columbia, 222 Math Science Building, Columbia, MO 65211, USA

[b]Department of Statistics, Purdue University, Math Science Building, West Lafayette, IN 47907, USA

Abstract

This paper begins with general background and motivation of the Bayesian Sequential Reliability Demonstration Test (BSRDT). Two different approaches to the BSRDT, based on posterior loss and predictive loss are presented. Three testing models are considered. Various risks and features are investigated. Detailed examination for the Weibull and related distributions are developed and summarized.

Key words and Phrases. Reliability demonstration test, Bayesian analysis, predictive distribution, exponential distribution, Weibull distribution, expected stopping time, producer's risk, consumer's risk.

1 Introduction

In order to produce a new product, engineers should carry out a series of experiments, which consists of prototype, development, qualification and reliability demonstration. Reliability demonstration is the last and a necessary step in the experimentation period. Reliability Demonstration Testing (RDT) is often used for the purpose of verifying whether a specified reliability has been achieved in a newly designed product. Based on a demonstration test, a decision is made to either accept the design and start formal production, or reject the design and send the product back for reengineering.

A serious problem with RDT is that a reliability test can be very expensive in terms of money and time, especially in the case of products that require very high reliability and have a long lifetime. A common solution is to take into consideration prior information, typically from engineering knowledge or knowledge of previous similar products, and to test in a sequential fashion.

[1]This research was supported by NSF grants DMS-8702620 and DMS-8923071.

There are several issues in the reliability demonstration. What are the suitable distributions of lifetimes for a product? How should we use the engineer's knowledge and knowledge of previous similar products? How many units need to be tested to reach a decision? How long does it take to make a decision? What are expected losses? What are reasonable testing procedures which are acceptable to both producers and consumers? These questions will be discussed in this paper.

It is well known that a Bayesian Sequential Test (BST) can be a useful tool, since it can take into consideration prior information, loss structure and the cost of testing. In many cases, use of a BST can significantly reduce the amount of testing required. However, BST's are often too complex to evaluate.

Several authors have thus proposed simple Bayesian Sequential Reliability Demonstration Tests (BSRDT). These simpler tests ignore any loss structure or costs of conducting the test. They are based solely on the way in which consecutive observations of failures modify the prior information on the parameter of interest to produce a posterior distribution for the parameter of interest. Testing continues until the posterior distribution is decisive, according to appropriate criterion, at which stage it terminates and a decision is taken on the quality of the product.

The first BSRDT was introduced by Schafer & Singpurwalla (1970). They introduced the following test procedure. One unit at a time is tested, where the lifetimes of units are independently identically exponentially distributed with mean θ. The unknown θ is assumed to have an inverse Gamma prior distribution. Choose a minimum acceptable value, say θ_1, and let $P_n = P(\theta \geq \theta_1|\text{data})$. The test is terminated when $P_n \geq 1 - \alpha_2$, in which case a decision to accept the product is made, or when $P_n \leq \alpha_1$, in which case a decision to reject the product is made. Schafer & Singpurwalla (1970) were primarily concerned with the acceptance probability of this procedure, and developed approximations for it. Some related computations and approximations for other risks were done in Schafer and Sheffield (1971) and Mann, Shafer and Singpurwalla (1974). The extreme difficulty of all computations of this type is discussed in Martz and Waller (1982); one of the major motivations of our recent work is to show how such computations can be done explicitly, in closed form.

The stopping rule of Schafer & Singpurwalla (1970) is discrete, in the sense that one can only stop the test when a failure occurs. This can be inefficient when observations are very expensive and/or have long lifetimes. Barnett (1972) proposed a continuous BSRDT plan for the exponential failure rate problem. By his method, one can stop the test at any time that enough information has accumulated. Again, however, closed form answers were not obtained.

Related work can be found in Chandra and Singpurwalla (1981), Epstein and Sobel (1953), Goel and Coppola (1979), Harris and Singpurwalla (1968, 1969), Lindley and Singpurwalla (1991a,b), MacFarland (1971), Martz and Waller (1979), Montagne and Singpurwalla (1984), Ray (1965), and Soland (1969).

Recently, two different approaches to the BSRDT have been considered for our purposes (see Sun and Berger (1991) and Sun (1991)). One of them, which is still called the BSRDT, is stimulated by the work of Barnett (1972). Testing continues until the posterior loss is decisive according to a desired criterion, at which time testing terminates and a decision made concerning the quality of the product. The other, the PSRDT (Predictive Sequential Reliability Demonstration Tests), is based on the predictive loss of future products. Those two procedures and their discrete versions will be described in Section 2.

In Section 3, two common testing models, with replacement and without replacement, and their generalization, the stepwise model, are considered. In Section 4, various risk criteria and other important features for the BSRDT and the PSRDT are presented. In Section 5, details about the Weibull and related distributions are developed and summarized.

2 Stopping and Decision Rules

Suppose that the underlying distribution of the new product is characterized by an unknown parameter $\theta(> 0)$. Larger θ provide larger mean time to failure, and θ is assumed to have a prior distribution $\pi(\theta)$.

2.1 The BSRDT

There are a variety of possible goals for sequential experimentation. The following BSRDT plan is the intersection of two goals.

1. Let θ_1 be the goal to begin production, in the sense that the experiment will stop and production begin if there is $100(1-\alpha_1)\%$ "confidence" that $\theta > \theta_1$.

2. Let θ_2 be the mature product goal , in the sense that the experiment will stop and the product will be rejected (sent back for reengineering) if there is $100(1-\alpha_2)\%$ "confidence" that $\theta < \theta_2$.

Here α_1 and α_2 are two usually somewhat small numbers, and $\theta_1 < \theta_2$ are two prespecified values. The region $\theta_1 < \theta < \theta_2$ is often called the indifference region.

The BSRDT also arises in formal decision models. Suppose that a product with small $\theta < (\theta_1)$ should be rejected and with large $\theta > (\theta_2)$ should be accepted. Let $l(\theta)$ be the loss for making a wrong decision, where $l(\theta)$ is nonincreasing and nondecreasing in $(0, \theta_1]$ and $[\theta_2, \infty)$, respectively, and $l(\theta) = 0$ for $\theta \in (\theta_1, \theta_2)$. The test will stop and production begin if the posterior loss of accepting the product ($\int_0^{\theta_1} l(\theta)\pi(\theta|\text{data})d\theta$) is small enough, and the test will stop and the product be rejected if the posterior loss of rejecting the product ($\int_{\theta_2}^{\infty} l(\theta)\pi(\theta|\text{data})d\theta$) is small enough. The BSRDT arises if $l(\theta)$ is constant on both $(0, \theta_1]$ and $[\theta_2, \infty)$.

It is easy to see that the testing plan is equivalent to

$$\begin{cases} \text{Stop and accept the product,} & \text{if } q^*(\alpha_1) > \theta_1; \\ \text{Stop and reject the product,} & \text{if } q^*(1-\alpha_2) \le \theta_2, \\ \text{Continue testing,} & \text{otherwise,} \end{cases} \tag{1}$$

where $q^*(\alpha)$ is the αth posterior quantile. More details about the BSRDT can be found from Sun and Berger (1991).

2.2 The PSRDT

For the BSRDT plans, inference is made based on the information about the unknown parameter in the distribution of time to failure. This is understandable and acceptable to an engineer. But from the viewpoint of a manager, it might be more natural to consider the time to failure of a future product. In fact, the purpose of reliability demonstration testing is to make a decision about a future product. Basically, it is a predictive problem. Thus we also consider an alternative sequential procedure, the PSRDT, which makes inference based on the predictive distribution of a new time to failure Z, given the current data. The idea is as follows. Choose the desired lifetime, t_2, and the minimum lifetime to begin production, t_1. A product will be rejected (accepted) if its time to fail is less then t_1 (more then t_2). Let $l(\cdot)$ be the loss function of making a wrong decision, where $l(z)$ is nondecreasing and nonincreasing in $(0, t_1]$ and $[t_2, \infty)$, respectively, and $l(z) = 0$ for $z \in (t_1, t_2)$. The plan is the intersection of following two goals:

1. The experiment will stop and production begin at time t if the predictive loss of accepting the product $(\int_0^{t_1} l(z) f(z|\text{data at t}) dz) < \alpha_1$.

2. The experiment will stop and product will be rejected at time t (sent back for reengineering) if the predictive loss of rejecting the product $(\int_{t_2}^{\infty} l(z) f(z|\text{data at t}) dz) < \alpha_2$.

Here α_1 and α_2 are two usually somewhat small numbers, and $f(z|\text{data at t})$ is the predictive p.d.f. For the 0-1 loss, the predictive BSRDT plan is equivalent to

$$\begin{cases} \text{Stop and accept the product at t,} & \text{if } P(Z < t_1|\text{data at t}) < \alpha_1, \\ \text{Stop and reject the product at t,} & \text{if } P(Z > t_2|\text{data at t}) < \alpha_2. \end{cases} \tag{2}$$

The BSRDT and the PSRDT describe above are actually using continuous stopping times. In contrast, there are discrete versions, the discrete BSRDT and the discrete PSRDT, in which the two design goals are examined only at each failure.

3 Testing Models

We now consider three testing models, with replacement, without replacement and stepwise models.

- *With Replacement.* Units are independently tested on m machines. Whenever a unit fails, it is replaced by a new unit and testing is continued until enough information has been obtained. This model includes the case in which m machines are tested themselves and, upon failure, a machine is repaired or rebuilt (immediately) so that the repaired machine is as good as new.

- *Without Replacement.* Units are independently tested on m machines without repair.

- *Stepwise Models.* Suppose that, at time 0, m (≥ 1) machines are used to test the items with replacement. If no decision is made at time $X_1(> 0)$, $m_1(\geq -1)$ more machines are put on test with replacement. Generally, if no decision is made at time X_j ($> X_{j-1}$), $m_j(\geq 0)$ more machines are put on test with replacement. The $X_j's$ are called the evaluation points.

The stepwise model generalizes the first two models. If $m_j = 0$ ($j \geq 0$) then the stepwise model is the model with replacement. If $m_j = -1(1 \leq j \leq m)$ and $m_j = 0(j \geq m)$ then the stepwise model is the model without replacement.

Note that one can "accept" when accumulated nonfailure time is large enough, and this could happen at any time, but rejection takes place only on the occurrence of a failure. Let $T_1 \leq T_2 \leq \cdots \leq T_n$ be the first n ordered failure times for all the machines. A natural choice of evaluation points is

$$X_j = T_j, \ i = 1, 2, \cdots. \tag{3}$$

This is reasonable, since we would typically consider more machines when the failure time has been too long to reject, but not long enough to accept.

4 Features of the BSRDT and the PSRDT

4.1 Risks

Let $\mathcal{A}$ and $\mathcal{R}$ denote the action (or, by an abuse of notation, the region) of accepting the product and the action of rejecting the product, respectively. Several risk criteria, defined in Chapter 10 of Martz and Waller (1982), can be used to measure the goodness of the BSRDT. The following names of these risks are borrowed from related conventions in quality control.

1. Classical Producer's Risk, $\gamma = P(\mathcal{R}|\theta_2)$, and Classical Consumer's Risk, $\delta = P(\mathcal{A}|\theta_1)$. Here γ is the probability that a product at the mature product goal will fail the BSRDT and δ is the probability that a product at the goal to begin production will pass the BSRDT. If the lifetime distribution has a monotone likelihood ratio, $P(\mathcal{A}|\theta)$ is monotonically increasing in θ. In this case, $P(\mathcal{R}|\theta) < \gamma$ for $\theta > \theta_2$, and $P(\mathcal{A}|\theta) < \delta$ for $\theta < \theta_1$.
2. Average Producer's Risk, $\tilde{\gamma} = P(\mathcal{R}|\theta \geq \theta_2)$, and Average Consumer's Risk, $\tilde{\delta} = P(\mathcal{A}|\theta \leq \theta_1)$. Here $\tilde{\gamma}$ is the probability of rejecting a good product and $\tilde{\delta}$ is the probability of accepting a bad product. Note that computation of these risks involves the prior.
3. Posterior Producer's Risk, $\gamma^* = P(\theta \geq \theta_2|\mathcal{R})$, and Posterior Consumer's Risk, $\delta^* = P(\theta \leq \theta_1|\mathcal{A})$. Here γ^* is the posterior probability that a rejected product is good, and δ^* is the posterior that an accepted product is bad.
4. Rejection probability, $P(\mathcal{R}) = \int_\Theta P(\mathcal{R}|\theta)\pi(\theta)d\theta$, and Acceptance probability, $P(\mathcal{A}) = 1 - P(\mathcal{R})$. Here $P(\mathcal{A})$ is the unconditional probability of the product passing the BSRDT.

Rejection probability and Acceptance probability can also be used for the PSRDT. The choice of criteria to evaluate the BSRDT is left to the user. For the fixed sample size problem, many papers are available concerning how to choose the criteria. For example, Balaban (1975) favors the mixed classical/Bayesian pair (γ, δ^*) to determine a Bayesian reliability demonstration test. Also see Easterling (1970), Schafer and Sheffield (1971), Schick and Drnas (1972), Goel and Joglekar (1976).

4.2 Other Design Criteria

Besides various risks, the following features are also important to design:

1. N_{TU}, the total sample size or the total number of testing units put on test;
2. T, the total time on test;
3. $N = N(T)$, the number of failed units at the time when testing stops.

Finding the expected stopping time, $E(T) = \int E(T|\theta)\pi(\theta)d\theta$, and the expected sample size, $E(N_{TU}) = \int E(N_{TU}|\theta)\pi(\theta)d\theta$, is important for design. The following relationship between the total sample size and the number of failures follows immediately from the definition of the stopping rule, and allows us to consider N instead of N_{TU}.

Theorem 4.1 Consider the stepwise model. For both the BSRDT and the PSRDT,

$$N_{TU} = m + \sum_{j=0}^{N-I(\mathcal{R})} (1 + m_j). \tag{4}$$

Here $I(\cdot)$ is the indicator function. □

Remark 4.1 1. Note that, for the model with replacement, $N_{TU} = m + N - I(\mathcal{R})$, and for the model without replacement, $N_{TU} = m - N + I(\mathcal{R})$.

2. For the stepwise model and the discrete versions of the BSRDT and PSRDT,

$$N_{TU} = m_0 + \sum_{j=0}^{N-1} (1 + m_j). \tag{5}$$

5 Details for the Weibull and Related Distributions

5.1 Failure Distributions

In this section, it is assumed that the lifetime probability density function of the product has the following form

$$f(x|\theta) = \frac{H'(x)}{Q(\theta)} \exp\left\{-\frac{H(x)}{Q(\theta)}\right\}, \quad t > 0. \tag{6}$$

Here $H(\cdot)$ is a known increasing function satisfying $H(0^+) = 0$ and $\lim_{x\to\infty} H(x) = \infty$, $Q(\cdot)$ is a known and strictly increasing function, and θ is the unknown characteristic life. The density of (6) is a special form of the exponential family and encompasses many common reliability distributions.

Example 5.1 If, in (6), $Q(x) \equiv H(x) = x^{\beta}, (x > 0)$ for some known positive constant β, the density becomes

$$f(x|\theta) = \frac{\beta x^{\beta-1}}{\theta^{\beta}} \exp\Big\{-\Big(\frac{x}{\theta}\Big)^{\beta}\Big\}, \quad t > 0, \tag{7}$$

which is the p.d.f. of the Weibull distribution, $\mathcal{W}(\theta, \beta)$. It is well known that the Weibull distribution encompasses both increasing (with $\beta > 1$) and decreasing (with $\beta < 1$) hazard rates, and has been successfully used to describe both initial failures and wearout failure (Von Alven (1964) and Lieblein & Zelen (1956)). It has been argued that when a system is composed of a number of components and failure is due to the most severe defect of a large number of possible defects, the Weibull distribution is often especially appropriate. It is found in practice to be suitable for data on failure strengths and also failure times. It is also one of the stable distributions of extreme value theory. The exponential and Raleigh distributions are obtained when $\beta = 1$ and 2, respectively.

Example 5.2 Assume that, for given θ, X_1, X_2 are independent random variables and X_i has reliability function $\exp\{-t^{\beta_i}/\theta^{\beta}\}$, where β, β_1, and β_2 are known positive constants. Then $\min\{X_1, X_2\}$ has the p.d.f.

$$f(x|\theta) = \frac{\beta_1 x^{\beta_1-1} + \beta_2 x^{\beta_2-1}}{\theta^{\beta}} \exp\Big\{-\frac{x^{\beta_1} + x^{\beta_2}}{\theta^{\beta}}\Big\}, \quad t > 0, \tag{8}$$

which is a special case of (6).

Example 5.3 Assume that, for given θ, $\{X_i\}_{i\geq 1}$ is a sequence of independent random variables and X_i has reliability function $\exp\Big\{-x^{\beta_1}/(i!\theta^{\beta})\Big\}$, where β_1 and β are known positive constants. Then $\inf_{n\geq 1} X_n$ has the p.d.f.

$$f(x|\theta) = \frac{\beta_1 e^{\beta_1 x}}{\theta^{\beta}} \exp\Big\{-(e^{\beta_1 x} - 1)/\theta^{\beta}\Big\}, \quad t > 0, \tag{9}$$

which is the truncated extreme value distribution, and again a special case of (6).

Example 5.4 Let $H(x) = \ln(x+1)\,(x > 0)$ in (6). Then the lifetimes t_{ij}, $i = 1, 2, \cdots, m$, $j = 1, 2, \cdots$, are *iid*. Pareto distributions with p.d.f.

$$f(x|\theta) = 1\Big/\Big[Q(\theta)(x+1)^{\frac{1}{Q(\theta)}+1}\Big], \quad t > 0. \tag{10}$$

From Billingsley (1986) ((21.9) on Page 282), $E(X) = \int_0^{\infty} P(X > x)dx$. It follows from the assumptions on $Q(\cdot)$ and $H(\cdot)$ that larger θ provide larger expected lifetime.

5.2 Prior, Posterior and Predictive Distributions

Prior information about the unknown parameter θ is assumed available in the form of a prior density function $\pi(\cdot)$. Schafer (1969) and Schafer and Sheffield (1971) observed that the inverse Gamma prior distributions are often reasonable for exponential failure problems. Here, the prior p.d.f. of θ for the family (6) will be assumed to belong to the conjugate family

$$\pi(\theta) = \frac{b^a}{\Gamma(a)} \frac{Q'(\theta)}{Q^{a+1}(\theta)} \exp\left\{-\frac{b}{Q(\theta)}\right\}, \text{ for } \theta > 0. \tag{11}$$

Note that then $Q(\theta)$ has an inverse gamma distribution, $\mathcal{IG}(a,b)$. Methods of choosing a and b can be found from Sun and Berger (1991).

Let $N(t)$ be the total number of failures at time t and define the adjusted total time on test by

$$\begin{aligned} V(t) &= \sum_{t_j:\text{ observed failure time before t}} H(t_j) \\ &+ \sum_{t_j^*:\text{ observed test time not failed by t}} H(t_j^*). \end{aligned} \tag{12}$$

Then the posterior p.d.f. of θ given data at time t is

$$\pi(\theta|\text{data at t}) = \frac{(V(t)+b)^{N(t)+a}}{\Gamma(N(t)+a)} \frac{Q'(\theta)}{Q^{N(t)+a+1}(\theta)} \exp\left\{-\frac{V(t)+b}{Q(\theta)}\right\},$$

for $\theta > 0$, i.e., $Q(\theta)$ has, *a posteriori*, an inverse gamma distribution, $\mathcal{IG}(N(t)+a, V(t)+b)$. In particular, it follows that the posterior α^{th} quantile is

$$q^*(\alpha) = Q^{-1}\Big(2(V(t)+b)/\chi^2_{2(N(t)+a)}(1-\alpha)\Big), \quad \text{for } 0 < \alpha < 1, \tag{13}$$

where $Q^{-1}(\cdot)$ is the inverse function of $Q(\cdot)$ and $\chi^2_j(1-\alpha)$ is the $(1-\alpha)^{th}$ quantile of the χ^2 distribution with j degrees of freedom.

Assume that Z, a future time to failure, is independent of current data. From Berger (1985) (page 157), the predictive survival function at time t is

$$\begin{aligned} P(Z > s|\text{data at t}) &= \int_0^\infty P(Z > s|\theta)\pi(\theta|\text{data at t})d\theta \\ &= \frac{(V(t)+b)^{N(t)+a}}{\Gamma(N(t)+a)} \int_0^\infty \frac{Q'(\theta)}{(Q(\theta))^{N(t)+a+1}} \exp\left\{-\frac{V(t)+H(s)+b}{Q(\theta)}\right\} d\theta \\ &= \left\{1 + \frac{H(s)}{V(t)+b}\right\}^{-N(t)-a}, \quad s > 0. \end{aligned}$$

This implies that the αth quantile of the predictive distribution is

$$q^Z(\alpha) = H^{-1}\Big(\big[V(t)+b\big]\big[(1-\alpha)^{-1/(N(t)+a)} - 1\big]\Big), \quad \textit{for } 0 < \alpha < 1, \tag{14}$$

where $H^{-1}(\cdot)$ is the inverse function of $H(\cdot)$.

5.3 Representations for the BSRDT and the PSRDT

The stopping (and decision) rules for the BSRDT and the PSRDT are equivalent to

$$\begin{cases} \text{Stop and accept the product at t,} & \text{if } V(t)+b > \frac{1}{2}Q(\theta_1)\chi^2_{2(N(t)+a)}(1-\alpha_1), \\ \text{Stop and reject the product at t,} & \text{if } V(t)+b \le \frac{1}{2}Q(\theta_2)\chi^2_{2(N(t)+a)}(\alpha_2). \end{cases} \tag{15}$$

and

$$\begin{cases} \text{Stop and accept the product at t,} & \text{if } V(t)+b > H(t_1)\Big/\Big\{(1-\alpha_1)^{-1/(N(t)+a)}-1\Big\}, \\ \text{Stop and reject the product at t,} & \text{if } V(t)+b \le H(t_2)\Big/\Big\{\alpha_2^{-1/(N(t)+a)}-1\Big\}. \end{cases} \tag{16}$$

respectively. As with certain classical sequential tests these procedures are "semicontinuous" (see Epstein and Sobel, 1955): one can "accept" when the continuous time of accumulated nonfailure is large enough, but can "reject" only on the (discrete) occurrence of a failure.

For $i = 0, 1, 2, \cdots$, let

$$c_i = \begin{cases} \frac{1}{2}Q(\theta_2)\chi^2_{2(a+i)}(\alpha_2), & \text{for the BSRDT,} \\ H(t_2)\Big/\Big\{\alpha_2^{-1/(a+i)}-1\Big\}, & \text{for the PSRDT,} \end{cases} \tag{17}$$

and

$$d_i = \begin{cases} \frac{1}{2}Q(\theta_1)\chi^2_{2(a+i)}(1-\alpha_1), & \text{for the BSRDT,} \\ H(t_1)\Big/\Big\{(1-\alpha_1)^{-1/(a+i)}-1\Big\}, & \text{for the PSRDT.} \end{cases} \tag{18}$$

Then the stopping (and decision) rules for the BSRDT and the PSRDT have the common form: Stop and accept (reject) the product at t, if $V(t) + b > d_{N(t)}$ $(\le c_{N(t)})$.

5.4 Summary about Evaluation of Important Features

Sun and Berger (1991) found closed form expressions for all the important features of the BSRDT defined in Section 4, for the model with replacement. Exact expressions for the expected sample size, various risks and the distribution of total number of failures were determined. Bounds on expected testing time were given. Based on a result about a Poisson process, the exact expected testing time for exponential failure was also found. Included in these risks and expected stopping times were frequentist versions, thereof, so that the results also provided frequentist answers for a class of interesting stopping rules.

For the discrete BSRDT, Sun (1991) found the exact formulas for the various risks, and the distributions of the number of failures and sample size. If only one unit at a time is tested, the expected testing time has a closed form. Some asymptotic properties of the discrete BSRDT plan were also discussed. For a general loss, the discrete BSRDT is asymptotically equivalent to the Bayes sequential test when the cost per observation is small enough.

For the stepwise model, Sun (1991) gave closed form expressions for all the risks, the expected sample size, and the distribution of total number of failures for the BSRDT and the PSRDT.

Appendix

Here are summaries of exact form expressions which are true for both the BSRDT and the PSRDT and all the three models. Their proofs can be found in Sun and Berger (1991) and Sun(1991). We use the notation and assumptions of Section 5.

A1. Technical Preliminaries

Since both α_1 and α_2 are small for the BSRDT and the PSRDT, it can be assumed that $\alpha_2 < 1 - \alpha_1$. Then for c_i and d_i defined by (17) and (18), respectively, it can be shown that $c_i \geq d_{i-1}$ when i is large enough, as long as $\theta_1 < \theta_2$ for the BSRDT or $t_1 < t_2$ for the PSRDT. Thus there is an i_0 such that

$$i_0 = \min\{i = 1, 2, \ldots : c_i \geq d_{i-1}\}. \tag{19}$$

For $n = 1, \cdots, i_0$, let

$$G_n \;=\; \{(y_1, \cdots, y_n) : y_j > 0, c_j - b < y_1 + \cdots + y_j \leq d_{j-1} - b, j = 1, \cdots, n\}, \tag{20}$$

let $\|G_0\| = 1$ and let $\|G_n\|$ denote the volume or Lebesgue measure of G_n $(n \geq 1)$. For $j = 1, 2, \cdots, i = 0, 1, 2, \cdots$, and $y \geq 0$, define

$$a_{ij}(y) = \begin{cases} (c_{i+j} \wedge d_{j-1}) \vee y - (c_j \vee y), & \text{if } i \geq 1, \\ d_{j-1} - c_j \vee y, & \text{if } i = 0, \end{cases} \tag{21}$$

where $x \vee y = \max(x, y)$ and $x \wedge y = \min(x, y)$. Then

$$a_{ij}(c_j) = \begin{cases} c_{i+j} \wedge d_{j-1} - c_j, & \text{if } i \geq 1, \\ d_{j-1} - c_j, & \text{if } i = 0. \end{cases} \tag{22}$$

For $n \geq 2$, define two sets of partitions of n by

$$\Psi_n = \Big\{(i_1, \cdots, i_k) : \sum_{j=1}^{k} i_j = n, \; k \geq 1, \; i_j \geq 1\Big\} \tag{23}$$

and

$$\Psi_n^* = \Big\{(i_1, \cdots, i_k) : (i_1, \cdots, i_k) \in \Psi_n, i_k \geq 2\Big\}. \tag{24}$$

For $1 \leq r \leq n, (i_1, \cdots, i_k) \in \Psi_n$, define

$$\begin{aligned} \rho_1(b; n; r; i_1, \cdots, i_k) \;&\equiv\; \rho_1(b; n; r; i_1, \cdots, i_k; c_j, d_{j-1}, 1 \leq j \leq n) \\ &= \begin{cases} \dfrac{a_{n-r,r}^{r}(b)}{r!}, & \text{if } k = 1, \\ \dfrac{a_{n-r,r}^{i_1}(c_r)}{\prod_{j=1}^{k} i_j!} \Big\{\prod_{j=2}^{k-1} a_{r,j}^{*}(c_{r,j}^{*})\Big\} a_{r,k}^{*}(b), & \text{if } k \geq 2, \end{cases} \end{aligned} \tag{25}$$

and for $2 \le l \le r \le n, (i_1, \cdots, i_k) \in \Psi_l^*$,

$$\rho_2(b;n;r;l;i_1,\cdots,i_k) \equiv \rho_2(b;n;r;l;i_1,\cdots,i_k;c_j,d_{j-1},1 \le j \le n)$$
$$= \begin{cases} \dfrac{a_{n-r,r}^l(d_{r-l})}{l!}\|G_{r-l}\|, & \text{if } k=1, \\ \dfrac{a_{n-r,r}^{i_1}(c_r)}{\prod_{j=1}^k i_j!}\Big\{\prod_{j=2}^{k-1} a_{r,j}^*(c_{r,j}^*)\Big\} a_{r,k}^*(d_{r-l})\|G_{r-l}\|, & \text{if } k \ge 2, \end{cases} \tag{26}$$

where $i_0 = 0$, $i_{(j)} = i_0 + i_1 + \cdots + i_j$, $a_{r,j}^*(\cdot) = a_{i_{j-1},r-i_{(j)}}^{i_j}(\cdot)$, $a_{ij}(\cdot)$ is given by (21), $c_{n,j}^* = c_{n-i_{(j)}}$, and $\prod_{j=2}^1 \cdot = 1$. For $2 \le n \le i_0$ and $1 \le r \le n$, let

$$\xi_{n,r} = \sum_{(i_1,\cdots,i_k)\in\Psi_r} \rho_1(b;n;r;i_1,\cdots,i_k) - \sum_{l=2}^r \sum_{(i_1,\cdots,i_k)\in\Psi_l^*} \rho_2(b;n;r;l;i_1,\cdots,i_k). \tag{27}$$

Lemma A1 The volume of G_n is $\|G_n\| = \xi_{n,n}$ $(n \ge 1)$.

A2. Exact form Expressions

Theorem A1 The rejection probability for given θ is

$$P(\mathcal{R} \mid \theta) = 1 - P(\mathcal{A} \mid \theta) = \sum_{n=0}^{i_0-1} \frac{\|G_n\|}{Q^n(\theta)} \exp\Big\{-\frac{d_n - b}{Q(\theta)}\Big\},\ \theta > 0,$$

and the unconditional rejection probability is

$$P(\mathcal{R}) = 1 - P(\mathcal{A}) = 1 - \sum_{n=0}^{i_0-1} \frac{b^a\Gamma(a+n)}{d_n^{a+n}\Gamma(a)}\|G_n\|.$$

Theorem A2 The cumulative distribution function of N for given θ is

$$P(N \le n|\theta) = \begin{cases} \exp\big\{-\frac{d_0-b}{Q(\theta)}\big\}, & \text{if } n = 0, \\ 1 - J_{\theta,n} + \sum_{i=n-1}^n \dfrac{\|G_i\|}{Q^i(\theta)} \exp\Big\{-\dfrac{d_i - b}{Q(\theta)}\Big\}, & \text{if } 1 \le n < i_0; \end{cases} \tag{28}$$

the expected number of failed units for given θ is

$$E(N \mid \theta) = \sum_{n=0}^{i_0-1} J_{\theta,n} - 2\sum_{n=0}^{i_0-2} \frac{\|G_n\|}{Q^n(\theta)} \exp\Big\{-\frac{d_n - b}{Q(\theta)}\Big\} - \frac{\|G_{i_0-1}\|}{Q^{i_0-1}(\theta)} \exp\Big\{-\frac{d_{i_0-1} - b}{Q(\theta)}\Big\}; \tag{29}$$

the marginal cumulative distribution of N is

$$P(N \le n) = \begin{cases} \left(\frac{b}{d_0}\right)^a, & \text{if } n = 0, \\ 1 - J_n + \sum_{i=n-1}^n \|G_i\| \dfrac{b^a\Gamma(a+i)}{d_i^{a+i}\Gamma(a)}, & \text{if } 1 \le n < i_0; \end{cases} \tag{30}$$

and the expected number of failed units is

$$E(N) = \sum_{n=0}^{i_0-1} J_n - 2\sum_{n=0}^{i_0-2} \|G_n\| \frac{b^a\Gamma(a+n)}{d_n^{a+n}\Gamma(a)} - \|G_{i_0-1}\| \frac{b^a\Gamma(a+i_0-1)}{d_{i_0-1}^{a+i_0-1}\Gamma(a)}, \tag{31}$$

where $\sum_0^{-1} \cdot = 0$, $J_{\theta,0} = J_0 = 1$,

$$J_{\theta,n} = \exp\{-\frac{c_n \vee b - b}{Q(\theta)}\} - \sum_{i=0}^{n-2} \frac{\exp\{-\frac{c_n \vee d_i - b}{Q(\theta)}\}}{Q^i(\theta)} \|G_i\| + \exp\{-\frac{c_n - b}{Q(\theta)}\} \sum_{r=1}^{n-1} \frac{\xi_{n,r}}{Q^r(\theta)}, \tag{32}$$

and

$$J_n = \frac{b^a}{(c_n \vee b)^a} - \sum_{i=0}^{n-2} \frac{b^a\Gamma(a+i)\|G_i\|}{(c_n \vee d_i)^{a+i}\Gamma(a)} + \sum_{r=1}^{n-1} \frac{b^a\Gamma(a+r)}{c_n^{a+r}\Gamma(a)} \xi_{n,r}, \tag{33}$$

for $n \geq 1$.

References

[1] Balaban, H.D. (1969). A Bayesian Approach to Reliability Demonstration. *Annals of Assurance Sciences* **8**, 497-506.

[2] Balaban, H.S. (1975). Reliability Demonstration: Purposes, Practice, and Value. *Proceedings 1975 Annual Reliability and Maintainability Symposium*, 246-248.

[3] Barlow, R.E., Bartholomew, D.J., Bremner, J.M., and Brunk, H.D (1972). *Statistical Inference under Order Restrictions.* John Wiley and Sons, New York.

[4] Barlow, R.E. and Proschan, F. (1965). *Mathematical Theory of Reliability.* John Wiley, New York.

[5] Barlow, R.E. and Proschan, F., (1975). *Statistical Theory of Reliability and Life Testing,* New York: Holt, Rinehart, and Winston

[6] Barnett, V.D. (1972). A Bayes Sequential Life Test. *Technometrics* **14**, 453-467.

[7] Berger, J.O. (1985). *Statistical Decision Theory and Bayesian Analysis, 2nd ed.* Springer, New York.

[8] Berger, J.O. and Sun, D. (1991). Bayesian Analysis for the Poly-Weibull Distribution. Technical Report **203**, Department of Statistics, The University of Michigan.

[9] Billingsley, P. (1986). *Probability and Measure, 2nd ed..* John Wiley Sons, New York.

[10] Bivens, G., Born, F., Caroli, J., and Hyle, R. (1987). Reliability Demonstration Technique for Fault Tolerant Systems. *Proceedings 1987 Annual Reliability and Maintainability Symposium*, 316-320.

[11] Blumenthal, S., Greenwood, J.A., and Herbach, L.H. (1984). Series Systems and Reliability Demonstration Tests. *Operations Research* **32**, 641-648.

[12] Bonis, A.J. (1966). Bayesian Reliability Demonstration Plans. *Annals of Reliability and Maintainability* **9** , 31-35.

[13] Canavos, G.C. and Tsokos, C.P. (1973). Bayesian Estimation of Life Parameters in the Weibull Distribution. *Operation Research* **21**, 755-763.

[14] Chandra, M. and Singpurwalla, N.D. (1981). Relationships Between Some Notions which are Common to Reliability Theory and Economics. *Mathematics of Operations Research* **6**, 113-121.

[15] Eastering, R.G. (1970). On the Use of Prior Distribution in Acceptance Sampling. *Annals of Reliability and Maintainability* **9**, 31-35.

[16] Eastering, R.G. (1975). Risk Quantification. *Proceedings 1975 Annual Reliability and Maintainability Symposium*, 249-250.

[17] Epstein, B. (1953). The Exponential Distribution and Its Role in Life Testing. *Industrial Quality Control* **15**, R5-9.

[18] Epstein, B. and Sobel, M. (1953). Life Testing. *J. Amer. Statist. Assoc.* **48**, 486-502.

[19] Epstein, B. and Sobel, M. (1955). Sequential Life Tests in the Exponential Case. *Ann. Math. Statist.* **26**, 82-93.

[20] Ghare, P.M. (1981). Sequential Tests Under Weibull Distribution. *Proceedings 1981 Annual Reliability and Maintainability Symposium*, 375-380.

[21] Goel, A.L. (1975). Panel Discussion on Some Recent Developments in Reliability Demonstration, Introductory Remarks. *Proceedings 1975 Annual Reliability and Maintainability Symposium*, 244-245.

[22] Goel, A.L. and Coppola, A. (1979). Design of reliability Acceptance Sampling Plans Based Upon Prior Distribution. *Proceedings 1979 Annual Reliability and Maintainability Symposium*, 34-38.

[23] Goel, A.L. and Joglekar, A.M. (1976). *Reliability Acceptance Sampling Plans Based upon Prior Distribution.* Technical Reports 76-1 to 76-5, Department of Industrial Engineering and Operations Research, Syracuse University, Syracuse, NY.

[24] Harris, C.M. and Singpurwalla, N.D. (1968). Life Distributions Derived from Stochastic Hazard Functions. *IEEE Transaction on Reliability* **R-17**, 70-79.

[25] Harris, C.M. and Singpurwalla, N.D. (1969). On Estimation in Weibull Distribution with Random Scale Parameters. *Naval Research Logistics Quarterly* **16**, 405-410.

[26] Joglekar, A.M. (1975). Reliability Demonstration Based on Prior Distribution - Sensitivity Analysis and Multi Sample Plans. *Proceedings 1975 Annual Reliability and Maintainability Symposium*, 251-252.

[27] Lindley, D.V., 1965. , Introduction to Probability and Statistics from a Bayesian Viewpoint: 2 Inference, C.U.P. Cambridge.

[28] Lindley, D.V. and Singpurwalla, N.D. (1991a). On the Amount of Evidence Needed to Reach Agreement between Adversaries. *J. Amer. Statist. Assoc.* **86**, p. To Appear.

[29] Lindley, D.V. and Singpurwalla, N.D. (1991b). Adversarial Life Testing. *submitted.*

[30] MacFarland, W.J. (1971). Sequential Analysis and Bayes Demonstration. *Proceedings 1971 Annual Reliability and Maintainability Symposium*, 24-38.

[31] Mann, N.R., Schafer, R.E., and Singpurwalla, N.D. (1974). *Methods for Statistical Analysis of Reliability and Life Data.* John Wiley Sons, New York.

[32] Martz, H.F. and Waller, R.A. (1979). A Bayesian Zero-Failure (BAZE) Reliability Demonstration Testing Procedure. *Journal of Quality Technology* **11**, 128-138.

[33] Martz, H.F. and Waller, R.A. (1982). *Bayesian Reliability Analysis.* John Wiley Sons, New York.

[34] Montagne, E.R. and Singpurwalla, N.D. (1985). Robustness of Sequential Exponential Life-Testing Procedures. *J. Amer. Statist. Assoc.* **80**, 715-719.

[35] Ray, S.N. (1965). Bounds on the Minimum Sample Size of a Bayes Sequential Procedure. *Ann. Math. Statist.* **39**, 859-878 .

[36] Schafer, R.E. (1975). Some Approaches to Bayesian Reliability Demonstration. *Proceedings 1975 Annual Reliability and Maintainability Symposium*, 253-254.

[37] Schafer, R.E. (1969). Bayesian Reliability Demonstration, Phase I–Data for the A Priori Distribution. *RADC-TR-69-389* , Rome Air Development Center, Rome, NY.

[38] Schafer, R.E. and Sheffield, T.S. (1971). Bayesian Reliability Demonstration, Phase II–Development of a Prior Distribution. *RADC-TR-71-139*, Rome Air Development Center, Rome, NY.

[39] Schafer, R.E. and Singpurwalla, N.D. (1970). A Sequential Bayes Procedure for Reliability Demonstration. *Naval Research Logistics Quarterly* **17**, 55-67.

[40] Schemee, J. (1975). Application of the Sequential t-Test to Maintainability Demonstration. *Proceedings 1975 Annual Reliability and Maintainability Symposium*, 239-243.

[41] Schick, G.J. and Drnas, T.M. (1972). Bayesian Reliability Demonstration. *AIIE Transactions* **4**, 92-102.

[42] Smith, R.L. and Naylor, J.C. (1987). A Comparison of Maximum Likelihood and Bayesian Estimators for the Three-parameter Weibull Distribution. *Appl. Statist.* **36**, no. 3, 358-369.

[43] Soland, R.M. and Sobel, M. (1953). An Essential Complete Class of Decision Functions for Certain Standard Sequential Problems. *Ann. Math. Statist.* **19**, 326-339.

[44] Soland, R.M. (1968). Bayesian Analysis of the Weibull Process with Unknown Scale Parameter and its Application to Acceptance Sampling. *IEEE Transactions on Reliability* **R-17**, 84-90.

[45] Soland, R.M., (1969). Bayesian Analysis of the Weibull Process with Unknown Scale Parameter and Shape Parameters, R-18, pp. 181-184.

[46] Sun, D. (1991). On Bayesian Sequential Reliability Demonstration Tests for Weibull and Related Distributions, Ph.D. Thesis, Department of Statistics, Purdue University, West Lafayette.

[47] Sun, D. and Berger, J.O. (1991). Bayesian Sequential Reliability for Weibull and Related Distributions, Technical Report **91-40c**, Department of Statistics, Purdue University, West Lafayette.

ADVANCES IN RELIABILITY
Edited by Asit P. Basu

Characterizations of a Family of Bivariate Exponential Distributions

KAI SUN and ASIT P. BASU

University of Missouri-Columbia, Columbia, MO 65211

In this paper a bivariate failure rate representation based on Cox's conditional failure rate is introduced, relationship between this failure rate and the bivariate loss of memory property is established, and characterizations of a family of bivariate exponential distributions, including the Freund model, the Marshall-Olkin model, the Block-Basu model, the Proschan-Sullo model, and the Friday-Patil model, are obtained.

KEY WORDS: bivariate exponential distribution; bivariate failure rate; bivariate loss of memory property.

1. INTRODUCTION

The univariate exponential distribution is a very useful model to study physical systems. To model complex systems, many bivariate exponential distributions have been proposed in the statistical literature since the 60's. A survey of some of these is given by Basu (1988). A family of bivariate exponential distributions, called the BEE family , consists of a number of well known distributions: the Freund model (1961), the Marshall-Olkin model (1967), the Block-Basu model (1974), the Proschan-Sullo model (1974), and the Friday-Patil model (1977). In this paper we attempt to find characterizations of the distributions of this family. Since a fundamental characterization of the univariate exponential distribution is constant failure rate, it is natural to hope that a bivariate exponential distribution would also have constant failure rate(s). There have been some different approaches to define failure rate in the bivariate case, for example, the bivariate failure rate $r(x,y) = f(x,y)/\bar{F}(x,y)$ discussed by Basu (1971) and Puri and Rubin (1974), and the vector hazard failure rate

$h(x,y) = (-\partial \log \bar{F}(x,y)/\partial x, -\partial \log \bar{F}(x,y)/\partial y)$ discussed by Johnson and Kotz (1975) and Marshall (1975). Here $\bar{F}(x,y)$ is the joint survival function P{X>x, Y>y}, with $\bar{F}(0,0)=1$. Unfortunately, the distributions of the BEE family do not have either constant bivariate failure rate or constant vector hazard rate.

Cox (1972) introduced a concept of conditional failure rate and a failure rate formulation for the absolutely continuous bivariate variable which views the bivariate lifetime as a point process. Cox's formulation is very useful and has been extended to the multivariate failure processes, see, e.g., Shaked and Shanthikumar (1988). However, little work has been done in characterizing bivariate distributions by using this formulation. In this paper we apply Cox's conditional failure rate to the case where singular parts exist in the bivariate distribution and simultaneous failures may occur in the two-component system, and we characterize the bivariate exponential distributions of the BEE family in terms of this failure rate. The relationship between constant failure rate and the bivariate loss of memory property is discussed in Section 2. Characterizations of bivariate exponential distributions are given in Section 3.

Failure of a two-component parallel system can be considered to consist of two stages: first one of the two components fails, then the remaining component fails. In a situation where one of the components has failed, it is useful to take advantage of that knowledge when considering the still surviving component's residual life. From this point of view, it seems reasonable to represent a system failure rate by using both the first stage failure rate, that is, the failure rate of min(X,Y), and the second stage failure rate, which is the conditional failure rate of one component given that the other component fails first. To this end, we define total failure rate of a bivariate random variable as follows.

Definition 1.1 Let (X,Y) be a nonnegative bivariate random variable having joint survival function $\bar{F}(x,y)$. If (X,Y) has a joint density function f(x,y) on the region $x \neq y$, the vector

$$(r(t),\ r_1(x|y) \text{ for } x>y>0,\ r_2(y|x) \text{ for } y>x>0)$$

is called the total failure rate of (X,Y) or of $\bar{F}$, where

$$r(t) = -d\log \bar{F}(t,t)/dt, \tag{1.1}$$

$$r_1(x|y) = f(x,y)/\int_x^{\infty} f(x,y)dx, \tag{1.2}$$

and

$$r_2(y|x) = f(x,y)/\int_y^{\infty} f(x,y)dy. \tag{1.3}$$

Notice that r(t) is the failure rate of min(X,Y), and that $r_1(x|y)$ is the conditional failure rate of X given X>Y and Y=y (Cox, 1972), in other words, it is the conditional failure rate of component A given that component B fails first at time y. $r_2(y|x)$ is defined similarly.

According to this definition, the following exponential distributions have constant total failure rates.

a) Freund (1961) distribution
The joint density is

$$\begin{aligned} f(x, y) &= \alpha'\beta\exp[-\alpha'x-(\alpha+\beta-\alpha')y], && \text{for } x>y>0, \\ &= \alpha\beta'\exp[-\beta'y-(\alpha+\beta-\beta')x], && \text{for } y>x>0, \end{aligned} \tag{1.4}$$

where α, β, α', $\beta'>0$.

Total failure rate = $(\alpha + \beta, \alpha', \beta')$.

b) Marshall-Olkin (1967) distribution (BVE)
The joint survival function is

$$\bar{F}(x,y) = \exp[-\lambda_1 x - \lambda_2 y - \lambda_{12}\max(x,y)], \quad \text{for } x > 0, y > 0, \tag{1.5}$$

where $\lambda_1, \lambda_2, \lambda_{12} > 0$, and $\lambda = \lambda_1 + \lambda_2 + \lambda_{12}$.

Total failure rate = $(\lambda, \lambda_1 + \lambda_{12}, \lambda_2 + \lambda_{12})$.

c) Block-Basu (1974) distribution (ACBVE)
The joint survival function is

$$\begin{aligned} \bar{F}(x,y) &= [\lambda/(\lambda_1 + \lambda_2)]\exp[-\lambda_1 x - \lambda_2 y - \lambda_{12}\max(x,y)] \\ &\quad - [\lambda_{12}/(\lambda_1 + \lambda_2)]\exp[-\lambda\max(x,y)], \quad \text{for } x > 0, y > 0, \end{aligned} \tag{1.6}$$

where $\lambda_1, \lambda_2, \lambda_{12} > 0$, and $\lambda = \lambda_1 + \lambda_2 + \lambda_{12}$.

Total failure rate= $(\lambda, \lambda_1 + \lambda_{12}, \lambda_2 + \lambda_{12})$.

d) Proschan-Sullo (1974) distribution

The joint survival function is

$$\bar{F}(x, y)=\alpha_0 \bar{F}_A(x, y)+(1-\alpha_0) \bar{F}_S(x, y), \qquad \text{for} \quad x>0, \quad y>0.$$

Here $\bar{F}_A$ is an absolutely continuous part of $\bar{F}$ and $\bar{F}_S$ is a singular part of $\bar{F}$, and the density of $\bar{F}_A$ is

$$f_A(x,y) = (\alpha_0)^{-1}\lambda_2(\lambda_1' + \lambda_{12})\exp[-(\lambda_1' + \lambda_{12})x - (\lambda - \lambda_1' - \lambda_{12})y], \quad \text{for } x > y > 0,$$

$$= (\alpha_0)^{-1}\lambda_1(\lambda_2' + \lambda_{12})\exp[-(\lambda_2' + \lambda_{12})y - (\lambda - \lambda_2' - \lambda_{12})x], \quad \text{for } y > x > 0,$$

(1.7)

$$\bar{F}_S(x,y) = \exp[-\lambda \max(x,y)], \quad \text{for } x>0, y>0, \tag{1.8}$$

where $\lambda_1, \lambda_2, \lambda_1', \lambda_2' > 0$, $\lambda_{12} \geq 0$, $\alpha_0 = (\lambda_1 + \lambda_2)/\lambda$, and $\lambda = \lambda_1 + \lambda_2 + \lambda_{12}$.

Total failure rate= $(\lambda, \lambda_1' + \lambda_{12}, \lambda_2' + \lambda_{12})$.

e) Friday-Patil (1977) distribution (BEE)

The joint survival function is

$$\bar{F}(x, y)=\alpha_0 \bar{F}_A(x, y)+(1-\alpha_0) \bar{F}_S(x, y), \quad \text{for } x>0, y>0.$$

Here $\bar{F}_A$ is an absolutely continuous part of $\bar{F}$ and $\bar{F}_S$ is a singular part of $\bar{F}$, the density of $\bar{F}_A$ is

$$f_A(x,y)=\alpha'\beta\exp[-\alpha'x-(\alpha+\beta-\alpha')y], \qquad \text{for } x>y>0,$$

$$=\alpha\beta'\exp[-\beta'y-(\alpha+\beta-\beta')x], \qquad \text{for } y>x>0, \tag{1.9}$$

and

$$\bar{F}_S(x,y) = \exp[-(\alpha + \beta)\max(x, y)], \qquad \text{for } x>0, y>0, \tag{1.10}$$

where α, β, α', $\beta'>0$, $0<\alpha_0 \leq 1$.

Total failure rate = $(\alpha + \beta, \alpha', \beta')$.

Remark 1.1 The joint distribution for the BEE given by (2.7) in Friday-Patil [7] cannot be reduced exactly to the Freund distribution when $\alpha_0=1$, since there are no constraints, $\alpha+\beta-\beta'\neq 0$ or $\alpha+\beta-\alpha'\neq 0$, in the latter distribution. In this paper the distribution for the BEE given by (1.9) and (1.10) has no constraint on the parameter space, α, β, α', $\beta'>0$, $0<\alpha_0\leq 1$, but if $\alpha+\beta-\alpha'=0$ ($\alpha+\beta-\beta'=0$) the marginal distribution of X (Y) is a mixture of an exponential distribution and a gamma distribution.

2. RELATIONSHIP BETWEEN CONSTANT TOTAL FAILURE RATE AND THE LMP

The loss of memory property (LMP) of Marshall-Olkin [10] is given by the equation $\bar{F}(x+t, y+t)=\bar{F}(x, y)\,\bar{F}(t, t)$, for all x, y, t>0. Since total failure rate describes explicitly the two-stage failures of a two-component system, derivations of bivariate exponential distributions based on constant total failure rate are often intuitively appealing. In this paper we assume that $\bar{F}$ is absolutely continuous on $x\neq y$, but $\bar{F}$ does not necessarily have absolute continuity everywhere. The following theorem is readily verified.

Theorem 2.1. Suppose that (X,Y) is a nonnegative bivariate random variable. If

a) total failure rate= $(\alpha+\beta, \alpha', \beta')$,

$P\{X>Y|\min(X,Y)=s\}=\beta/(\alpha+\beta)$,

$$P\{X<Y|\min(X,Y)=s\}=\alpha/(\alpha+\beta), \tag{2.1}$$

then (X,Y) has the Freund distribution;

b) total failure rate= $(\lambda, \lambda_1+\lambda_{12}, \lambda_2+\lambda_{12})$,

$$P\{X>Y|\min(X,Y)=s\}=\lambda_2/\lambda,\quad P\{X<Y|\min(X,Y)=s\}=\lambda_1/\lambda, \tag{2.2}$$

then (X,Y) has the BVE distribution;

c) total failure rate= $(\lambda, \lambda_1+\lambda_{12}, \lambda_2+\lambda_{12})$,

$P\{X>Y|\min(X,Y)=s\}=\lambda_2/(\lambda_1+\lambda_2)$,

$$P\{X<Y|\min(X,Y)=s\}=\lambda_1/(\lambda_1+\lambda_2), \tag{2.3}$$

then (X,Y) has the ACBVE distribution;

d) total failure rate = $(\alpha + \beta, \alpha', \beta')$,

$P\{X>Y|\min(X,Y)=s\} = \alpha_0\beta/(\alpha+\beta)$,

$P\{X<Y|\min(X,Y)=s\} = \alpha_0\alpha/(\alpha+\beta)$, (2.4)

with $\alpha', \beta' > (\alpha+\beta)(1-\alpha_0)$,

then (X,Y) has the Proschan-Sullo distribution;

e) total failure rate = $(\alpha + \beta, \alpha', \beta')$,

$P\{X>Y|\min(X,Y)=s\} = \alpha_0\beta/(\alpha+\beta)$,

$P\{X<Y|\min(X,Y)=s\} = \alpha_0\alpha/(\alpha+\beta)$, (2.5)

then (X,Y) has the BEE distribution;

where $\alpha, \beta, \alpha', \beta' > 0$, $0 < \alpha_0 \le 1$, $\lambda_1, \lambda_2, \lambda_{12} > 0$, and $\lambda = \lambda_1 + \lambda_2 + \lambda_{12}$.

Since the Freund distribution, the BVE, the ACBVE, the Proschan-Sullo distribution and the BEE possess the LMP as well as the constant total failure rate property, one might expect that the constant total failure rate property characterizes the LMP. The following examples prove the contrary case.

Example 2.1. Suppose that Z, U and V are mutually independent exponentially distributed random variables. Let

$$X = Z + UI_{[Z>1]},$$

and

$$Y = Z + VI_{[Z\le 1]}.$$

It is easy to show that (X,Y) has a constant total failure rate but it does not satisfy the LMP.

Example 2.2. Suppose that Z is an exponentially distributed random variable. Let

$$X = Z I_{[Z>1]} + (Z+1) I_{[Z\le 1]} ,$$

$$Y = (Z+1) I_{[Z>1]} + Z I_{[Z\le 1]} ,$$

and

$$S(x,y)=.5P\{X>x,Y>y\}+.5P\{X>y,Y>x\}.$$

From Theorem 3.1 of Block (1977), it follows that S(x,y) has the LMP. But it can be readily shown that S(x,y) does not have a constant total failure rate.

The following theorem, however, proves that under appropriate conditions, the constant total failure rate property implies the LMP.

Theorem 2.2. Let (X,Y) be a nonnegative bivariate variable and assume that both $P\{X>Y|\min(X,Y)=s\}$ and $P\{X<Y|\min(X,Y)=s\}$ are constant. Then the constant total failure rate property implies the LMP.

Proof. Let total failure rate=(a_1, a_2, a_3), $P\{X>Y|\min(X,Y)=s\}=a_4$, and $P\{X<Y|\min(X,Y)=s\}=a_5$, where $a_i>0$, $i=1,\ldots,5$; $a_4+a_5 \le 1$.

Using a one-one transformation,

$$a_0=a_4+a_5,\ a=a_1a_5/(a_4+a_5),\ b=a_1a_4/(a_4+a_5),$$
$$a'=a_2,\quad b'=a_3, \tag{2.6}$$

we obtain

total failure rate=(a+b, a', b'),

$$P\{X>Y|\min(X,Y)=s\}=a_0b/(a+b),$$

and

$$P\{X<Y|\min(X,Y)=s\}=a_0a/(a+b).$$

From Theorem 2.1, this yields the BEE. Hence (X,Y) has the LMP.

The converse of the above theorem does not hold in general. To prove a partial characterization of the LMP, consider the following Lemma.

Lemma 2.1. If functional forms of the marginal densities of (X,Y) are specified as

$$f_1(x)=p_1a_1\exp(-a_1x)+q_1q\exp(-qx) \tag{2.7}$$

or

$$f_1(x)=[(q-a_1)+qa_1x]\exp(-qx), \tag{2.8}$$

and

$$f_2(y)=p_2a_2\exp(-a_2y)+q_2q\exp(-qy) \tag{2.9}$$

or

$$f_2(y)=[(q-a_2)+qa_2y]\exp(-qy), \tag{2.10}$$

where $p_i+q_i=1$, $p_i \neq 0$, $a_i>0$, $i=1,2$; $a_i \neq q>0$, $i=1,2$, and $E[\min(X,Y)]=1/q$, then the constant failure rate property implies that both $P\{X>Y|\min(X,Y)=s\}$ and $P\{X<Y|\min(X,Y)=s\}$ are constant.

Proof. Let the total failure rate be (q, q_1, q_2), where $q, q_1, q_2>0$, and let $f(x,y)$ be the density function of (X,Y) on $x \neq y$, that is,

$$f(x,y) = \partial^2 P(X \neq Y, X > x, Y > y)/\partial x \partial y. \tag{2.11}$$

Let

$$P\{X>Y|\min(X,Y)=s\}=g_1(s),$$

$$P\{X<Y|\min(X,Y)=s\}=g_2(s),$$

and

$$P\{X=Y|\min(X,Y)=s\}=g_3(s). \tag{2.12}$$

Then

$$\begin{aligned} f(x,y)&=qq_1g_1(y)\exp[-(q-q_1)y-q_1x], \quad \text{for } x>y>0,\\ &=qq_2g_2(x)\exp[-(q-q_2)x-q_2y], \quad \text{for } y>x>0. \end{aligned} \tag{2.13}$$

Noting that

$$\lim_{\Delta\to 0}(\Delta)^{-1}P\{X=Y, x+\Delta > X \geq x\}$$

$$= \lim_{\Delta\to 0}(\Delta)^{-1}P\{X=Y, x+\Delta > \min(X,Y) \geq x\}$$

$$= \lim_{\Delta\to 0}(\Delta)^{-1}P\{x+\Delta > \min(X,Y) \geq x\}P\{X=Y|x+\Delta > \min(X,Y) \geq x\}$$

$$= \theta \exp(-\theta x)g_3(x),$$

we have

$$f_1(x) = \int_0^x \theta\theta_1 g_1(y)\exp[-(\theta-\theta_1)y-\theta_1 x]dy$$

$$+ \int_x^\infty \theta\theta_2 g_2(x)\exp[-(\theta-\theta_2)x-\theta_2 y]dy + \theta g_3(x)\exp(-\theta x)$$

$$= qq_1\exp(-q_1x)\int_0^x g_1(y)\exp[-(\theta-\theta_1)y]dy + q[1-g_1(x)]\exp(-qx). \quad (2.14)$$

From the assumption (2.7), we have

$$p_1a_1\exp(-a_1x)+q_1q\exp(-qx)$$

$$= \theta\theta_1\exp(-\theta_1 x)\int_0^x g_1(y)\exp[-(\theta-\theta_1)y]dy + \theta[1-g_1(x)]\exp(-\theta x) \quad (2.15)$$

From this functional equation, it follows that $g_1(x)$ is continuous and differentiable. Differentiating with respect to x in the above equation, we have

$$p_1a_1(a_1-q_1)q^{-1}\exp[-(a_1-q)x]+q_1(q-q_1)=(q-q_1)-qg_1(x)+g_1'(x). \quad (2.16)$$

Solving this differential equation, we obtain

$$g_1(x)=c\cdot\exp(qx)+p_1(q-q_1)q^{-1}-p_1(a_1-q_1)q^{-1}\exp[(q-a_1)x], \tag{2.17}$$

where c is constant.

Since

$p_1\exp(-a_1x)+q_1\exp(-qx)=P\{X>x\}>P\{X>x,Y>x\}=\exp(-qx)$,

for some $x>0$, we know that $a_1<q$.

From $0\leq g_1(x)\leq 1$, for any $x>0$, we have $c=0$ and $a_1-q_1=0$.

Hence

$$g_1(x)=p_1(q-q_1)q^{-1}. \tag{2.18}$$

From the assumption (2.8), we have

$$[(q-a_1)+qa_1x]\exp(-qx)$$

$$=\theta\theta_1\exp(-\theta_1x)\int_0^x g_1(y)\exp[-(\theta-\theta_1)y]dy+\theta(1-g_1(x))\exp(-\theta x) \tag{2.19}$$

Using arguments similar to those given above we can show that $g_1(x)$ is constant. Hence from either (2.7) or (2.8) we prove that $g_1(x)$ is constant.

Similarly, we can prove that $g_2(x)$ is constant from either (2.9) or (2.10).

The following theorem shows that under certain conditions constant total failure rate is equivalent to the LMP.

Theorem 2.3. If functional forms of the marginal densities of $\bar{F}$ are specified as in Lemma 2.1, then $\bar{F}$ has a constant failure rate if and only if $\bar{F}$ has the LMP.

Proof. The "only" part follows from Lemma 2.1 and Theorem 2.2.

To prove the "if" part, assume that the LMP holds. Then $r(t)=q$,

and for $x>y>0$,

$$P\{X>x|Y=y\}=[q\bar{F}_1(x-y)-f_1(x-y)]\exp(-qy)[f_2(y)]^{-1}.$$

From the assumption (2.7), for x>y>0,

$$P\{X>x|Y=y)\}=[f_2(y)]^{-1}\exp(-qy)\{q[p_1\exp(-a_1(x-y))+q_1\exp(-q(x-y))]$$
$$-[p_1a_1\exp(-a_1(x-y))+q_1q\exp(-q(x-y))]\}$$

$$=[f_2(y)]^{-1}\exp(-qy)(qp_1-a_1p_1)\exp[-a_1(x-y)], \tag{2.20}$$

$$f(x|y)=a_1[f_2(y)]^{-1}\exp(-qy)(qp_1-a_1p_1)\exp[-a_1(x-y)]. \tag{2.21}$$

Hence

$$r_1(x|y)=f(x|y)/P\{X>x|Y=y\}=a_1.$$

From the assumption (2.8), for x>y>0,

$$P\{X>x|Y=y\}=[f_2(y)]^{-1}\exp(-qx)\{q[1+a_1(x-y)]-[(q-a_1)+qa_1(x-y)]\}$$
$$=[f_2(y)]^{-1}a_1\exp(-qx),$$

$$f(x|y)=q[f_2(y)]^{-1}a_1\exp(-qx),$$

$$r_1(x|y)=f(x|y)/P\{X>x|Y=y\}=q.$$

Hence from either (2.7) or (2.8) we prove that $r_1(x|y)$ is constant. Similarly, we can prove that $r_2(y|x)$ is constant.

3. CHARACTERIZATIONS OF FIVE BIVARIATE EXPONENTIAL DISTRIBUTIONS

Friday and Patil (1977) study the relationships among the five distributions in Section 1 and show that the BEE distribution includes the other four distributions as special cases. In this section we discuss characterizations of the distributions in this family. Since the BVE is the most important bivariate exponential distribution, we start our discussion with the BVE.

Theorem 3.1. (X,Y) has the BVE distribution if and only if (X,Y) has the exponential marginals and a constant total failure rate.
Proof. Follows from Theorem 2.3. (set $q_1=q_2=0$) and the properties of the BVE.

Theorem 3.2. The following conditions are equivalent.

(1) $\bar{F}$ is the BEE distribution.

(2) $\bar{F}$ has a constant total failure rate and functional forms of the marginal densities of $\bar{F}$ are specified as in Lemma 2.1.

(3) $\bar{F}$ has a constant total failure rate, and both $P\{X>Y|\min(X,Y)=s\}$ and $P\{X<Y|\min(X,Y)=s\}$ are constant.

Proof.

(1) $\Rightarrow$(2)

Follows from the properties of the BEE.

(2) $\Rightarrow$(3)

Follows from Lemma 2.1.

(3) $\Rightarrow$(1)

It has been shown in the proof of Theorem 2.2.

Theorem 3.2 has the following corollaries.

Corollary 3.1. The following conditions are equivalent.

(1) $\bar{F}$ is the Proschan-Sullo distribution.

(2) $\bar{F}$ has a constant total failure rate, (a_1, a_2, a_3), and functional forms of the marginal densities of $\bar{F}$ are specified as in Lemma 2.1, with $a_2, a_3 > a_1P\{X=Y\}$.

(3) $\bar{F}$ has a constant total failure rate, (a_1, a_2, a_3), and $P\{X>Y|\min(X,Y)=s\}=a_4$, $P\{X<Y|\min(X<Y)=s\}=a_5$, with $a_i>a_1(1-a_4-a_5)$, i=2, 3, where $a_i>0$, i=1,...,5 and $a_4+a_5 \leq 1$.

Corollary 3.2. The following conditions are equivalent.

(1) $\bar{F}$ is the Freund distribution.

(2) $\bar{F}$ is absolutely continuous having a constant total failure and functional forms of the marginal densities of $\bar{F}$ are specified as in Lemma 2.1.

(3) $\bar{F}$ has a constant total failure rate, $P\{X>Y)|\min(X<Y)=s\}=a_4$, $P\{X<Y|\min(X,Y)=s\}=a_5$, with $a_4+a_5=1$, where $a_i>0$, $i=4, 5$.

Corollary 3.3. The following conditions are equivalent.

(1) $\bar{F}$ is the BVE distribution.

(2) $\bar{F}$ has a constant total failure rate and exponential marginals.

(3) $\bar{F}$ has a constant total failure rate, (a_1, a_2, a_3), and $P\{X>Y|\min(X,Y)=s\}=a_4$, $P\{X<Y|\min(X,Y)=s\}=a_5$ with $a_i<a_1<a_2+a_3$, $i=2,3$, $a_4a_1=(a_1-a_2)$, and $a_5a_1=(a_1-a_3)$; where $a_i>0$, $i=1,...,5$.

Corollary 3.4. The following conditions are equivalent.

(1) $\bar{F}$ is the ACBVE distribution.

(2) $\bar{F}$ has a constant total failure rate, (a_1, a_2, a_3), and functional forms of the marginal densities of $\bar{F}$ are specified as in (2.7) and (2.9), with $p_1=p_2=a_1(2a_1-a_2-a_3)^{-1}>1$.

(3) $\bar{F}$ has a constant total failure rate, (a_1, a_2, a_3), and $P\{X>Y|\min(X,Y)=s\}=a_4$, $P\{X<Y|\min(X,Y)=s\}=a_5$ with $a_i<a_1<a_2+a_3$, $i=2,3$, $a_4(2a_1-a_2-a_3)=(a_1-a_2)$, and $a_5(2a_1-a_2-a_3)=(a_1-a_3)$; where $a_i>0$, $i=1,...,5$.

4. CONCLUDING REMARKS

The constant total failure rate plays a role similar to the LMP in the theorems in Section 3. Although the constant total failure rate, like the bivariate LMP, without some additional assumptions cannot characterize any specific exponential distribution, it is an important characteristic of a distribution. In fact, a total failure rate with $P\{X>Y|\min(X,Y)=s\}$ and $P\{X<Y|\min(X,Y)=s\}$ uniquely characterize the corresponding joint distribution and,

therefore, it should be also useful in modeling two-component systems and developing statistical methods.

Another consideration is that the concept of total failure rate has the potential for introducing a new notion of aging which would be more intuitive and practical in comparison with the usual definitions of the bivariate increasing failure rate class by using the survival function.

Acknowledgement

This research was partially sponsored by the Air Force Office of Scientific Research, Air Force System Command, USAF, under grant number AFOSR F49620-92-J-0371.

REFERENCES

[1] Basu, A. P. (1971). Bivariate failure rate. J. Amer. Statist. Assoc. **66**, 103-104.

[2] Basu, A. P. (1988). Multivariate exponential distributions and their applications in reliability. In *Handbook of Statistics, Vol.7.* (P. R. Krishnaiah and C. R. Rao, eds.), pp. 467-477. Elsevier, Amsterdam.

[3] Block, H. W. (1977). A characterization of a bivariate exponential distribution. Ann. Statist. **5**, 808-812.

[4] Block, H. W. and Basu, A. P. (1974). A continuous bivariate exponential extension. J. Amer. Statist. Assoc. **69**, 1031-1037.

[5] Cox, D. R. (1972). Regression models and life-tables. J. R. Statist. Soc. B **34**, 187-220.

[6] Freund, J. (1961). A bivariate extension of the exponential distribution. J. Amer. Statist. Assoc. **56**,971-977.

[7] Friday, D. S. and Patil, G. P. (1977). A bivariate exponential model with applications to reliability and computer generation of random variables. In *Theory and Applications of Reliability, Volume I.* (C. P. Tsokos and I. Shimi, eds.), pp. 527-549.Academic Press, New York.

[8] Johnson, N. L. and Kotz, S. (1975). A vector multivariate hazard rate. J. Mult. Anal. **5**, 53-66.

[9] Marshall, A. W. (1975). Some comments on the hazard gradient. Stoch. Proc. Appl. **3**, 293-300.

[10] Marshall, A. W. and Olkin, I. (1967). A multivariate exponential distribution. J. Amer. Statist. Assoc. **62**, 30-44.

[11] Proschan, F. and Sullo, P. (1974). Estimating the parameters of a bivariate exponential distribution in several sampling situations. In *Reliability and Biometry* (F. Proschan and R. J. Serfling, eds.), pp. 423-440. SIAM, Philadelphia.

[12] Puri, P. S. and Rubin, H. (1974). On a characterization of the family of distributions with constant multivariate failure rates. Ann. Probab. **2**, 738-740.

[13] Shaked, M. and Shanthikumar, J. G. (1988). Multivariate conditional hazard rates and the MIFRA and MIFR properties. J. Appl. Prob. **25**, 150-168.

ADVANCES IN RELIABILITY
Edited by Asit P. Basu

Bayesian Reliability of Stress-Strength Systems

R.D. Thompson[a] and A.P. Basu[b]

[a]School of Business Administration, University of Michigan, Ann Arbor, Michigan 48109

[b]Department of Statistics, University of Missouri-Columbia, Columbia, Missouri 65211

Abstract

Posterior densities of the reliabilities of exponential stress-strength systems are derived using subjective and non-informative prior distributions specified on the system reliability itself, rather than on expected stress and expected strength (the usual approach). The method of Berger and Bernardo (1989) is employed to derive a reference posterior for the reliability of a simple stress-stress system in the presence of nuisance parameters. This posterior is shown to be identical to the Jeffreys posterior.

1. INTRODUCTION

In its most basic formulation, a so-called stress-strength system consists of a component, a stress applied to it, and the rule that the component will "fail" (*i.e.*, cease to properly function) if and only if the magnitude X of the applied stress exceeds the component's inherent strength Y. This system is motivated in Birnbaum's seminal article (1956, p. 14):

> If structural components of a mechanism are mass produced, the strength at failure Y of each single component (equals stress at which this component will fail) may be considered a random variable. The component is installed in an assembly and exposed to a stress which reaches its maximum value X, again a random variable. If $Y < X$, then the component will fail in use. In this situation, $p = \Pr\{Y < X\}$ is the probability that failure will occur because, due to chance, a component with relatively low strength was paired off with a high stress. It clearly is of interest to estimate this probability.

This basic definition of a stress-strength system has been extended to include a component suffering multiple stresses, a system consisting of multiple components each of which suffers a (the same) single stress, and a system consisting of multiple components each of which suffers its own stress. Johnson (1988) and Basu (1977, 1981) survey the considerable literature that pertains to inference on the reliability of these systems, much of which is written from a "classical" (*i.e.*, "frequentist") statistics perspective.

This research is concerned with Bayesian investigations of stress-strength systems. Section §2 reviews models and reliabilities of important stress-strength systems, including exponential stress-strength systems. Section §3 discusses Bayesian estimation of these reliabilities. Posterior distributions are derived with respect to subjective and non-informative priors defined on these reliabilities. In particular, a reference posterior is derived for Birnbaum's (1956) simple stress-strength system using an approach described in Berger and Bernardo (1989). This reference posterior is shown to be identical to the Jeffreys posterior.

2. STRESS-STRENGTH MODELS

2.1. Reliability of General Stress-Strength Systems

To aid in the clarity of this presentation, the following stress-strength models are defined.

MODEL I. *Simple Stress-Strength System.* Let X denote the magnitude of a stress that is applied to a component having strength Y. The system functions if $Y > X$, and does not function otherwise. Then, system reliability is given by

$$\eta_1 \equiv P(Y > X) = \iint_{\{(x,y)\,:\,y>x\}} f_{X,Y}(x,y)dydx = \int_{-\infty}^{\infty} \overline{F}_{Y|X}(x|x)dF_X(x) \tag{1a}$$

where $F_{X,Y}$ and $f_{X,Y}$ denote the joint cdf and joint pdf of the bivariate random variable (X,Y), $F_{X|Y}(\cdot|y)$ and $f_{X|Y}(\cdot|y)$ denote the conditional cdf and conditional pdf of X given $Y=y$, F_X is the marginal cdf of X, and $\overline{F}_{Y|X} \equiv 1 - F_{Y|X}$ is the conditional survivor function of Y given X. In the special case that X and Y are independent, (1a) reduces to

$$\eta_1 \equiv P(Y > X) = \int_{-\infty}^{\infty} \overline{F}_Y(x)dF_X(x), \tag{1b}$$

where $\overline{F}_Y \equiv 1 - F_Y$. ▲

MODEL II. *Single Component and* p *Stresses.* Let $X_1,\cdots,X_p$ denote the magnitudes of p stresses that are applied to a component having strength Y. The system functions if Y exceeds at least k $(k \le p)$ of the X_i's, and does not function otherwise. If the X_i's are iid and independent of Y, then system reliability is given by

$$\eta_2 = \sum_{j=k}^{p} \binom{p}{j} \int_{-\infty}^{\infty} [F_X(u)]^j [\overline{F}_X(u)]^{p-j} dF_Y(u). \quad ▲ \tag{2}$$

MODEL III. p *Components and a Single Stress.* Let $Y_1,\cdots,Y_p$ denote the respective strengths of the p components of a system suffering a single stress of magnitude X. The system functions if at least k $(k \le p)$ of the Y_i's exceed X, and does not function otherwise. If the Y_i's are iid and independent of X, then system reliability is given by

$$\eta_3 = \sum_{j=k}^{p} \binom{p}{j} \int_{-\infty}^{\infty} [F_Y(u)]^{p-j} [\bar{F}_Y(u)]^j \, dF_X(u). \ \blacktriangle \tag{3}$$

MODEL IV. p *Components and* p *Respective Stresses*. Let $Y_1, \cdots, Y_p$ denote the respective strengths of the components of a p-component system. Let X_i denote the magnitude of the stress applied to the system's i^{th} component $(i = 1, \cdots, p)$. The system functions if at least k $(k \le p)$ of the Y_i's exceed its respective X_i, and does not function otherwise. If the (X_i, Y_i) pairs are iid, then system reliability is given by

$$\eta_4 = \sum_{j=k}^{p} \binom{p}{j} [P(Y > X)]^j [P(Y < X)]^{p-j} = \sum_{j=k}^{p} \binom{p}{j} \eta_1{}^j (1 - \eta_1)^{p-j}. \ \blacktriangle \tag{4}$$

2.2. Reliability of Exponential Stress-Strength Models

The negative exponential distribution has proven to be extremely useful in the areas of mathematical reliability and life testing and has been used extensively in Bayesian inference re stress-strength systems. This distribution is used in the current investigation.

Let $f(\,\cdot\,|\theta)$ denote the density of a negative exponential random variable with expectation θ $(\theta > 0)$—that is,

$$f(u|\theta) \equiv \frac{1}{\theta} e^{-u/\theta} \cdot 1_{(0,\infty)}(u). \tag{5}$$

Then, given expected strength ϕ and expected stress θ $(\theta, \phi > 0)$, the following results are immediate: If $X, X_1, \cdots, X_p \sim \text{iid} f(x|\theta)$ and $Y, Y_1, \cdots, Y_p \sim \text{iid} f(y|\phi)$ independently, the system reliabilities (1b) through (4) associated with stress-strength Models I—IV are

$$\eta_1 = \frac{\phi}{\theta + \phi} = \frac{1}{1 + \lambda}, \tag{6}$$

$$\eta_2 = \frac{\Gamma(p+1) \cdot \Gamma\left(p + \frac{1}{\lambda} + 1 - k\right)}{\Gamma(p+1-k) \cdot \Gamma\left(p + \frac{1}{\lambda} + 1\right)}, \tag{7}$$

$$\eta_3 = 1 - \frac{\Gamma(k+\lambda)\Gamma(p+1)}{\Gamma(p+\lambda+1)\Gamma(k)}, \tag{8}$$

and

$$\eta_4 = \sum_{j=k}^{p} \binom{p}{j} \frac{\lambda^j}{(1+\lambda)^p}, \tag{9}$$

where $\lambda = \theta/\phi$ is the so-called stress-to-strength ratio, and $\Gamma(u) = \int_0^\infty t^{u-1} e^{-t} dt$ is the gamma function. Basu and Tarmast (1987) derive expressions (6) through (9) as special cases of expressions (1b) through (4) when stress and strength distributions satisfy the Lehmann alternatives—that is, when X and Y have proportional failure rates.

3. BAYESIAN INFERENCE WITH EXPONENTIAL STRESS-STRENGTH SYSTEMS

3.1. Data Collection

The usual approach to generating data to be used for statistical inference on (6) through (9) involves observing a sample of m iid stresses $\mathbf{X} = (X_1,\cdots,X_m)$ and the iid strengths $\mathbf{Y} = (Y_1,\cdots,Y_n)$ of an independent sample of n components. This approach was employed in Enis and Geisser (1971), the first Bayesian investigation of stress-strength systems, and has several compelling motivations (see Birnbaum (1956, p. 13) and Bhattacharyya and Johnson (1974, p. 966), for example). This standard sampling approach is adopted in this paper.

Data have been generated *via* other approaches, also. Although these approaches will not be pursued in this paper, they are described briefly, below. The most straightforward approach would involve placing several stress-strength systems on test and observing the number of systems that function or "survive". In a variation of this survivor-count approach, n replicates or cohorts each consisting of m components are subjected to n respective random stresses. The respective numbers $\mathbf{N} = (N_1,\cdots,N_n)$ of survivors, rather than the values of the n stresses and nm component strengths, are observed. This approach is due to Bhattacharyya (1977) and is employed by Choi and Kim (1983) in a Bayesian investigation of (8). For the special case in which the values of the stresses can be set prior to testing, quantal response data can be generated, provided the stress distribution is known. (In a non-Bayesian investigation, Mazumdar (1970, p. 159) motivates the important case in which the stress distribution is known completely.) Johnstone (1983) employs this approach in a Bayesian investigation of (6).

Recently Thompson (1992) has investigated (6) through (9) in the case of censored stress-strength data—that is, when both the value of a random stress and the resulting *state* of the component, but not necessarily the component's *strength*, can be determined.

3.2. Bayesian Inference Under Subjective Priors

3.2.1. CONJUGATE POSTERIORS FOR η_i

Bayes inferences on system reliabilities (6) through (9) necessarily involve prior beliefs about the η_i. These beliefs have historically been induced from prior densities defined on ϕ and θ, an approach that is appropriate whenever prior opinions exist separately about stresses and component strengths. Typically, independent inverted gamma priors are specified for ϕ and θ; Raiffa and Schlaifer's (1961) well-known result that the inverted gamma is conjugate for (5) yields rich classes of conjugate priors for these reliabilities. The inverted gamma density is given by

$$g(u|\alpha,\beta) = \frac{\beta^{\alpha}}{\Gamma(\alpha)} \cdot \frac{1}{u^{\alpha+1}} e^{-\beta/u} \cdot 1_{(0,\infty)}(u), \tag{10}$$

where α and β $(\alpha,\beta > 0)$ are hyperparameters. The following result is immediate.

Theorem 1. *Given the hyperparameters* α, β, δ, *and* γ $(\alpha,\beta,\delta,\gamma > 0)$, *suppose expected stress* θ *and expected component strength* ϕ *have independent inverted gamma priors* $\theta \sim g(\theta|\alpha,\beta)$ *and* $\phi \sim g(\phi|\delta,\gamma)$. *Then, if* $X_1,\cdots,X_m \sim$ iid $f(x|\theta)$ *and* $Y_1,\cdots,Y_n \sim$ iid $f(y|\phi)$ *are two independent samples of exponential stresses and component strengths, the corresponding independent posterior densities are* $\theta \sim g(\theta|\alpha+m,\beta+m\bar{x})$ *and* $\phi \sim g(\phi|\delta+n,\gamma+n\bar{y})$, *and the conjugate posterior* $\pi_C(\eta_1|\mathrm{D})$ *for the system reliability* η_1 *is*

$$\pi_C(u|\mathrm{D}) = \pi(u;\alpha+m,\beta+m\bar{x},\delta+n,\gamma+n\bar{y}), \tag{11}$$

where

$$\pi(u;\alpha,\beta,\delta,\gamma) = \frac{(\beta/\gamma)^{\alpha}}{\mathrm{B}(\alpha,\delta)} \cdot \frac{u^{\alpha-1}(1-u)^{\delta-1}}{\left[1-(1-\beta/\gamma)u\right]^{\alpha+\delta}} \cdot 1_{(0,1)}(u), \tag{12}$$

$\bar{x} = \sum_{i=1}^{m} x_i$ *and* $\bar{y} = \sum_{i=1}^{n} y_i$ *are realized sample means, and* $\mathrm{B}(\alpha,\delta) = \Gamma(\alpha)\Gamma(\delta)/\Gamma(\alpha+\delta)$ *is the beta function.*

Proof. Straightforward. ▲

The straightforward derivation of this result is found in Thompson (1991). Enis and Geisser (1971), using independent gamma priors (rather than inverted gamma priors) and the parameterization of the exponential density for which the gamma is conjugate, indicate an essentially analogous result. The values of the hyperparameters α, β, δ, and γ are specified to reflect prior opinion about either η_1 or ϕ and θ. In the special case that $\beta = \gamma$, (12) is the beta prior (14).

The utility of the conjugate prior (12) should be emphasized: If an investigator's prior opinion is solely in terms of the *system*, the hyperparameters α, β, δ, and γ would to specified so that (12) will reflect this opinion. The traditional perspective is that an investigator will have opinions solely in terms of components and system stresses, in which case α, β, δ, and γ are specified so that the inverted gamma priors for ϕ and θ will reflect these opinions.

Analogous closed-form expressions for the posterior densities of system reliabilities (7) through (9) are not available. However, these posteriors can be simulated easily using the posterior of λ which, readily derived from (11), is shown in Basu and Tarmast (1987) to be

$$\varphi_C(\lambda|\mathrm{D}) = \varphi(\lambda; \alpha + m, \beta + m\bar{x}, \delta + n, \gamma + n\bar{y}), \tag{13}$$

where

$$\varphi(u; \alpha, \beta, \delta, \gamma) = \frac{(\beta/\gamma)^{\alpha}}{\mathrm{B}(\alpha,\delta)} \cdot \frac{u^{\delta-1}}{(u+\beta/\gamma)^{\alpha+\delta}} \cdot 1_{(0,\infty)}(u).$$

The posterior density (11) of η_1 and the simulated posteriors of the other η_i derived from (13) can be used to obtain point estimates for these stress-strength system reliabilities. Typically, however, these estimates cannot be obtained in closed form. For example, the simplified expressions for Bayes estimates of (6) through (9) derived in Basu and Tarmast (1987) under squared-error loss involve integrals which must be numerically evaluated. Bayes estimates that correspond to asymmetric loss functions, ostensibly more appropriate when estimating system reliability, also typically cannot be expressed in closed form. (See Thompson and Basu (1992), Basu and Ebrahimi (1992), and Thompson (1991) for discussions re asymmetric loss functions appropriate for estimating system reliability.)

3.2.2. ALTERNATIVE SUBJECTIVE POSTERIORS FOR η_1

Priors specified for θ and ϕ induce not only a prior for η_1, but also a prior for any parameter ω for which (η_1, ω) is a one-to-one transformation of (θ, ϕ). This suggests the following alternative method for obtaining a posterior for η_1: An investigator would specify a joint prior for η_1 and ω and obtain the marginal posterior for η_1 directly from the resulting joint posterior. If prior opinions exist separately for η_1 and ω, this joint prior would simply be the product of appropriate independent marginal priors specified for η_1 and ω.

To admit maximum flexibility for capturing subjective prior beliefs about the system reliability η_1, the prior for η_1 should belong to a rich parametric class of distributions (*e.g.*, the beta distribution). How the prior for ω might be specified depends, however, on the nature of ω: If ω is not meaningful—for example, if ω is defined simply for convenience—a vague or non-informative prior would be preferred for what would be a "nuisance" parameter; alternately, if ω is meaningful, its prior should be chosen from among some rich parametric class.

The utility of this alternative method for deriving a subjective posterior for η_1 is apparent if, for example, prior opinions are held not only for the system reliability η_1 but also for the expected system stress θ. This alternative method would allow a Bayesian investigation of η_1 to incorporate both of these disparate sources of prior opinion, both of which influence the marginal posterior distribution of η_1. (In this example, $\omega \equiv \theta$.)

The following result is an application of this method. Thompson (1992) contains other similar results.

Theorem 2. *Suppose the prior density of* η_1 *and* $\omega = \theta + \phi$ *is given by* $\pi_1(\eta_1) \cdot \pi_2(\omega)$, *where, given* α, β, *and* δ $(\alpha, \beta, \delta > 0)$,

$$\pi_1(u) = \frac{1}{\mathrm{B}(\alpha,\beta)} u^{\alpha-1}(1-u)^{\beta-1} \cdot 1_{(0,1)}(u) \tag{14}$$

and

$$\pi_2(u) \propto \frac{1}{u^{\delta}} \cdot 1_{(0,\infty)}(u). \tag{15}$$

Then, if $X_1, \cdots, X_m \sim$ iid $f(x|\theta)$ *and* $Y_1, \cdots, Y_n \sim$ iid $f(y|\phi)$ *are two independent samples of exponential stresses and component strengths, the posterior* $\pi_S(\eta_1|\mathrm{D})$ *of* η_1 *is given by*

$$\pi_S(u|\mathrm{D}) \propto \frac{u^{\alpha+\delta+m-1}(1-u)^{\beta+\delta+n-1}}{\left[1-\left(1-\frac{m\bar{x}}{n\bar{y}}\right)u\right]^{\delta+m+n-1}} \cdot 1_{(0,1)}(u). \tag{16}$$

Proof. Straightforward. ▲

In the above theorem, $\omega = \theta + \phi$ is a nuisance parameter, and its prior (15) is intuitively non-informative if either $\delta = 0$ or $\delta = 1$. Expression (14) is the beta density.

3.3. Bayesian Estimation Under Non-Informative Priors

If a prior reflects vague or non-existent prior opinions, then the resulting posterior can be used as a baseline reference for posteriors obtained from subjective priors. Non-informative priors are not unique, however, and neither are reference posteriors. This section describes several reference posteriors for system reliability η_1.

If the parameters of the beta distribution (14) satisfy $\alpha = \beta = 1$, then a uniform distribution results. Thus, Theorem 2 provides an intuitive reference posterior for the system reliability η_1 if $\alpha = \beta = \delta = 1$ (alternately, if $\alpha = \beta = 1$ and $\delta = 0$) is substituted into (16), as these parameter values reflect vague or non-existent prior beliefs about both η_1 and the nuisance parameter $\omega = \theta + \phi$.

Theorem 1 provides a different intuitive reference posterior for η_1. The inverted gamma prior (10) for θ is non-informative for θ when its hyperparameters satisfy $\alpha = \beta = 0$. (The prior is non-informative or "information-less" in several senses, including those of Jeffreys (1961), Box and Tiao (1973), and Zellner (1984).) Hence, the posterior that results when $\alpha = \beta = \delta = \gamma = 0$ is substituted into (11) will reflect vague or non-existent prior beliefs about the expected stress θ, the expected strength ϕ, and, intuitively, the system reliability η_1. (Since the derivation of this "reference" posterior for η_1 involves transforming priors that are non-informative for θ and ϕ, it may in fact not reflect vague or non-existent prior beliefs about η_1.)

Other intuitive reference posteriors for η_1 might be derived by first expressing the joint distribution of the samples **X** and **Y** in terms of η_1 and a parameter ω (*via* a one-to-one reparameterization of θ and ϕ) and then deriving a non-informative joint prior for η_1 and ω *via* a standard recipe (*e.g.*, Jeffreys). Marginal posteriors for η_1 obtained from such joint posteriors are *intuitively* reference priors only, since, they might not reflect vague or non-existent prior beliefs about η_1. (See Dawid, Stone, and Zidek (1973), for example, for a discussion of this so-called "marginalization paradox.")

Berger and Bernardo (1989) refines a procedure proposed in Bernardo (1979) which provides a reference prior for a parameter in the presence of a nuisance parameter. This procedure is employed below to derive a reference prior for the system reliability η_1.

Lemma. *If* $X_1,\cdots,X_m \sim$ iid $f(x|\theta)$ *and* $Y_1,\cdots,Y_n \sim$ iid $f(y|\phi)$ *are two independent samples of exponential stresses and component strengths, the Jeffreys posterior for* η_1 *is equal to* (11) *with* $\alpha=\beta=\delta=\gamma=0$.

Proof. Reparameterization of the joint distribution of the samples **X** and **Y** yields the following likelihood function of $\eta=\phi/(\theta+\phi)$ and $\omega=\theta+\phi$

$$\mathrm{lik}(\eta,\omega|\mathbf{x},\mathbf{y})=\frac{\beta^{\alpha}\gamma^{\delta}}{\Gamma(\alpha)\Gamma(\delta)}\cdot\frac{e^{-\left\{\frac{n\bar{y}[1-(1-m\bar{x}/n\bar{y})\eta]}{\eta(1-\eta)}\cdot\frac{1}{\omega}\right\}}}{\eta^{n}(1-\eta)^{m}\omega^{m+n}}\cdot 1_{(0,1)\times(0,\infty)}(\eta,\omega),$$

from which the expected Fisher information matrix is readily seen to be

$$I(\eta,\omega)=\left(I_{ij}\right)(\eta,\omega)=\begin{pmatrix}\dfrac{n}{\eta^2}+\dfrac{m}{(1-\eta)^2} & \dfrac{1}{\omega}\left(\dfrac{n}{\eta}+\dfrac{m}{1-\eta}\right)\\ (\text{sym.}) & \dfrac{m+n}{\omega^2}\end{pmatrix}. \tag{17}$$

Then the Jeffreys multivariate prior for η and ω is

$$\pi(\eta,\omega)\propto \det{}^{1/2} I(\eta,\omega)\propto\frac{1}{\eta(1-\eta)\omega}, \tag{18}$$

which is readily seen to lead to (11) with $\alpha=\beta=\delta=\gamma=0$. ▲

Theorem 3. *The Berger-Bernardo reference posterior for the system reliability* η_1 *is identical to the Jeffreys reference posterior for* η_1.

Proof. The algorithm, described in Berger and Bernardo (1989), proceeds as follows:

Step 1. Let $\pi(\omega|\eta)$ be the usual (Jeffreys) reference prior for ω with η given, defined by $\pi(\omega|\eta)\propto\sqrt{\det I_{22}(\eta,\omega)}$: From (17),

$$\pi(\omega|\eta) \propto \frac{1}{\omega} \cdot 1_{(0,\infty)}(\omega).$$

Step 2. Choose a sequence $\Lambda_1 \subset \Lambda_2 \subset \cdots$ of subsets of the parameter space Λ for (η,ω), such that $\bigcup_i \Lambda_i = \Lambda$ and $\pi(\omega|\eta)$ has finite mass on $\Omega_{i,\eta} = \{\omega : (\eta,\omega) \in \Lambda_i\}$ for all η. Then normalize $\pi(\omega|\eta)$ on each $\Omega_{i,\eta}$, obtaining $p_i(\omega|\eta) = K_i(\eta)\pi(\omega|\eta) \cdot 1_{\Omega_{i,\eta}}(\omega)$ where $K_i(\eta) = \left\{\int_{\Omega_{i,\eta}} \pi(\omega|\eta)d\omega\right\}^{-1}$:

$\Lambda = (0,1) \times (0,\infty)$, $\Lambda_i = (0,1) \times (1/i,i)$, and $\Omega_{i,\eta} = (1/i,i)$;

$$1/K_i(\eta) = \int_{1/i}^{i} \frac{1}{\omega} d\omega = 2\ln i \ ;$$

$$p_i(\omega|\eta) = \frac{1}{2\omega\ln i} \cdot 1_{(1/i,i)}(\omega).$$

Step 3. Find the marginal reference prior for η with respect to $p_i(\omega|\eta)$. This is $\pi_i(\eta) \propto \exp\left\{\frac{1}{2}\int_{\Omega_{i,\eta}} p_i(\omega|\eta) \cdot \ln[\det I(\eta,\omega)/\det I_{22}(\eta,\omega)]d\omega\right\}$, assuming the integral exists:

$$\pi_i(\eta) \propto \exp\left\{\frac{1}{2}\int_{1/i}^{i} \frac{1}{(2\ln i)\omega} \ln\left(\frac{\omega^2}{\omega^2\eta^2(1-\eta)^2}\right)d\omega\right\} = \frac{1}{\eta(1-\eta)}.$$

Step 4. Define the reference prior for (η,ω) when ω is a nuisance parameter by $\pi(\eta,\omega) \propto \lim_{i\to\infty}\left\{\frac{K_i(\eta)\pi_i(\eta)}{K_i(\eta_0)\pi_i(\eta_0)}\right\}\pi(\omega|\eta)$, assuming the limit exists; here η_0 is any fixed point:

$$\pi(\eta,\omega) \propto \lim_{i\to\infty}\left\{\frac{2(\ln i)\eta_0(1-\eta_0)}{2(\ln i)\eta(1-\eta)}\right\}\frac{1}{\omega} \cdot 1_{(0,1)\times(0,\infty)}(\eta,\omega) \propto \frac{1}{\omega\eta(1-\eta)} \cdot 1_{(0,1)\times(0,\infty)}(\eta,\omega)$$

Since this prior is identical to (18), the theorem follows from the lemma. ▲

4. SUMMARY

There remain relatively few Bayesian investigations among the extensive literature pertaining to stress-strength systems. This paper reviewed several important stress-strength systems that have appeared since the original Birnbaum (1956) system, described their reliabilities in the context of independent negative exponential stress and strength distributions, and derived posterior distributions for them when independent samples of iid system stresses and iid component strengths are available. These posteriors were derived assuming that prior beliefs will be represented by prior distributions specified on the reliabilities themselves rather than on the expected stress and expected strength, the usual and perhaps less practical approach. These posterior distributions are derived from several subjective and non-informative prior distributions. Finally, a reference posterior for the reliability of a simple stress-strength system was developed using the Berger-Bernardo (1989) approach and shown to coincide with the Jeffreys posterior, but differ from a reference posterior derived from a non-data-generated non-informative prior.

References

Basu, A.P. (1977). Estimation of reliability in the stress-strength model. *Proceedings of the Twenty-Second Conference on the Design of Experiments in Army Research, Development, and Testing* **22**, 97-110.

Basu, A.P., and Ebrahimi, N. (1991). Bayesian approach to life testing and reliability estimation using asymmetric loss function. *Journal of Statistical Planning and Inference* **29**, 21-31.

Basu, A.P., and Tarmast, G. (1987). Reliability of a complex system from Bayesian viewpoint. In *Probability and Bayesian Statistics* (R. Viertl, ed.), pp. 31-38. Plenum Publishing Corporation.

Basu, A.P. (1981). The estimation of $P(X<Y)$ for distributions useful in life testing. *Naval Research Logistics Quarterly* **28**, 383-392.

Berger, J.O. (1985). *Statistical Decision Theory and Bayesian Analysis, 2nd Edition.* New York: Springer-Verlag, Inc.

Berger, J.O. and J.M. Bernardo (1989). Estimating a product of means: Bayesian analysis with reference priors. *J. Amer. Statist. Assoc.* **84** (405) 200-207.

Bernardo, J.M. (1979). Reference posterior distributions for Bayesian inference. *J. Royal Statist. Soc., Series B* **41** (2) 113-147.

Bhattacharyya, G.K. (1977). Reliability estimation from survivor count data in a stress-strength setting. *Journal of the Indian Association for Productivity, Quality, and Reliability* **2**, 1-16.

Bhattacharyya, G.K., and Johnson, R.A. (1974). Estimation of reliability in a multicomponent stress-strength model. *J. Amer. Statist. Assoc.* **69** (348), 966-970.

Birnbaum, Z.M. (1956). On a use of the Mann-Whitney statistic. In *Proceedings of the Third Berkeley symposium on Mathematical Statistics and Probability, Volume I: Contributions to the Theory of Statistics and Probability*, pp. 13-17. Berkeley, California: University of California Press.

Box, G.E.P., and Tiao, G.C. (1973). *Bayesian Inference in Statistical Analysis*. Reading, Massachusetts: Addison-Wesley Pub. Co., Inc.

Choi, S.S., and Kim, J.J. (1983). A Bayes reliability estimation from life test in a stress-strength model. *J. Korean Statist. Soc.* **12** (1), 1-9.

Dawid, A.P.; Stone, N.; and Zidek, J.V. (1973). Marginalization paradoxes in Bayesian and structural inference. *Journal of the Royal Statistical Society* **B35** (2) 189-233.

Enis, P., and Geisser, S. (1971). Estimation of the probability that Y<X. *J. Amer. Statist. Assoc.* **66** (333), 162-168.

Jeffreys, H. (1983, c1961). *Theory of Probability* (3rd Edition). London: Oxford University Press.

Johnson, R.A. (1988). Stress-strength models for reliability. In *Handbook of Statistics, Volume 7: Quality Control and Reliability* (P.R. Krishnaiah and C.R. Rao, eds.), pp. 27-54. New York: Elsevier Science Pub. Co., Inc.

Johnstone, M.A. (1983). Bayesian estimation of reliability in the stress-strength context. *Journal of the Washington Academy of Sciences* **73** (4), 140-150.

Mazumdar, M. (1970). Some estimates of reliability using interference theory. *Naval Research Logistics Quarterly* **17**, 159-165.

Raiffa, H., and Schlaifer, R. (1961). *Applied Statistical Decision Theory*. Boston: Harvard University Press.

Thompson, R.D. (1991). *Bayesian Estimation of System Reliability Under Asymmetric Loss*. PhD dissertation. University of Missouri-Columbia.

Thompson, R.D. (1992). Bayesian estimation of stress-strength reliability using censored data. Research Manuscript. University of Michigan.

Thompson, R.D., and Basu, A.P. (1992). Asymmetric loss functions for estimating system reliability. Research manuscript. University of Michigan.

Zellner, A. (1984). A note on the maximal data information prior (MDIP) for the parameters of the Weibull density function. Research Manuscript. University of Chicago.

ADVANCES IN RELIABILITY
Edited by Asit P. Basu

An efficient test for increasing failure rate average distribution under random censoring

Ram C. Tiwari[a] and Jyoti N. Zalkikar[b]

[a]Department of Mathematics, University of North Carolina at Charlotte, Charlotte, NC 28223, USA

[b]Department of Statistics, Florida International University, Miami, FL 33199, USA

Abstract

A continuous life distribution function F is an increasing failure rate average (IFRA) if $\{\overline{F}(bx)\} \geq \{\overline{F}(x)\}^b$, $0 < b < 1$, $x \geq 0$, where $\overline{F} \equiv 1 - F$. Under a model of random censorship, we consider the problem of testing H_0: F is exponential, versus H_1: F is IFRA, but not exponential. This paper introduces a process, called an IFRA process, $\{\zeta_n(b);\ 0 \leq b \leq 1\}$, where

$$\zeta_n(b) = \int_0^\infty \overline{F}_n(bx)dF_n(x),$$

and $\overline{F}_n$ is the Kaplan-Meier estimator of $\overline{F}$. The proposed test statistic is $J_n^c := \int_0^1 \zeta_n(b)db$. Most of the tests that have been studied in the literature depend on the nuisance parameter b. The proposed test does not have this disadvantage. The asymptotic normality of J_n^c is established and asymptotically distribution - free tests are obtained by using estimators for the null standard deviations. The efficiency loss due to censoring for the Weibull family of IFRA alternatives is studied. The test is applied to the survival data of Hollander and Proschan (1979).

1. INTRODUCTION

A life distribution function (d.f.) F is said to be increasing failure rate average (IFRA) if $-\ell n\,\overline{F}(x)$ is star shaped for $x > 0$, or equivalently if and only if

$$\overline{F}(bx) \geq \{\overline{F}(x)\}^b \text{ for all } x \geq 0,\ 0 < b < 1, \tag{1.1}$$

where $\overline{F}(x) = 1 - F(x)$ (see, e.g., Barlow and Proschan (1975), p.84). The equality in (1.1) holds if and only if F is an exponential distribution. The dual class of decreasing failure rate average distributions (DFRA) can be defined analogously by reversing the direction of inequalty in (1.1). The IFRA class of distributions plays an important role in reliability investigations. It is the smallest class of probability distributions which contains the exponential distributions and, is closed under the formation of coherent systems (Barlow and Proschan (1975)).

We consider the problem of testing the null hypothesis

H_0: $F(x) = 1 - \exp(-x/\mu)$, $x \geq 0$ (μ unspecified),

versus

H_1: F(x) is IFRA, but not exponential.

In the uncensored model, where we get a complete sample, tests for H_0 vs H_1 have been proposed by Deshpande (1983), Kochar (1985), Link (1989), Bandyopadhyay and Basu (1989), and Jammalamadaka, et. al. (1990). For randomly censored model see Gerlach (1989), Tiwari, et. al. (1989) and Wells and Tiwari (1991).

To describe a randomly censored model, let $X_1, \ldots, X_n$ denote independent identically distributed (i.i.d.) random variables (r.v.'s) having a common continuous d.f. F. Independent of the X_i's, let $Y_1, \ldots, Y_n$ also denote i.i.d.r.v.'s having a common continuous d.f. H, which is unknown and is treated as a nuisance parameter. The Y_i's are treated as the random times to the right censorship. Finally, let

$$Z_i = X_i \wedge Y_i \text{ and } \delta_i = 1[X_i \le Y_i] \text{ for } i = 1, \ldots, n, \tag{1.2}$$

where $a \wedge b = \min\{a, b\}$ and $1[A]$ denotes the indicator function of the set A. For testing H_0 against H_1 on the basis of the randomly censored data $\{(Z_i, \delta_i), i = 1, \ldots, n\}$ the tests are of the form

$$\int_0^\infty \Psi(\bar{F}_n(bx))dF_n(x) \tag{1.3}$$

where Ψ is a suitable weight function, and F_n is the Kaplan-Meier (Kaplan and Meier (1958)) estimator of F defined by

$$\bar{F}_n(x) := 1 - F_n(x) = \prod_{\{i:\, Z_{(i)} \le x\}} \left(\frac{n-i}{n-i+1}\right)^{\delta_{(i)}}, \tag{1.4}$$

where $Z_{(1)} < \ldots < Z_{(n)}$ denote the ordered values of Z_i's and $\delta_{(1)}, \ldots, \delta_{(n)}$ are the δ_i's corresponding to $Z_{(1)}, \ldots, Z_{(n)}$, respectively. The tests in (1.3) reject H_0 in favor of H_1 for large values. One disadvantage of the tests in (1.3) is that they depend on the nuisance parameter b.

To overcome this problem, in Section 2, we first study the process

$$\zeta_n(b) = \int_0^\infty \bar{F}_n(bx)dF_n(x),\ 0 \le b \le 1, \tag{1.5}$$

and then, for testing H_0 versus H_1, propose the test statistic

$$J_n^c := \int_0^1 \zeta_n(b)db \tag{1.6}$$

The asymptotic normality of J_n^c is established, and a consistent estimator of the null asymptotic variance is provided.

In Section 3, the efficacy of the test is computed and the efficiency loss due to censoring is studied for the Weibull alternatives in the proportional censoring model. This section also contains an application of the proposed test to the survival data of Hollander and Proschan (1979).

2. THE $\zeta_n(b)$ PROCESS AND THE TEST

Let $\bar{K}(t) = \bar{F}(t)\bar{H}(t)$, and $\{\phi(t);\ 0 \le t < \infty\}$ be a Gaussian process with mean zero and covariance kernel defined by

$$\mathrm{Cov}(\phi(s),\ \phi(t)) = \overline{F}(s)\overline{F}(t) \int_0^s \{\overline{K}(u)\overline{F}(u)\}^{-1} dF(u), \text{ for } 0 \le s < t < \infty. \tag{2.1}$$

We impose the following assumptions on the d.f.'s F and H:

(A.1) Both F and H have a common support $[0, \infty)$,

(A.2) $\int_0^\infty \{\overline{H}(u)\}^{-1} dF(u) < \infty$.

Then Theorem 2.1 of Gill (1983) implies that the weak convergence of $\phi_n(x) := n^{\frac{1}{2}}\{F_n(x \wedge Z_{(n)}) - F(x \wedge Z_{(n)})\}$ to $\phi(x)$ on $[0, \infty)$ holds as $n \to \infty$. Let,

$$\varsigma(b) = \int_0^\infty \overline{F}(bx)dF(x),\ 0 \le b \le 1. \tag{2.2}$$

From (1.5) and (2.2), we have

$$\sqrt{n}[\varsigma_n(b) - \varsigma(b)] = \int_0^\infty \sqrt{n}[\overline{F}_n(bx) - \overline{F}(bx)]dF_n(x) - \int_0^\infty \sqrt{n}[\overline{F}_n(bx) - \overline{F}(bx)]dF(x)$$

$$+ \int_0^\infty \sqrt{n}[(\overline{F}_n(bx) - \overline{F}(bx)) - (\overline{F}_n(\tfrac{x}{b}) - \overline{F}(\tfrac{x}{b}))]dF(x)$$

$$= [\int_0^\infty \phi_n(bx)dF_n(x) - \int_0^\infty \phi_n(bx)dF(x)]$$

$$+ \int_0^\infty [\phi_n(bx) - \phi_n(\tfrac{x}{b})]dF(x). \tag{2.3}$$

The first term in (2.3) goes to zero in probability as $n \to \infty$, and, thus, the weak convergence of $\sqrt{n}[\varsigma_n(b) - \varsigma(b)]$, $0 \le b \le 1$ is identical to that of the process $\int_0^\infty \left[\phi_n(bx) - \phi_n(\tfrac{x}{b})\right] dF(x),\ 0 \le b \le 1.$

Let D[0, 1] be the class of real valued, bounded and right continuous functions defined on [0, 1] with finite left limits at each $x \in [0, 1]$. D[0, ∞] is a metric space with Skorohod metric. Then we have the following theorem. [The proof of this theorem will be presented in a separate article.]

Theorem 2.1. The process $\xi_n(b) := \sqrt{n}[\varsigma_n(b) - \varsigma(b)]$, $0 \le b \le 1$, converges weakly to a mean zero Gaussian process $\xi(b) := \int_0^\infty (\phi(bx) - \phi(\tfrac{x}{b}))dF(x)$, $0 \le b \le 1$, with covariance kernel given by

$$\mathrm{Cov}(\varsigma(b_1),\ \varsigma(b_2)) = \int_0^\infty\int_0^\infty E[(\phi(b_1x) - \phi\tfrac{x}{b_1})(\phi(b_2y) - \phi\tfrac{y}{b_2})]dF(x)dF(y),\quad 0 \le b_1,$$

$b_2 \leq 1$,

where ϕ is defined as in (2.1).

Define,

$$\Delta(F) = \int_0^1 \int_0^\infty \overline{F}(bx)dF(x)db = \int_0^1 \varsigma(b)db$$

$$\overset{H_0}{=} \int_0^1 \frac{1}{(1+b)} db = \log_e 2.$$

The from Theorem 2.1 we have the following.

Corollary 2.2. Under Assumptions (A.1) and (A.2), $n^{\frac{1}{2}}\{J_n^c - \Delta(F)\}$ coverges in distribution to a normal r.v. with mean zero and finite variance

$$\sigma^2(F) = \int_0^1 \int_0^1 \int_0^\infty \int_0^\infty E[(\phi(b_1 x) - \phi(\tfrac{x}{b_1}))(\phi(b_2 y) - \phi(\tfrac{y}{b_2}))]dF(x)dF(y)db_1 db_2 \quad (2.4)$$

Corollary 2.3. Under H_0 and Assumptions (A.1) and (A.2), $n^{\frac{1}{2}}\{J_n^c - \log_e 2\}$ converges in distribution to a normal r.v. with mean zero and variance σ^2 given by

$$\sigma^2 = \int_0^1 \int_0^1 \int_0^1 f(z, b_1, b_2)\{\overline{K}(-b_1 \mu \ell n z)\}^{-1} dz db_1 db_2$$

$$+ \int_0^1 \int_0^1 \int_0^1 f(z, b_2, b_1)\{\overline{K}(-b_2 \mu \ell n z)\}^{-1} dz db_1 db_2, \quad (2.5)$$

where

$$f(z, b_1, b_2) = \left\{ \frac{b_1 b_2 z^{(2b_1 + \frac{b_1}{b_2})}}{(b_2+1)(b_1+b_2+2b_1 b_2)} - \frac{b_1 b_2 z^{(b_1 b_2 + 2b_1)}}{(b_1+1)(b_2+1)} \right.$$

$$\left. + \frac{b_1^2 b_2 z^{b_1(b_1+b_2+2)-1}}{(b_1+1)(b_1+b_2+2)} \right\}.$$

An alternative proof of Corollary 2.2 can be given along the lines of Kumazawa (1987) using the following lemma. Define, $F^{-1}(s) = \inf\{x: \ F(x) \geq s\}$. Lemma 2.4 shows that the statistical function $\Delta(F)$ is Hadamard differentiable (see Fernholz (1983) and Gill (1983) for details). For the continuous d.f. F,

$$\Delta(F) = \int_0^1 \int_0^\infty \overline{F}(bx)dF(x)db$$

$$= \int_0^1 \int_0^1 [1 - \overline{F}(bF^{-1}(s))]dsdb.$$

Lemma 2.4. Let τ be the functional induced on D[0, 1] by $\tau(G) = \Delta(G \circ F)$ for G in D[0, 1]. Then τ is Hadamard differentiable at I with derivative

$$\tau_I'(G) = -\int_0^1 \int_0^1 G \circ F(bF^{-1}(s)db$$

$$+ \int_0^1 \int_0^1 G \circ F(F^{-1}(s)/b)db,$$

where $I(u) = u$ for $0 \leq u \leq 1$.

Proof. For fixed F define,

$\gamma_1(G)(s) = F^{-1} \circ G^{-1}(s)$,

$\gamma_2(V)(s, b) = bV(\dot{s})$,

$\gamma_3(U, G)(s, b) = 1 - G \circ F[U(s, b)]$,

and

$$\gamma_4(U) = \int_0^1 \int_0^1 U(x, y)dxdy,$$

for G in D[0, 1], V in $L^1[0, 1]$, U in $L^1[0, 1] \times L^1[0, 1]$ and $0 \leq s, u \leq 1$. Then $\tau(G) = \gamma_4 \circ \gamma_3(\gamma_2 \circ \gamma_1, G)$, and it follows from Proposition 6.1.1 and 6.1.6 of Fernholz (1983) that τ is differentiable at I and some calculation yields the derivative $\tau_1'(G)$. □

Corollary 2.3 shows that the null asymptotic variance σ^2 depends on the unknown mean μ, and the censoring d.f. H. In order to perform the test based on J_n^c we estimate σ^2. We can write σ^2 as

$$\sigma^2 = \int_0^1\int_0^1\int_0^\infty \frac{1}{b_1} f(\bar{F}(\tfrac{x}{b_1}), b_1, b_2)\bar{F}(\tfrac{x}{b_1})\{\bar{K}(x)\}^{-2} dH^*(x)db_1db_2$$

$$+ \int_0^1\int_0^1\int_0^\infty \frac{1}{b_2} f(\bar{F}(\tfrac{x}{b_2}), b_2, b_1)\bar{F}(\tfrac{x}{b_2}) \{\bar{K}(x)\}^{-2} dH^*(x)db_1db_2,$$

where $H^*(x) = P(Z_1 \leq x, \delta_1 = 1)$. Define,

$\bar{K}_n(x) = \frac{1}{n} \sum_{i=1}^n 1[Z_i > x]$, $H_n^*(x) = \frac{1}{n} \sum_{i=1}^n 1[Z_i \leq x, \delta_i = 1]$.

We estimate σ^2 by

$$\hat{\sigma}_n^2 = \int_0^1\int_0^1\int_0^\infty [\frac{f(\bar{F}_n(\frac{x}{b_1}-), b_1, b_2)\bar{F}_n(\frac{x}{b_1}-)}{b_1} + \frac{f(\bar{F}_n(\frac{x}{b_2}-), b_2, b_1)}{b_2}$$

$$\bar{F}_n(\tfrac{x}{b_2}-)]\{\bar{K}_n(x-)\}^{-2} dH_n^*(x)db_1db_2$$

$$= \frac{1}{n} \sum_{i:\ \delta_i=1}^n \int_0^1\int_0^1 [\frac{1}{b_1} f(\bar{F}_n(\tfrac{Z_i}{b_1})-), b_1, b_2)\bar{F}_n(\tfrac{Z_i}{b}-)$$

$$+ \frac{1}{b_2} f(\bar{F}_n(\tfrac{Z_i}{b_2}-), b_2, b_1)\bar{F}_n(\tfrac{Z_i}{b_2}-)]\{\bar{K}_n(Z_i-)\}^{-2} db_1db_2$$

Under H_0, and Assumptions (A.1) and (A.2) the consistency of $\hat{\sigma}_n^2$ can be proved along the lines of Lemma 2.4 of Kumazawa (1987).

Another consistent estimator of σ^2 can be obtained by replacing μ and $\bar{K}$ in (2.5) by $\hat{\mu}_n = (\sum_{i=1}^n \delta_i)^{-1} \sum_{i=1}^n Z_i$ and $\bar{K}_n$, respectively. The proof of this can be given using the results of Chen et al. (1983) and Tiwari et al. (1989).

From above discussions, and from Corollary 2.3 it follows that the tests rejecting H_0 in favor of H_1 for large values of $\sqrt{n}\ [J_n^c - \log_e 2]/\hat{\sigma}_n$ are consistent against all continuous IFRA alternatives.

3. EFFICIENCY LOSS DUE TO CENSORING

We shall consider the alternatives $\{F_{\lambda_n}\}$ with $\lambda_n = \lambda_0 + \frac{\gamma}{\sqrt{n}}$, $\gamma > 0$ and $F_{\lambda_0}(x) := 1 - \exp(-x)$. The alternatives are in the Weibull family of d.f.'s: $F_\lambda(x) = 1 - \exp(-x^\lambda)$, $\lambda > 1$. The calculations yield

$$\mu'(\lambda_0) = -\int_0^1 \frac{b \log_e b}{(1+b)^2}\, db \qquad (3.1)$$

For the proportional hazard censoring model, $\overline{H}(x) = \overline{F}^\beta(x)$, $0 \le \beta < 1$, we have

$$\sigma^2(\beta) = \int_0^1 \int_0^1 [f(b_1, b_2) + f(b_2, b_1)]db_1 db_2$$

where

$$f(b_1, b_2) = \left\{ \frac{b_1 b_2}{(b_2+1)(b_1+b_2+2b_1b_2)(2b_1+\frac{b_1}{b_2}-b_1(1+\beta)+1)} \right.$$

$$- \frac{b_1 b_2}{(b_1+1)(b_2+1)(b_1b_2+2b_1-b_1(1+\beta)+1)}$$

$$\left. + \frac{b_1^2 b_2}{(b_1+1)(b_1+b_2+2)(b_1+b_2+2)-b_1(1+\beta))} \right\}. \qquad (3.2)$$

The Pitman efficacy of J_n^c is defined by

$$e(\beta) = \{\mu'(\lambda_0)\}^2/\sigma^2(\beta),\ 0 \le \beta < 1 \qquad (3.3)$$

where $\mu'(\lambda_0)$ and σ^2 are given by (3.1) and (3.2), respectively. Table 1 gives the efficacy of J_n^c for values of $\beta = 0(.1).9$.

Table 1
Efficacy of J_n^c Test

β	$e(\beta)$
0.0	1.3870
0.1	1.3276
0.2	1.2638
0.3	1.1955
0.4	1.1226
0.5	1.0450
0.6	0.9629
0.7	0.8763
0.8	0.7855
0.9	0.6906

The decrease in efficacy of J_n^c as β increases indicates the efficiency loss as the censoring becoems heavier.

We now apply the J_n^c tests to the survival data of Hollander and Proschan (1979). The data corresponds to 211 prostate cancer patients of whom 90 died of prostrate cancer, 105 died of other diseases, and 16 were still alive at the closing date. Those observations corresponding to deaths due to other causes and those corresponding to 16 survivors are treated as censored observations. $Z_{(n)}$ is treated as death so that $\delta_{(n)} = 1$. We consider the proportional hazards censoring model where $\overline{H} \equiv \overline{F}^{\beta}$ for some $\beta \geq 0$. The data consist of 57% of censored observations and corresponding β is approximately 0.7. The computations yield $J_{211}^c = 0.7214184$, $\sigma^2(0.7) = 0.0190848$, and $\Delta(F) \overset{H_0}{=} \log_e 2$. [Note that under proportional hazards censoring model, $\sigma^2(\beta)$ given by (3.2) can be computed numerically and hence we do not have to use its estimate $\hat{\sigma}_n^2$.] The one-sided p-value is 0.0015 indicating that IFRA model is preferrable to an exponential model at 5% level of significance.

The calculation of J_n^c was done using the representation

$$J_n^c = \int_0^1 (\Sigma_{j=1}^n [\overline{F}_n(bZ_{(j)})][\overline{F}_n(Z_{(j-1)}) - \overline{F}_n(Z_{(j)})])db.$$

IMSL Subroutines were used for computing the Kaplan-Meier estimator $\overline{F}_n$, and for the integrating w.r.t. b.

4. REFERENCES

Bandyopadhyay, D. and Basu, A.P. (1989). "A note on tests for exponentiality by Deshpande", *Biometrika*, 76, 403-405.

Barlow, R.E. and Proschan, E. (1975). *Statistical Theory of Reliability and Life Testing*, Holt Rinehart and Winston.

Billingsley, P. (1968). *Convergence of Probability Measures*, John Wiley & Sons, Inc., New York.

Chen, Y.Y.; Hollander, M.; and Langberg, N.A. (1983). "Testing whether new is better than used with randomly censored data," *Ann. Statist., 11*, 267-274.

Deshpande, J.V. (1983). "A class of tests of exponentiality against increasing failure rate average alternative," *Biometrika, 70*, 514-518.

Fernholz, L.T. (1983). *von-Mises Calculus for Statistical Functionals. Lecture Notes in Statist., 19*, Springer, New York.

Gerlach, B. (1989). "Tests for increasing failure rate average with randomly right censored data," 287-295.

Gill, R. (1983). "Large sample behavior of the product-limit estimator on the whole line," *Ann. Statist., 11*, 49-58.

Hollander, M. and Proschan, F. (1979). "Testing to determine the underlying distribution using randomly censored data," *Biometrics, 35*, 393-401.

Jammalamadaka, S.R.; Tiwari, R.C. and Zalkikar, J.N. (1990). "Testing for exponentiality against IFRA alternatives using a U-statistic process," *Revista Brasileria de Probabilidade Estatistica, 4*, 147-159.

Kaplan, E.L. and Meier, P. (1958). "Nonparametric estimation from incomplete observations," *J. Amer. Statist. Assoc., 53,* 457-481.

Kochar, S.C. (1985). "Testing exponentiality against monotone failure rate average," *Commun. Statist. - Theor. Meth. A, 14*, 381-392.

Kumazawa, Y. (1987). "On testing whether new is better than used using randomly censored data," *Ann. Statist., 15*, 420-426.

Link, W.A. (1989). "Testing for exponentiality against monotone failure rate average alternatives," *Commun. Statist. Theor. - Meth., 18*, 3009-3017.

Tiwari, R.C.; Jammalamadaka, S.R.; and Zalkikar, J.N. (1989). "Testing an increasing failure rate average distribution with censored data." *Statistics, 20*, 279-286..

Wells, M.T. and Tiwari, R.C. (1991). "A class of tests for an increasing failure-rate-average distribution with randomly right-censored data," *IEEE Transactions on Reliability,* vol. 40, 152-156.

ADVANCES IN RELIABILITY
Edited by Asit P. Basu

On the problem of masked system life data

John S. Usher

Department of Industrial Engineering, University of Louisville, Louisville, KY 40292

Abstract

This paper reviews some of the current research being done to estimate component reliability from system life data. In particular, we focus on the problem of masked data, i.e., data where the time of system failure is known but the exact failing component is not. We discuss procedures for finding MLEs under various component life distribution assumptions for the case of masked data. In addition, we present an iterative method for obtaining the component reliability estimates. We conclude with a discussion of areas for further study.

1. INTRODUCTION

Reliability engineers are often interested in estimating the reliability of each of a system's components through analysis of the system life data. By observing the time and cause of system failure, it is relatively easy to estimate the reliability of any component for which sufficient failures are observed, by assuming a series system and using a competing risks model. The data required for this type of analysis is the time to failure of the system and the exact component causing failure. See Nelson [1] for an overview of models and methods for using the competing risks model under a variety of life distribution assumptions. The component reliability estimates obtained from analysis of system life data are invaluable because they reflect the actual reliability of the components *after* they have been installed into a system. This runs contrary to the usual types of component reliability estimates obtained from individual component life tests. Analysts often use the estimates to make decisions regarding such things as, warranty requirements, spares provisioning, effects of design changes on a current system, and feasibility of new configurations of the same types of components, to name but a few.

Usher, Alexander, and Thompson [2] utilize the competing risks approach to estimate the reliability of electronic components at IBM. They first develop an extensive database system to store and retrieve system life data about time to system failure and cause of failure based upon 40 critical component types. Each of these component types is assumed to have a Weibull life distribution. They then find maximum likelihood estimates (MLEs) of the Weibull parameters and use these to predict the reliability of new computer systems under development. In their model, they assume that the exact cause of system failure failure is known.

However, we have begun to see an increased emphasis on maintainability aspects of systems which has resulted in more modular construction. While this facilitates fast repair, thus effectively increasing the availability of the system, it yields a lack of information on the exact cause of system failure. That is, a repair technician may replace a module that contains many components, and never know the exact failing component. This results in confounded information on the cause of system failure. Consider a set of field data for computers systems. Upon analysis, you would likely find that the data consists of the time to failure of the system and two types of failure-cause information, (1) exact cause of failure, e.g., resistors, diodes, connectors, transformers, and (2) module failures, circuit cards or packs, power supply, I/O, disk drive etc. This second type of failure-cause data represents partial information on the cause of failure since these items contain many subcomponents, e.g., resistors, diodes, connectors, etc. That is, we have isolated the exact cause of failure down to some subset of components. We refer to this as *masked data* since the cause of failure is masked from our view.

In this paper we present some models and methods for estimating component reliability from a combination of masked and non-masked data. That is, we assume that we observe system failure times, some exact causes of system failure and some masked causes of system failure. In particular, we address the estimation problem for components assumed to have exponential and Weibull life distributions using the method of maximum likelihood and an iterative graphical approach. We find that, except in few special cases, the ML estimators are not available in closed form. The iterative graphical procedure yields good estimators with little computation time, but the statistical properties of the estimators have yet to be investigated.

2. MAXIMUM LIKELIHOOD ESTIMATORS

Consider a series system of J components. Assume that n such system are being observed. Let T_{ij} represent the random lifelength of component j in system i, $i=1,2,\ldots,n$ and $j=1,2,\ldots J$. Let the lifelength of component j be described by the density function $f_j(t)$ with some parameter vector $\underline{\theta}_j$. Let $R_j(t)$ denote the corresponding reliability function for component j. We assume that T_{ij}, $i=1,2,\ldots,n$, $j=1,2,\ldots,J$ are s-independent random variables with T_{1j}, $T_{2j},\ldots$, T_{nj} being identically distributed for each j. Under the series system assumption, the lifelength of system i is:

$$X_i = min\left\{T_{i1}, T_{i2}, \ldots, T_{iJ}\right\}$$

Let δ_i be the indicator variable to denote if system i is censored, that is,

$$\delta_i = \begin{cases} 0 \text{ if system } i \text{ is censored} \\ 1 \text{ if system } i \text{ is failed} \end{cases}$$

Upon failure of system i, some subset, S_i, of the J components within the system is known to contain the failing component. Note that if S_i contains a single component, then the cause of system failure is known exactly, otherwise the cause is masked. We assume that masking occurs independently of the cause of failure. Also note that S_i could contain all J components, in which case the cause of failure would be completely unknown. Thus the observable quantities from the n systems are $\{x_i, \delta_i, S_i\}$, for $i=1,2,\ldots, n$.

Guess, Usher and Hodgson [3] found that the likelihood function for the masked data case can be expressed as:

$$L = \prod_{i=1}^{n} \left(\left(\sum_{j \varepsilon S_i} \left\{ f_j(x_i) \prod_{\substack{l=1 \\ l \neq j}}^{J} R_l(x_i) \right\} \right)^{\delta_i} \cdot \left(\prod_{l=1}^{J} R_l(x_i) \right)^{1-\delta_i} \right) \tag{1}$$

To help clarify this complex expression, consider the following simple example of a test of two identical systems consisting of three components in series, i.e., $J=3$ and $n=2$. Assume that System 1 was observed for 20 hours without failure. This would yield, $(x_1=20,\ \delta_1=0,\ S_i=\{\ \})$. Further assume that System 2 failed at time 13 hours and the cause was isolated down to only components 2 or 3. This would yield $(x_2=13,\ \delta_2=1,\ S_i=\{2,3\})$. Then the likelihood function for this sample data would be found as:

$$L = R_1(20)\, R_2(20)\, R_3(20) \cdot \Big(f_2(13)\, R_1(13)\, R_2(13) + f_3(13)\, R_1(13)\, R_2(13) \Big)$$

The values of the parameter vector $\underline{\theta}_j$ that maximize L in equation (1), or equivalently the log-likelihood, $\mathcal{L} = ln(L)$, yield the MLEs denoted as $\hat{\underline{\theta}}_j$. These can be found by solving the set of log-likelihood equations:

$$\frac{\partial \mathcal{L}}{\partial \underline{\theta}_j} = 0 \qquad j=1,2,\ldots J.$$

3. THE EXPONENTIAL CASE

3.1. Two-Component Case

Miyakawa [4] first considered the masked data situation for a 2-component system. Assume that n, 2-component series systems are placed on test. Assume that each component lifelength is exponentially distributed with failure rate λ_j, $j=1,2$. For ease of exposition, we introduce some additional notation. Let n_1 and n_2, denote the number of failures where $S_i=\{1\}$, $S_i=\{2\}$ respectively. Let n_{12} denote the number of masked observations, i.e., where $S_i=\{1,2\}$. The log-likelihood is then given as:

$$\mathcal{L} = \left(-\sum_{i=1}^{n} x_i \right) (\lambda_1+\lambda_2) + n_1 ln(\lambda_1) + n_2 ln(\lambda_2) + n_{12} ln(\lambda_1+\lambda_2)$$

Partial differentiation yields the following likelihood equations:

$$\frac{\partial \mathcal{L}}{\partial \lambda_1} = 0 = -\sum_{i=1}^{n} x_i + \frac{n_1}{\lambda_1} + \frac{n_{12}}{\lambda_1+\lambda_2}$$

$$\frac{\partial \mathcal{L}}{\partial \lambda_1} = 0 = -\sum_{i=1}^{n} x_i + \frac{n_2}{\lambda_2} + \frac{n_{12}}{\lambda_1+\lambda_2}$$

The closed-form solutions to these are easily found to be:

$$\hat{\lambda}_1 = \left(\sum_{i=1}^{n} x_i \right)^{-1} \left(n_1 + n_{12} \frac{n_1}{n_1+n_2} \right)$$

$$\hat{\lambda}_2 = \left(\sum_{i=1}^{n} x_i \right)^{-1} \left(n_2 + n_{12} \frac{n_2}{n_1+n_2} \right). \qquad (2)$$

Note that here, as well as in subsequent sections, that when the sample does not contain any masked observations the estimators reduce to the standard MLEs for the failure rate parameter λ_j, i.e., the ratio of the total number of failures of component j to the total time on test.

Miyakawa points out that these estimators can be used in the more general case of a J-component series system where J is some positive integer. That is, one simply considers a particular component as "component 1" and all others (since they are in series) as "component 2". The cause of system failure is then either known to be component 1 or component 2 (one of the others in the system) or else is completely unknown.

The 2-component model when applied to the J-component case, however, implicitly assumes that when the cause of system failure is unknown, all J components can be suspected as being the true cause, i.e., the cause of system failure is completely unknown.

3.2. Three-Component Case

Usher and Hodgson [5] extended the results in (2) by considering a $J=3$ component system, and their results can be easily extended to consider any number of
components. Assume that n 3-component series systems are observed. Let n_1, n_2, and n_3 denote the number of failures where $S_i=\{1\}$, $S_i=\{2\}$, and $S_i=\{3\}$ respectively. Let n_{12}, n_{13}, and n_{23} denote the number of partially-masked observations where $S_i=\{1,2\}$, $S_i=\{1,3\}$, and $S_i=\{2,3\}$ respectively. Let n_{123} denote the number of completely masked observations where $S_i=\{1,2,3\}$. The log-likelihood is then given as:

$$\mathcal{L} = \left(-\sum_{i=1}^{n} x_i \right) (\lambda_1+\lambda_2+\lambda_3) + n_1 ln(\lambda_1) + n_2 ln(\lambda_2) + n_3 ln(\lambda_3) + n_{12} ln(\lambda_1+\lambda_2) +$$
$$+ n_{13} ln(\lambda_1+\lambda_3) + n_{23} ln(\lambda_2+\lambda_3) + n_{123} ln(\lambda_1+\lambda_2+\lambda_3).$$

Partial differentiation yields the following likelihood equations:

$$\lambda_1 = \left(\sum_{i=1}^{n} x_i \right)^{-1} \left(n_1 + n_{12} \frac{\lambda_1}{\lambda_1+\lambda_2} + n_{13} \frac{\lambda_1}{\lambda_1+\lambda_3} + n_{123} \frac{\lambda_1}{\lambda_1+\lambda_2+\lambda_3} \right)$$

$$\lambda_2 = \left(\sum_{i=1}^{n} x_i \right)^{-1} \left(n_2 + n_{12} \frac{\lambda_2}{\lambda_1+\lambda_2} + n_{23} \frac{\lambda_2}{\lambda_2+\lambda_3} + n_{123} \frac{\lambda_2}{\lambda_1+\lambda_2+\lambda_3} \right) \quad (3)$$

$$\lambda_3 = \left(\sum_{i=1}^{n} x_i \right)^{-1} \left(n_3 + n_{13} \frac{\lambda_3}{\lambda_1+\lambda_3} + n_{23} \frac{\lambda_3}{\lambda_2+\lambda_3} + n_{123} \frac{\lambda_3}{\lambda_1+\lambda_2+\lambda_3} \right)$$

Note that these equations have an intuitive appeal. That is the standard MLE for λ_j (in the non-masked case) is just the number of failures observed for component j (n_j), divided by the total time on test. Here we see that the the masked failures involving component j are weighted by the failure rates of the components. For example, consider the number of failures masked as the set {1,2}, given by the quantity n_{12}. The fraction of these failures attributed to component 2 is given by $\lambda_2/(\lambda_1+\lambda_2)$. Note that extending these equations to $J>3$ components is a straightforward process of considering all possible combinations of component subsets.

In the 3-component exponential case, closed form MLEs for the λ_j are intractable. But, these equations are of the form $\underline{\lambda} = T(\underline{\lambda})$ where $\underline{\lambda} = [\lambda_1\ \lambda_2\ \lambda_3]^T$. A common technique used to approximate solutions to this type of equation set is to define an iteration process by the equation:

$$\underline{\lambda}^{(i+1)} = T(\underline{\lambda}^{(i)}) \quad (4)$$

That is, an initial approximation $\underline{\lambda}^{(0)}$ is obtained (guessed), then $\underline{\lambda}^{(1)} = T(\underline{\lambda}^{(0)})$, $\underline{\lambda}^{(2)} = T(\underline{\lambda}^{(1)})$, ... form successive approximations to the solution of (4). This procedure, when applied to the exponential case, converges rapidly. It is extremely simple to implement and yield solutions even with poor starting values.

3.3. Other Cases

Usher and Hodgson [5] also present assumptions under which solutions for the 3-component exponential case are available in closed form. For example, consider the case where $n_{12} = n_{13} = n_{23} = 0$ and n_1, n_2, n_3, and $n_{123} \geq 0$. This case represents a situation where the cause of failure is either known or completely unknown. This could apply to a situation where only part of a sample of systems have been failure analyzed to determine the exact cause of each system failure. For, the remainder of the sample, no failure analysis has been performed. Under this simplifying assumption we find closed form estimators as:

$$\hat{\lambda}_1 = \left(\sum_{i=1}^{n} x_i \right)^{-1} \left(n_1 + n_{123} \frac{n_1}{n_1+n_2+n_3} \right)$$

$$\hat{\lambda}_2 = \left(\sum_{i=1}^{n} x_i \right)^{-1} \left(n_2 + n_{123} \frac{n_2}{n_1+n_2+n_3} \right) \tag{5}$$

$$\hat{\lambda}_3 = \left(\sum_{i=1}^{n} x_i \right)^{-1} \left(n_3 + n_{123} \frac{n_3}{n_1+n_2+n_3} \right)$$

This set of solutions should be recognized as the natural extension of Miyakawa's 2-component model for the 3-component case. Other simple closed forms are possible, but may be of limited value in actual applications.

4. OTHER ESTIMATORS

As a result of the problem's intractable nature, (even in the exponential case) engineers are forced to use numerical procedures to obtain MLEs of component reliabilities from masked data. While there are many non-linear numerical optimization algorithms available, see Reklaitis, Ravindran, and Ragsdell [6], they can be time-consuming and cumbersome to apply. Most require good initial starting points to guarantee convergence. For a good discussion of convergence problems encountered in the use of the ML method in multiparameter situations, see Mantel and Myers [7]. In addition, the algorithms are generally computationally intensive. This can result in extremely long solution times for large problems in a real-world setting. In addition, the faster gradient methods, such as Newton-Raphson, require first and second partial derivatives of the likelihood function. For the masking case, these are not available in closed form, therefore requiring even more numerical computation.

The overall result is that MLEs, with their desirable statistical properties (invariance, asymptotic efficiency and normality, etc.), may be expensive to obtain. Industry experience suggests that many engineers are interested in simpler solution procedures that generate "good" estimates at reasonable cost. The widespread use of graphical (probability plotting) techniques is a good example of this phenomenon. See Jensen and Petersen [8] for numerous examples of Weibull probability plots. In a sense, these graphical techniques gain the advantage of computational ease with the sacrifice of some statistical power. See Lawless [9] and Cox [10] for a review and discussion of graphical procedures and their properties.

Usher and Guess [11] address the problem of obtaining parametric estimates of component reliabilities from masked system life data. They propose an iterative solution procedure that is analogous to traditional probability plotting. That is, they first find nonparametric reliability estimates (plotting points) for each component in the system. These nonparametric estimates account for the effects of masking. To this, they then fit an assumed parametric distribution, e.g., Weibull. The fitting process is performed using least-squares, the "graphical step". The resulting parametric estimates are next used to obtain revised plotting points for the nonparametric reliability function. These are then used to obtain revised parametric estimates. The process is repeated iteratively until convergence is reached.

We now give an overview of their iterative method. We assume the same notation as before. In addition, let $p_j(x_i)$ denote the conditional probability that component j is the true cause of system failure at time x_i. We make the reasonable assumption that each component has a smooth and continuous hazard function, $h_j(x)$ and estimate $p_j(x_i)$ as:

$$\hat{p}_j(x_i) = \frac{\hat{h}_j(x_i)}{\sum_{l \in S_i} \hat{h}_l(x_i)} \tag{6}$$

That is, $\hat{p}_j(x_i)$ represents the fractional contribution of component j's hazard rate to the total hazard rate of all components suspected as being the cause of failure. Note that when the cause is known to be component j, $S_i=\{ j \}$, $\hat{p}_j(x_i)=1$, and $\hat{p}_k(x_i)=0$ for all $k \neq j$.

Now, suppose that the $\hat{p}_j(x_i)$ have been obtained for all i (for j fixed). As an extension to the product limit estimator, consider the following nonparametric estimator for the reliability of component j:

$$\hat{R}_j^*(x) = \prod_{x_i \leq x} \left(\frac{q_i - \hat{p}_j(x_i)}{q_i} \right) \tag{7}$$

where q_i denotes the number of systems still functioning at time x_i and $\hat{p}_j(x_i)$ is as given in (6). The term in parentheses represents the estimated conditional reliability of component j given survival to time x_i.

When the cause of failure is known to be component j at time x_i we find that $\hat{p}_j(x_i)=1$. This shows that (7) reduces to the standard product limit estimator in the complete data case. Also note that if component j is not the cause of failure at time x_i, i.e., $\hat{p}_j(x_i)=0$, then the conditional reliability term (in parentheses) appropriately reduces to 1. That is, x_i simply acts as a censoring time for component j.

The values of $\hat{R}_j^*(x)$ yield plotting points that provide parametric estimates either graphically or through the method of least squares. However, we must first estimate $p_j(x_i)$, $j=1,2,\ldots,J$, and $i=1,2,\ldots, n$ using (6). We see that this in turn requires estimates of $h_j(x_i)$ $j=1,2,\ldots J$, and $i=1,2,\ldots, n$. These estimated hazard rates are functions of $\hat{\underline{\theta}}_j$, $j=1,2,\ldots,J$. The parameter estimates, $\hat{\underline{\theta}}_j$ are estimated by fitting the parametric reliability function, R_j, to the nonparametric plotting points given by $(x_i, \hat{R}_j^*(x))$ $i=1, 2, \ldots, n$. This apparent closed-loop problem can be solved through the use of an iterative procedure as follows:

1. From the sample of n observations, select only those where the cause of

failure is known. (We assume, of course, that at least two observations with known cause of failure are available.)

2. Find the initial nonparametric estimates, (plotting points), $R_j^*(x)$, at the selected failure times, using (7). I.e., using only the x_i with $S_i=\{k\}$ for some k, find the point $(x_i, \hat{R}_j^*(x))$ with $\hat{p}_j(x_i)=1$ if $S_i=\{j\}$ and $\hat{p}_j(x_i)=0$ for $S_i=\{k\}$ and $k \neq j$.
3. Find $\underline{\hat{\theta}}_j$ by fitting (graphically or by least squares) the parametric reliability function R_j to the nonparametric plotting points from step 2 (first iteration) or from step 6 (subsequent iterations).
4. Calculate $\hat{h}_j(x_i)$, $j=1,2,...,J$, and $i=1,2,...,n$ using $\underline{\hat{\theta}}_j$ from step 3.
5. Find updated values of $\hat{p}_j(x_i)$, $j=1,2,...,J$, and $i=1,2,...,n$ using (6).
6. Find updated estimates, $\hat{R}_j^*(x)$, $j=1,2,...J$, and $i=1,2,...,n$ using (7).
7. Go to step 3 and repeat until $\underline{\hat{\theta}}_j$ converges, $j=1,2,...,J$.

The major advantage of the approach is its ability to yield good estimates with much less computation than the method of maximum likelihood. For a small numerical example, Usher and Guess find good Weibull parameter estimates with only 4% of the computation time required to find comparable MLEs. This advantage would become even more important in the analysis of complex systems of many components.

5. CONCLUSIONS

We have presented a review of several methods for estimating component reliabilities from masked system life test data. We consider that the cause of system failure can be isolated to some subset of the system's components. This subset approach allows us to consider the full range of possible information on the cause of system failure. But, as with any topic, there is always more to be done. In particular we cite some topics that deserve more consideration.

1. The models presented here explicitly assume that masking occurs independently of the cause of failure. Clearly there are systems where certain component failures make it difficult to determine the cause of failure, thus yielding masked system data. Whereas other component failures may yield a system fault that makes failure cause easy to identify. The effect of this dependency should be investigated.

2. The problem of determining the true cause of failure from a masked observation could be investigated under a Bayesian framework. In this way the analyst could utilize prior information/engineering judgment in estimating the true cause of failure.

3. The "shape" of the likelihood function in n-dimensional space for various distributional assumptions needs to be investigated. This would help determine the most effective method for maximizing the log-likelihood function to obtain MLEs in large scale (100-200 parameters) problems. We are currently investigating the use of global optimization strategies to this problem. Preliminary results indicate that the

likelihood function is very flat at the optimum and does not have multiple peaks. Thus, gradient based procedures could be cost effective, if closed form derivatives of the likelihood function can be found.

4. The iterative estimators, based upon a modified product-limit estimator seem to yield good results at very little expense, but their properties are unknown. It would be of interest to determine how the level of masking affects the quality of these estimators.

5. The concept of estimating detailed component reliabilities from system life data is appealing for many industries, e.g., automotive, aircraft, appliance, military equipment, etc. The approach does, however, require the development of a large database of system life data. Thus, there is a need for effective database design strategies that can be successfully applied to real-world applications.

6. REFERENCES

[1] Nelson, W. (1982). *Applied Life Data Analysis,* Wiley, New York.

[2] Usher, J.S., S. M. Alexander, and J. D. Thompson (1990). "System Reliability Prediction Based On Historical Data", *Quality and Reliability Engineering International,* Vol. 6., No. 3, pp 209-218.

[3] Guess, F. M., J. S. Usher, and T. J. Hodgson, (1991). "Estimating system and component reliabilities under partial information on the cause of failure", *Journal of Statistical Planning and Inference,* Vol. 29, pp 75-85.

[4] Miyakawa, M. (1984). "Analysis of incomplete data in competing risks model", *IEEE Trans. Reliab.* , R-33, No. 4, pp 293-296.

[5] Usher, J. S. and T. J. Hodgson, (1988) "Maximum likelihood analysis of component reliability using masked system life test data", *IEEE Trans. Reliab.,* Vol. 37, No. 5, pp 550-555.

[6] Reklaitis, G. V., A. Ravindran and K.M. Ragsdell, (1983). *Engineering Optimization: Methods and Applications,* John Wiley, New York.

[7] Mantel, N. and M. Myers, (1971). "Problems of convergence of maximum likelihood iterative procedures in multiparameter situations", *JASA,* Vol. 66, No. 335, pp 484-491.

[8] Jensen, F. and Petersen, N. E., (1982). *Burn-In,* Wiley, New York.

[9] Lawless, J.F., (1982). *Statistical Models and Methods for Lifetime Data,* Wiley, New York.

[10] Cox, D. R., (1978). "Some remarks on the role in statistics of graphical procedures" , *Appl. Stat.,* 27, pp. 4-9.

[11] Usher, J. S., and F. M. Guess, (1989). "An Iterative Approach for Estimating Component Reliability From Masked System Life Data", *Quality and Reliability Engineering International,* Vol. 5, pp 257-261.

ADVANCES IN RELIABILITY
Edited by Asit P. Basu

Group replacement policies that incorporate statistical learning

John G. Wilson*

Department of Operations Research, Case Western Reserve University, Cleveland, OH 44106-7235

* Supported in part by National Science Foundation grant # DDM-8910378.

Abstract

Suppose that n identical production machines are operating in parallel. A downtime cost is incurred for each failed machine. Repairing or replacing machines involves fixed costs and unit replacement costs. In the literature on Group Replacement Policies for such a system, it is generally assumed that the parameters of the underlying failure time distributions are known. The approach of this article is novel in that parameters are estimated from data obtained while operating the machines and from prior quality control information. An easily implemented group replacement policy is introduced. This article considers the specific case where the failure times are independent identically distributed exponential random variables and the prior is gamma. For this case, the results are numerically tractable and do not involve numerical integrations.

Key words: reliability, Bayesian updating, exponentially failing machines, gamma priors.

1. INTRODUCTION

Consider a system of n machines operating in parallel. Downtime costs are incurred whenever a failed machine is not replaced or repaired. Both fixed and unit costs are incurred whenever a machine is repaired. In common with much of the literature, assume that attention is restricted to group replacement and repair policies. In most of the literature on Group Replacement Policies, the objective function is average cost per unit time. This makes sense in many applications since one expects to operate the machines over a long horizon. Two forms of group replacement policy, the m-failure and T-age policy, have received much attention in the literature. An m-failure policy calls for replacement or repair of the system to as good as new at the time of the mth failure. For examples of this approach see Rade (1976), Nakagawa (1979), Okumoto and Elsayed (1983), Assaf and Shanthikumar (1987) and Wilson and Benmerzouga (1990). A T-age policy calls for replacing the system after T time units have elapsed (see, e.g., Barlow and Hunter (1960), Boland (1982), Boland and Proschan (1982)). Ritchken and Wilson (1990), Nakagawa (1983) and Yeh (1988) have provided results for the combination policy which calls for replacement after the system has experienced m failures or at time T, whichever occurs first. Wilson and Benmerzouga (1992) consider a class of adaptive policies that include m-failure, T-age and (m,T) policies as special cases. All of these approaches consider only the expected cost associated with policies. Wilson (1992) provides a procedure for calculating the variance of the cost per unit time associated with group maintenance policies.

Each time a batch of n machines is purchased, a fixed cost of c_0 is incurred. This fixed cost includes the cost of buying n new machines and other costs, such as paperwork etc., involved in the purchase. The salvage value of i functioning but used machines is given by the function $c_s(i)$. Each failed machine is assumed to have zero salvage value. Each failed machine incurs a downtime cost of c_d per unit time.

Assume that the failure time distributions are i.i.d. exponential random variables with parameter λ. Optimality results and algorithms have only recently appeared for the case of known λ - see Assaf and Shanthikumar (1987) and Wilson and Benmerzouga (1990). Thus, this seems an appropriate assumption with which to start an investigation of Bayesian policies. Assumption of an exponential failure time with known parameter implies that there is no reliability growth or degradation, i.e. the failure rate is constant.

Assume that the machines are bought in batches and that $\pi(\lambda)$ is the prior density for the distribution for λ among batches. This density can represent the quality control information from the supplier of the machines. It is assumed that the supplier has a quality control program in operation and that $\pi(\lambda)$ remains the same from purchase to purchase. For the case of known λ, m-failure policies are optimal (see Assaf and Shanthikumar 1987). However, these policies have some managerially unpalatable consequences. Suppose, for instance, that n=100 and the 51-failure policy is optimal. This policy treats the following two situations as being equivalent: no machines fail for the first 365 days and 51 fail on day 366; 50 machines fail on day 1 and the 51st machine fails on day 366. Most managers would prefer a policy that prescribes different actions for the above two situations. The policy analysed in this paper is called a (t,m) policy, where t and m are parameters determined by the decision maker. When a failure occurs, the expected posterior mean of λ is calculated. If this mean is larger than t (which would imply an unreliable system), then the system is replaced. Otherwise wait until the next failure and recompute the expected posterior mean. If replacement has not occurred by the time of the mth failure, then replace the system. The advantage of this policy is that it includes m-failure policies as a special case and allows the manager to replace the system early if this particular batch looks like being unreliable. This policy utilises both the information contained in the prior distribution and the sufficient statistic for λ.

The major result in this paper is given by the Corollary to Theorem 2. There it is shown that the cost per unit time associated with a (t,m) policy can be written in closed form for the case of gamma prior information.

2. (t,m) POLICIES

The prior density $\pi(\cdot)$ is assumed to be gamma with parameters $\alpha>1$ and $\beta>0$, i.e.

$$\pi(\lambda) = \frac{\beta^{\alpha}}{\Gamma(\alpha)} \lambda^{\alpha-1} e^{-\beta\lambda}, \quad \lambda>0.$$

Let τ_i denote the time of the ith machine to fail. Then, at the time of the ith failure, $s_i=(n-i+1)\tau_i + \sum_{j=1}^{i-1}\tau_j$ is a sufficient statistic for the parameter λ. Thus, this quantity and the fact that i machines have failed are the crucial quantities on which decisions should be based. The smaller the value of the sufficient statistic, the more evidence there is that the system is unreliable. The posterior mean for λ at the time of the ith failure is a function of the sufficient statistic. The larger this posterior mean, the more evidence that the system in unreliable. Thus, it makes sense to replace the system whenever this quantity is large.

For a given $t\geq0$ and $m\in\{1,...,n\}$, define the (t,m) policy as follows: for $1\leq i<m$ replace the system at the ith failure if $E[\lambda|s_i]>t$; otherwise replace the system at the mth failure.

At the time of the ith failure $E[\lambda|s_i]>t$ if and only if $\frac{\alpha+i}{\beta+s_i}>t$.

Thus the (t,m) policy calls for replacement at the first failure where $s_i<a_i(t) \equiv \frac{\alpha+i}{t} - \beta$ or at the mth failure, whichever occurs first. Thus, implementation of a given (t,m) policy is straightforward since one simply compares the value of s_i to $a_i(t)$.

From the above description, it is clear that the posterior distribution of λ is utilised when deciding on whether or not to replace the system. If the decision is not to replace, then one would like to replace the prior distribution for λ with the posterior distribution. However, since an average cost per unit time criterion is being used, this is mathematically equivalent to assuming λ is fixed and then integrating with respect to the prior distribution. This removes the need to formally replace the prior distribution for λ with the posterior distribution.

3. THE OBJECTIVE FUNCTION ASSOCIATED WITH A (t,m) POLICY

Throughout this section, the quantities t and m are fixed. Thus, for notational convenience $a_i(t)$ will be written as a_i.

Assume that failure time parameter is known to be λ. Then define $T(t,m;\lambda)$ to be the expected cycle length when the (t,m) policy is used. Let $D(t,m;\lambda)$ and $S(t,m;\lambda)$ denote the expected downtime costs and salvage costs over one cycle when the (t,m) policy is used. The times at which machines are replaced form renewal points for the system. Thus, the expected cost per unit time associated with the (t,m) policy is given by

$$C(t,m) \equiv \frac{c_0 - E[S(t,m;\lambda)] + E[D(t,m;\lambda)]}{E[T(t,m;\lambda)]},$$

where the expectations are taken with respect to the prior distribution. The goal is to obtain tractable expressions for the expected values in the above expression.

Define $p_i(\lambda)$ to be the probability that the system is replaced after the ith failure if the parameter of the exponential distribution is known to be λ, i.e., $p_i(\lambda)=P[s_1 \geq a_1, s_2 \geq a_2, \ldots, s_i \geq a_i]$. If group replacement is not made at the ith failure, then there are n-i functioning machines and the time to the next failure is an exponential random variable with parameter $(n-i)\lambda$. Thus, the expected time from the ith to the (i+1)st failure is given by $\{(n-i)\lambda\}^{-1}$ and the expected downtime cost over this period is given by $c_d i\{(n-i)\lambda\}^{-1}$. Thus, when the failure time parameter is fixed at the value λ, the expected length of the cycle, the expected downtime costs and the expected salvage value are given by the following:

$$T(t,m;\lambda) = \sum_{i=1}^{m} \frac{p_{i-1}(\lambda)}{(n-i+1)\lambda}$$

$$D(t,m;\lambda) = c_d \sum_{i=1}^{m} \frac{(i-1)p_{i-1}(\lambda)}{(n-i+1)\lambda}$$

$$S(t,m;\lambda) = \sum_{i=1}^{m} c_s (n-i) \{p_{i-1}(\lambda) - p_i(\lambda)\}. \tag{1}$$

Thus, computation of the objective function requires an explicit expression for the quantity $p_i(\lambda)$. First, some preliminary definitions are required.

Define a sequence of constants $\{c_j\}$ by the following recursive procedure:

$$c_0 = 1$$

$$c_j = -\sum_{i=0}^{j-1} c_i \frac{(a_j)^{j-i}}{(j-i)!} \quad \text{for } j=1,2,\ldots \tag{2}$$

Now define a sequence of constants $\{d_j(k)\}$ by the following recursive procedure:

$$d_j(k) = \begin{cases} 1 \text{ for } j=0 \\ \sum_{i=0}^{j} c_i \frac{(a_k)^{j-i}}{(j-i)!} \quad \text{for } j=1,2,\ldots,k-2 \\ \sum_{i=0}^{k-2} c_i \frac{(a_k)^{k-1-i} - (a_{k-1})^{k-1-i}}{(k-1-i)!} \quad \text{for } j=k-1 \end{cases} \tag{3}$$

The following theorem is the key to obtaining a tractable expression for the objective function. The proof is provided in Wilson and Benmerzouga (1991).

Theorem 1

Assume the (t,m) policy is being used. For $1 \le k < m$, and failure time parameter equal to λ, the probability of replacing after the kth failure is given by

$$p_k(\lambda) = e^{-\lambda a_k} \sum_{j=0}^{k-1} d_j(k)\lambda^j.$$

It is shown in Wilson and Benmerzouga (1991), that the above theorem holds even for the case of arbitrary (nonconjugate) prior distributions. However, as will be shown in this article, assumption of gamma prior information considerably simplifies the results. For this case, Lemmas 1 and 2 provide closed form, nonrecursive, expressions for the constants c_j and $d_j(k)$.

Lemma 1

The constants $\{c_j\}$ can be written as

$$c_j = \begin{cases} 1 \quad \text{for } j=0 \\ (-1)^j \left(\frac{\alpha+j}{t}-\beta\right) \left(\frac{\alpha}{t}-\beta\right)^{j-1} (j!)^{-1} \end{cases} \tag{4}$$

Proof

The proof will be by induction. From definition (2),

$$c_1 = -c_0 a_1$$

$$= -\left(\frac{\alpha+1}{t} - \beta\right).$$

Thus, the lemma is true for j=1.

For the induction assumption, assume that

$$c_i = (-1)^i\left(\frac{\alpha+i}{t} - \beta\right)\left(\frac{\alpha}{t} - \beta\right)^{i-1}(i!)^{-1}, \quad \text{for } i=1,\ldots,j-1.$$

The goal is to show that the statement remains true for i=j.

From definition (2) and the induction hypothesis:

$$c_j = -\left\{\frac{(a_j)^j}{j!} + \sum_{i=1}^{j-1} c_i \frac{(a_j)^{j-i}}{(j-i)!}\right\}$$

$$= \frac{-1}{j!}\left\{(a_j)^j + \sum_{i=1}^{j-1}(-1)^i\binom{j}{i}a_j^{j-i}\left[\left(\frac{\alpha}{t} - \beta\right)^i + \left(\frac{\alpha}{t} - \beta\right)^{i-1}\frac{i}{t}\right]\right\}. \tag{5}$$

Note, from the Binomial Theorem, that

$$\sum_{i=1}^{j-1}(-1)^i\binom{j}{i}a_j^{j-i}\left(\frac{\alpha}{t} - \beta\right)^i = \left\{a_j - \left(\frac{\alpha}{t} - \beta\right)\right\}^j - (a_j)^j - (-1)^j\left(\frac{\alpha}{t} - \beta\right)^j$$

$$= \left(\frac{j}{t}\right)^j - (a_j)^j - (-1)^j\left(\frac{\alpha}{t} - \beta\right)^j. \tag{6}$$

Again using the Binomial Theorem, note that

$$\sum_{i=1}^{j-1}(-1)^i\binom{j}{i}a_j^{j-i}\left(\frac{\alpha}{t} - \beta\right)^{i-1}\frac{i}{t} = \frac{\partial}{\partial\alpha}\sum_{i=1}^{j-1}(-1)^i\binom{j}{i}a_j^{j-i}\left(\frac{\alpha}{t} - \beta\right)^i$$

$$= \frac{\partial}{\partial\alpha}\left\{\left[a_j - \left(\frac{\alpha}{t} - \beta\right)\right]^j - (a_j)^j - (-1)^j\left(\frac{\alpha}{t} - \beta\right)^j\right\}$$

$$= -\frac{j}{t}\left[a_j - \left(\frac{\alpha}{t} - \beta\right)\right]^{j-1} - (-1)^j\frac{j}{t}\left(\frac{\alpha}{t} - \beta\right)^{j-1}$$

$$= -\left(\frac{j}{t}\right)^j - (-1)^j\frac{j}{t}\left(\frac{\alpha}{t} - \beta\right)^{j-1}. \tag{7}$$

Use (6) and (7) in expression (5) to obtain

$$c_j = -\frac{1}{j!}\left\{-(-1)^j\left(\frac{\alpha}{t} - \beta\right)^{j-1}\left(\frac{\alpha+j}{t} - \beta\right)\right\},$$

the required result. ■

Lemma 2

The constants $\{d_j(k)\}$ can be written as

$$d_j(k) = \frac{k^j - j\,k^{j-1}}{j!t^j} \quad \text{for } j=1,2,\ldots,k-1$$

Proof

First it will be shown that

$$\sum_{i=0}^{j} c_i \frac{(a_r)^{j-i}}{(j-i)!} = \frac{r^j - jr^{j-1}}{j!\;t^j}, \tag{8}$$

for r = k and r = k-1. The validity of this result for r = k and definition (3) proves the statement of the lemma for j = 1,2,...,k-2. The validity of (7) for r = k-1 will be used to establish the truth of the lemma for j = k-1. Insert expression (4) for c_i and rearrange terms to obtain

$$\sum_{i=0}^{j} c_i \frac{(a_r)^{j-i}}{(j-i)!} = \frac{(a_r)^j}{j!} + \frac{1}{j!}\sum_{i=1}^{j}\binom{j}{i}(-1)^i\left(\frac{\alpha}{t} - \beta\right)^i\left(\frac{\alpha+r}{t} - \beta\right)^{j-i}$$

$$+ \frac{1}{tj!}\sum_{i=1}^{j} i\binom{j}{i}(-1)^i\left(\frac{\alpha}{t} - \beta\right)^{i-1}\left(\frac{\alpha+r}{t} - \beta\right)^{j-i}. \tag{9}$$

Using the Binomial Theorem:

$$\sum_{i=1}^{j}\binom{j}{i}(-1)^i\left(\frac{\alpha}{t} - \beta\right)^i\left(\frac{\alpha+r}{t} - \beta\right)^{j-i} = \left(\frac{r}{t}\right)^j - \left(\frac{\alpha+r}{t} - \beta\right)^j. \tag{10}$$

Again use the Binomial Theorem to see that

$$\sum_{i=1}^{j} i\binom{j}{i}(-1)^i\left(\frac{\alpha}{t}-\beta\right)^{i-1}\left(\frac{\alpha+r}{t}-\beta\right)^{j-i} = \frac{\partial}{\partial y}\left[\sum_{i=1}^{j}\binom{j}{i}(-1)^i y^i\left(\frac{\alpha+r}{t}-\beta\right)^{j-i}\right]\Bigg|_{y=\frac{\alpha}{t}-\beta}$$

$$= \frac{\partial}{\partial y}\left[\left(\frac{\alpha+r}{t}-\beta-y\right)^j-\left(\frac{\alpha+r}{t}-\beta\right)^j\right]\Bigg|_{y=\frac{\alpha}{t}-\beta}$$

$$= j\left(\frac{r}{t}\right)^{j-1}. \qquad (11)$$

The proof of (8) is completed by using (10), (11) and the fact that $a_r = \frac{\alpha+r}{t} - \beta$ in expression (9).

On putting r=k into (8), the lemma is seen to be true for j=1,2,...,k-2. From definition (3),

$$d_{k-1}(k) = \left\{\sum_{i=0}^{k-1} c_i \frac{(a_k)^{k-1-i}}{(k-1-i)!} - c_{k-1}\right\} - \left\{\sum_{i=0}^{k-1} c_i \frac{(a_{k-1})^{k-1-i}}{(k-1-i)!} - c_{k-1}\right\}.$$

Now apply (8) to the summations in the above expression to obtain

$$d_{k-1}(k) = \frac{k^{k-1} - (k-1)k^{k-2}}{(k-1)!\ t^{k-1}} - \frac{(k-1)^{k-1}-(k-1)(k-1)^{k-2}}{(k-1)!\ t^{k-1}}$$

$$= \frac{k^{k-1} - (k-1)k^{k-2}}{(k-1)!\ t^{k-1}},$$

the required result. ■

The results of Theorem 1 and Lemmas 1 and 2 will now be used in (1) to obtain expressions for the various expected values in the objective function. First, some preliminary definitions are required.

Let the notation $(x)_k$ denote the product $x(x-1)\ldots(x-k+1)$. Define constants $r_1(m)$, $r_2(m)$ and $r_3(m)$ as follows:

$$r_1(m) = \sum_{k=-1}^{m-3} \frac{\beta^\alpha}{(k+1)!}(\alpha+k-1)_k \sum_{i=k+2}^{m-1} (n-i)^{-1}(\alpha+i)^{-(\alpha+k)}(i^{k+1}-(k+1)i^k)$$

$$r_2(m) = \sum_{k=-1}^{m-3} \frac{\beta^\alpha}{(k+1)!}(\alpha+k-1)_k \sum_{i=k+2}^{m-1} i(n-i)^{-1}(\alpha+i)^{-(\alpha+k)}(i^{k+1}-(k+1)i^k)$$

$$r_3(m) = \sum_{k=-1}^{m-3} \frac{\beta^\alpha}{(k+1)!}(\alpha+k+1)_{k+1} \sum_{i=k+2}^{m-1} (\alpha+i)^{-(\alpha+k+1)}(i^{k+1}-(k+1)i^k)[c_s(n-i-1)-c_s(n-i)]$$

Theorem 2

Suppose a (t,m) policy is being used.

a) The expected preposterior length of a cycle is given by

$$E[T(t,m;\lambda)] = \frac{\beta}{n(\alpha-1)} + r_1(m)t^{\alpha-1}.$$

b) The expected preposterior downtime cost is given by

$$E[D(t,m;\lambda)] = c_d\, r_2(m)\, t^{\alpha-1}.$$

c) The expected preposterior salvage value is given by

$$E[S(t,m;\lambda)] = c_s(n-1) + r_3(m)\, t^{\alpha}.$$

Proof of (a)

From (1) and Theorem 1, the expected cycle length when using the (t,m) policy is given by

$$E[T(t,m;\lambda)] = E\left[\sum_{i=1}^{m} \frac{P_{i-1}(\lambda)}{(n-i+1)\lambda}\right]$$

$$= E\left[\frac{1}{n\lambda} + \sum_{i=2}^{m} \frac{e^{-\lambda a_{i-1}}}{(n-i+1)\lambda} \sum_{j=0}^{i-2} d_j(i-1)\lambda^j\right]. \qquad (12)$$

Interchange summation operations and integrate with respect to the prior gamma density to obtain

$$E[T(t,m;\lambda)] = E\left[\frac{1}{n\lambda} + \sum_{k=-1}^{m-3} \sum_{i=k+2}^{m-1} \frac{d_{k+1}(i)}{n-i} \lambda^k e^{-\lambda a_i}\right]$$

$$= \frac{\beta}{n(\alpha-1)} + \sum_{k=-1}^{m-3} \sum_{i=k+2}^{m-1} \frac{d_{k+1}(i)}{n-i} \left(\frac{t}{\alpha+i}\right)^{\alpha+k} \beta^{\alpha}(\alpha+k-1)_k.$$

Now use Lemma 2 and rearrange terms:

$$E[T(t,m;\lambda)] = \frac{\beta}{n(\alpha-1)} + t^{\alpha-1}\beta^{\alpha} \sum_{k=-1}^{m-3} \frac{(\alpha+k-1)_k}{(k+1)!} \sum_{i=k+2}^{m-1} \frac{i^{k+1}-(k+1)i^k}{(n-i)(\alpha+i)^{\alpha+k}},$$

the required result. ∎

Proof of (b)

From (1) the expected downtime cost in a cycle can be written as

$$E[D(t,m;\lambda)] = c_d\, E\left[\sum_{i=1}^{m} \frac{(i-1)P_{i-1}(\lambda)}{(n-i+1)\lambda}\right].$$

The only difference between this expression and (12) is the presence of c_d and the factor (i-1). Thus, proceed exactly as in Theorem 2(a) to obtain:

$$E[D(t,m;\lambda)] = c_d\, E\left[\frac{(1-1)}{n\lambda} + \sum_{i=2}^{m} \frac{(i-1)e^{-\lambda a_{i-1}}}{(n-i+1)\lambda} \sum_{j=0}^{i-2} d_j(i-1)\lambda^j\right]$$

$$= c_d\, E\left[\sum_{k=-1}^{m-3} \sum_{i=k+2}^{m-1} \frac{id_{k+1}(i)}{n-i} \lambda^k e^{-\lambda a_i}\right]$$

$$= c_d\, t^{\alpha-1}\beta^{\alpha} \sum_{k=-1}^{m-3} \frac{(\alpha+k-1)_k}{(k+1)!} \sum_{i=k+2}^{m-1} \frac{i[i^{k+1}-(k+1)i^k]}{(n-i)(\alpha+i)^{\alpha+k}},$$

the required result. ■

Proof of (c)

Rearrange terms in the expression for $S(t,m;\lambda)$ provided in (1), note that $p_0(\lambda)\equiv 0$ and $p_m(\lambda)\equiv 0$ to obtain $E[S(t,m;\lambda)]$ = $c_s(n-1)-E\left[\sum_{i=2}^{m}\{c_s(n-i+1)\ -c_s(n-i)\}\ p_{i-1}(\lambda)\right]$. Use Theorem 1, interchange summation operations and integrate with respect to the prior density to obtain

$$E[S(t,m;\lambda)] = c_s(n-1)-E\left[\sum_{j=2}^{m}\{c_s(n-i+1)-c_s(n-i)\}\ e^{-\lambda a_{i-1}} \sum_{j=0}^{i-2} d_j(i-1)\lambda^j\right]$$

$$= c_s(n-1)-E\left[\sum_{k=-1}^{m-3} \sum_{i=k+2}^{m-1} \{c_s(n-i)-c_s(n-i-1)\}d_{k+1}(i)\lambda^{k+1}e^{-\lambda a_i}\right]$$

$$= c_s(n-1)- \sum_{k=-1}^{m-3} \sum_{i=k+2}^{m-1}\{c_s(n-i)-c_s(n-i-1)\}d_{k+1}(i)\left(\frac{t}{\alpha+i}\right)^{\alpha+k+1} \beta^{\alpha}(\alpha+k)_{k+1}$$

Use Lemma 2 and rearrange terms to obtain

$$E[S(t,m;\lambda)]=c_s(n-1)-t^{\alpha}\beta^{\alpha}\sum_{k=-1}^{m-3}\frac{(\alpha+k)_{k+1}}{(k+1)!}\sum_{i=k+2}^{m-1}\frac{c_s(n-i)-c_s(n-i-1)}{(\alpha+i)^{\alpha+k+1}}(i^{k+1}-(k+1)i^k),$$

the required result. ■

Corollary

The expected cost per unit time from using the (t,m) policy is given by

$$C(t,m) = \frac{c_0-c_s(n-1) + c_d r_2(m)t^{\alpha-1} - r_3(m)t^{\alpha}}{\beta n^{-1}(\alpha-1)^{-1} + r_1(m)t^{\alpha-1}}$$

The form of the objective function given above is particularly appealing. The parameter m only appears in the easily computed constants $r_1(m)$, $r_2(m)$, and $r_3(m)$. The parameter t only appears in two places in numerator and one in the denominator that forms the expression for C(t,m). Thus, finding the optimal (t,m) policy is a relatively straightforward numerical task. For a given m, the t which minimises C(t,m) can be found by investigating the solutions to $\frac{\partial}{\partial t}C(t,m) = 0$. Alternatively, the value can be found by inspection simply by plotting C(t,m) as a function of t. Repeat this procedure for all values of m between 1 and n and choose the policy with the smallest overall cost.

If it is decided to perform replacement whenever a failure occurs, the expected cost per unit time equals $\alpha\beta^{-1}[c_0-c_s(n-1)]$. By comparing this quantity with $\inf_{t,m} C(t,m)$, management can determine the cost of requiring that the system always be replaced whenever a failure occurs. An interesting future research question is to evaluate the performance of replacement policies (other than the ones considered in this paper) when prior information is available.

The tractability of the problems in this paper result in large part from the mathematical properties of the exponential distribution. Future research will address the question of how the results vary if distributions other than an exponential are used.

4. CONCLUSIONS

A new class of group replacement policies has been introduced. This class of policies includes the well known m-failure policies as a special case. This policy allows the incorporation of two kinds of statistical information: prior information before the machines are operated and information gathered during the operation of the machines. This article has investigated in detail the case of exponentially failing machines with a gamma prior distribution. It has been shown that computation of the optimal policy is numerically straightforward and does not require techniques such as numerical integration. The resulting policy is easy to implement. The policy allows the manager to replace the machines earlier than an m-failure policy would allow if the data suggest that the machines are unreliable.

5. ACKNOWLEDGEMENT

Thanks are due to an anonymous referee for helpful comments and suggestions.

6. REFERENCES

Assaf, D. and Shanthikumar, J.G. (1987), "Optimal Group Maintenance Policies with Continuous and Periodic Inspections", *Management Science*, 33, 1440-1452.

Barlow, R. and Hunter, L.C. (1960), "Optimum Preventive Maintenance Policies", *Operations Research*, 8, 90-100.

Boland, P.J. (1982), "Periodic Replacement with Minimal Repair Costs Vary with Time", *Naval Research Logistics Quarterly*, 29, 541-546.

Boland, P.J. and Proschan, F. (1982), "Periodic Replacement with Increasing Minimal Repair Costs at Failure", *Operations Research*, 30, 1183-1189.

Nakagawa, T. (1979), "Replacement Problems of a Parallel System in Random Environment", *Journal of Applied Probability*, 16, 203-205.

——————— (1983), "Optimal Number of Failures Before Repair Time", *IEEE Transactions on Reliability*, 32, 115-116.

Okumoto, K. and Elsayed, E.A. (1983), "An Optimal Group Maintenance Policy", *Naval Research Logistics Quarterly*, 30, 667-674.

Rade, L. (1976), "Reliability Systems in a Random Environment", *Journal of Applied Probability*, 13, 407-410.

Ritchken, P. and Wilson J.G. (1990), "(m,T) Group Maintenance Policies", *Management Science*, 36, 632-639.

Wilson J.G. (1992), "Variance Reducing Group Maintenance Policies", to appear in *Management Science*.

Wilson, J.G. and Benmerzouga, A. (1990), "Optimal m-Failure Policies with Random Repair Time", *Operations Research Letters*, 9, 203-209.

——————— (1991), "Bayesian Group Replacement Policies", to appear in *Operations Research*.

——————— (1992), "A General Adaptive Group Maintenance Policy", to appear in *IEEE Transactions on Reliability*.

Yeh, L. (1988), "A Note on the Optimal Replacement Time of Damaged Devices", *Advances in Applied Probability*, 20, 479-482.

ADVANCES IN RELIABILITY
Edited by Asit P. Basu

An eavesdrop on a chance encounter of like minded ghosts

N. D. Singpurwalla

Department of Operations Research, George Washington University, Washington, DC 20052, United States of America

Abstract

The material given here formed the basis of an after dinner talk given by the author at the "International Research Conference on Reliability" at Columbia, Missouri in June 1991. Its aim is to convey some thoughts on the foundational issues, current state of affairs and future directions in statistical reliability within the relaxed atmosphere of a conference banquet.

The author would like to take this opportunity to thank Professor Asit Basu for his tireless efforts - over several years - to promote statistical aspects of reliability via his own research and conferences such as this which have had the unique flavor of participation by a diverse group of persons from academia, government and industry. He would also like to thank Professor Barlow for his comments on certain technical aspects of this talk.

LOCATION: Above The Memorial Union at UM Columbia.

TIME: After the Panel Discussion at the Reliability Conference

CAST OF CHARACTERS

Ghost No. 1: HAROLD, with a CAMBRIDGE Accent.

Ghost No. 2: JIMMY, with a DETROIT Accent.

Ghost No. 3: BRUNO, with an ITALIAN Accent.

SCENE 1: Introductions & Salutations:

Ghosts 1 & 2.

Jimmy: Greetings, Sir Harold, what brings you to the BAYESIAN BUSHES?

Harold: Well Professor, it is the SADDAMS OF STATISTICS!

The conference in Reliability was advertised in the universe, and the US Government (through NSF) has given funds to peddle influence in heaven, which it now views as a third world cosmos .

And what made you come here?

Jimmy: I received an invitation from A Basu, but misunderstood it as being from Dee Basu.

At about this juncture ghost 3 arrives on the scene

Harold & Jimmy: Greetings Bruno, welcome to Missouri, how come you are here at this down to earth Conference?

Bruno: Basu asked the Vatican to sponsor the Conference and invited the Pope to be a moderator,

PROVIDED HE PAY THE REGISTRATION FEE.

The Pope being a man of certainty, asked for advice from God,

WHO ALWAYS PLAYS DICE WITH MEN,

but

ACTS LIKE A BAYESIAN.

HE said:

"Remember John Paul, what I told you when you pontificated on birth control

"WHEN YOU DON'T PLAYETH THE GAME YOU DON'T MAKETH THE RULES!

God then recommended Fabio Spizzichino, who told HIM that he is in a state of SHANTI because he has had a

REVELATION FROM SAINT DICK OF BERKELEY

on

THE MEANING OF AGEING.

The SAINT Says

"TO AGE IS TO BE SCHUR-CONCAVE."

Jimmy: Fabio must be accumulating a lot of FREQUENT FLYER mileage.

Bruno: Jimmy, You know how much I abhor the word FREQUENT!

Every time I hear it, I have a SPIT FIT.

Harold: Then you must have done a lot of spitting here

This place has too many frequentists.

SCENE 2: On Reliability Books

Harold: I have nothing against the frequentist view of probability. What I object to is the frequentist's violation of the LIKELIHOOD PRINCIPLE.

Bruno: What a pity; it has been over 50 years since I wrote on SUBJECTIVE PROBABILITY, and these reliabilists have yet to catch on.

Jimmy: But you wrote about it in Italian, and moreover, you tended to ramble a lot - actually, you still do. Furthermore, you dwelt on philosophic issues.

How do you expect these simple minded cowboys with no taste for good Italian food and wine, not to mention sex, understand deep philosophical issues?

Harold: Even their professors really did not understand sex.

Jimmy: To them computing MLE's, is like

HAVING SEX WITH THE LIKELIHOOD.

They forget the PRIOR, which INDUCES ROMANCE.

Bruno: Look who is talking; you wrote a book in English on "The Foundations of Statistics," but made it unintelligible.

It is no wonder it sells now for a pittance. About 10000 LIRA.

Jimmy: But so is Harold's book on Probability (which Dennis has been reading and rereading), is relatively cheap.

Bruno: What about the book by this Minnesota kid, whose name reminds me of American fast food?

Jimmy: You mean my namesake Berger?

Bruno: Oh Yes! I was thinking of BERGER KING.

Jimmy: Yes, he is indeed like a king, but his book sells cheap. Besides his publisher is giving away complementary copies, and the Soviets are grabbing it because they think that it is food for the mind.

Harold: What I can't understand is why the reliability books cost so much. For example the books by

PARLOW and BROSCHAN

PAIN SELSON

GERHARD LAWFUL, and

ARCHER and GOLDFINGER.

Jimmy: The Parlow-Broschan costs a lot because it has the distinction of being

THE MOST REFERENCED BUT LEAST USED

book in reliability.

The Archer-Goldfinger book is something you take along to read on a plane trip to Missouri. They write with the style of Stephen King.

DETAILS ON SEX, BUT NO PRINCIPLES
ON AFFAIRS.

The Selson book costs a lot because it is obsolete and has the potential of being

A COLLECTOR'S ITEM

The LAWFUL book costs a lot because it is in one sense AWFUL; it declares on the front cover that it is basically

LAWLESS

Harold: And what about the Bayesian book by the two East Coast Cowboys,

ARTZ and WAVER?

Jimmy: Now that is a masterpiece. It is pricy, because it sells well; and it sells well because

THE BAYESIANS THINK IT IS BAYESIAN

and

THE FREQUENTISTS KNOW BETTER.

So what matters is what the customers think, not what the authors think.

For that matter authors of books don't even have to think; all they need to know is plagiarize.

Bruno: As I have said,

IT IS ALL IN THE MIND!

Harold: I discovered a long time ago, that if you want to get rich writing books, make sure that they have the ILLUSION OF OBJECTIVITY like those of my colleague Ronald whose book on

"Statistical Methods for Research Workers"

sells like McDonalds (billions and billions) and not like Berger King (only thousands and thousands).

SCENE 3: **<u>On The Basu - Broschan Futility Function</u>**

Bruno: It looks as if the Bayesians are using the TAGUCHI METHOD; they are minimizing societal loss, by writing books that
DIMINISH ROYALTY BUT INCREASE LOYALTY

Jimmy: Be careful what you say. Don't draw any analogies between the Bayesians and Taguchi. The frequentists will criticize the Bayesians for ignoring

"INTERACTIONS."

Harold: The Bayesians should only REACT, and not bother to INTERACT? In any case why is Taguchi not at this conference?

Jimmy: Taguchi is a very controversial man.

THE BELL-BOYS LOVE HIM

and

THE BOX-BOYS CURSE HIM

The former think that he is the biggest coming since the BREAK - UP of statistics by KUTEY.

The latter feel that he is stealing their GURU'S thunder.

Harold: But the Guru's place in history is assured with his prescription for

HOW TO KEEP AN EYE ON ROBUST POSTERIORS

Jimmy: And Bruno, you wrote in the Berkeley symposium that

ROBUST POSTERIORS ARE MADE BY GOD

whereas

PRIORS AND LIKELIHOODS ARE MADE BY MEN

Bruno: How come Frank is not here.

Jimmy: Do you mean FRANK PROSCHAN or FRANK PLUMPTON?

Bruno: Who the hell is Proschan?

Harold: Proschan is the wierdo who wrote two books and several papers on reliability, some having a Bayesian flavor without knowing why he did it.

His utility has a strange form; it goes like this:

"ANOTHER PAPER ANOTHER DOLLAR"

Jimmy: Ramsey is not here because Tom advised him against it.

Tom Bayes did not come because he is offended that a reliabilist of Indian origin has had the gall to name his dog BAYES
and

THE DOG BAYES BIT THE RELIABLE BAYESIAN

Bruno: And what about LaPlace?

Harold: Does the organizer of this conference know that it was LaPlace, who, strained as an engineer, pioneered the use of INVERSE PROBABILITY which Tom is now famous for?

Jimmy: I doubt it; Asit has an utility function which is opposite to that of Proschan.

I call it FUTILES and and it goes as follows:

"ANOTHER CONFERENCE ANOTHER PAPER!"

SCENE 4: <u>On Exchangeability</u>

Bruno: The American academics are a strange lot.

They label a conference international if there is one Italian, one Englishman, a few Chinese and many Indians.

Harold: And the US Government funds such a conference because they know that in the eyes of the public

AMERICAN - INDIANS AND INDIAN - AMERICANS ARE EXCHANGEABLE!

Bruno: I wonder if there is a REPRESENTATION THEOREM there.

Jimmy: Only if there has been some MIXING of the two.

Harold: I guess that once such a theorem is discovered Dercy, The McCartherite will screw it up by arguing that the number of

INDIAN - AMERICANS IS GOING TO INFINITY

whereas, the number of

AMERICAN - INDIANS IS GOING TO ZERO.

SCENE 5: <u>The Dry and Wet Ages of Statistics in Reliability</u>

Jimmy: It appears that the DRY AGE OF STATISTICS is coming to an end.

Bruno: It lasted for a long time, almost 70 years, from 1920 until 1990.

Harold: Thanks to the three of us, and my student DENNIS we kept the subject moist.

Jimmy: What a shame your other student SIR DAVID LEAN missed the boat.

Bruno: He was fair, and unlike the TECHNOMETRIKA MAFIOSO did not screw up PERESTROIKA.

Harold: Recently, he did publish a paper on

INFORMATION by AYATOLLAH SOOFI AND AYATOLLAH EBRAHIMI.

Jimmy: The wet ages are upon us; The Bayesians are having fun in Valencia, with a lot of Sangria.

Bruno: Even some frequentists go there! For them

EVERYTHING IS MAJORIZED BY SANGRIA.

Harold: Must I apologize for my Cambridge colleague for turning the clock backwards and requiring that we

HAVE AFFAIRS ONLY WITH THE LIKELIHOOD.

Bruno: As an Italian, I prefer a little infidelity

I LOVE THE LIKELIHOOD BUT FEEL PASSIONATE ABOUT THE PRIOR;

THE PRIOR IS LIKE ADULTERY IN THE MIND. IT IS ALWAYS THERE,

WHEREAS THE LIKELIHOOD MAY BE FLAT AND NEVER ATTAIN A CLIMAX.

Jimmy: To have a coherent marriage, we need both

A SPOUSE, LIKE THE LIKELIHOOD,

and

OTHER SIGNIFICANT, LIKE THE PRIOR.

Harold: The non-Bayesians are like members of closed clubs:

THEY DON'T ADMIT MISTRESSES AND CONSEQUENTLY HAVE A ROUTINE FREQUENTIST LIFE, INFINITELY OFTEN.

THEY STICK TO THIS ROUTINE CONFIDENTALLY UNTIL INFINITY DOES THEM APART.

Jimmy: And in the process of confidentally sampling they get

INFATUATED WITH CONFIDENCE LIMITS

Harold: Do you know the case of the two members of the GAUSSIAN FAMILY?

Bruno: No I don't.

Harold: Each had different means but the same variance, and interest was focused on estimating the ratio of means.
The confidence limits are [– infinity to + infinity], but the probability of coverage is not 1.

Bruno: A damn stupid result. I say curse on the house of confidence limits.

Jimmy: Even Jacob de Wolf condemned confidence limits!

Bruno: Another contributor to the dry ages of statistics is Jerzey Polanski.
He should have become a Pope instead of becoming

THE LATTER - DAY BISHOP OF BERKELEY.

Jimmy: Be careful what you say: Is he not the one who drew attention of your work to the Anglo-American establishment?

Harold: THE SAMURAI OF STATISTICS, BIG ED did British statistics a favor by seducing Jerzey out of the UK and into the USA.

Now Big Ed dismisses Jerzey's work as irresponsive.

Personally, I would have liked to have him continue in the UK, but not at Cambridge!

His feud with Fisher kept Sir Ronald off my back, and allowed me to sprinkle champagne on the dry ages of statistics.

Jimmy: Ronald's affair with p-values has recently been shattered by

BERGER AND THE BERGER - BOYS.

Bruno: Despite all this, the reliability folk continue to be attracted to p-values, Type 1 & 2 Errors and the

FALLACIOUS BELIEFS THAT:

* TRUE RELIABILITY EXISTS OUTSIDE THE MIND

* THAT FAILURE MODELS REALLY EXIST IN THE MIND OF GOD WHO IS MERCILESS IN NOT DIVULGING THEM TO THE POOR FREQUENTISTS.

* THAT ERROR DISTRIBUTIONS IN REGRESSION, TIME SERIES, AND ANOVA ARE TRULY GAUSSIAN OR LOGNORMAL, OR SOMETHING ELSE.

Jimmy: Talking about God I am reminded of the Hindu Philosophy that

EVEN GOD EXISTS IN THE MIND, OR THAT PEOPLE ARE A PART OF GOD.

Bruno: Why then are so many statisticians of Indian origin frequentist?

Jimmy: Either because they made a pact with the

MEPHISTOPHELES OF ACADEMIC TENURE

or

THEY ARE ATTRACTED TO INFINITIES.

which is also a tenet of Hindu Philosophy.

Harold: I guess the frequentists need Type 1 & 2 errors for

RELIABILITY DEMONSTRATION TESTING

and

P-VALUES FOR GOODNESS OF FIT TESTING OF HYPOTHESIZED FAILURE MODELS.

In any case, what alternatives do you propose?

Jimmy:

* Instead of making decisions using Type 1 & 2 Errors we should assess costs and make decisions that maximize expected utilities.

  This is something that both Big Ed and Gengichi also advocate.

* Errors in Regression and Time Series should be based on Bruno's

  CONSIDERATIONS OF ROBUSTNESS.

* Failure Models should be initially chosen based on Bruno's

  (SUBJECTIVE) PRINCIPLE OF INDIFFERENCE AND ITS VARIANTS.

* The chosen models could be updated in the light of data, a suggestion also made by Box.

* Competing failure models should be compared with respect to their predictive capabilities via

  POSTERIOR ODDS.

* Competing models in Time Series should be compared via

  PREQUENTIAL LIKELIHOOD RATIOS.

* Finally focus on Reliability and Quality Control problems should be shifted away from inference about Greek Symbols, created by the minds of Greeks to

  DECISION MAKING

  and

  PREDICTIVISM.

Harold: I am not a subjectivist. Can you enlighten me about the "Principle of Indifference?"

Jimmy: Will be glad to do it.

1. A sequence of binary variables $\{X_i\}$, $i=1,\ldots,\infty$, can be represented as a mixture over p of coin tossing processes with parameter p, iff for each n, given

   $$\sum^{n} X_i = t,$$

   the sequence $X_1,\ldots, X_n$ is judged uniform over the $\binom{n}{t}$ sequences of t 1's and (n-t) 0's.

2. A sequence of variables $\{X_i\}$, $i=1,\ldots, 2$ can be represented as a mixture over μ and σ^2 of i.i.d. $\mathcal{N}(\mu, \sigma^2)$, iff for each n, given

   $$U_n = \sum^{n} X_i \text{ and } V_n = (\sum^{n} X_i^2)^{1/2},$$

   the distribution of $X_1,\ldots, X_n$ is uniform over the (n-2) sphere in $\mathbb{R}^n$; the sphere has center U_n and radius V_n.

3. A sequence of variables $\{X_i\}$, $i=1,\ldots, \infty$ can be represented as a mixture over θ of i.i.d. uniforms over $[0, \theta]$, iff for each n, given $M_n = \max(X_1,\ldots, X_n)$ the X_i's are uniform over $[0, M_n]$.

4. A sequence of variables $\{X_i\}$, $X_i \geq 0$, $i = 1,\ldots, \infty$, can be represented as a mixture over λ of i.i.d. exponentials with parameter λ, iff for each n, given

   $$\sum^{n} X_i = S_n,$$

   $X_1, \ldots, X_n$ is uniform over the simplex S_n and $X_i \geq 0$.

   Thus indifference is the judgment of uniformity over an appropriate space, and it is such judgments that give rise to the well known probability models, not goodness of fit tests. Such ideas were also noted by the Physicists such as Maxwell, Boltzman, Einstein, etc.

Bruno: I cannot disagree with you. I guess that what is needed for the future welfare of Reliability and Quality Control is a thorough

RE-EDUCATION OF THE RELIABILISTS!

Harold: Perhaps the next conference would be a

DE-PROGRAMMING CAMP FOR OLD RASCALS.

Harold,
Jimmy
& Bruno: See you next time at the George Washington University for a truly Bayesian Reliability Conference to be organized by

SAINT THOMAS of MAZZUCHI

and

SULTAN SOYER of ISTANBUL.

AUTHOR INDEX